Multivariable Dynamic Calculus on Time Scales

Martin Bohner · Svetlin G. Georgiev

Multivariable Dynamic Calculus on Time Scales

Springer

Martin Bohner
Department of Mathematics and Statistics
Missouri University of Science
and Technology
Rolla, MO
USA

Svetlin G. Georgiev
Faculty of Mathematics and Informatics
Sofia University St. Kliment Ohridski
Sofia
Bulgaria

ISBN 978-3-319-47619-3 ISBN 978-3-319-47620-9 (eBook)
DOI 10.1007/978-3-319-47620-9

Library of Congress Control Number: 2016955534

Printed on acid-free paper

This Springer imprint is published by Springer Nature
The registered company is Springer International Publishing AG
The registered company address is: Gewerbestrasse 11, 6330 Cham, Switzerland

Dedicated to the Memory of
Professor Gusein Shirin Guseinov
(1951–2015)

Dedicated to the Memory of

Professor Gusein Sh. Guseinov

(1951–2015)

Preface

The theory of time scales was introduced by Stefan Hilger in his Ph.D. thesis [31] in 1988 (supervised by Bernd Aulbach) in order to unify continuous and discrete analysis and to extend the continuous and discrete theories to cases "in between." Since then, research in this area of mathematics has exceeded by far a thousand publications, and numerous applications to literally all branches of science such as statistics, biology, economics, finance, engineering, physics, and operations research have been given. For an introduction to *single-variable* time scales calculus and its applications, we refer the reader to the monograph [21] by Bohner and Peterson.

In this book, we offer the reader an overview of recent developments of *multivariable* time scales calculus. The book is primarily intended for senior undergraduate students and beginning graduate students of engineering and science courses. Students in mathematical and physical sciences will find many sections of direct relevance. This book contains nine chapters, and each chapter consists of results with their proofs, numerous examples, and exercises with solutions. Each chapter concludes with a section featuring advanced practical problems with solutions followed by a section on notes and references, explaining its context within existing literature. Altogether, the book contains 123 definitions, 230 theorems including corollaries, lemmas, and propositions, 275 examples, and 239 exercises including advanced practical problems.

The first three chapters deal with single-variable time scales calculus. Many of the presented results including their proofs are extracted from [21]. Chapter 1 introduces the most fundamental concepts related to time scales, namely the forward and backward jump operators and the graininess. In addition, the induction principle on time scales is given. Chapter 2 deals with differential calculus for single-variable functions on time scales. The basic definition of delta differentiation is due to Hilger. Many examples on differentiation in various time scales are included, as well as the Leibniz formula for the nth derivative of a product of two functions. Mean value results are presented that will be used later on in the book in the multivariable case. Several versions of the chain rule are included. Sufficient conditions for a local maximum and a minimum are given. Moreover, sufficient conditions for convexity and concavity of single-variable functions are presented. A sufficient condition for complete delta differentiability of single-variable functions is given, and the geometric sense of differentiability is discussed and illustrated. In Chapter 3, the main concepts for regulated, rd-continuous, and pre-differentiable functions are introduced. The indefinite integral and the Riemann delta integral are defined and many of their properties are deduced. Hilger's complex plane is introduced. Some elementary functions such as the exponential functional, hyperbolic functions, and trigonometric functions are defined and their properties are given. Moreover, Taylor's formula and L'Hôpital's rule are presented. Improper integrals of the first and the second kind are introduced and studied.

The next two chapters discuss sequences and series of functions as well as parameter-dependent integrals. The results of these chapters are adopted from [36, 40]. In Chapter 4, the Dini theorem, the Cauchy criterion for uniform convergence of a function series, the Weierstraß M-test for uniform convergence of a function series, the Abel test, and the Dirichlet test are presented and numerous examples are given. Chapter 5 introduces and studies both normal parameter-dependent integrals and improper parameter-dependent integrals of the first kind.

The final four chapters deal with multivariable time scales calculus. The presented results including their proofs are n-dimensional analogues of the two-dimensional results given in [8, 9, 14, 17]. Chapter 6 is devoted to partial differentiation on time scales. Definitions for partial derivatives and completely delta differentiable functions are given. Some sufficient conditions for differentiability are presented. The chain rule and some of the properties of implicit functions are given. The directional derivative is introduced. In Chapter 7, multiple Riemann integrals over rectangles and over more general sets are introduced. Many of their properties are given. Mean value results are presented. Chapter 8 defines the length of time scale curves. Line integrals of the first kind and of the second kind are introduced. Moreover, Green's formula is derived. Chapter 9 deals with surface integrals. Many of their properties are given.

The aim of this book is to present a clear and well-organized treatment of the concept behind the development of mathematics as well as solution techniques. The text material of this book is presented in a readable and mathematically solid format. Many practical problems are given, displaying the power of multivariable dynamic calculus on time scales.

Many of the results presented in this book are based on the work by one of the authors (Martin Bohner) and Professor Gusein Shirin Guseinov, who unexpectedly passed away on March 20, 2015. Both authors have attended the memorial conference for Professor Guseinov at Atılım University in Ankara, Turkey, July 11–13, 2016, and they have decided there to dedicate this book to the memory of Professor Guseinov.

Finally, the authors would like to thank Rasheed Al-Salih, Mehdi Nategh, and Dr. Özkan Öztürk for a careful reading of the manuscript. The authors would also like to thank Professor Mohtar Kirane (University of La Rochelle, France) for bringing the subject of multivariable calculus on time scales to the attention of the second author. Moreover, the authors are grateful to the production staff at Springer.

Rolla, USA Martin Bohner
Paris, France Svetlin G. Georgiev
December 2016

Contents

Chapter 1
Time Scales

1.1 Forward and Backward Jump, Graininess

Definition 1.1 A *time scale* is an arbitrary nonempty closed subset of the real numbers.

We denote a time scale by the symbol $\mathbb{T}$. We suppose that a time scale $\mathbb{T}$ has the topology that it inherits from the real numbers with the standard topology.

Example 1.2 $[1, 2]$, $\mathbb{R}$, and $\mathbb{N}$ are time scales.

Example 1.3 $[a, b)$, $(a, b]$, and (a, b) are not time scales if $a < b$.

Example 1.4 The set

$$\{1, 2, 3\} \cup [4, 5] \cup \{11, 12, 13\}$$

is a time scale.

Exercise 1.5 Prove that the following sets are time scales.

1. $h\mathbb{Z} = \{hk : k \in \mathbb{Z}\}$, $h \in \mathbb{R}$,
2. $\mathbb{P}_{a,b} = \bigcup_{k=0}^{\infty}[k(a+b), k(a+b)+a]$, $a > 0$,
3. $\overline{q^{\mathbb{Z}}} = \{q^k : k \in \mathbb{Z}\} \cup \{0\}$, $q > 1$,
4. $\mathbb{N}_0^n = \{k^n : k \in \mathbb{N}_0\}$, $n \in \mathbb{N}$,
5. $\{H_n : n \in \mathbb{N}_0\}$, where H_n are the so-called harmonic numbers defined by

$$H_0 = 0 \quad \text{and} \quad H_n = \sum_{k=1}^{n} \frac{1}{k} \quad \text{for} \quad n \in \mathbb{N}.$$

6. $C = \bigcap_{n=0}^{\infty} K_n$, where $K_0 = [0, 1]$, K_1 is obtained by removing the open "middle third" of K_0, i.e., the open interval $\left(\frac{1}{3}, \frac{2}{3}\right)$, K_2 is obtained by removing the two open middle thirds of K_1, i.e., the open intervals $\left(\frac{1}{9}, \frac{2}{9}\right)$ and $\left(\frac{7}{9}, \frac{8}{9}\right)$ from K_1, and so on. The set C is called the Cantor set.

M. Bohner and S.G. Georgiev, *Multivariable Dynamic Calculus on Time Scales*,
DOI 10.1007/978-3-319-47620-9_1

Definition 1.6 For $t \in \mathbb{T}$, we define the *forward jump operator* $\sigma : \mathbb{T} \to \mathbb{T}$ by

$$\sigma(t) = \inf\{s \in \mathbb{T} : s > t\}.$$

We note that $\sigma(t) \geq t$ for any $t \in \mathbb{T}$.

Definition 1.7 For $t \in \mathbb{T}$, we define the *backward jump operator* $\rho : \mathbb{T} \to \mathbb{T}$ by

$$\rho(t) = \sup\{s \in \mathbb{T} : s < t\}.$$

We note that $\rho(t) \leq t$ for any $t \in \mathbb{T}$.

Definition 1.8 We set

$$\inf \emptyset = \sup \mathbb{T}, \quad \sup \emptyset = \inf \mathbb{T}.$$

Definition 1.9 For $t \in \mathbb{T}$, we define the following.

1. If $\sigma(t) > t$, then t is called *right-scattered.*
2. If $t < \sup \mathbb{T}$ and $\sigma(t) = t$, then t is called *right-dense.*
3. If $\rho(t) < t$, then t is called *left-scattered.*
4. If $t > \inf \mathbb{T}$ and $\rho(t) = t$, then t is called *left-dense.*
5. If t is left-scattered and right-scattered at the same time, then t is said to be *isolated.*
6. If t is left-dense and right-dense at the same time, then t is said to be *dense.*

Example 1.10 Let

$$\mathbb{T} = \{\sqrt{2n+1} : n \in \mathbb{N}\}.$$

If $t = \sqrt{2n+1}$ for some $n \in \mathbb{N}$, then

$$n = \frac{t^2 - 1}{2}$$

and

$$\sigma(t) = \inf\{l \in \mathbb{N} : \sqrt{2l+1} > \sqrt{2n+1}\} = \sqrt{2n+3} = \sqrt{t^2+2},$$

$$\rho(t) = \sup\{l \in \mathbb{N} : \sqrt{2l+1} < \sqrt{2n+1}\} = \sqrt{2n-1} = \sqrt{t^2-2}$$

for $n \geq 2$, $n \in \mathbb{N}$,

$$\rho(\sqrt{3}) = \sup \emptyset = \inf \mathbb{T} = \sqrt{3}$$

for $n = 1$. Because

$$\sqrt{t^2-2} < t < \sqrt{t^2+2} \quad \text{for} \quad n \geq 2,$$

we conclude that every point $\sqrt{2n+1}$, $n \in \mathbb{N} \setminus \{1\}$, is right-scattered and left-scattered, i.e., every point $\sqrt{2n+1}$, $n \geq 2$, $n \in \mathbb{N}$, is isolated. Since

$$\rho(\sqrt{3}) = \sqrt{3} < \sqrt{5} = \sigma(\sqrt{3}),$$

we have that $\sqrt{3}$ is right-scattered.

Example 1.11 Let

$$\mathbb{T} = \left\{ \frac{1}{2n} : n \in \mathbb{N} \right\} \cup \{0\}.$$

Let $t \in \mathbb{T}$ be arbitrarily chosen.

1. $t = \frac{1}{2}$. Then

$$\sigma\left(\frac{1}{2}\right) = \inf\left\{s \in \mathbb{T} : s > \frac{1}{2}\right\} = \inf \emptyset = \sup \mathbb{T} = \frac{1}{2}.$$

   Moreover,

$$\rho\left(\frac{1}{2}\right) = \sup\left\{s \in \mathbb{T} : s < \frac{1}{2}\right\} = \frac{1}{4} < \frac{1}{2},$$

   i.e., $\frac{1}{2}$ is left-scattered.
2. $t = \frac{1}{2n}$, $n \in \mathbb{N}$, $n \geq 2$. Then

$$\sigma\left(\frac{1}{2n}\right) = \inf\left\{s \in \mathbb{T} : s > \frac{1}{2n}\right\} = \frac{1}{2(n-1)} > \frac{1}{2n}$$

   and

$$\rho\left(\frac{1}{2n}\right) = \sup\left\{s \in \mathbb{T} : s < \frac{1}{2n}\right\} = \frac{1}{2(n+1)} < \frac{1}{2n}.$$

   Therefore, all points $\frac{1}{2n}$, $n \in \mathbb{N}$, $n \geq 2$, are right-scattered and left-scattered, i.e., all points $\frac{1}{2n}$, $n \in \mathbb{N}$, $n \geq 2$, are isolated.
3. $t = 0$. Then

$$\rho(0) = \sup\{s \in \mathbb{T} : s < 0\} = \sup \emptyset = \inf \mathbb{T} = 0.$$

   Moreover,

$$\sigma(0) = \inf\{s \in \mathbb{T} : s > 0\} = 0,$$

   i.e., 0 is right-dense.

Example 1.12 Let

$$\mathbb{T} = \left\{ \frac{n}{3} : n \in \mathbb{N}_0 \right\}.$$

Let $t = \frac{n}{3}$, $n \in \mathbb{N}_0$, be arbitrarily chosen.

1. $n \in \mathbb{N}$. Then
$$\sigma\left(\frac{n}{3}\right) = \inf\left\{s \in \mathbb{T} : \ s > \frac{n}{3}\right\} = \frac{n+1}{3} > \frac{n}{3}$$
and
$$\rho\left(\frac{n}{3}\right) = \sup\left\{s \in \mathbb{T} : \ s < \frac{n}{3}\right\} = \frac{n-1}{3} < \frac{n}{3}.$$
Therefore, all points $\frac{n}{3}$, $n \in \mathbb{N}$, are right-scattered and left-scattered, i.e., all points $\frac{n}{3}$, $n \in \mathbb{N}$, are isolated.
2. $n = 0$. Then
$$\rho(0) = \sup\{s \in \mathbb{T} : \ s < 0\} = \sup\emptyset = \inf\mathbb{T} = 0.$$
Moreover,
$$\sigma(0) = \inf\{s \in \mathbb{T} : \ s > 0\} = \frac{1}{3} > 0,$$
i.e., 0 is right-scattered.

Exercise 1.13 Classify each point
$$t \in \mathbb{T} = \{\sqrt[3]{2n-1} : \ n \in \mathbb{N}_0\}$$
as left-dense, left-scattered, right-dense, or right-scattered.

Solution Each point $\sqrt[3]{2n-1}$, $n \in \mathbb{N}$ is isolated, -1 is right-scattered.

Exercise 1.14 Find σ and ρ for
$$\mathbb{T} = \{H_n : \ n \in \mathbb{N}_0\}.$$

Solution $\sigma(H_n) = H_{n+1}$, $n \in \mathbb{N}_0$ and $\rho(H_n) = H_{n-1}$, $n \in \mathbb{N}$, $\rho(H_0) = H_0$.

Exercise 1.15 Find σ and ρ for

1. $\mathbb{T} = h\mathbb{Z}$, $h > 0$,
2. $\mathbb{T} = h\mathbb{Z}$, $h < 0$,
3. $\mathbb{T} = \mathbb{R}$,
4. $\mathbb{T} = \mathbb{Z}$,
5. $\mathbb{T} = \mathbb{N}^k$, $k \in \mathbb{N}$,
6. $\mathbb{T} = q^{\mathbb{Z}} \cup \{0\}$, $q > 1$,
7. $\mathbb{T} = q^{\mathbb{N}_0}$, $q > 1$,
8. $\mathbb{T} = p^{\mathbb{N}_0} \cup \{0\}$, $p \in (0, 1)$,

Solution 1. $\sigma(t) = t + h$ and $\rho(t) = t - h$,
2. $\sigma(t) = t - h$ and $\rho(t) = t + h$,

3. $\sigma(t) = \rho(t) = t$,
4. $\sigma(t) = t + 1$, $\rho(t) = t - 1$,
5. $\sigma(t) = \left(1 + \sqrt[k]{t}\right)^k$ for $t \in \mathbb{T}$ and $\rho(t) = \left(\sqrt[k]{t} - 1\right)^k$ for $t \in \mathbb{T} \setminus \{1\}$, $\rho(1) = 1$,
6. $\sigma(t) = qt$ and $\rho(t) = \frac{t}{q}$,
7. $\sigma(t) = qt$ for $t \in q^{\mathbb{N}_0}$ and $\rho(t) = \frac{t}{q}$ for $t \in q^{\mathbb{N}}$, $\rho(1) = 1$,
8. $\sigma(t) = \frac{t}{p}$ for $t \in p^{\mathbb{N}} \cup \{0\}$, $\sigma(1) = 1$ and $\rho(t) = pt$ for $t \in p^{\mathbb{N}_0}$, $\rho(0) = 0$.

Example 1.16 Let $t \in \mathbb{T}$ be isolated with $\sigma(t) = t + 1$ and $\rho(t) = t - 1$. Consider

$$A = (\sigma(t))^2 - 2\sigma(t)\rho(t) + (\rho(t))^2 + 3\rho(t) - \sigma(t), \quad t \in \mathbb{T}.$$

Then

$$\begin{aligned} A &= (t+1)^2 - 2(t+1)(t-1) + (t-1)^2 + 3(t-1) - (t+1) \\ &= t^2 + 2t + 1 - 2t^2 + 2 + t^2 - 2t + 1 + 3t - 3 - t - 1 \\ &= 2t. \end{aligned}$$

Exercise 1.17 Simplify

$$A = (\sigma(t))^2 + (\sigma(t) + \rho(t))^3 - t\sigma(t) + \rho(t)$$

if

1. $t \in \mathbb{T}$ is isolated with $\sigma(t) = t + 1$ and $\rho(t) = t - 1$.
2. $t \in \mathbb{T}$ is right-scattered and left-dense with $\sigma(t) = t + 1$.
3. $t \in \mathbb{T}$ is right-dense and left-scattered with $\rho(t) = t - 1$.
4. $t \in \mathbb{T}$ is dense.

Solution 1. Since $\sigma(t) = t + 1$ and $\rho(t) = t - 1$, we have

$$\begin{aligned} A &= (t+1)^2 + (t+1+t-1)^3 - t(t+1) + t - 1 \\ &= t^2 + 2t + 1 + 8t^3 - t^2 - t + t - 1 \\ &= 8t^3 + 2t. \end{aligned}$$

2. Since $\sigma(t) = t + 1$ and $\rho(t) = t$, we have

$$\begin{aligned} A &= (t+1)^2 + (t+1+t)^3 - t(t+1) + t \\ &= t^2 + 2t + 1 + 8t^3 + 12t^2 + 6t + 1 - t^2 - t + t \\ &= 8t^3 + 12t^2 + 8t + 2. \end{aligned}$$

3. Since $\sigma(t) = t$ and $\rho(t) = t - 1$, we have

$$\begin{aligned} A &= t^2 + (t + t - 1)^3 - t^2 + t - 1 \\ &= t^2 + (2t - 1)^3 - t^2 + t - 1 \\ &= t^2 + 8t^3 - 12t^2 + 6t - 1 - t^2 + t - 1 \\ &= 8t^3 - 12t^2 + 7t - 2. \end{aligned}$$

4. Since $\sigma(t) = \rho(t) = t$, we have

$$\begin{aligned} A &= t^2 + (t + t)^3 - t^2 + t \\ &= 8t^3 + t. \end{aligned}$$

Exercise 1.18 Let $t \in \mathbb{T}$ be right-dense and left-scattered with $\rho(t) = t - 1$. Simplify

$$A = \frac{\sigma(t) + t + (\rho(t))^2 - 2\sigma(t)\rho(t)}{\sigma(t) + \rho(t) + 2}, \quad t \in \mathbb{T}, \quad \sigma(t) + \rho(t) + 2 \neq 0.$$

Solution We have

$$\sigma(t) = t, \quad \rho(t) = t - 1$$

and

$$\sigma(t) + \rho(t) + 2 = t + t - 1 + 2 = 2t + 1.$$

Hence, we get

$$\begin{aligned} A &= \frac{t + t + (t - 1)^2 - 2t(t - 1)}{2t + 1} \\ &= \frac{2t + t^2 - 2t + 1 - 2t^2 + 2t}{2t + 1} \\ &= \frac{-t^2 + 2t + 1}{2t + 1} \quad \text{for} \quad t \neq -\frac{1}{2}. \end{aligned}$$

Exercise 1.19 Let $t \in \mathbb{T}$ be isolated with $\sigma(t) = t + 1$ and $\rho(t) = t - 1$. Simplify

$$A = (\sigma(t))^2 + 2\sigma(t)(t - \rho(t)) - t^2 + (\rho(t))^2,$$

$$B = \sigma(t) - \rho(t) - 3,$$

$$C = A^2 + B - B^2, \quad t \in \mathbb{T}.$$

Solution Since $\sigma(t) = t + 1$ and $\rho(t) = t - 1$, we have

$$\begin{aligned} A &= (t+1)^2 + 2(t+1)(t-t+1) - t^2 + (t-1)^2 \\ &= t^2 + 2t + 1 + 2t + 2 - t^2 + t^2 - 2t + 1 \\ &= t^2 + 2t + 4, \\ B &= \sigma(t) - \rho(t) - 3 \\ &= t + 1 - (t-1) - 3 \\ &= t + 1 - t + 1 - 3 \\ &= -1, \\ C &= (t^2 + 2t + 4)^2 - 1 - (-1)^2 \\ &= t^4 + 4t^2 + 16 + 4t^3 + 8t^2 + 16t - 1 - 1 \\ &= t^4 + 4t^3 + 12t^2 + 16t + 14. \end{aligned}$$

Exercise 1.20 Let $t \in \mathbb{T}$ be right-scattered and left-dense with $\sigma(t) = t + 1$. Simplify

$$A = (\sigma(t))^3 - 3(\sigma(t))^2\rho(t) + (\rho(t))^2 - t,$$

$$B = (\sigma(t))^2 - (\rho(t))^2,$$

$$C = 2A + B - (A + B)^2.$$

Solution Since $\sigma(t) = t + 1$ and $\rho(t) = t$, we have

$$A = -2t^3 - 2t^2 - t + 1,$$

$$B = 1 + 2t,$$

$$C = -4t^6 - 8t^5 + 8t^3 + 3t^2 - 4t - 1.$$

Exercise 1.21 For $t \in \mathbb{T}$, simplify

$$A = \frac{1}{(\sigma(t))^2 - \sigma(t)} + \frac{2}{1 - (\sigma(t))^2} + \frac{1}{(\sigma(t))^2 + \sigma(t)}, \quad \sigma(t) \notin \{-1, 0, 1\}.$$

Solution For $t \in \mathbb{T}$ and $\sigma(t) \notin \{-1, 0, 1\}$, we have

$$\begin{aligned}
A &= \frac{1}{\sigma(t)(\sigma(t) - 1)} - \frac{2}{(\sigma(t) - 1)(\sigma(t) + 1)} + \frac{1}{\sigma(t)(\sigma(t) + 1)} \\
&= \frac{\sigma(t) + 1 - 2\sigma(t)}{\sigma(t)(\sigma(t) - 1)(\sigma(t) + 1)} + \frac{1}{\sigma(t)(\sigma(t) + 1)} \\
&= -\frac{\sigma(t) - 1}{\sigma(t)(\sigma(t) - 1)(\sigma(t) + 1)} + \frac{1}{\sigma(t)(\sigma(t) + 1)} \\
&= -\frac{1}{\sigma(t)(\sigma(t) + 1)} + \frac{1}{\sigma(t)(\sigma(t) + 1)} \\
&= 0.
\end{aligned}$$

Exercise 1.22 For $t \in \mathbb{T}$, simplify

$$A = \frac{1}{6\sigma(t) + 3} - \frac{\frac{16}{3}\sigma(t) + 3}{8(\sigma(t))^2 - 2} + \frac{1}{2\sigma(t) - 1}, \quad \sigma(t) \notin \left\{-\frac{1}{2}, \frac{1}{2}\right\}.$$

Solution For $t \in \mathbb{T}$, $\sigma(t) \notin \left\{-\frac{1}{2}, \frac{1}{2}\right\}$, we have

$$\begin{aligned}
A &= \frac{1}{3(2\sigma(t) + 1)} - \frac{\frac{16}{3}\sigma(t) + 3}{2(4(\sigma(t))^2 - 1)} + \frac{1}{2\sigma(t) - 1} \\
&= \frac{1}{3(2\sigma(t) + 1)} - \frac{\frac{16}{3}\sigma(t) + 3}{2(2\sigma(t) - 1)(2\sigma(t) + 1)} + \frac{1}{2\sigma(t) - 1} \\
&= \frac{4\sigma(t) - 2 - 16\sigma(t) - 9}{6(2\sigma(t) - 1)(2\sigma(t) + 1)} + \frac{1}{2\sigma(t) - 1} \\
&= \frac{-12\sigma(t) - 11}{6(2\sigma(t) + 1)(2\sigma(t) - 1)} + \frac{1}{2\sigma(t) - 1}
\end{aligned}$$

$$= \frac{-12\sigma(t) - 11 + 12\sigma(t) + 6}{6(4(\sigma(t))^2 - 1)}$$

$$= \frac{5}{6(1 - 4(\sigma(t))^2)}.$$

Exercise 1.23 For $t \in \mathbb{T}$, simplify

$$A = \frac{1 - \frac{3(\sigma(t))^2}{1-(\sigma(t))^2}}{\frac{\sigma(t)}{\sigma(t)-1} + 1}, \quad \sigma(t) \notin \left\{-1, \frac{1}{2}, 1\right\}.$$

Solution For $t \in \mathbb{T}$ and $\sigma(t) \notin \left\{-1, \frac{1}{2}, 1\right\}$, we have

$$A = \frac{\frac{1-(\sigma(t))^2-3(\sigma(t))^2}{1-(\sigma(t))^2}}{\frac{\sigma(t)+\sigma(t)-1}{\sigma(t)-1}}$$

$$= \frac{\frac{4(\sigma(t))^2-1}{(\sigma(t))^2-1}}{\frac{2\sigma(t)-1}{\sigma(t)-1}}$$

$$= \frac{(2\sigma(t) - 1)(2\sigma(t) + 1)}{(\sigma(t) - 1)(\sigma(t) + 1)} \cdot \frac{\sigma(t) - 1}{2\sigma(t) - 1}$$

$$= \frac{2\sigma(t) + 1}{\sigma(t) + 1}.$$

Exercise 1.24 For $t \in \mathbb{T}$, simplify

1. $\frac{(\sigma(t))^2-\sigma(t)-6}{(\sigma(t))^2-4} - \frac{\sigma(t)-1}{2-\sigma(t)} - 2$, $\sigma(t) \notin \{-2, 2\}$,
2. $\frac{6}{5\sigma(t)-20} + \frac{5-\sigma(t)}{(\sigma(t))^2-8\sigma(t)+16}$, $\sigma(t) \neq 4$,
3. $\frac{2}{\sigma(t)+4} + \frac{\sigma(t)-9}{16-(\sigma(t))^2} - \frac{\sigma(t)-3}{(\sigma(t))^2-8\sigma(t)+16}$, $\sigma(t) \notin \{-4, 4\}$,
4. $\frac{2\rho(t)-1}{2\rho(t)} - \frac{1}{2\rho(t)-4(\rho(t))^2} - \frac{2\rho(t)}{2\rho(t)-1}$, $\rho(t) \notin \left\{-\frac{1}{2}, 0, \frac{1}{2}\right\}$,
5. $\frac{\frac{\rho(t)}{6-3\rho(t)} + \frac{\rho(t)}{\rho(t)+2} + \frac{4\rho(t)}{(\rho(t))^2-4}}{\frac{\rho(t)-4}{\rho(t)-2}}$, $\rho(t) \notin \{-2, 2, 4\}$,
6. $\left(\frac{2}{\rho(t)} - \frac{4}{\rho(t)+2}\right) \cdot \left(2 + \frac{(\rho(t))^2+4}{\rho(t)-2}\right)$, $\rho(t) \notin \{-2, 0, 2\}$.

Solution 1. 0,
2. $\frac{\sigma(t)+1}{5(\sigma(t)-4)^2}$,
3. $\frac{4(2-\sigma(t))}{(\sigma(t)+4)(\sigma(t)-4)^2}$,
4. $-\frac{1}{\rho(t)}$,
5. $\frac{2\rho(t)}{3(\rho(t)-4)}$,
6. -2.

Definition 1.25 The *graininess function* $\mu : \mathbb{T} \to [0, \infty)$ is defined by

$$\mu(t) = \sigma(t) - t \quad \text{for all} \quad t \in \mathbb{T}.$$

Example 1.26 Let $\mathbb{T} = \{2^{n+1} : n \in \mathbb{N}\}$. Assume $t = 2^{n+1} \in \mathbb{T}$ for some $n \in \mathbb{N}$. Then

$$\sigma(t) = \inf\{2^{l+1} : 2^{l+1} > 2^{n+1},\ l \in \mathbb{N}\} = 2^{n+2} = 2t.$$

Hence,

$$\mu(t) = \sigma(t) - t = 2t - t = t, \quad \text{i.e.,} \quad \mu\left(2^{n+1}\right) = 2^{n+1}.$$

Example 1.27 Let $\mathbb{T} = \{\sqrt{n+1} : n \in \mathbb{N}\}$. Assume $t = \sqrt{n+1}$ for some $n \in \mathbb{N}$. Then $n = t^2 - 1$ and

$$\sigma(t) = \inf\left\{\sqrt{l+1} : \sqrt{l+1} > \sqrt{n+1},\ l \in \mathbb{N}\right\} = \sqrt{n+2} = \sqrt{t^2+1}.$$

Hence,

$$\mu(t) = \sigma(t) - t = \sqrt{t^2+1} - t, \quad \text{i.e.,} \quad \mu(\sqrt{n+1}) = \sqrt{n+2} - \sqrt{n+1}.$$

Example 1.28 Let $\mathbb{T} = \{\frac{n}{2} : n \in \mathbb{N}_0\}$. Assume $t = \frac{n}{2}$ for some $n \in \mathbb{N}_0$. Then $n = 2t$ and

$$\sigma(t) = \inf\left\{\frac{l}{2} : \frac{l}{2} > \frac{n}{2},\ l \in \mathbb{N}_0\right\} = \frac{n+1}{2} = t + \frac{1}{2}.$$

Hence,

$$\mu(t) = \sigma(t) - t = t + \frac{1}{2} - t = \frac{1}{2}, \quad \text{i.e.,} \quad \mu\left(\frac{n}{2}\right) = \frac{1}{2}.$$

Example 1.29 Suppose that $\mathbb{T}$ consists of finitely many different points $t_1, t_2, \ldots, t_k$. Without loss of generality, we can assume that

$$t_1 < t_2 < \ldots < t_k.$$

For $i \in \{1, \ldots, k-1\}$, we have

$$\sigma(t_i) = \inf\{t_l \in \mathbb{T} : t_l > t_i, l \in \{1, \ldots, k\}\} = t_{i+1}.$$

Hence,

$$\mu(t_i) = t_{i+1} - t_i, \quad i \in \{1, \ldots, k-1\}.$$

Also,

$$\sigma(t_k) = \inf\{t_l \in \mathbb{T} : t_l > t_k, l \in \{1, \ldots, k\}\} = \inf \emptyset = \sup \mathbb{T} = t_k.$$

Therefore,

$$\mu(t_k) = \sigma(t_k) - t_k = t_k - t_k = 0.$$

From here,

$$\sum_{i=1}^{k} \mu(t_i) = \sum_{i=1}^{k-1} \mu(t_i) + \mu(t_k) = \sum_{i=1}^{k-1} (t_{i+1} - t_i) = t_k - t_1.$$

Exercise 1.30 Find μ for $\mathbb{T} = \left\{\sqrt[3]{n+2} : n \in \mathbb{N}_0\right\}$.

Solution $\mu\left(\sqrt[3]{n+2}\right) = \sqrt[3]{n+3} - \sqrt[3]{n+2}$.

Exercise 1.31 Find μ if

1. $\mathbb{T} = h\mathbb{Z}, h > 0,$
2. $\mathbb{T} = h\mathbb{Z}, h < 0,$
3. $\mathbb{T} = \mathbb{R},$
4. $\mathbb{T} = \mathbb{Z},$
5. $\mathbb{T} = \mathbb{N}^k, k \in \mathbb{N},$
6. $\mathbb{T} = q^{\mathbb{Z}} \cup \{0\}, q > 1,$
7. $\mathbb{T} = q^{\mathbb{N}_0}, q > 1,$
8. $\mathbb{T} = p^{\mathbb{N}_0} \cup \{0\}, p \in (0, 1),$

Solution 1. $\mu(t) = h,$
2. $\mu(t) = -h,$
3. $\mu(t) = 0,$
4. $\mu(t) = 1,$
5. $\mu(t) = \left(1 + \sqrt[k]{t}\right)^k - t,$
6. $\mu(t) = (q-1)t,$
7. $\mu(t) = (q-1)t,$
8. $\mu(t) = \frac{1-p}{p} t, t \in \mathbb{T} \setminus \{1\}, \mu(1) = 0.$

Exercise 1.32 Find μ and σ if

$$\frac{2}{\mu(t) - 3} + \frac{3}{\mu(t) + 3} + \frac{7}{9 - \mu^2(t)} = 0, \quad \mu(t) \neq 3.$$

Solution We have

$$\begin{aligned} 0 &= 2(\mu(t) + 3) + 3(\mu(t) - 3) - 7 \\ &= 2\mu(t) + 6 + 3\mu(t) - 9 - 7 \\ &= 5\mu(t) - 10, \end{aligned}$$

so that

$$\mu(t) = 2.$$

Hence,

$$\sigma(t) = \mu(t) + t = 2 + t.$$

Exercise 1.33 Find μ and σ if

$$\frac{3}{(\mu(t))^2 + \mu(t) - 2} = \frac{1}{\mu(t)(\mu(t)-1)^2} + \frac{3}{\mu(t)(\mu(t)-3)}, \quad \mu(t) \notin \{0, 1, 3\}.$$

Solution For $\mu(t) \notin \{0, 1, 3\}$, we have

$$\frac{3}{(\mu(t)+2)(\mu(t)-1)} = \frac{1}{\mu(t)(\mu(t)-1)^2} + \frac{3}{\mu(t)(\mu(t)-3)},$$

whereupon

$$\begin{aligned}
0 &= 3\mu(t)(\mu(t)-1)(\mu(t)-3) - \Big((\mu(t)+2)(\mu(t)-3) + 3(\mu(t)-1)^2(\mu(t)+2)\Big) \\
&= 3\mu(t)((\mu(t))^2 - 4\mu(t) + 3) \\
&\quad - \Big((\mu(t))^2 - \mu(t) - 6 + 3((\mu(t))^2 - 2\mu(t) + 1)(\mu(t)+2)\Big) \\
&= 3(\mu(t))^3 - 12(\mu(t))^2 + 9\mu(t) \\
&\quad - \Big((\mu(t))^2 - \mu(t) - 6 + 3(\mu(t))^3 + 6(\mu(t))^2 - 6(\mu(t))^2 - 12\mu(t) + 3\mu(t) + 6\Big) \\
&= -13(\mu(t))^2 + 19\mu(t),
\end{aligned}$$

so that

$$\mu(t) = \frac{19}{13} \quad \text{and} \quad \sigma(t) = \mu(t) + t = \frac{19}{13} + t.$$

Exercise 1.34 Find μ and σ if

$$\sqrt{\mu(t) + \sqrt{\mu(t)+11}} - \sqrt{\mu(t) - \sqrt{\mu(t)+11}} = 2.$$

Solution From the given equation, we get

$$\mu(t) - \sqrt{\mu(t)+11} \geq 0,$$

whereupon

$$\mu(t) \geq \frac{1+3\sqrt{5}}{2}.$$

Then

$$0 = \left(\sqrt{\mu(t) + \sqrt{\mu(t) + 11}} - \sqrt{\mu(t) - \sqrt{\mu(t) + 11}}\right)^2 - 4$$

$$= \mu(t) + \sqrt{\mu(t) + 11} - 2\sqrt{(\mu(t))^2 - \mu(t) - 11} + \mu(t) - \sqrt{\mu(t) + 11} - 4$$

$$= -2\sqrt{(\mu(t))^2 - \mu(t) - 11} - 4 + 2\mu(t),$$

whence

$$0 = (\mu(t))^2 - \mu(t) - 11 - 4 + 4\mu(t) - (\mu(t))^2$$

$$= 3\mu(t) - 15$$

and

$$\mu(t) = 5.$$

Hence,

$$\sigma(t) = \mu(t) + t = 5 + t.$$

Exercise 1.35 Find μ and σ if

1. $\frac{2\mu(t)-5}{\mu(t)-2} + 1 - \frac{3\mu(t)-5}{\mu(t)-1} = 0$, $\mu(t) \notin \{1, 2\}$,
2. $3\mu(t) - \frac{3\mu^2(t)+2}{\mu(t)+5} = 4$,
3. $\frac{3\mu(t)+9}{3\mu(t)-1} + \frac{2\mu(t)-13}{2\mu(t)+5} = 2$, $\mu(t) \neq \frac{1}{3}$,
4. $\frac{5\mu(t)+13}{5\mu(t)+4} + \frac{6\mu(t)-4}{3\mu(t)-1} = 3$, $\mu(t) \neq \frac{1}{3}$,
5. $\frac{\mu(t)-2}{2\mu(t)+6} + \frac{\mu(t)+3}{3\mu(t)-6} = \frac{5}{6}$, $\mu(t) \neq 2$,
6.
$$\frac{\mu(t)}{2(\mu(t))^2 + 12\mu(t) + 10} + \frac{3\mu(t) + 1}{4(\mu(t))^2 + 16\mu(t) - 20} - \frac{\mu(t) + 34}{(\mu(t))^3 + 5(\mu(t))^2 - \mu(t) - 5} = 0, \quad \mu(t) \neq 1,$$
7.
$$\frac{1}{2\mu(t) + 3} - \frac{1}{(\mu(t))^2 - 16} + \frac{1}{2(\mu(t))^2 + 11\mu(t) + 12} - \frac{\mu(t) - 8}{2(\mu(t))^3 + 3(\mu(t))^2 - 32\mu(t) - 48} = 0, \quad \mu(t) \neq 4$$
8. $\frac{(\mu(t))^2+\mu(t)-7}{\mu(t)} - \frac{35\mu(t)}{(\mu(t))^2+\mu(t)-7} = 2$, $\mu(t) \notin \left\{0, \frac{\sqrt{29}-1}{2}\right\}$,

9. $2\left(1+\frac{9}{\mu(t)}\right)+3\left(\frac{\mu(t)+9}{\mu(t)}\right)^2=56,\ \mu(t)\neq 0,$
10. $\left(\mu(t)+\frac{8}{\mu(t)}\right)^2+\mu(t)=42-\frac{8}{\mu(t)},\ \mu(t)\neq 0,$
11. $(\mu(t))^2-4\mu(t)+\sqrt{(\mu(t))^2-4\mu(t)+4}=8,$
12. $2(\mu(t))^2+\mu(t)+\sqrt{2(\mu(t))^2+\mu(t)+4}=26,$
13. $3(\mu(t))^2+\mu(t)-\sqrt{(\mu(t))^2+\mu(t)+2}=4,$
14. $(\mu(t))^2-\mu(t)-\sqrt{3(\mu(t))^2-3\mu(t)+13}=5,$
15. $\sqrt{4(\mu(t))^2+3\mu(t)+14}-\sqrt{4(\mu(t))^2+3\mu(t)+3}=1.$

Solution 1. $3, 3+t,$
2. $2, 2+t,$
3. $2, 2+t,$
4. $1, 1+t,$
5. $12, 12+t,$
6. $\frac{27}{5}, \frac{27}{5}+t,$
7. $5, 5+t,$
8. $1, 1+t, 7, 7+t,$
9. $3, 3+t,$
10. $2, 2+t, 4, 4+t,$
11. $5, 5+t,$
12. $3, 3+t,$
13. $1, 1+t,$
14. $4, 4+t,$
15. $2, 2+t.$

Definition 1.36 For $f:\mathbb{T}\to\mathbb{R}$, we define the *forward shift* $f^\sigma:\mathbb{T}\to\mathbb{R}$ by

$$f^\sigma(t)=f(\sigma(t))\quad\text{for any}\quad t\in\mathbb{T},\quad\text{i.e.,}\quad f^\sigma=f\circ\sigma.$$

Example 1.37 Let $\mathbb{T}=\{t=2^{n+2}:n\in\mathbb{N}\}$, $f(t)=t^2+t-1$. Then

$$\sigma(t)=\inf\left\{2^{l+2}:\ 2^{l+2}>2^{n+2},\ l\in\mathbb{N}\right\}=2^{n+3}=2t.$$

Hence,

$$f^\sigma(t)=f(\sigma(t))=(\sigma(t))^2+\sigma(t)-1=(2t)^2+2t-1=4t^2+2t-1,\quad t\in\mathbb{T}.$$

Example 1.38 Let $\mathbb{T}=\{t=\sqrt{n+3}:n\in\mathbb{N}\}$, $f(t)=t+3$, $t\in\mathbb{T}$. Then $n=t^2-3$ and

$$\sigma(t)=\inf\{\sqrt{l+3}:\ \sqrt{l+3}>\sqrt{n+3},\ l\in\mathbb{N}\}=\sqrt{n+4}=\sqrt{t^2+1}.$$

Hence,

$$f(\sigma(t))=\sigma(t)+3=\sqrt{t^2+1}+3.$$

Example 1.39 Let $\mathbb{T} = \left\{\frac{1}{n} : n \in \mathbb{N}\right\} \cup \{0\}$, $f(t) = t^3 - t$, $t \in \mathbb{T}$.

1. If $t = \frac{1}{n}$ for $n \in \mathbb{N} \setminus \{1\}$, then $n = \frac{1}{t}$ and

$$\sigma(t) = \inf\left\{s \in \mathbb{T} : \ s > \frac{1}{n}\right\} = \frac{1}{n-1} = \frac{t}{1-t}$$

so that

$$f(\sigma(t)) = (\sigma(t))^3 - \sigma(t) = \left(\frac{t}{1-t}\right)^3 - \frac{t}{1-t} = \frac{t^3}{(1-t)^3} - \frac{t}{1-t}$$
$$= \frac{t^3 - t(1-t)^2}{(1-t)^3} = \frac{t^3 - t(1-2t+t^2)}{(1-t)^3}$$
$$= \frac{t^3 - t + 2t^2 - t^3}{(1-t)^3} = \frac{t(2t-1)}{(1-t)^3}.$$

2. If $t = 1$, then

$$\sigma(1) = \inf\{s \in \mathbb{T} : \ s > 1\} = \inf \emptyset = \sup \mathbb{T} = 1$$

and

$$f(\sigma(1)) = (\sigma(1))^3 - \sigma(1) = 1 - 1 = 0.$$

3. If $t = 0$, then

$$\sigma(0) = \inf\{s \in \mathbb{T} : \ s > 0\} = 0$$

and

$$f(\sigma(0)) = (\sigma(0))^3 - \sigma(0) = 0.$$

Exercise 1.40 Let $\mathbb{T} = \left\{t = \sqrt[3]{n+2} : n \in \mathbb{N}\right\}$, $f(t) = 1 - t^3, t \in \mathbb{T}$. Find $f(\sigma(t))$, $t \in \mathbb{T}$.

Solution $-t^3$.

Exercise 1.41 Determine the conditions for t under which we have both

$$\sigma(\rho(t)) = t \quad \text{and} \quad \rho(\sigma(t)) = t.$$

Solution 1. If t is isolated, then

$$\sigma(\rho(t)) = t \quad \text{and} \quad \rho(\sigma(t)) = t.$$

2. If t is right-scattered and left-dense, then

$$\sigma(\rho(t)) = \sigma(t) \neq t \quad \text{and} \quad \rho(\sigma(t)) = t.$$

3. If t is right-dense and left-scattered, then

$$\rho(\sigma(t)) = \rho(t) \neq t \quad \text{and} \quad \sigma(\rho(t)) = t.$$

4. If t is dense, then

$$\rho(\sigma(t)) = \rho(t) = t \quad \text{and} \quad \sigma(\rho(t)) = \sigma(t) = t.$$

Definition 1.42 We assume that $a \leq b$. We define the interval $[a, b]$ in $\mathbb{T}$ by

$$[a, b] = [a, b]_{\mathbb{T}} = \{t \in \mathbb{T} : a \leq t \leq b\}.$$

Open intervals, half-open intervals and so on are defined accordingly.

Definition 1.43 We define the set

$$\mathbb{T}^{\kappa} = \begin{cases} \mathbb{T} \setminus (\rho(\sup \mathbb{T}), \sup \mathbb{T}] & \text{if} \quad \sup \mathbb{T} < \infty \\ \\ \mathbb{T} & \text{otherwise.} \end{cases}$$

Example 1.44 Let $\mathbb{T} = \left\{\frac{1}{n} : n \in \mathbb{N}\right\} \cup \{0\}$. Then $\sup \mathbb{T} = 1$ and

$$\rho(1) = \sup \{s \in \mathbb{T} : \ s < 1\} = \frac{1}{2}.$$

Therefore,

$$\mathbb{T}^{\kappa} = \mathbb{T} \setminus \left(\frac{1}{2}, 1\right] = \left\{\frac{1}{n} : n \in \mathbb{N} \setminus \{1\}\right\} \cup \{0\}.$$

Example 1.45 Let $\mathbb{T} = \{2n : n \in \mathbb{N}\}$. Then $\sup \mathbb{T} = \infty$ and $\mathbb{T}^{\kappa} = \mathbb{T}$.

Example 1.46 Let $\mathbb{T} = \left\{\frac{1}{n^2+3} : n \in \mathbb{N}\right\} \cup \{0\}$. Then $\sup \mathbb{T} = \frac{1}{4} < \infty$ and

$$\rho\left(\frac{1}{4}\right) = \sup \left\{s \in \mathbb{T} : \ s < \frac{1}{4}\right\} = \frac{1}{7}.$$

Hence,

$$\mathbb{T}^{\kappa} = \mathbb{T} \setminus \left(\frac{1}{7}, \frac{1}{4}\right] = \left\{\frac{1}{n^2 + 3} \ : \ n \in \mathbb{N} \setminus \{1\}\right\} \cup \{0\}.$$

Example 1.47 Let $[a, b]$ be an interval in $\mathbb{T}$ and let $b \in \mathbb{T}$ be left-dense. Then

$$\sup[a, b] = b,$$

and since b is left-dense, we have $\rho(b) = b$. Hence,

$$[a, b]^\kappa = [a, b] \setminus (b, b] = [a, b] \setminus \emptyset = [a, b].$$

Example 1.48 Let $[a, b]$ be an interval in $\mathbb{T}$ and let $b \in \mathbb{T}$ be left-scattered. Then

$$\sup[a, b] = b,$$

and since b is left-scattered, we have $\rho(b) < b$. We assume that there exists $c \in (\rho(b), b]$, $c \in \mathbb{T}$, $c \neq b$. Then $\rho(b) < c < b$, which contradicts the definition of $\rho(b)$. Therefore,

$$[a, b]^\kappa = [a, b] \setminus (\rho(b), b] = [a, b).$$

Exercise 1.49 Let $\mathbb{T} = \left\{\frac{1}{2n+1} : n \in \mathbb{N}\right\} \cup \{0\}$. Find $\mathbb{T}^\kappa$.

Solution $\left\{\frac{1}{2n+1} : n \in \mathbb{N} \setminus \{1\}\right\} \cup \{0\}$.

1.2 Induction Principle

Theorem 1.50 (Induction Principle) *Let $t_0 \in \mathbb{T}$ and assume that*

$$\{S(t) : t \in [t_0, \infty)\}$$

is a family of statements satisfying

(i) $S(t_0)$ is true.
(ii) If $t \in [t_0, \infty)$ is right-scattered and $S(t)$ is true, then $S(\sigma(t))$ is true.
(iii) If $t \in [t_0, \infty)$ is right-dense and $S(t)$ is true, then there exists a neighbourhood U of t such that $S(s)$ is true for all $s \in U \cap (t, \infty)$.
(iv) If $t \in (t_0, \infty)$ is left-dense and $S(s)$ is true for $s \in [t_0, t)$, then $S(t)$ is true.

Then $S(t)$ is true for all $t \in [t_0, \infty)$.

Proof Let

$$S^* = \{t \in [t_0, \infty) : S(t) \text{ is not true}\}.$$

We assume that $S^* \neq \emptyset$. Let $\inf S^* = t^*$. Because $\mathbb{T}$ is closed, we have $t^* \in \mathbb{T}$.

1. If $t^* = t_0$, then $S(t^*)$ is true.
2. If $t^* \neq t_0$ and $t^* = \rho(t^*)$, then, using (iv), we get that $S(t^*)$ is true.
3. If $t^* \neq t_0$ and $\rho(t^*) < t^*$, then $\rho(t^*)$ is right-scattered. Since $S(\rho(t^*))$ is true, we get that $S(t^*)$ is true.

Consequently, $t^* \notin S^*$. If we suppose that t^* is right-scattered, then, using that $S(t^*)$ is true and (ii), we conclude that $S(\sigma(t^*))$ is true, which is a contradiction. From the definition of t^*, it follows that $t^* \neq \sup \mathbb{T}$. Since t^* is not right-scattered and $t^* \neq \sup \mathbb{T}$, we obtain that t^* is right-dense. Because $S(t^*)$ is true, using (iii), there exists a neighbourhood U of t^* such that $S(s)$ is true for all $s \in U \cap (t^*, \infty)$, which is a contradiction. Consequently, $S^* = \emptyset$. □

Theorem 1.51 (Dual Version of Induction Principle) *Let $t_0 \in \mathbb{T}$ and assume that*

$$\{S(t) : \ t \in (-\infty, t_0]\}$$

is a family of statements satisfying

(i) $S(t_0)$ *is true.*
(ii) *If $t \in (-\infty, t_0]$ is left-scattered and $S(t)$ is true, then $S(\rho(t))$ is true.*
(iii) *If $t \in (-\infty, t_0]$ is left-dense and $S(t)$ is true, then there exists a neighbourhood U of t such that $S(s)$ is true for all $s \in U \cap (-\infty, t)$.*
(iv) *If $t \in (-\infty, t_0)$ is right-dense and $S(s)$ is true for $s \in (t, t_0)$, then $S(t)$ is true.*

Then $S(t)$ is true for all $t \in (-\infty, t_0]$.

Proof Let

$$S^* = \{t \in (-\infty, t_0] : \ S(t) \quad \text{is not true} \quad \}.$$

Assume that $S^* \neq \emptyset$. Note that $t_0 \notin S^*$. Let

$$t^* = \sup S^*.$$

1. If $t^* = t_0$, then $S(t^*)$ is true.
2. If $t^* \neq t_0$ and $t^* = \sigma(t^*)$, then, using *iv*, we obtain that $S(t^*)$ is true.
3. If $t^* \neq t_0$ and $t^* < \sigma(t^*)$, then $\sigma(t^*)$ is left-scattered, $\rho(\sigma(t^*)) = t^*$, and $S(\sigma(t^*))$ is true. Therefore, we conclude that $S(t^*)$ is true.

Therefore, $t^* \notin S^*$. If t^* is left-scattered, then, using (ii), we get that $S(\rho(t^*))$ is true, which is a contradiction. If t^* is left-dense, then, using (iii), there exists a neighbourhood U of t^* such that $S(s)$ is true for all $s \in U \cap (-\infty, t^*)$, which is a contradiction. Consequently, $S^* = \emptyset$. □

Example 1.52 Let $\mathbb{T} = \mathbb{N}$. We will prove that

$$\frac{1}{3}(s^2 - r^2) \leq \frac{1}{7}(s^3 - r^3) \quad \text{for every} \quad r \leq s, \quad s, r \in \mathbb{T}. \tag{1.1}$$

Let $\varepsilon > 0$ be arbitrarily chosen. We will prove that

$$S(t) : \ \frac{1}{3}(t^2 - r^2) \leq \frac{1}{7}(t^3 - r^3) + \varepsilon(t - r) \tag{1.2}$$

is a true statement for all $t \in [r, s]$.

1. The statement $S(r)$ is trivially satisfied.

2. Let t be right-scattered. Let $S(t)$ be true for $t \in [r, s)$. Then $\sigma(t) = t + 1$. Note that
$$9t^2 - 5t - 4 \geq 0.$$

Hence,
$$14t + 7 \leq 9t^2 + 9t + 3,$$

whereupon
$$\frac{1}{3}(2t + 1) \leq \frac{1}{7}(3t^2 + 3t + 1). \tag{1.3}$$

Since $S(t)$ is true, we have that
$$\begin{aligned}\frac{1}{3}(t^2 - r^2) &\leq \frac{1}{7}(t^3 - r^3) + \varepsilon(t - r)\\ &\leq \frac{1}{7}(t^3 - r^3) + \varepsilon(t + 1 - r).\end{aligned}$$

From the last inequality and from (1.3), we get
$$\frac{1}{3}(t^2 + 2t + 1 - r^2) \leq \frac{1}{7}(t^3 + 3t^2 + 3t + 1 - r^3) + \varepsilon(t + 1 - r),$$

i.e.,
$$\frac{1}{3}\left((t + 1)^2 - r^2\right) \leq \frac{1}{7}\left((t + 1)^3 - r^3\right) + \varepsilon(t + 1 - r),$$

i.e.,
$$\frac{1}{3}\left((\sigma(t))^2 - r^2\right) \leq \frac{1}{7}\left((\sigma(t))^3 - r^3\right) + \varepsilon(\sigma(t) - r).$$

Therefore, $S(\sigma(t))$ holds.

Hence, by the induction principle, we conclude that (1.2) is true for all $t \in [r, s]$. Because $\varepsilon > 0$ was arbitrarily chosen, we conclude that (1.1) holds for all $r \leq s$, $r, s \in \mathbb{T}$.

Exercise 1.53 Let $\mathbb{T} = 3^{\mathbb{N}}$. Prove that
$$\frac{1}{4}(s^2 - r^2) \leq \frac{1}{13}(s^3 - r^3) \quad \text{for every} \quad r \leq s, \quad s, r \in \mathbb{T}.$$

1.3 Advanced Practical Problems

Problem 1.54 Classify each point $t \in \mathbb{T} = \{\sqrt[4]{7n} : n \in \mathbb{N}_0\}$ as left-dense, left-scattered, right-dense or right-scattered.

Solution Each point $t = \sqrt[4]{7n}$, $n \in \mathbb{N}$, is isolated, $t = 0$ is right-scattered.

Problem 1.55 Let $t \in \mathbb{T}$ be isolated so that $\sigma(t) = t + 1$ and $\rho(t) = t - 1$. Simplify

$$A = (\sigma(t))^2 + (\rho(t))^3 + 1,$$

$$B = A - (\sigma(t) + \rho(t)) + (\sigma(t))^2,$$

$$C = A + 2B.$$

Solution $A = t^3 - 2t^2 + 5t + 1$, $B = t^3 - t^2 + 5t + 2$, $C = 3t^3 - 4t^2 + 15t + 5$.

Problem 1.56 For $t \in \mathbb{T}$, simplify

1. $\frac{\frac{\sigma(t)}{\sigma(t)-1} - \frac{\sigma(t)+1}{\sigma(t)}}{\frac{\sigma(t)-1}{\sigma(t)} - \frac{\sigma(t)}{\sigma(t)+1}} \left((\sigma(t))^2 + (\sigma(t))^3 - \frac{\sigma(t)}{1 - \frac{\sigma(t)}{\sigma(t)+1}} \right)$, $\sigma(t) \notin \{-1, 0, 1\}$,
2. $\frac{\frac{1}{1 + \frac{\sigma(t)}{1 - \frac{\sigma(t)}{\sigma(t)+2}}}}{\frac{\frac{1}{1-\sigma(t)} + \frac{1}{1+\sigma(t)}}{\frac{1}{1-\sigma(t)} - \frac{1}{1+\sigma(t)}}}$, $\sigma(t) \notin \{-2, -1, 0, 1\}$,
3. $\frac{\frac{\rho(t)-4}{4\rho(t)} + \frac{\rho(t)-12}{4\rho(t)-16} - \frac{\rho(t)+4}{4\rho(t)-(\rho(t))^2}}{\frac{1}{2\rho(t)}}$, $\rho(t) \notin \{0, 4\}$.

Solution 1. $-\sigma(t)(\sigma(t) + 1)^2$,
2. $\frac{2\sigma(t)}{(\sigma(t))^2 + 2\sigma(t) + 2}$,
3. $\rho(t) - 4$.

Problem 1.57 Let $\mathbb{T} = \left\{ \sqrt[4]{n+7} : n \in \mathbb{N}_0 \right\}$. Find μ.

Solution $\mu\left(\sqrt[4]{n+7}\right) = \sqrt[4]{n+8} - \sqrt[4]{n+7}$.

Problem 1.58 Find μ and σ if

1. $\frac{\mu(t)+5}{5\mu(t)-20} + \frac{\mu(t)-4}{4\mu(t)+20} = \frac{9}{20}$, $\mu(t) \neq 4$,
2. $\frac{1+4\mu(t)}{2} = \frac{1-4\mu(t)}{1+6\mu(t)} + \frac{1+6\mu(t)}{3}$,
3. $\frac{2\mu(t)-1}{5\mu(t)-1} = \frac{2\mu(t)+1}{4} - \frac{3\mu(t)-1}{6}$, $\mu(t) \neq \frac{1}{5}$,
4. $\frac{1}{(\mu(t)+1)^2} + \frac{4}{\mu(t)(1+\mu(t))^2} = \frac{5}{2\mu(t)(\mu(t)+1)}$, $\mu(t) \neq 0$,
5. $\frac{4\mu(t)-1}{(\mu(t))^2-4} + \frac{4}{2-\mu(t)} + \frac{1}{\mu(t)+2} = 0$, $\mu(t) \neq 2$,
6. $\frac{(\mu(t))^2-2\mu(t)+5}{\mu(t)-1} - \frac{20\mu(t)-20}{(\mu(t))^2-2\mu(t)+5} = 1$, $\mu(t) \neq 1$,
7. $\frac{21}{(\mu(t))^2-4\mu(t)+10} - (\mu(t))^2 + 4\mu(t) = 6$,

8. $7\left(\mu(t)+\frac{1}{\mu(t)}\right)-2\left((\mu(t))^2+\frac{1}{\mu^2(t)}\right)=9,\ \mu(t)\neq 0,$
9. $(\mu(t))^2+\frac{4}{(\mu(t))^2}+6\left(\mu(t)+\frac{2}{\mu(t)}\right)=23,\ \mu(t)\neq 0,$
10. $((\sigma(t))^2-3\sigma(t)+1)^2=8(\sigma(t))^2-24\sigma(t)+41,$
11. $\sqrt{15(\mu(t))^2+6\mu(t)+4}-\sqrt{10(\mu(t))^2+4\mu(t)+2}=1,$
12. $\sqrt{5(\mu(t))^2+3\mu(t)+1}+\sqrt{5(\mu(t))^2+3\mu(t)+8}=7,$
13. $\sqrt{3(\mu(t))^2-5\mu(t)+4}-\sqrt{3(\mu(t))^2-5\mu(t)-3}=1,\ \mu(t)\geq\frac{5+\sqrt{61}}{6},$
14. $\sqrt{\frac{\mu(t)+5}{\mu(t)}}+4\sqrt{\frac{\mu(t)}{\mu(t)+5}}=4,\ \mu(t)\neq 0,$
15. $(\mu(t))^2+\mu(t)=\sqrt{9-6\mu(t)+(\mu(t))^2}.$

Solution 1. $40, 40+t,$
2. $\frac{1}{6}, \frac{1}{6}+t,$
3. No solutions,
4. $1, 1+t,$
5. $11, 11+t,$
6. $2, 2+t, 5, 5+t,$
7. $1, 1+t, 3, 3+t,$
8. $\frac{1}{2}, \frac{1}{2}+t, 2, 2+t,$
9. $1, 1+t, 2, 2+t,$
10. $5, 5-t, -2, -2-t,$
11. $1, 1+t,$
12. $1, 1+t,$
13. $3, 3+t,$
14. $\frac{5}{3}, \frac{5}{3}+t,$
15. $1, 1+t.$

Problem 1.59 Let $\mathbb{T}=2^{\mathbb{N}}$ and define $f:\mathbb{T}\to\mathbb{R}$ by

$$f(t)=t^3-4t^2+t.$$

Find $f^\sigma(t)$.

Solution $8t^3-16t^2+2t.$

Problem 1.60 Let $\mathbb{T}=\left\{\frac{1}{3n+5}:n\in\mathbb{N}\right\}\cup\{0\}$. Find $\mathbb{T}^\kappa$.

Solution $\left\{\frac{1}{3n+5}:n\in\mathbb{N}\setminus\{1\}\right\}\cup\{0\}.$

Problem 1.61 Let $\mathbb{T}=3\mathbb{N}$. Prove that

$$s^2-r^2\leq\frac{1}{6}(s^3-r^3)\quad\text{for every}\quad r\leq s,\quad s,r\in\mathbb{T}.$$

1.4 Notes and References

The theory of time scales was introduced by Stefan Hilger in his PhD thesis [31] in 1988 (supervised by Bernd Aulbach) in order to unify continuous and discrete analysis. The first publications on this subject are Hilger [32] and Aulbach and Hilger [4, 5]. Important examples of time scales are given in this chapter. Such examples contain of course $\mathbb{R}$ (the set of all real numbers) and $\mathbb{Z}$ (the set of all integers), the set of all integer multiples of a number $h > 0$, the set of all integer powers of a number $q > 1$ (including 0), unions of disjoint closed intervals, and the Cantor set. Classifications of points on time scales are considered in this chapter. The basic definitions of forward and backward jump operators are due to Hilger. In this chapter, many examples of jump operators on various time scales are given. The graininess function, which is the distance from a point to the "next" point on the right, is introduced in this chapter. The induction principle (Theorem 1.50) on time scales is essentially contained in Dieudonné [26], and it can be found together with its proof in [21, Theorem 1.7]. All results presented in this chapter are taken from Bohner and Peterson [21].

Chapter 2
Differential Calculus of Functions of One Variable

2.1 Differentiable Functions of One Variable

Definition 2.1 Assume that $f : \mathbb{T} \to \mathbb{R}$ is a function and let $t \in \mathbb{T}^\kappa$. We define $f^\Delta(t)$ to be the number, provided it exists, with the property that for any $\varepsilon > 0$, there exists a neighbourhood U of t, $U = (t - \delta, t + \delta) \cap \mathbb{T}$ for some $\delta > 0$, such that

$$|f(\sigma(t)) - f(s) - f^\Delta(t)(\sigma(t) - s)| \le \varepsilon|\sigma(t) - s| \quad \text{for all} \quad s \in U.$$

We call $f^\Delta(t)$ the *delta* or *Hilger* derivative of f at t. We say that f is *delta* or *Hilger differentiable*, shortly *differentiable*, in $\mathbb{T}^\kappa$ if $f^\Delta(t)$ exists for all $t \in \mathbb{T}^\kappa$. The function $f^\Delta : \mathbb{T} \to \mathbb{R}$ is said to be the *delta derivative* or *Hilger derivative*, shortly *derivative*, of f in $\mathbb{T}^\kappa$.

Remark 2.2 If $\mathbb{T} = \mathbb{R}$, then the delta derivative coincides with the classical derivative.

Theorem 2.3 *The delta derivative is well defined.*

Proof Let $t \in \mathbb{T}^\kappa$. Suppose $f_1^\Delta(t)$ and $f_2^\Delta(t)$ are such that

$$|f(\sigma(t)) - f(s) - f_1^\Delta(t)(\sigma(t) - s)| \le \frac{\varepsilon}{2}|\sigma(t) - s|$$

and

$$|f(\sigma(t)) - f(s) - f_2^\Delta(t)(\sigma(t) - s)| \le \frac{\varepsilon}{2}|\sigma(t) - s|$$

for any $\varepsilon > 0$ and any s belonging to a neighbourhood U of t, $U = (t - \delta, t + \delta) \cap \mathbb{T}$ for some $\delta > 0$. Hence, if $s \ne \sigma(t)$, then

M. Bohner and S.G. Georgiev, *Multivariable Dynamic Calculus on Time Scales*,
DOI 10.1007/978-3-319-47620-9_2

$$
\begin{aligned}
|f_1^{\Delta}(t) - f_2^{\Delta}(t)| &= \left| f_1^{\Delta}(t) - \frac{f(\sigma(t)) - f(s)}{\sigma(t) - s} + \frac{f(\sigma(t)) - f(s)}{\sigma(t) - s} - f_2^{\Delta}(t) \right| \\
&\leq \left| f_1^{\Delta}(t) - \frac{f(\sigma(t)) - f(s)}{\sigma(t) - s} \right| + \left| \frac{f(\sigma(t)) - f(s)}{\sigma(t) - s} - f_2^{\Delta}(t) \right| \\
&= \frac{|f(\sigma(t)) - f(s) - f_1^{\Delta}(t)(\sigma(t) - s)|}{|\sigma(t) - s|} + \frac{|f(\sigma(t)) - f(s) - f_2^{\Delta}(t)(\sigma(t) - s)|}{|\sigma(t) - s|} \\
&\leq \frac{\varepsilon}{2} + \frac{\varepsilon}{2} \\
&= \varepsilon.
\end{aligned}
$$

Since $\varepsilon > 0$ was arbitrarily chosen, we conclude that

$$
f_1^{\Delta}(t) = f_2^{\Delta}(t).
$$

This completes the proof. □

Remark 2.4 Let us assume that $\sup \mathbb{T} < \infty$ and $f^{\Delta}(t)$ is defined at a point $t \in \mathbb{T} \setminus \mathbb{T}^{\kappa}$ with the same definition as given in Definition 2.1. Then the unique point $t \in \mathbb{T} \setminus \mathbb{T}^{\kappa}$ is $\sup \mathbb{T}$. Hence, for any $\varepsilon > 0$, there is a neighbourhood $U = (t - \delta, t + \delta) \cap (\mathbb{T} \setminus \mathbb{T}^{\kappa})$, for some $\delta > 0$, such that

$$
f(\sigma(t)) = f(s) = f(\sigma(\sup \mathbb{T})) = f(\sup \mathbb{T}), \quad s \in U.
$$

Therefore, for any $\alpha \in \mathbb{R}$ and $s \in U$, we have

$$
\begin{aligned}
|f(\sigma(t)) - f(s) - \alpha(\sigma(t) - s)| &= |f(\sup \mathbb{T}) - f(\sup \mathbb{T}) - \alpha(\sup \mathbb{T} - \sup \mathbb{T})| \\
&\leq \varepsilon |\sigma(t) - s|,
\end{aligned}
$$

i.e., any $\alpha \in \mathbb{R}$ is the delta derivative of f at the point $t \in \mathbb{T} \setminus \mathbb{T}^{\kappa}$.

Example 2.5 Let $f(t) = \alpha \in \mathbb{R}$. We will prove that $f^{\Delta}(t) = 0$ for any $t \in \mathbb{T}^{\kappa}$. Indeed, for $t \in \mathbb{T}^{\kappa}$ and for any $\varepsilon > 0$, $s \in (t - 1, t + 1) \cap \mathbb{T}$ implies

$$
|f(\sigma(t)) - f(s) - 0(\sigma(t) - s)| = |\alpha - \alpha| = 0 \leq \varepsilon |\sigma(t) - s|.
$$

Example 2.6 Let $f(t) = t$, $t \in \mathbb{T}$. We will prove that $f^{\Delta}(t) = 1$ for any $t \in \mathbb{T}^{\kappa}$. Indeed, for $t \in \mathbb{T}^{\kappa}$ and for any $\varepsilon > 0$, $s \in (t - 1, t + 1) \cap \mathbb{T}$ implies

$$|f(\sigma(t)) - f(s) - 1(\sigma(t) - s)| = |\sigma(t) - s - (\sigma(t) - s)|$$

$$= 0 \leq \varepsilon|\sigma(t) - s|.$$

Example 2.7 Let $f(t) = t^2$, $t \in \mathbb{T}$. We will prove that $f^\Delta(t) = \sigma(t) + t$, $t \in \mathbb{T}^\kappa$. Indeed, for $t \in \mathbb{T}^\kappa$ and for any $\varepsilon > 0$, $s \in (t - \varepsilon, t + \varepsilon) \cap \mathbb{T}$ implies $|t - s| < \varepsilon$ and

$$|f(\sigma(t)) - f(s) - (\sigma(t) + t)(\sigma(t) - s)|$$

$$= |(\sigma(t))^2 - s^2 - (\sigma(t) + t)(\sigma(t) - s)|$$

$$= |(\sigma(t) - s)(\sigma(t) + s) - (\sigma(t) + t)(\sigma(t) - s)|$$

$$= |\sigma(t) - s||t - s|$$

$$\leq \varepsilon|\sigma(t) - s|.$$

Exercise 2.8 Let $f(t) = \sqrt{t}$, $t \in \mathbb{T}$, $t > 0$. Prove that

$$f^\Delta(t) = \frac{1}{\sqrt{t} + \sqrt{\sigma(t)}} \quad \text{for all} \quad t \in \mathbb{T}^\kappa \cap (0, \infty).$$

Exercise 2.9 Let $f(t) = t^3$, $t \in \mathbb{T}$. Prove that

$$f^\Delta(t) = (\sigma(t))^2 + t\sigma(t) + t^2 \quad \text{for all} \quad t \in \mathbb{T}^\kappa.$$

Theorem 2.10 *Assume $f : \mathbb{T} \to \mathbb{R}$ is a function and let $t \in \mathbb{T}^\kappa$. Then we have the following.*

1. *If f is differentiable at t, then f is continuous at t.*
2. *If f is continuous at t and t is right-scattered, then f is differentiable at t with*

$$f^\Delta(t) = \frac{f(\sigma(t)) - f(t)}{\mu(t)}.$$

3. *If t is right-dense, then f is differentiable at t iff the limit*

$$\lim_{s \to t} \frac{f(t) - f(s)}{t - s}$$

exists as a finite number. In this case,

$$f^{\Delta}(t) = \lim_{s \to t} \frac{f(t) - f(s)}{t - s}.$$

4. *If f is differentiable at t, then the "simple useful formula"*

$$f(\sigma(t)) = f(t) + \mu(t) f^{\Delta}(t)$$

holds.

Proof 1. Assume that f is differentiable at $t \in \mathbb{T}^{\kappa}$. Let $\varepsilon \in (0, 1)$ be arbitrarily chosen. Set

$$\varepsilon^* = \frac{\varepsilon}{1 + |f^{\Delta}(t)| + 2\mu(t)}.$$

Since f is differentiable at t, there exists a neighbourhood U of t such that

$$|f(\sigma(t)) - f(s) - f^{\Delta}(t)(\sigma(t) - s)| \leq \varepsilon^* |\sigma(t) - s|.$$

Hence, for all $s \in U \cap (t - \varepsilon^*, t + \varepsilon^*)$, we have

$$\begin{aligned}
|f(t) - f(s)| &= |f(t) + f(\sigma(t)) - f(s) - f^{\Delta}(t)(\sigma(t) - s) \\
&\quad - f(\sigma(t)) + f^{\Delta}(t)(\sigma(t) - s)| \\
&\leq |f(\sigma(t)) - f(s) - f^{\Delta}(t)(\sigma(t) - s)| \\
&\quad + |f(\sigma(t)) - f(t) - f^{\Delta}(t)(\sigma(t) - s)| \\
&\leq \varepsilon^* |\sigma(t) - s| + |f(\sigma(t)) - f(t) - f^{\Delta}(t)(\sigma(t) - t) \\
&\quad + f^{\Delta}(t)(s - t)| \\
&\leq \varepsilon^* |\sigma(t) - s| + |f(\sigma(t)) - f(t) - f^{\Delta}(t)(\sigma(t) - t)| \\
&\quad + |f^{\Delta}(t)||s - t| \\
&\leq \varepsilon^* |\sigma(t) - s| + \varepsilon^* \mu(t) + \varepsilon^* |f^{\Delta}(t)| \\
&= \varepsilon^* \left(|\sigma(t) - s| + \mu(t) + |f^{\Delta}(t)| \right) \\
&= \varepsilon^* \left(|\sigma(t) - t + t - s| + \mu(t) + |f^{\Delta}(t)| \right)
\end{aligned}$$

$$
\begin{aligned}
&\le \varepsilon^* \left(\sigma(t) - t + |t-s| + \mu(t) + |f^{\Delta}(t)|\right)\\
&= \varepsilon^* \left(2\mu(t) + |t-s| + |f^{\Delta}(t)|\right)\\
&\le \varepsilon^*(2\mu(t) + \varepsilon^* + |f^{\Delta}(t)|)\\
&\le \varepsilon^*(1 + 2\mu(t) + |f^{\Delta}(t)|)\\
&= \varepsilon,
\end{aligned}
$$

which completes the proof.

2. Assume that f is continuous at t and t is right-scattered. By continuity, we have

$$
\begin{aligned}
\lim_{s\to t} \frac{f(\sigma(t)) - f(s)}{\sigma(t) - s} &= \frac{f(\sigma(t)) - f(t)}{\sigma(t) - t}\\
&= \frac{f(\sigma(t)) - f(t)}{\mu(t)}.
\end{aligned}
$$

Therefore, for any $\varepsilon > 0$, there exists a neighbourhood U of t such that

$$
\left| \frac{f(\sigma(t)) - f(s)}{\sigma(t) - s} - \frac{f(\sigma(t)) - f(t)}{\mu(t)} \right| \le \varepsilon
$$

for all $s \in U$, i.e.,

$$
\left| f(\sigma(t)) - f(s) - \frac{f(\sigma(t)) - f(t)}{\mu(t)}(\sigma(t) - s) \right| \le \varepsilon |\sigma(t) - s|
$$

for all $s \in U$. Hence,

$$
f^{\Delta}(t) = \frac{f(\sigma(t)) - f(t)}{\mu(t)}.
$$

3. Assume that t is right-dense. Let $\varepsilon > 0$ be arbitrarily chosen. Then f is differentiable at t iff there is a neighbourhood U of t such that

$$
|f(t) - f(s) - f^{\Delta}(t)(t-s)| \le \varepsilon |t-s| \quad \text{for all} \quad s \in U,
$$

i.e., iff

$$
\left| \frac{f(t) - f(s)}{t - s} - f^{\Delta}(t) \right| \le \varepsilon \quad \text{for all} \quad s \in U,
$$

i.e., iff

$$\lim_{s\to t}\frac{f(t)-f(s)}{t-s}=f^{\Delta}(t).$$

4. Assume that f is differentiable at t.
   a. If t is right-dense, then $\sigma(t)=t$, $\mu(t)=0$ and
   $$f(\sigma(t))=f(t)=f(t)+\mu(t)f^{\Delta}(t).$$
   b. If t is right-scattered, then
   $$f^{\Delta}(t)=\frac{f(\sigma(t))-f(t)}{\mu(t)},$$
   whereupon
   $$f(\sigma(t))=f(t)+\mu(t)f^{\Delta}(t).$$

This completes the proof. □

Example 2.11 Let

$$\mathbb{T}=\left\{\frac{1}{2n+1}:n\in\mathbb{N}_0\right\}\cup\{0\}$$

and $f(t)=\sigma(t)$ for $t\in\mathbb{T}$. We will find $f^{\Delta}(t)$ for $t\in\mathbb{T}^{\kappa}=\mathbb{T}\setminus\{1\}$. For

$$t\in\mathbb{T}^{\kappa}\setminus\{0\},\quad t=\frac{1}{2n+1},\quad n=\frac{1-t}{2t},\quad n\in\mathbb{N},$$

we have

$$\begin{aligned}\sigma(t)&=\inf\left\{s\in\mathbb{T}:\ s>\frac{1}{2n+1}\right\}\\&=\frac{1}{2n-1}\\&=\frac{1}{2\frac{1-t}{2t}-1}\\&=\frac{t}{1-2t}>t,\end{aligned}$$

i.e., any point $t=\frac{1}{2n+1}$, $n\in\mathbb{N}$, is right-scattered. At these points,

$$\begin{aligned}
f^{\Delta}(t) &= \frac{f(\sigma(t)) - f(t)}{\sigma(t) - t} \\
&= \frac{\sigma(\sigma(t)) - \sigma(t)}{\sigma(t) - t} \\
&= 2\frac{(\sigma(t))^2}{(1 - 2\sigma(t))(\sigma(t) - t)} \\
&= 2\frac{\left(\frac{t}{1-2t}\right)^2}{\left(1 - \frac{2t}{1-2t}\right)\left(\frac{t}{1-2t} - t\right)} \\
&= 2\frac{\frac{t^2}{(1-2t)^2}}{\frac{1-4t}{1-2t}\frac{2t^2}{1-2t}} \\
&= 2\frac{t^2}{2t^2(1 - 4t)} \\
&= \frac{1}{1 - 4t}.
\end{aligned}$$

Let now $t = 0$. Then

$$\sigma(0) = \inf\{s \in \mathbb{T} : s > 0\} = 0.$$

Consequently, $t = 0$ is right-dense. Also,

$$\lim_{h \to 0} \frac{\sigma(h) - \sigma(0)}{h} = \lim_{h \to 0} \frac{\frac{h}{1-2h} - 0}{h} = \lim_{h \to 0} \frac{1}{1 - 2h} = 1.$$

Therefore, $f^{\Delta}(0) = 1$. Altogether, $f^{\Delta}(t) = \frac{1}{1-4t}$ for all $t \in \mathbb{T}^{\kappa}$.

Example 2.12 Let $\mathbb{T} = \{n^2 : n \in \mathbb{N}_0\}$ and $f(t) = t^2$, $g(t) = \sigma(t)$ for $t \in \mathbb{T}$. We will find $f^{\Delta}(t)$ and $g^{\Delta}(t)$ for $t \in \mathbb{T}^{\kappa} = \mathbb{T}$. For $t \in \mathbb{T}$, $t = n^2$, $n = \sqrt{t}$, $n \in \mathbb{N}_0$, we have

$$\sigma(t) = \inf\{l^2 : l^2 > n^2,\ l \in \mathbb{N}_0\} = (n + 1)^2 = (1 + \sqrt{t})^2 > t.$$

Therefore, all points of $\mathbb{T}$ are right-scattered. We note that f and g are continuous functions in $\mathbb{T}$. Hence,

$$\begin{aligned}
f^{\Delta}(t) &= \frac{f(\sigma(t)) - f(t)}{\sigma(t) - t} \\
&= \frac{(\sigma(t))^2 - t^2}{\sigma(t) - t} \\
&= \sigma(t) + t
\end{aligned}$$

$$= \left(1+\sqrt{t}\right)^2 + t$$

$$= t + 2\sqrt{t} + 1 + t$$

$$= 1 + 2\sqrt{t} + 2t$$

and

$$\begin{aligned} g^{\Delta}(t) &= \frac{g(\sigma(t)) - g(t)}{\sigma(t) - t} \\ &= \frac{\sigma(\sigma(t)) - \sigma(t)}{\sigma(t) - t} \\ &= \frac{(1+\sqrt{\sigma(t)})^2 - \sigma(t)}{\sigma(t) - t} \\ &= \frac{\sigma(t) + 2\sqrt{\sigma(t)} + 1 - \sigma(t)}{\sigma(t) - t} \\ &= \frac{1 + 2\sqrt{\sigma(t)}}{\sigma(t) - t} \\ &= \frac{1 + 2(1+\sqrt{t})}{(1+\sqrt{t})^2 - t} \\ &= \frac{3 + 2\sqrt{t}}{1 + 2\sqrt{t}}. \end{aligned}$$

Example 2.13 Let $\mathbb{T} = \left\{\sqrt[4]{2n+1} : n \in \mathbb{N}_0\right\}$ and $f(t) = t^4$ for $t \in \mathbb{T}$. We will find $f^{\Delta}(t)$ for $t \in \mathbb{T}$. For $t \in \mathbb{T}$, $t = \sqrt[4]{2n+1}$, $n = \frac{t^4-1}{2}$, $n \in \mathbb{N}_0$, we have

$$\begin{aligned} \sigma(t) &= \inf\{\sqrt[4]{2l+1} : \ \sqrt[4]{2l+1} > \sqrt[4]{2n+1}, \ l \in \mathbb{N}_0\} \\ &= \sqrt[4]{2n+3} \\ &= \sqrt[4]{t^4+2} > t. \end{aligned}$$

Therefore, every point of $\mathbb{T}$ is right-scattered. We note that the function f is continuous in $\mathbb{T}$. Hence,

$$\begin{aligned} f^{\Delta}(t) &= \frac{f(\sigma(t)) - f(t)}{\sigma(t) - t} \\ &= \frac{(\sigma(t))^4 - t^4}{\sigma(t) - t} \\ &= (\sigma(t))^3 + t(\sigma(t))^2 + t^2\sigma(t) + t^3 \\ &= \sqrt[4]{(t^4+2)^3} + t^2\sqrt[4]{t^4+2} + t\sqrt{t^4+2} + t^3. \end{aligned}$$

Exercise 2.14 Let $\mathbb{T} = \left\{\sqrt[5]{n+1} : n \in \mathbb{N}_0\right\}$ and $f(t) = t + t^3$ for $t \in \mathbb{T}$. Find $f^{\Delta}(t)$ for $t \in \mathbb{T}^{\kappa}$.

Solution $1 + \sqrt[5]{(t^5+1)^2} + t\sqrt[5]{t^5+1} + t^2$.

Example 2.15 Let $\mathbb{T} = \mathbb{Z}$ and f be differentiable at t. Note that all points of t are right-scattered and $\sigma(t) = t + 1$. Therefore,

$$\begin{aligned} f^{\Delta}(t) &= \frac{f(\sigma(t)) - f(t)}{\sigma(t) - t} \\ &= \frac{f(t+1) - f(t)}{t+1-t} \\ &= f(t+1) - f(t) \\ &= \Delta f(t), \end{aligned}$$

where Δ is the usual forward difference operator.

Exercise 2.16 Let $\mathbb{T} = h\mathbb{Z}$ for some $h > 0$. Prove that

$$\begin{aligned} t &= \frac{1}{2} f_2^{\Delta}(t) - \frac{h}{2}, \\ t^2 &= \frac{1}{3} f_3^{\Delta}(t) - \frac{h}{2} f_2^{\Delta}(t) + \frac{1}{6}h^2, \\ t^3 &= \frac{1}{4} f_4^{\Delta}(t) - \frac{1}{2} h f_3^{\Delta}(t) + \frac{1}{4} h^2 f_2^{\Delta}(t), \\ t^4 &= \frac{1}{5} f_5^{\Delta}(t) - \frac{1}{2} h f_4^{\Delta}(t) + \frac{1}{3} h^2 f_3^{\Delta}(t) - \frac{1}{30} h^4, \\ t^5 &= \frac{1}{6} f_6^{\Delta}(t) - \frac{1}{2} h f_5^{\Delta}(t) + \frac{5}{12} h^2 f_4^{\Delta}(t) - \frac{1}{12} h^4 f_2^{\Delta}(t), \end{aligned}$$

where $f_i(t) = t^i$, $i = 1, \ldots, 6$, $t \in \mathbb{T}$.

Solution We note that all points of $\mathbb{T}$ are right-scattered and $\sigma(t) = t + h$, $\mu(t) = h$, $t \in \mathbb{T}$. Then

$$\begin{aligned}
\frac{1}{2} f_2^{\Delta}(t) - \frac{h}{2} &= \frac{1}{2} \frac{(\sigma(t))^2 - t^2}{\sigma(t) - t} - \frac{h}{2} \\
&= \frac{1}{2} \frac{(\sigma(t) - t)(\sigma(t) + t)}{\sigma(t) - t} - \frac{h}{2} \\
&= \frac{1}{2}(\sigma(t) + t) - \frac{h}{2} \\
&= \frac{1}{2}(2t + h) - \frac{h}{2} \\
&= t + \frac{h}{2} - \frac{h}{2} \\
&= t,
\end{aligned}$$

$$\begin{aligned}
\frac{1}{3} f_3^{\Delta}(t) - \frac{h}{2} f_2^{\Delta}(t) + \frac{1}{6}h^2 &= \frac{1}{3} \frac{(\sigma(t))^3 - t^3}{\sigma(t) - t} - \frac{h}{2} \frac{(\sigma(t))^2 - t^2}{\sigma(t) - t} + \frac{1}{6}h^2 \\
&= \frac{1}{3} \frac{(\sigma(t) - t)((\sigma(t))^2 + t\sigma(t) + t^2)}{\sigma(t) - t} \\
&\quad - \frac{h}{2} \frac{(\sigma(t) - t)(\sigma(t) + t)}{\sigma(t) - t} + \frac{1}{6}h^2 \\
&= \frac{1}{3}((\sigma(t))^2 + t\sigma(t) + t^2) - \frac{h}{2}(\sigma(t) + t) + \frac{1}{6}h^2 \\
&= \frac{1}{3}\left((t + h)^2 + t(t + h) + t^2\right) - \frac{h}{2}(t + h + t) + \frac{1}{6}h^2 \\
&= \frac{1}{3}(t^2 + 2ht + h^2 + t^2 + ht + t^2) - \frac{h}{2}(2t + h) + \frac{1}{6}h^2 \\
&= \frac{1}{3}(3t^2 + 3ht + h^2) - ht - \frac{h^2}{2} + \frac{1}{6}h^2 \\
&= t^2 + ht + \frac{h^2}{3} - ht - \frac{h^2}{3}
\end{aligned}$$

$$= t^2,$$

$$\frac{1}{4}f_4^{\Delta}(t) - \frac{1}{2}hf_3^{\Delta}(t) + \frac{1}{4}h^2 f_2^{\Delta}(t) = \frac{1}{4}\frac{(\sigma(t))^4 - t^4}{\sigma(t) - t} - \frac{1}{2}h\frac{(\sigma(t))^3 - t^3}{\sigma(t) - t}$$

$$+ \frac{1}{4}h^2\frac{(\sigma(t))^2 - t^2}{\sigma(t) - t}$$

$$= \frac{1}{4}\frac{(\sigma(t) - t)((\sigma(t))^3 + t(\sigma(t))^2 + t^2\sigma(t) + t^3)}{\sigma(t) - t}$$

$$- \frac{1}{2}h\frac{(\sigma(t) - t)((\sigma(t))^2 + t\sigma(t) + t^2)}{\sigma(t) - t}$$

$$+ \frac{1}{4}h^2\frac{(\sigma(t) - t)(\sigma(t) + t)}{\sigma(t) - t}$$

$$= \frac{1}{4}((\sigma(t))^3 + t(\sigma(t))^2 + t^2\sigma(t) + t^3) - \frac{1}{2}h((\sigma(t))^2 + t\sigma(t) + t^2)$$

$$+ \frac{1}{4}h^2(\sigma(t) + t)$$

$$= \frac{1}{4}\left((t + h)^3 + t(t + h)^2 + t^2(t + h) + t^3\right)$$

$$- \frac{1}{2}h\left((t + h)^2 + t(t + h) + t^2\right) + \frac{1}{4}h^2(t + h + t)$$

$$= \frac{1}{4}\left(t^3 + 3t^2h + 3th^2 + h^3 + t(t^2 + 2th + h^2) + t^3 + ht^2 + t^3\right)$$

$$- \frac{1}{2}h(t^2 + 2ht + h^2 + t^2 + ht + t^2) + \frac{1}{4}h^2(2t + h)$$

$$= \frac{1}{4}(3t^3 + 4t^2h + 3th^2 + h^3 + t^3 + 2t^2h + th^2)$$

$$- \frac{1}{2}h(3t^2 + 3ht + h^2) + \frac{1}{2}th^2 + \frac{1}{4}h^3$$

$$= \frac{1}{4}(4t^3 + 6t^2h + 4th^2 + h^3) - \frac{3}{2}ht^2 - \frac{3}{2}h^2t - \frac{1}{2}h^3 + \frac{1}{2}h^2t + \frac{1}{4}h^3$$

$$= t^3 + \frac{3}{2}ht^2 + h^2t + \frac{1}{4}h^3 - \frac{3}{2}ht^2 - \frac{3}{2}h^2t - \frac{1}{2}h^3 + \frac{1}{2}h^2t + \frac{1}{4}h^3$$

$$= t^3,$$

$$\frac{1}{5}f_5^{\Delta} - \frac{h}{2}f_4^{\Delta} + \frac{1}{3}h^2 f_3^{\Delta} - \frac{1}{30}h^4 = \frac{1}{5}\frac{(\sigma(t))^5 - t^5}{\sigma(t) - t}$$

$$-\frac{h}{2}\frac{(\sigma(t))^4 - t^4}{\sigma(t) - t} + \frac{1}{3}h^2\frac{(\sigma(t))^3 - t^3}{\sigma(t) - t} - \frac{1}{30}h^4$$

$$= \frac{1}{5}\frac{(\sigma(t) - t)((\sigma(t))^4 + t(\sigma(t))^3 + t^2(\sigma(t))^2 + t^3\sigma(t) + t^4)}{\sigma(t) - t}$$

$$-\frac{h}{2}\frac{(\sigma(t) - t)((\sigma(t))^3 + t(\sigma(t))^2 + t^2\sigma(t) + t^3)}{\sigma(t) - t}$$

$$+\frac{1}{3}h^2\frac{(\sigma(t) - t)((\sigma(t))^2 + t\sigma(t) + t^2)}{\sigma(t) - t} - \frac{1}{30}h^4$$

$$= \frac{1}{5}\left((t+h)^4 + t(t+h)^3 + t^2(t+h)^2 + t^3(t+h)_+ t^4\right)$$

$$-\frac{h}{2}\left((t+h)^3 + t(t+h)^2 + t^2(t+h) + t^3\right)$$

$$+\frac{1}{3}h^2\left((t+h)^2 + t(t+h) + t^2\right) - \frac{1}{30}h^4$$

$$= \frac{1}{5}\Big(t^4 + 4t^2h^2 + h^4 + 4t^3h + 2t^2h^2 + 4th^3 + t(t^3 + 3t^2h + 3th^2 + h^3)$$

$$+ t^2(t^2 + 2th + h^2) + t^4 + ht^3 + t^4\Big)$$

$$-\frac{h}{2}\left(t^3 + 3t^2h + 3th^2 + h^3 + t(t^2 + 2th + h^2) + t^3 + ht^2 + t^3\right)$$

$$+\frac{1}{3}h^2(t^2 + 2th + h^2 + t^2 + ht + t^2) - \frac{1}{30}h^4$$

$$= \frac{1}{5}(t^4 + 6t^2h^2 + h^4 + 4t^3h + 4th^3 + t^4 + 3t^3h + 3t^2h^2 + th^3$$

$$
\begin{aligned}
&+t^4+2t^3h+t^2h^2+2t^4+ht^3)\\
&-\frac{h}{2}(3t^3+4t^2h+3th^2+h^3+t^3+2t^2h+th^2)\\
&+\frac{1}{3}h^2(3t^2+3ht+h^2)-\frac{1}{30}h^4\\
&=\frac{1}{5}(5t^4+10t^3h+10t^2h^2+5th^3+h^4)-\frac{h}{2}(4t^3+6t^2h+4th^2+h^3)\\
&+h^2t^2+h^3t+\frac{1}{3}h^4-\frac{1}{30}h^4\\
&=t^4+2t^3h+2t^2h^2+th^3+\frac{1}{5}h^4-2ht^3-3t^2h^2-2th^3-\frac{h^4}{2}\\
&+h^2t^2+h^3t+\frac{1}{3}h^4-\frac{1}{30}h^4\\
&=t^4.
\end{aligned}
$$

The last equality is left to the reader for exercise.

Exercise 2.17 Let $\mathbb{T}=q^{\mathbb{N}}$ for some $q>1$. Prove

$$
\begin{aligned}
t &= \frac{1}{1+q}f_2^{\Delta}(t),\\
t^2 &= \frac{1}{1+q+q^2}f_3^{\Delta}(t),\\
t^3 &= \frac{1}{1+q+q^2+q^3}f_4^{\Delta}(t),\\
t^4 &= \frac{1}{1+q+q^2+q^3+q^4}f_5^{\Delta}(t),\\
t^5 &= \frac{1}{1+q+q^2+q^3+q^4+q^5}f_6^{\Delta}(t),
\end{aligned}
$$

where $f_i(t)=t^i$, $i=1,\ldots,6$, $t\in\mathbb{T}$.

Theorem 2.18 *Assume $f,g:\mathbb{T}\to\mathbb{R}$ are differentiable at $t\in\mathbb{T}^{\kappa}$. Then we have the following.*

1. *The sum* $f + g : \mathbb{T} \to \mathbb{R}$ *is differentiable at* t *with*

$$(f + g)^{\Delta}(t) = f^{\Delta}(t) + g^{\Delta}(t).$$

2. *For any constant* α, $\alpha f : \mathbb{T} \to \mathbb{R}$ *is differentiable at* t *with*

$$(\alpha f)^{\Delta}(t) = \alpha f^{\Delta}(t).$$

3. *The product* $fg : \mathbb{T} \to \mathbb{R}$ *is differentiable at* t*, and the "product rule"*

$$(fg)^{\Delta}(t) = f^{\Delta}(t)g(t) + f(\sigma(t))g^{\Delta}(t) = f(t)g^{\Delta}(t) + f^{\Delta}(t)g(\sigma(t))$$

holds.

4. *If* $g(t)g(\sigma(t)) \neq 0$*, then the quotient* $\frac{f}{g} : \mathbb{T} \to \mathbb{R}$ *is differentiable at* t*, and the "quotient rule"*

$$\left(\frac{f}{g}\right)^{\Delta}(t) = \frac{f^{\Delta}(t)g(t) - f(t)g^{\Delta}(t)}{g(t)g(\sigma(t))}$$

holds.

Proof 1. Let $\varepsilon > 0$ be arbitrarily chosen. Since f and g are differentiable at t, there exist neighbourhoods U_1 and U_2 of t so that

$$|f(\sigma(t)) - f(s) - f^{\Delta}(t)(\sigma(t) - s)| \leq \frac{\varepsilon}{2}|\sigma(t) - s| \quad \text{for all} \quad s \in U_1$$

and

$$|g(\sigma(t)) - g(s) - g^{\Delta}(t)(\sigma(t) - s)| \leq \frac{\varepsilon}{2}|\sigma(t) - s| \quad \text{for all} \quad s \in U_2.$$

Hence, for $s \in U_1 \cap U_2$, we have

$$\begin{aligned}
&|f(\sigma(t)) + g(\sigma(t)) - f(s) - g(s) - (f^{\Delta}(t) + g^{\Delta}(t))(\sigma(t) - s)| \\
&\leq |f(\sigma(t)) - f(s) - f^{\Delta}(t)(\sigma(t) - s)| + |g(\sigma(t)) - g(s) - g^{\Delta}(t)(\sigma(t) - s)| \\
&\leq \frac{\varepsilon}{2}|\sigma(t) - s| + \frac{\varepsilon}{2}|\sigma(t) - s| \\
&= \varepsilon|\sigma(t) - s|,
\end{aligned}$$

which completes the proof.

2. Let $\alpha \neq 0$. Assume that $\varepsilon > 0$ is arbitrarily chosen. Since f is differentiable at t, there exists a neighbourhood U of t such that

$$|f(\sigma(t)) - f(s) - f^{\Delta}(t)(\sigma(t) - s)| \le \frac{\varepsilon}{|\alpha|}|\sigma(t) - s| \quad \text{for all} \quad s \in U.$$

Hence, for $s \in U$, we have

$$|\alpha f(\sigma(t)) - \alpha f(s) - \alpha f^{\Delta}(t)(\sigma(t) - s)|$$

$$= |\alpha||f(\sigma(t)) - f(s) - f^{\Delta}(t)(\sigma(t) - s)|$$

$$\le |\alpha|\frac{\varepsilon}{|\alpha|}$$

$$= \varepsilon,$$

which completes the proof.

3. Let $\varepsilon > 0$ be arbitrarily chosen. Let also $[a, b] \subset \mathbb{T}$ be such that $t, \sigma(t) \in [a, b]$. Set

$$M = \max_{t \in [a,b]} |f(t)|$$

and

$$\varepsilon^* = \frac{\varepsilon}{1 + M + |g(\sigma(t))| + |g^{\Delta}(t)|}.$$

Since f is differentiable at t, there exists a neighbourhood U_1 of t such that

$$|f(\sigma(t)) - f(s) - f^{\Delta}(t)(\sigma(t) - s)| \le \varepsilon^*|\sigma(t) - s| \quad \text{for any} \quad s \in U_1.$$

Since g is differentiable at t, there exists a neighbourhood U_2 of t such that

$$|g(\sigma(t)) - g(s) - g^{\Delta}(t)(\sigma(t) - s)| \le \varepsilon^*|\sigma(t) - s| \quad \text{for any} \quad s \in U_2.$$

Since f is differentiable at t, f is continuous at t. Therefore, there exists a neighbourhood U_3 of t such that

$$|f(t) - f(s)| \le \varepsilon^* \quad \text{for any} \quad s \in U_3.$$

Then, for any $s \in U_1 \cap U_2 \cap U_3 \cap [a, b]$, we get

$$|f(\sigma(t))g(\sigma(t)) - f(s)g(s) - (f^{\Delta}(t)g(\sigma(t)) + f(t)g^{\Delta}(t))(\sigma(t) - s)|$$

$$= |f(\sigma(t))g(\sigma(t)) - f(s)g(\sigma(t)) - f^{\Delta}(t)g(\sigma(t))(\sigma(t) - s)$$

$$+ f(s)g(\sigma(t)) + f^{\Delta}(t)g(\sigma(t))(\sigma(t) - s) + g^{\Delta}(t)f(s)(\sigma(t) - s)$$

$$
\begin{aligned}
&- (f^{\Delta}(t)g(\sigma(t)) + f(t)g^{\Delta}(t))(\sigma(t) - s) - f(s)g(s) - g^{\Delta}(t)f(s)(\sigma(t) - s)| \\
&= |(f(\sigma(t)) - f(s) - f^{\Delta}(t)(\sigma(t) - s))g(\sigma(t)) \\
&\quad + f(s)(g(\sigma(t)) - g(s) - g^{\Delta}(t)(\sigma(t) - s)) - (f(t) - f(s))g^{\Delta}(t)(\sigma(t) - s)| \\
&\leq |g(\sigma(t))||f(\sigma(t)) - f(s) - f^{\Delta}(t)(\sigma(t) - s)| \\
&\quad + |f(s)||g(\sigma(t)) - g(s) - g^{\Delta}(t)(\sigma(t) - s)| + |f(t) - f(s)||g^{\Delta}(t)||\sigma(t) - s| \\
&\leq \varepsilon^*|g(\sigma(t))||\sigma(t) - s| \\
&\quad + \varepsilon^*|f(s)||\sigma(t) - s| + \varepsilon^*|g^{\Delta}(t)||\sigma(t) - s| \\
&\leq \varepsilon^*(|g(\sigma(t))| + M + |g^{\Delta}(t)|)|\sigma(t) - s| \\
&\leq \varepsilon|\sigma(t) - s|,
\end{aligned}
$$

which completes the proof.

4. Let $\varepsilon > 0$ be arbitrarily chosen. Since f and g are differentiable at t, they are continuous at t. Because $g(t) \neq 0$, there exists a neighbourhood U of t such that

$$|g(s)| \geq m_1 > 0 \quad \text{for all} \quad s \in U,$$

for some constant m_1. Let $[a, b] \subset \mathbb{T}$ be such that $\sigma(t), t \in [a, b]$. Set

$$M_1 = \sup_{t\in[a,b]} |f(t)|, \quad M_2 = \sup_{t\in[a,b]} |g(t)|$$

and

$$\varepsilon^* = \varepsilon \frac{m_1|g(\sigma(t))g(t)|}{1 + 2M_1M_2 + |g^{\Delta}(t)|}.$$

Since f and g are continuous at t, there exists a neighbourhood U_1 of t such that

$$|f(t)g(s) - f(s)g(t)| \leq \varepsilon^* \quad \text{and} \quad |g(s) - g(t)| \leq \varepsilon^*$$

for any $s \in U_1$. Since f is differentiable at t, there exists a neighbourhood U_2 of t such that

$$|f(\sigma(t)) - f(s) - f^{\Delta}(t)(\sigma(t) - s)| \leq \varepsilon^*|\sigma(t) - s| \quad \text{for any} \quad s \in U_2.$$

Since g is differentiable at t, there exists a neighbourhood U_3 of t such that

$$|g(\sigma(t)) - g(s) - g^{\Delta}(t)(\sigma(t) - s)| \leq \varepsilon^*|\sigma(t) - s| \quad \text{for any} \quad s \in U_3.$$

Hence, for $s \in U_1 \cap U_2 \cap U_3 \cap U \cap [a, b]$, we get

$$\left| \frac{f(\sigma(t))}{g(\sigma(t))} - \frac{f(s)}{g(s)} - \frac{f^{\Delta}(t)g(t) - f(t)g^{\Delta}(t)}{g(t)g(\sigma(t))}(\sigma(t) - s) \right|$$

$$= \frac{1}{|g(\sigma(t))g(s)g(t)|} \Big| f(\sigma(t))g(t)g(s) - f(s)g(t)g(\sigma(t))$$

$$- (f^{\Delta}(t)g(t) - f(t)g^{\Delta}(t))g(s)(\sigma(t) - s) \Big|$$

$$= \frac{1}{|g(\sigma(t))g(s)g(t)|} \Big| (f(\sigma(t)) - f(s) - f^{\Delta}(t)(\sigma(t) - s))g(t)g(s)$$

$$+ f(s)g(t)g(s) + f^{\Delta}(t)g(t)g(s)(\sigma(t) - s)$$

$$- (g(\sigma(t)) - g(s) - g^{\Delta}(t)(\sigma(t) - s))f(s)g(t)$$

$$- g(s)f(s)g(t) - f(s)g^{\Delta}(t)g(t)(\sigma(t) - s)$$

$$- f^{\Delta}(t)g(t)g(s)(\sigma(t) - s) + f(t)g^{\Delta}(t)g(s)(\sigma(t) - s) \Big|$$

$$= \frac{1}{|g(\sigma(t))g(s)g(t)|} \Big| (f(\sigma(t)) - f(s) - f^{\Delta}(t)(\sigma(t) - s))g(t)g(s)$$

$$- (g(\sigma(t)) - g(s) - g^{\Delta}(t)(\sigma(t) - s))f(s)g(t)$$

$$+ g^{\Delta}(t)(f(t)g(s) - f(s)g(t))(\sigma(t) - s) \Big|$$

$$\leq \frac{1}{|g(\sigma(t))g(s)g(t)|} \Big(|f(\sigma(t)) - f(s) - f^{\Delta}(t)(\sigma(t) - s)||g(t)||g(s)|$$

$$+ |g(\sigma(t)) - g(s) - g^{\Delta}(t)(\sigma(t) - s)||f(s)||g(t)|$$

$$+|g^{\Delta}(t)||f(t)g(s)-f(s)g(t)||\sigma(t)-s|\Big)$$

$$\leq \frac{1}{|g(\sigma(t))g(s)g(t)|}(\varepsilon^* M_1 M_2|\sigma(t)-s|+\varepsilon^*|\sigma(t)-s|M_1 M_2$$

$$+|g^{\Delta}(t)|\varepsilon^*|\sigma(t)-s|)$$

$$\leq \varepsilon^* \frac{1}{|g(\sigma(t))g(s)g(t)|}(1+2M_1M_2+|g^{\Delta}(t)|)|\sigma(t)-s|$$

$$\leq \varepsilon^* \frac{1}{m_1|g(\sigma(t))g(t)|}(1+2M_1M_2+|g^{\Delta}(t)|)|\sigma(t)-s|$$

$$=\varepsilon|\sigma(t)-s|,$$

which completes the proof. □

Example 2.19 Let $f, g, h : \mathbb{T} \to \mathbb{R}$ be differentiable at $t \in \mathbb{T}^{\kappa}$. Then

$$(fgh)^{\Delta}(t) = ((fg)h)^{\Delta}(t)$$

$$= (fg)^{\Delta}(t)h(t) + (fg)(\sigma(t))h^{\Delta}(t)$$

$$= (f^{\Delta}(t)g(t) + f(\sigma(t))g^{\Delta}(t))h(t) + f^{\sigma}(t)g^{\sigma}(t)h^{\Delta}(t)$$

$$= f^{\Delta}(t)g(t)h(t) + f^{\sigma}(t)g^{\Delta}(t)h(t) + f^{\sigma}(t)g^{\sigma}(t)h^{\Delta}(t).$$

Example 2.20 Let $f : \mathbb{T} \to \mathbb{R}$ be differentiable at $t \in \mathbb{T}^{\kappa}$. Then

$$\left(f^2\right)^{\Delta}(t) = (ff)^{\Delta}(t)$$

$$= f^{\Delta}(t)f(t) + f(\sigma(t))f^{\Delta}(t)$$

$$= f^{\Delta}(t)(f^{\sigma}(t) + f(t)).$$

Also,

$$\begin{aligned}
(f^3)^\Delta(t) &= (ff^2)^\Delta(t) \\
&= f^\Delta(t)f^2(t) + f(\sigma(t))(f^2)^\Delta(t) \\
&= f^\Delta(t)f^2(t) + f^\sigma(t)f^\Delta(t)(f^\sigma(t) + f(t)) \\
&= f^\Delta(t)(f^2(t) + f(t)f^\sigma(t) + (f^\sigma)^2(t)).
\end{aligned}$$

We assume that

$$(f^n)^\Delta(t) = f^\Delta(t)\sum_{k=0}^{n-1} f^k(t)(f^\sigma)^{n-1-k}(t)$$

for some $n \in \mathbb{N}$. We will prove that

$$(f^{n+1})^\Delta(t) = f^\Delta(t)\sum_{k=0}^{n} f^k(t)(f^\sigma)^{n-k}(t).$$

Indeed,

$$\begin{aligned}
(f^{n+1})^\Delta(t) &= (ff^n)^\Delta(t) \\
&= f^\Delta(t)f^n(t) + f^\sigma(t)(f^n)^\Delta(t) \\
&= f^\Delta(t)f^n(t) + f^\Delta(t)f^\sigma(t)(f^{n-1}(t) + f^{n-2}(t)f^\sigma(t) \\
&\quad + \cdots + f(t)(f^\sigma)^{n-2}(t) + (f^\sigma)^{n-1}(t))f^\sigma(t) \\
&= f^\Delta(t)\left(f^n(t) + f^{n-1}(t)f^\sigma(t) + f^{n-2}(t)(f^\sigma)^2(t) + \cdots + (f^\sigma)^n(t)\right) \\
&= f^\Delta(t)\sum_{k=0}^{n} f^k(t)(f^\sigma)^{n-k}(t).
\end{aligned}$$

Example 2.21 Now, we consider $f(t) = (t-a)^m$ for $a \in \mathbb{R}$ and $m \in \mathbb{N}$. We set

$$h(t) = t - a.$$

Then

$$h^\Delta(t) = 1.$$

By Example 2.20, we get

$$\begin{aligned}(h^m)^{\Delta}(t) &= h^{\Delta}(t)\sum_{k=0}^{m-1} h^k(t)(h^{\sigma})^{m-1-k}(t)\\ &= \sum_{k=0}^{m-1}(t-a)^k(\sigma(t)-a)^{m-1-k}.\end{aligned}$$

Let now $g(t) = \frac{1}{f(t)}$. Then

$$g^{\Delta}(t) = -\frac{f^{\Delta}(t)}{f(\sigma(t))f(t)},$$

whereupon

$$\begin{aligned}\left(\frac{1}{h^m}\right)^{\Delta}(t) &= -\frac{1}{(\sigma(t)-a)^m(t-a)^m}\sum_{k=0}^{m-1}(t-a)^k(\sigma(t)-a)^{m-1-k}\\ &= -\sum_{k=0}^{m-1}\frac{1}{(t-a)^{m-k}}\frac{1}{(\sigma(t)-a)^{k+1}}.\end{aligned}$$

Exercise 2.22 Let $t \in \mathbb{T}$ be right-scattered. Find $f^{\Delta}(t)$ for

$$f(t) = 2\sin t + t^2 - 3t^3.$$

Solution We have

$$\begin{aligned}f^{\Delta}(t) &= 2\frac{\sin\sigma(t) - \sin t}{\mu(t)} + \frac{(\sigma(t)-t)(\sigma(t)+t)}{\mu(t)}\\ &\quad -3\frac{(\sigma(t)-t)((\sigma(t))^2 + t\sigma(t) + t^2)}{\mu(t)}\\ &= 4\frac{\sin\frac{\sigma(t)-t}{2}\cos\frac{\sigma(t)+t}{2}}{\mu(t)} + \sigma(t) + t - 3(\sigma(t))^2 - 3t\sigma(t) - 3t^2\\ &= 4\frac{\sin\frac{\mu(t)}{2}\cos\frac{\sigma(t)+t}{2}}{\mu(t)} + \sigma(t) + t - 3(\sigma(t))^2 - 3t\sigma(t) - 3t^2\\ &= 4\frac{\sin\frac{\mu(t)}{2}\cos\frac{\sigma(t)+t}{2}}{\mu(t)} - 3(\sigma(t))^2 + \sigma(t)(1-3t) + t - 3t^2.\end{aligned}$$

Exercise 2.23 Let $t \in 2^{\mathbb{N}_0}$. Find $f^{\Delta}(t)$ for

$$f(t) = \frac{t^3 + t^2 - 2t}{2t^2 + 3t + 1}.$$

Solution Here, $\sigma(t) = 2t$, $\mu(t) = t$, $t \in \mathbb{T}$. Then

$$\begin{aligned}
f^{\Delta}(t) &= \frac{\left((\sigma(t))^2 + t\sigma(t) + t^2 + t + \sigma(t) - 2\right)(2t^2 + 3t + 1)}{(2t^2 + 3t + 1)(2(\sigma(t))^2 + 3\sigma(t) + 1)} \\
&\quad - \frac{(t^3 + t^2 - 2t)(2\sigma(t) + 2t + 3)}{(2t^2 + 3t + 1)(2(\sigma(t))^2 + 3\sigma(t) + 1)} \\
&= \frac{(4t^2 + 2t^2 + t^2 + t + 2t - 2)(2t^2 + 3t + 1) - (t^3 + t^2 - 2t)(4t + 2t + 3)}{(2t^2 + 3t + 1)(8t^2 + 6t + 1)} \\
&= \frac{(7t^2 + 3t - 2)(2t^2 + 3t + 1) - (t^3 + t^2 - 2t)(6t + 3)}{(2t^2 + 3t + 1)(8t^2 + 6t + 1)} \\
&= \frac{14t^4 + 21t^3 + 7t^2 + 6t^3 + 9t^2 + 3t - 4t^2 - 6t - 2}{(2t^2 + 3t + 1)(8t^2 + 6t + 1)} \\
&\quad - \frac{6t^4 + 3t^3 + 6t^3 + 3t^2 - 12t^2 - 6t}{(2t^2 + 3t + 1)(8t^2 + 6t + 1)} \\
&= \frac{8t^4 + 18t^3 + 21t^2 + 3t - 2}{(2t^2 + 3t + 1)(8t^2 + 6t + 1)}.
\end{aligned}$$

Exercise 2.24 Let $\mathbb{T} = 3^{\mathbb{N}_0}$. Find $f^{\Delta}(t)$ for

$$f(t) = \frac{1}{4}t^4 - \frac{1}{3}t^3 - t^2 + 1.$$

Solution Here, $\sigma(t) = 3t$, $\mu(t) = 2t$, $t \in \mathbb{T}$. Then

$$\begin{aligned}
f^{\Delta}(t) &= \frac{1}{4}\frac{(\sigma(t))^4 - t^4}{\sigma(t) - t} - \frac{1}{3}\frac{(\sigma(t))^3 - t^3}{\sigma(t) - t} - \frac{(\sigma(t))^2 - t^2}{\sigma(t) - t} \\
&= \frac{1}{4}((\sigma(t))^3 + t(\sigma(t))^2 + t^2\sigma(t) + t^3) \\
&\quad - \frac{1}{3}((\sigma(t))^2 + t\sigma(t) + t^2) - (\sigma(t) + t) \\
&= \frac{1}{4}(27t^3 + 9t^3 + 3t^3 + t^3) - \frac{1}{3}(9t^2 + 3t^2 + t^2) - 4t \\
&= 10t^3 - \frac{13}{3}t^2 - 4t.
\end{aligned}$$

Exercise 2.25 Let $\mathbb{T} = \mathbb{Z}$. Find $f^{\Delta}(t)$ for

$$f(t) = \frac{1}{3}t^3 - \frac{5}{2}t^2 - 4t.$$

Solution $t^2 - 4t - \frac{37}{6}$.

Exercise 2.26 Let $\mathbb{T} = \{\sqrt[4]{n} : n \in \mathbb{N}\}$. Find $f^{\Delta}(t)$ for

$$f(t) = \frac{1}{3}t^3 - 2t^2 - 3t + 2.$$

Solution Let $t \in \mathbb{T}$, $t = \sqrt[4]{n}$ for some $n \in \mathbb{N}$. Then $n = t^4$ and

$$\begin{aligned}\sigma(t) &= \inf\{\sqrt[4]{l} : \sqrt[4]{l} > \sqrt[4]{n},\ l \in \mathbb{N}\}\\ &= \sqrt[4]{n+1}\\ &= \sqrt[4]{t^4+1}.\end{aligned}$$

Hence,

$$\begin{aligned}f^{\Delta}(t) &= \frac{1}{3}\frac{(\sigma(t))^3 - t^3}{\sigma(t) - t} - 2\frac{(\sigma(t))^2 - t^2}{\sigma(t) - t} - 3\\ &= \frac{1}{3}((\sigma(t))^2 + t\sigma(t) + t^2) - 2(\sigma(t) + t) - 3\\ &= \frac{1}{3}\left(\sqrt{t^4+1} + t\sqrt[4]{t^4+1} + t^2\right) - 2\left(\sqrt[4]{t^4+1} + t\right) - 3\\ &= \left(\frac{1}{3}t - 2\right)\sqrt[4]{t^4+1} + \frac{1}{3}\sqrt{t^4+1} + \frac{1}{3}t^2 - 2t - 3.\end{aligned}$$

Definition 2.27 Let $f : \mathbb{T} \to \mathbb{R}$ and $t \in (\mathbb{T}^{\kappa})^{\kappa} = \mathbb{T}^{\kappa^2}$. We define the second derivative of f at t, provided it exists, by

$$f^{\Delta^2} = f^{\Delta\Delta} = \left(f^{\Delta}\right)^{\Delta} : \mathbb{T}^{\kappa^2} \to \mathbb{R}.$$

Similarly, we define higher-order derivatives $f^{\Delta^n} : \mathbb{T}^{\kappa^n} \to \mathbb{R}$.

Example 2.28 Let $\mathbb{T} = \mathbb{Z}$. We will find $f^{\Delta\Delta}(t)$ for

$$f(t) = t^2 - 3t + 1.$$

We have $\sigma(t) = t + 1$ and

$$\begin{aligned} f^{\Delta}(t) &= \frac{(\sigma(t))^2 - t^2}{\sigma(t) - t} - 3\frac{\sigma(t) - t}{\sigma(t) - t} \\ &= \sigma(t) + t - 3 \\ &= t + 1 + t - 3 \\ &= 2t - 2. \end{aligned}$$

Hence,

$$f^{\Delta\Delta}(t) = 2.$$

Example 2.29 Let $\mathbb{T} = 3^{\mathbb{N}_0}$. We will find $f^{\Delta^3}(t)$ for

$$f(t) = t^3 + 3t^2 + t + 1.$$

Here $\sigma(t) = 3t$, $t \in \mathbb{T}$. Then

$$\begin{aligned} f^{\Delta}(t) &= \frac{(\sigma(t))^3 - t^3}{\sigma(t) - t} + 3\frac{(\sigma(t))^2 - t^2}{\sigma(t) - t} + \frac{\sigma(t) - t}{\sigma(t) - t} \\ &= (\sigma(t))^2 + t\sigma(t) + t^2 + 3(\sigma(t) + t) + 1 \\ &= 9t^2 + 3t^2 + t^2 + 12t + 1 \\ &= 13t^2 + 12t + 1, \end{aligned}$$

$$\begin{aligned} f^{\Delta\Delta}(t) &= 13\frac{(\sigma(t))^2 - t^2}{\sigma(t) - t} + 12\frac{\sigma(t) - t}{\sigma(t) - t} \\ &= 13(\sigma(t) + t) + 12 \\ &= 52t + 12. \end{aligned}$$

Hence,

$$f^{\Delta^3}(t) = 52.$$

Example 2.30 Let $\mathbb{T} = 2^{\mathbb{N}_0}$. We will find $f^{\Delta\Delta}(t)$ for

$$f(t) = \frac{t^2 + 1}{t + 1}.$$

Here $\sigma(t) = 2t$, $t \in \mathbb{T}$. Then

$$\begin{aligned}
f^{\Delta}(t) &= \frac{\frac{(\sigma(t))^2-t^2}{\sigma(t)-t}(t+1) - (t^2+1)\frac{\sigma(t)-t}{\sigma(t)-t}}{(t+1)(2t+1)} \\
&= \frac{(\sigma(t)+t)(t+1) - t^2 - 1}{2t^2+t+2t+1} \\
&= \frac{3t(t+1) - t^2 - 1}{2t^2+3t+1} \\
&= \frac{3t^2+3t-t^2-1}{2t^2+3t+1} \\
&= \frac{2t^2+3t-1}{2t^2+3t+1}.
\end{aligned}$$

Hence,

$$\begin{aligned}
f^{\Delta\Delta}(t) &= \frac{1}{(2t^2+3t+1)(8t^2+6t+1)}\left(\left(2\frac{(\sigma(t))^2-t^2}{\sigma(t)-t} + 3\frac{\sigma(t)-t}{\sigma(t)-t}\right)(2t^2+3t+1)\right. \\
&\quad \left. -(2t^2+3t-1)\left(2\frac{(\sigma(t))^2-t^2}{\sigma(t)-t} + 3\frac{\sigma(t)-t}{\sigma(t)-t}\right)\right) \\
&= \frac{(2(\sigma(t)+t)+3)(2t^2+3t+1) - (2t^2+3t-1)(2(\sigma(t)+t)+3)}{(2t^2+3t+1)(8t^2+6t+1)} \\
&= \frac{(6t+3)(2t^2+3t+1) - (2t^2+3t-1)(6t+3)}{(2t^2+3t+1)(8t^2+6t+1)} \\
&= \frac{(6t+3)(2t^2+3t+1-2t^2-3t+1)}{(2t^2+3t+1)(8t^2+6t+1)} \\
&= \frac{6(2t+1)}{(2t^2+3t+1)(8t^2+6t+1)}.
\end{aligned}$$

Exercise 2.31 Let $\mathbb{T} = 3^{\mathbb{N}_0}$. Find $f^{\Delta\Delta}(t)$ for

$$f(t) = \frac{t+2}{t+3}.$$

Solution $-\frac{4}{9}\frac{1}{(t+3)(3t^2+4t+1)}$.

Theorem 2.32 (Leibniz Formula) *Let $S_k^{(n)}$ be the set consisting of all possible strings of length n, containing exactly k times σ and $n-k$ times Δ. If*

$$f^{\Lambda} \quad \textit{exists for all} \quad \Lambda \in S_k^{(n)},$$

then

$$(fg)^{\Delta^n} = \sum_{k=0}^{n} \left(\sum_{\Lambda \in S_k^{(n)}} f^{\Lambda} \right) g^{\Delta^k}. \tag{2.1}$$

Proof We will use induction.

1. $n = 1$. Then

$$S_0^{(1)} = \Delta, \quad S_1^{(1)} = \sigma.$$

   Hence,

$$\begin{aligned}
\sum_{k=0}^{1} \left(\sum_{\Lambda \in S_k^{(1)}} f^{\Lambda} \right) g^{\Delta^k} &= \sum_{\Lambda \in S_0^{(1)}} f^{\Lambda} g + \sum_{\Lambda \in S_1^{(1)}} f^{\Lambda} g^{\Delta} \\
&= f^{\Delta} g + f^{\sigma} g^{\Delta} \\
&= (fg)^{\Delta},
\end{aligned}$$

   i.e., the assertion holds for $n = 1$.
2. Assume that the assertion is valid for some $n \in \mathbb{N}$. We will prove that

$$(fg)^{\Delta^{n+1}} = \sum_{k=0}^{n+1} \left(\sum_{\Lambda \in S_k^{(n+1)}} f^{\Lambda} \right) g^{\Delta^k}.$$

   Using (2.1), we have

$$\begin{aligned}
(fg)^{\Delta^{n+1}} &= \left((fg)^{\Delta^n} \right)^{\Delta} \\
&= \left(\sum_{k=0}^{n} \left(\sum_{\Lambda \in S_k^{(n)}} f^{\Lambda} \right) g^{\Delta^k} \right)^{\Delta} \\
&= \sum_{k=0}^{n} \left(\left(\sum_{\Lambda \in S_k^{(n)}} f^{\Lambda} \right) g^{\Delta^k} \right)^{\Delta} \\
&= \sum_{k=0}^{n} \left(\left(\sum_{\Lambda \in S_k^{(n)}} f^{\Lambda\sigma} \right) g^{\Delta^{k+1}} + \left(\sum_{\Lambda \in S_k^{(n)}} f^{\Lambda\Delta} \right) g^{\Delta^k} \right)
\end{aligned}$$

$$= \sum_{k=0}^{n} \left(\sum_{\Lambda \in S_k^{(n)}} f^{\Lambda\sigma} \right) g^{\Delta^{k+1}} + \sum_{k=0}^{n} \left(\sum_{\Lambda \in S_k^{(n)}} f^{\Lambda\Delta} \right) g^{\Delta^k}$$

$$= \sum_{k=1}^{n+1} \left(\sum_{\Lambda \in S_{k-1}^{(n)}} f^{\Lambda\sigma} \right) g^{\Delta^k} + \sum_{k=0}^{n} \left(\sum_{\Lambda \in S_k^{(n)}} f^{\Lambda\Delta} \right) g^{\Delta^k}$$

$$= \sum_{k=1}^{n} \left(\sum_{\Lambda \in S_{k-1}^{(n)}} f^{\Lambda\sigma} \right) g^{\Delta^k} + \left(\sum_{\Lambda \in S_n^{(n)}} f^{\Lambda\sigma} \right) g^{\Delta^{n+1}}$$

$$+ \sum_{k=1}^{n} \left(\sum_{\Lambda \in S_k^{(n)}} f^{\Lambda\Delta} \right) g^{\Delta^k} + \left(\sum_{\Lambda \in S_0^{(n)}} f^{\Lambda\Delta} \right) g$$

$$= \sum_{k=1}^{n} \left(\sum_{\Lambda \in S_{k-1}^{(n)}} f^{\Lambda\sigma} + \sum_{\Lambda \in S_k^{(n)}} f^{\Lambda\Delta} \right) g^{\Delta^k}$$

$$+ \left(\sum_{\Lambda \in S_n^{(n)}} f^{\Lambda\sigma} g^{\Delta^{n+1}} + \sum_{\Lambda \in S_0^{(n)}} f^{\Lambda\Delta} g \right)$$

$$= \sum_{k=1}^{n} \left(\sum_{\Lambda \in S_k^{(n+1)}} f^{\Lambda} \right) g^{\Delta^k} + \left(\sum_{\Lambda \in S_{n+1}^{(n+1)}} f^{\Lambda} \right) g^{\Delta^{n+1}} + \left(\sum_{\Lambda \in S_0^{(n+1)}} f^{\Lambda} \right) g$$

$$= \sum_{k=0}^{n+1} \left(\sum_{\Lambda \in S_k^{(n+1)}} f^{\Lambda} \right) g^{\Delta^k}.$$

The proof is complete. □

Example 2.33 Let f and μ be differentiable in $\mathbb{T}^\kappa$. Then

$$f^{\Delta\sigma} = \frac{f^{\sigma^2} - f^\sigma}{\mu^\sigma} \quad \text{and} \quad f^{\sigma\Delta} = \frac{f^{\sigma^2} - f^\sigma}{\mu}.$$

Therefore,

$$f^{\sigma\Delta} = \frac{f^{\Delta\sigma}\mu^\sigma}{\mu} = \frac{f^{\Delta\sigma}\mu(1+\mu^\Delta)}{\mu} = (1+\mu^\Delta) f^{\Delta\sigma}.$$

Also,

$$f^{\sigma^2\Delta} = \frac{f^{\sigma^3} - f^{\sigma^2}}{\mu} \quad \text{and} \quad f^{\sigma\Delta\sigma} = \frac{f^{\sigma^3} - f^{\sigma^2}}{\mu^\sigma}.$$

Therefore,

$$f^{\sigma\Delta\sigma} = \frac{f^{\sigma^2\Delta}\mu}{\mu^\sigma} = \frac{f^{\sigma^2\Delta}\mu}{\mu(1+\mu^\Delta)} = \frac{f^{\sigma^2\Delta}}{1+\mu^\Delta},$$

whereupon

$$f^{\sigma^2\Delta} = (1+\mu^\Delta)f^{\sigma\Delta\sigma}.$$

We have

$$\begin{aligned} f^{\Delta\sigma^2} &= \frac{f^{\sigma^3} - f^{\sigma^2}}{\mu^{\sigma^2}} \\ &= \frac{f^{\sigma^2\Delta}\mu}{\mu^{\sigma^2}} \\ &= \frac{\mu f^{\sigma^2\Delta}}{\mu(1+\mu^\Delta)(1-\mu^{\Delta\sigma})} \\ &= \frac{f^{\sigma^2\Delta}}{(1+\mu^\Delta)(1+\mu^{\Delta\sigma})}, \end{aligned}$$

from where

$$f^{\sigma^2\Delta} = (1+\mu^\Delta)(1+\mu^{\Delta\sigma})f^{\Delta\sigma^2}.$$

Example 2.34 Let

$$\mathbb{T} = \left\{ t_n = \left(\frac{1}{2}\right)^{2^n} : n \in \mathbb{N}_0 \right\} \cup \{0, -1\}.$$

Then

$$\sigma(t_n) \to 0 = \sigma(0) \quad \text{as} \quad n \to \infty.$$

Therefore, σ is continuous at 0. Also,

$$\begin{aligned} \lim_{s\to 0} \frac{\sigma(\sigma(s)) - \sigma(s)}{\sigma(s) - s} &= \lim_{s\to 0} \frac{\sigma(s) - \sqrt{\sigma(s)}}{s - \sqrt{s}} \\ &= \lim_{s\to 0} \frac{\sqrt{s} - \sqrt[4]{s}}{s - \sqrt{s}} \end{aligned}$$

$$= \lim_{s\to 0} \frac{1}{\sqrt{s}+\sqrt[4]{s}}$$

$$= \infty.$$

Consequently, σ is not differentiable at 0.

2.2 Mean Value Theorems

Let $\mathbb{T}$ be a time scale and $a, b \in \mathbb{T}$, $a < b$. Let $f : \mathbb{T} \to \mathbb{R}$ be a function.

Theorem 2.35 *If f is delta differentiable at t, then there exists a function g, defined in a neighbourhood U of t with*

$$\lim_{s\to t} g(s) = g(t) = 0,$$

such that

$$f(\sigma(t)) = f(s) + \left(f^{\Delta}(t) + g(s)\right)(\sigma(t) - s) \tag{2.2}$$

for all $s \in U$.

Proof Define

$$g(s) = \begin{cases} \frac{f(\sigma(t))-f(s)}{\sigma(t)-s} - f^{\Delta}(t) & \text{if } s \neq \sigma(t),\ t \in \mathbb{T} \\ \\ 0, & \text{if } s = \sigma(t),\ t \in \mathbb{T}. \end{cases} \tag{2.3}$$

Solving (2.3) for $f(\sigma(t))$ yields (2.2). Since f is differentiable at t, we have that f is continuous at t. Hence, g is continuous at t and

$$\lim_{s\to t} g(s) = g(t) = 0.$$

The proof is complete. □

Theorem 2.36 *Suppose that f has a delta derivative at each point of $[a, b]$. If*

$$f(a) = f(b),$$

then there exist points $\xi_1, \xi_2 \in [a, b]$ such that

$$f^{\Delta}(\xi_2) \le 0 \le f^{\Delta}(\xi_1).$$

Proof Since f is delta differentiable at each point of $[a, b]$, f is continuous on $[a, b]$. Therefore, there exist $\xi_1, \xi_2 \in [a, b]$ such that

$$m = \min_{t \in [a,b]} f(t) = f(\xi_1) \quad \text{and} \quad M = \max_{t \in [a,b]} f(t) = f(\xi_2).$$

Because $f(a) = f(b)$, we assume that $\xi_1, \xi_2 \in [a, b)$.

1. If $\sigma(\xi_1) > \xi_1$, then
$$f^{\Delta}(\xi_1) = \frac{f(\sigma(\xi_1)) - f(\xi_1)}{\sigma(\xi_1) - \xi_1} \geq 0.$$
2. If $\sigma(\xi_1) = \xi_1$, then
$$f^{\Delta}(\xi_1) = \lim_{t \to \xi_1} \frac{f(t) - f(\xi_1)}{t - \xi_1} \geq 0.$$
3. If $\sigma(\xi_2) > \xi_2$, then
$$f^{\Delta}(\xi_2) = \frac{f(\sigma(\xi_2)) - f(\xi_2)}{\sigma(\xi_2) - \xi_2} \leq 0.$$
4. If $\sigma(\xi_2) = \xi_2$, then
$$f^{\Delta}(\xi_2) = \lim_{t \to \xi_2} \frac{f(t) - f(\xi_2)}{t - \xi_2} \leq 0.$$

This completes the proof. □

Example 2.37 Let $\mathbb{T} = \mathbb{Z}$, $f(t) = t^2$. We will find $\xi_1, \xi_2 \in (-3, 3)$ such that

$$f^{\Delta}(\xi_2) \leq 0 \leq f^{\Delta}(\xi_1).$$

We have $\sigma(t) = t + 1$, $\mu(t) = 1$, $f(-3) = f(3) = 9$ and

$$f^{\Delta}(t) = \sigma(t) + t = 2t + 1.$$

Hence,

$$f^{\Delta}(\xi_2) = 2\xi_2 + 1 \leq 0 \quad \text{and} \quad f^{\Delta}(\xi_1) = 2\xi_1 + 1 \geq 0$$

hold for $\xi_1 \in \{0, 1, 2\}$ and $\xi_2 \in \{-2, -1\}$.

Example 2.38 Let $\mathbb{T} = 2^{\mathbb{N}_0}$, $f(t) = t^3 - \frac{73}{9}t^2 + 2$. We will find $\xi_1, \xi_2 \in (1, 8)$ such that

$$f^{\Delta}(\xi_2) \leq 0 \leq f^{\Delta}(\xi_1).$$

We have

$$f(1) = 1 - \frac{73}{9} + 2 = 3 - \frac{73}{9} = -\frac{46}{9}$$

and

$$f(8) = 512 - \frac{73}{9}64 + 2 = \frac{72 - 73}{9}64 + 2 = -\frac{64}{9} + 2. = -\frac{46}{9}.$$

Therefore, $f(1) = f(8)$. Also, $\sigma(t) = 2t$, $\mu(t) = t$ and

$$\begin{aligned} f^{\Delta}(t) &= (\sigma(t))^2 + t\sigma(t) + t^2 - \frac{73}{9}(\sigma(t) + t) \\ &= 4t^2 + 2t^2 + t^2 - \frac{73}{9}(3t) \\ &= 7t^2 - \frac{73}{3}t. \end{aligned}$$

Hence,

$$f^{\Delta}(\xi_2) = 7\xi_2^2 - \frac{73}{3}\xi_2 \leq 0 \quad \text{for} \quad \xi_2 = 2$$

and

$$f^{\Delta}(\xi_1) = 7\xi_1^2 - \frac{73}{3}\xi_1 \geq 0 \quad \text{for} \quad \xi_1 = 4.$$

Example 2.39 Let $\mathbb{T} = 3^{\mathbb{N}_0} \cup \{0\}$, $f(t) = t^2 - 9t + 3$. We will find $\xi_1, \xi_2 \in (0, 9)$ so that

$$f^{\Delta}(\xi_2) \leq 0 \leq f^{\Delta}(\xi_1).$$

Here, $\sigma(t) = 3t$, $t \neq 0$, $\sigma(0) = 1$,

$$f(0) = 3 \quad \text{and} \quad f(9) = 3.$$

Then

$$f^{\Delta}(t) = \sigma(t) + t - 9 = 3t + t - 9 = 4t - 9.$$

Hence,

$$f^{\Delta}(\xi_2) = 4\xi_2 - 9 \leq 0 \quad \text{for} \quad \xi_2 = 1$$

and

$$f^{\Delta}(\xi_1) = 4\xi_1 - 9 \geq 0 \quad \text{for} \quad \xi_1 = 3.$$

Exercise 2.40 Let $\mathbb{T} = \mathbb{Z}$, $f(t) = t^3 - 16t + 1$. Find $\xi_1, \xi_2 \in (-4, 4)$ so that

$$f^{\Delta}(\xi_2) \leq 0 \leq f^{\Delta}(\xi_1).$$

Solution $\xi_2 \in \{-2, -1, 0, 1\}$ and $\xi_1 \in \{-3, 2, 3\}$.

Theorem 2.41 (Mean Value Theorem) *Suppose that f is continuous on $[a, b]$ and has a delta derivative at each point of $[a, b)$. Then there exist $\xi_1, \xi_2 \in [a, b)$ such that*

$$f^{\Delta}(\xi_1)(b-a) \leq f(b) - f(a) \leq f^{\Delta}(\xi_2)(b-a). \tag{2.4}$$

Proof Consider the function ϕ defined on $[a, b]$ by

$$\phi(t) = f(t) - f(a) - \frac{f(b) - f(a)}{b-a}(t-a).$$

Then ϕ is continuous on $[a, b]$ and has a delta derivative at each point of $[a, b)$. Also, $\phi(a) = \phi(b) = 0$. Hence, there exist $\xi_1, \xi_2 \in [a, b)$ such that

$$\phi^{\Delta}(\xi_1) \leq 0 \leq \phi^{\Delta}(\xi_2),$$

i.e.,

$$f^{\Delta}(\xi_1) - \frac{f(b) - f(a)}{b-a} \leq 0 \leq f^{\Delta}(\xi_2) - \frac{f(b) - f(a)}{b-a},$$

and thus, we get (2.4). □

Example 2.42 Let $\mathbb{T} = \mathbb{Z}$, $f(t) = t^2 + 2t$, $[a, b] = [-2, 4]$. Then $\sigma(t) = t + 1$,

$$f^{\Delta}(t) = \sigma(t) + t + 2 = t + 1 + t + 2 = 2t + 3.$$

Since

$$f(4) = 24 \quad \text{and} \quad f(-2) = 0,$$

we have

$$2\xi_1 + 3 \leq \frac{24}{6} \leq 2\xi_2 + 3,$$

i.e.,

$$2\xi_1 + 3 \leq 4 \quad \text{and} \quad 2\xi_2 + 3 \geq 4,$$

so that

$$\xi_1 \leq \frac{1}{2} \quad \text{and} \quad \xi_2 \geq \frac{1}{2},$$

and therefore

$$\xi_1 \in \{-2, -1, 0\} \quad \text{and} \quad \xi_2 \in \{1, 2, 3\}.$$

Example 2.43 Let $\mathbb{T} = \mathbb{Z}$, $f(t) = e^t$. Then $\sigma(t) = t + 1$ and

$$f^{\Delta}(t) = \frac{e^{\sigma(t)} - e^t}{\sigma(t) - t} = e^{t+1} - e^t.$$

Hence, for every $t > 1$, there exist $\xi_1, \xi_2 \in (0, t)$ such that

$$e^{\xi_1+1} - e^{\xi_1} \le \frac{e^t - 1}{t} \le e^{\xi_2+1} - e^{\xi_2}.$$

Example 2.44 Let

$$\mathbb{T} = 2^{\mathbb{N}_0} \cup \{0\} \quad \text{and} \quad f(t) = \sin t.$$

Then, for every $t > 0$, we have $\sigma(t) = 2t$ and

$$\begin{aligned} f^{\Delta}(t) &= \frac{f(\sigma(t)) - f(t)}{\sigma(t) - t} \\ &= \frac{\sin(2t) - \sin t}{2t - t} \\ &= \frac{2}{t} \sin \frac{t}{2} \cos \frac{3t}{2}. \end{aligned}$$

Hence, for every $t > 1$, there exist $\xi_1, \xi_2 \in (0, t)$ such that

$$\frac{2}{\xi_1} \sin \frac{\xi_1}{2} \cos \frac{3\xi_1}{2} \le \frac{\sin t}{t} \le \frac{2}{\xi_2} \sin \frac{\xi_2}{2} \cos \frac{3\xi_2}{2}.$$

Exercise 2.45 Let $\mathbb{T} = 2^{\mathbb{N}_0} \cup \{0\}$. Prove that, for every $t > 1$, there exist $\xi_1, \xi_2 \in (0, t)$ so that

$$7\xi_1^2 + 3\xi_1 \le t^2 + t \le 7\xi_2^2 + 3\xi_2.$$

Solution Use the function $f(t) = t^3 + t^2$.

Corollary 2.46 *Let f be a continuous function on $[a, b]$ that has a delta derivative at each point of $[a, b)$. If $f^{\Delta}(t) = 0$ for all $t \in [a, b)$, then f is a constant function on $[a, b]$.*

Proof For every $t \in [a, b]$, using (2.4), we have that there exist $\xi_1, \xi_2 \in [a, b)$ such that

$$0 = f^{\Delta}(\xi_1)(t - a) \le f(t) - f(a) \le f^{\Delta}(\xi_2)(t - a) = 0,$$

i.e., $f(t) = f(a)$. □

Corollary 2.47 *Let f be a continuous function on $[a, b]$ that has a delta derivative at each point of $[a, b)$. Then f is increasing, decreasing, nondecreasing and nonincreasing on $[a, b]$ if $f^{\Delta}(t) > 0$, $f^{\Delta}(t) < 0$, $f^{\Delta}(t) \ge 0$, $f^{\Delta}(t) \le 0$ for all $t \in [a, b)$, respectively.*

Proof Let $f^{\Delta}(t) > 0$ for any $t \in [a, b]$. Then, for any $t_1, t_2 \in [a, b]$, $t_1 < t_2$, there exists $\xi \in (t_1, t_2)$ such that

$$f(t_1) - f(t_2) \le f^{\Delta}(\xi)(t_1 - t_2) < 0,$$

i.e., $f(t_1) < f(t_2)$. The cases $f^{\Delta}(t) < 0$, $f^{\Delta}(t) \ge 0$ and $f^{\Delta}(t) \le 0$ are left to the reader for exercise. □

Example 2.48 Let $\mathbb{T} = \mathbb{Z}$ and $f(t) = t^3 - 2t^2 - t$. Then $\sigma(t) = t + 1$ and

$$\begin{aligned} f^{\Delta}(t) &= (\sigma(t))^2 + t\sigma(t) + t^2 - 2(\sigma(t) + t) - 1 \\ &= (t+1)^2 + t(t+1) + t^2 - 2(t+1+t) - 1 \\ &= t^2 + 2t + 1 + t^2 + t + t^2 - 4t - 2 - 1 \\ &= 3t^2 - t - 2. \end{aligned}$$

Hence,

$$f^{\Delta}(t) = 0 \quad \text{iff} \quad t \in \left\{-\frac{2}{3}, 1\right\}.$$

Therefore,

$$f^{\Delta}(t) \ge 0 \quad \text{for} \quad t \in \left(-\infty, -\frac{2}{3}\right] \cup [1, \infty)$$

and

$$f^{\Delta}(t) \le 0 \quad \text{for} \quad t \in \left[-\frac{2}{3}, 1\right].$$

Consequently, f is increasing in $(-\infty, -1] \cup [1, +\infty)$ and f is decreasing in $[-1, 1]$.

Example 2.49 Let $\mathbb{T} = 2^{\mathbb{N}_0}$. We will investigate the monotonicity of the function

$$f(t) = \frac{1-2t}{1+t}.$$

Here, $\sigma(t) = 2t$. Then

$$\begin{aligned} f^{\Delta}(t) &= \frac{-2(1+t) - (1-2t)}{(1+t)(1+2t)} \\ &= \frac{-2 - 2t - 1 + 2t}{(1+t)(1+2t)} \\ &= -\frac{3}{(1+t)(1+2t)}. \end{aligned}$$

Therefore, the function f is decreasing for all $t \in \mathbb{T}$.

Example 2.50 Let $\mathbb{T} = \mathbb{Z}$. We will investigate where the function

$$f(t) = 4t^3 - 21t^2 + 18t + 20$$

is increasing and decreasing. Here, $\sigma(t) = t + 1$ and

$$\begin{aligned}
f^{\Delta}(t) &= 4((\sigma(t))^2 + t\sigma(t) + t^2) - 21(\sigma(t) + t) + 18 \\
&= 4((t+1)^2 + t(t+1) + t^2) - 21(t + 1 + t) + 18 \\
&= 4(t^2 + 2t + 1 + t^2 + t + t^2) - 21(2t + 1) + 18 \\
&= 12t^2 + 12t + 4 - 42t - 21 + 18 \\
&= 12t^2 - 30t + 1.
\end{aligned}$$

Hence,

$$f^{\Delta}(x) = 0 \quad \text{for} \quad x \in \left\{ \frac{15 + \sqrt{213}}{12}, \frac{15 - \sqrt{213}}{12} \right\}.$$

Therefore, $f^{\Delta}(t) \geq 0$ for $t \in (-\infty, 0] \cup [3, \infty)$ and $f^{\Delta}(t) \leq 0$ for $t \in [1, 2]$.

Exercise 2.51 Let $\mathbb{T} = \mathbb{Z}$. Investigate where the function

$$f(t) = t^3 - 4t^2 - 4t + 5$$

is increasing and decreasing.

Solution f is increasing for $t \in (-\infty, -1] \cup [3, \infty)$ and f is decreasing for $t \in [0, 2]$.

2.3 Chain Rules

Theorem 2.52 (Chain Rule) *Assume $g : \mathbb{R} \to \mathbb{R}$ is continuous, $g : \mathbb{T} \to \mathbb{R}$ is delta differentiable on $\mathbb{T}^{\kappa}$, and $f : \mathbb{R} \to \mathbb{R}$ is continuously differentiable. Then there exists $c \in [t, \sigma(t)]$ with*

$$(f \circ g)^{\Delta}(t) = f'(g(c))g^{\Delta}(t). \tag{2.5}$$

Proof Fix $t \in \mathbb{T}^{\kappa}$.

1. If t is right-scattered, then

$$(f \circ g)^{\Delta}(t) = \frac{f(g(\sigma(t))) - f(g(t))}{\mu(t)}.$$

If $g(t) = g(\sigma(t))$, then

$$(f \circ g)^{\Delta}(t) = 0 \quad \text{and} \quad g^{\Delta}(t) = 0,$$

and so (2.5) holds for any $c \in [t, \sigma(t)]$. Assume that $g(\sigma(t)) \neq g(t)$. Then, by the mean value theorem,

$$\begin{aligned}(f \circ g)^{\Delta}(t) &= \frac{f(g(\sigma(t))) - f(g(t))}{g(\sigma(t)) - g(t)} \frac{g(\sigma(t)) - g(t)}{\mu(t)} \\ &= f'(\xi)g^{\Delta}(t),\end{aligned}$$

where ξ is between $g(t)$ and $g(\sigma(t))$. Since $g : \mathbb{R} \to \mathbb{R}$ is continuous, there exists $c \in [t, \sigma(t)]$ such that $g(c) = \xi$.

2. If t is right-dense, then

$$\begin{aligned}(f \circ g)^{\Delta}(t) &= \lim_{s \to t} \frac{f(g(t)) - f(g(s))}{t - s} \\ &= \lim_{s \to t} \frac{f(g(t)) - f(g(s))}{g(t) - g(s)} \frac{g(t) - g(s)}{t - s} \\ &= \lim_{s \to t} \left(f'(\xi_s) \frac{g(t) - g(s)}{t - s} \right),\end{aligned}$$

where ξ_s is between $g(s)$ and $g(t)$. By the continuity of g, we get that $\lim_{s \to t} \xi_s = g(t)$. Therefore,

$$(f \circ g)^{\Delta}(t) = f'(g(t))g^{\Delta}(t).$$

This completes the proof. □

Example 2.53 Let $\mathbb{T} = \mathbb{Z}$, $f(t) = t^3 + 1$, $g(t) = t^2$. We have that $g : \mathbb{R} \to \mathbb{R}$ is continuous, $g : \mathbb{T} \to \mathbb{R}$ is delta differentiable on $\mathbb{T}^{\kappa}$, $f : \mathbb{R} \to \mathbb{R}$ is continuously differentiable, $\sigma(t) = t + 1$. Then

$$g^{\Delta}(t) = \sigma(t) + t$$

and

$$(f \circ g)^{\Delta}(1) = f'(g(c))g^{\Delta}(1) = 3g^2(c)(\sigma(1) + 1) = 9c^4. \tag{2.6}$$

Here, $c \in [1, \sigma(1)] = [1, 2]$. Also,

$$(f \circ g)(t) = f(g(t)) = g^3(t) + 1 = t^6 + 1$$

so that

$$(f \circ g)^{\Delta}(t) = (\sigma(t))^5 + t(\sigma(t))^4 + t^2(\sigma(t))^3 + t^3(\sigma(t))^2 + t^4\sigma(t) + t^5$$

and

$$(f \circ g)^{\Delta}(1) = (\sigma(1))^5 + (\sigma(1))^4 + (\sigma(1))^3 + (\sigma(1))^2 + \sigma(1) + 1 = 63.$$

By (2.6), we get

$$63 = 9c^4, \quad \text{so} \quad c^4 = 7, \quad \text{so} \quad c = \sqrt[4]{7} \in [1, 2].$$

Example 2.54 Let $\mathbb{T} = \{2^n : n \in \mathbb{N}_0\}$, $f(t) = t + 2$, $g(t) = t^2 - 1$. We note that $g : \mathbb{T} \to \mathbb{R}$ is delta differentiable, $g : \mathbb{R} \to \mathbb{R}$ is continuous and $f : \mathbb{R} \to \mathbb{R}$ is continuously differentiable. For $t \in \mathbb{T}$, $t = 2^n$, $n \in \mathbb{N}_0$, $n = \log_2 t$, we have

$$\sigma(t) = \inf \left\{2^l : 2^l > 2^n, l \in \mathbb{N}_0\right\} = 2^{n+1} = 2t > t.$$

Therefore, all points of $\mathbb{T}$ are right-scattered. Since $\sup \mathbb{T} = \infty$, we have that $\mathbb{T}^\kappa = \mathbb{T}$. Also, for $t \in \mathbb{T}$, we have

$$(f \circ g)(t) = f(g(t)) = g(t) + 2 = t^2 - 1 + 2 = t^2 + 1$$

and

$$(f \circ g)^{\Delta}(t) = \sigma(t) + t = 2t + t = 3t.$$

Hence,

$$(f \circ g)^{\Delta}(2) = 6. \tag{2.7}$$

Now, using Theorem 2.52, we get that there exists $c \in [2, \sigma(2)] = [2, 4]$ such that

$$(f \circ g)^{\Delta}(2) = f'(g(c))g^{\Delta}(2) = g^{\Delta}(2) = \sigma(2) + 2 = 4 + 2 = 6. \tag{2.8}$$

From (2.7) and (2.8), we find that for every $c \in [2, 4]$,

$$(f \circ g)^{\Delta}(2) = f'(g(c))g^{\Delta}(2).$$

Example 2.55 Let $\mathbb{T} = \left\{3^{n^2} : n \in \mathbb{N}_0\right\}$, $f(t) = t^2 + 1$, $g(t) = t^3$. We note that $g : \mathbb{R} \to \mathbb{R}$ is continuous, $g : \mathbb{T} \to \mathbb{R}$ is delta differentiable and $f : \mathbb{R} \to \mathbb{R}$ is continuously differentiable. For $t \in \mathbb{T}$, $t = 3^{n^2}$, $n \in \mathbb{N}_0$, $n = \left(\log_3 t\right)^{\frac{1}{2}}$, we have

$$\begin{aligned}\sigma(t) &= \inf\left\{3^{l^2} : 3^{l^2} > 3^{n^2},\ l \in \mathbb{N}_0\right\}\\ &= 3^{(n+1)^2}\\ &= 3\cdot 3^{n^2}\cdot 3^{2n}\\ &= 3t3^{2(\log_3 t)^{\frac{1}{2}}} > t.\end{aligned}$$

Consequently, all points of $\mathbb{T}$ are right-scattered. Also, $\sup\mathbb{T} = \infty$. Then $\mathbb{T}^\kappa = \mathbb{T}$. Hence, for $t \in \mathbb{T}$, we have

$$(f \circ g)(t) = f(g(t)) = g^2(t) + 1 = t^6 + 1$$

and

$$(f \circ g)^\Delta(t) = (\sigma(t))^5 + t(\sigma(t))^4 + t^2(\sigma(t))^3 + t^3(\sigma(t))^2 + t^4\sigma(t) + t^5.$$

Thus,

$$\begin{aligned}(f \circ g)^\Delta(1) =&(\sigma(1))^5 + (\sigma(1))^4 + (\sigma(1))^3 + (\sigma(1))^2 + \sigma(1) + 1\\ =&3^5 + 3^4 + 3^3 + 3^2 + 3 + 1 \qquad (2.9)\\ =&364.\end{aligned}$$

From Theorem 2.52, it follows that there exists $c \in [1, \sigma(1)] = [1, 3]$ such that

$$(f \circ g)^\Delta(1) = f'(g(c))g^\Delta(1) = 2g(c)g^\Delta(1) = 2c^3g^\Delta(1). \qquad (2.10)$$

Because all points of $\mathbb{T}$ are right-scattered, we have

$$g^\Delta(1) = (\sigma(1))^2 + \sigma(1) + 1 = 9 + 3 + 1 = 13.$$

By (2.10), we find

$$(f \circ g)^\Delta(1) = 26c^3.$$

From the last equation and from (2.9), we obtain

$$364 = 26c^3, \quad \text{so} \quad c^3 = \frac{364}{26} = 14, \quad \text{so} \quad c = \sqrt[3]{14}.$$

Exercise 2.56 Let $\mathbb{T} = \mathbb{Z}$, $f(t) = t^2 + 2t + 1$, $g(t) = t^2 - 3t$. Find a constant $c \in [1, \sigma(1)]$ such that

$$(f \circ g)^{\Delta}(1) = f'(g(c))g^{\Delta}(1).$$

Solution Any $c \in [1, 2]$.

Theorem 2.57 (Chain Rule) *Let $f : \mathbb{R} \to \mathbb{R}$ be continuously differentiable and suppose $g : \mathbb{T} \to \mathbb{R}$ is delta differentiable. Then $f \circ g : \mathbb{T} \to \mathbb{R}$ is delta differentiable, and the formula*

$$(f \circ g)^{\Delta}(t) = \left\{ \int_0^1 f'(g(t) + h\mu(t)g^{\Delta}(t))\mathrm{d}h \right\} g^{\Delta}(t) \tag{2.11}$$

holds.

Proof Note that

$$\begin{aligned} f(g(\sigma(t))) - f(g(s)) &= \int_{g(s)}^{g(\sigma(t))} f'(y)\mathrm{d}y \\ &= (g(\sigma(t)) - g(s)) \int_0^1 f'(h(g(\sigma(t))) + (1-h)g(s))\mathrm{d}h. \end{aligned}$$

Let $\varepsilon > 0$ and $t \in \mathbb{T}^{\kappa}$. Set

$$\varepsilon^* = \frac{\varepsilon}{1 + 2\int_0^1 |f'(hg(\sigma(t)) + (1-h)g(t))|\mathrm{d}h}.$$

Since g is differentiable at t, there exists a neighbourhood U_1 of t such that

$$|g(\sigma(t)) - g(s) - g^{\Delta}(t)(\sigma(t) - s)| \le \varepsilon^* |\sigma(t) - s| \quad \text{for all} \quad s \in U_1.$$

Since f' is continuous on $\mathbb{R}$, using that

$$\begin{aligned} |hg(\sigma(t)) + (1-h)g(s) - (hg(\sigma(t)) + (1-h)g(t))| &= (1-h)|g(t) - g(s)| \\ &\le |g(t) - g(s)| \end{aligned}$$

for $h \in [0, 1]$, there exists a neighbourhood U_2 of t such that

$$|f'(hg(\sigma(t)) + (1-h)g(s)) - f'(hg(\sigma(t)) + (1-h)g(t))| \le \frac{\varepsilon}{2(\varepsilon^* + |g^{\Delta}(t)|)}$$

for all $s \in U_2$. Let

$$U = U_1 \cap U_2, \quad s \in U.$$

We put

$$\alpha = hg(\sigma(t)) + (1-h)g(s) \quad \text{and} \quad \beta = hg(\sigma(t)) + (1-h)g(t).$$

Then, using the choice of ε, we have

$$|f'(\alpha) - f'(\beta)| < 1, \quad \text{i.e.,} \quad |f'(\alpha)| < 1 + |f'(\beta)|.$$

Also,

$$\begin{aligned}
&\left|(f \circ g)(\sigma(t)) - (f \circ g)(s) - (\sigma(t) - s)g^{\Delta}(t)\int_0^1 f'(\beta)\mathrm{d}h\right| \\
&= \left|(g(\sigma(t)) - g(s))\int_0^1 f'(\alpha)\mathrm{d}h - (\sigma(t) - s)g^{\Delta}(t)\int_0^1 f'(\beta)\mathrm{d}h\right| \\
&= \left|(g(\sigma(t)) - g(s) - (\sigma(t) - s)g^{\Delta}(t))\int_0^1 f'(\alpha)\mathrm{d}h\right. \\
&\quad \left. + (\sigma(t) - s)g^{\Delta}(t)\int_0^1 (f'(\alpha) - f'(\beta))\mathrm{d}h\right| \\
&\le |g(\sigma(t)) - g(s) - (\sigma(t) - s)g^{\Delta}(t)|\int_0^1 |f'(\alpha)|\mathrm{d}h \\
&\quad + |\sigma(t) - s||g^{\Delta}(t)|\int_0^1 |f'(\alpha) - f'(\beta)|\mathrm{d}h \\
&\le \varepsilon^*|\sigma(t) - s|\int_0^1 |f'(\alpha)|\mathrm{d}h + |\sigma(t) - s||g^{\Delta}(t)|\int_0^1 |f'(\alpha) - f'(\beta)|\mathrm{d}h \\
&\le \varepsilon^*|\sigma(t) - s|\int_0^1 |f'(\alpha)|\mathrm{d}h \\
&\quad + \left[\varepsilon^* + |g^{\Delta}(t)|\right]|\sigma(t) - s|\int_0^1 |f'(\alpha) - f'(\beta)|\mathrm{d}h
\end{aligned}$$

$$\leq \frac{\varepsilon}{2}|\sigma(t)-s| + \frac{\varepsilon}{2}|\sigma(t)-s|$$

$$= \varepsilon|\sigma(t)-s|.$$

Therefore, $f \circ g$ is differentiable at t, and its derivative satisfies (2.11). □

Example 2.58 Let $\mathbb{T} = \mathbb{Z}$, $f(x) = \frac{1}{1+x^2}$, $g(t) = t+1$. Note that $f : \mathbb{R} \to \mathbb{R}$ is continuously differentiable and $g : \mathbb{T} \to \mathbb{R}$ is delta differentiable. We have

$$f'(x) = -\frac{2x}{(1+x^2)^2}, \quad \mu(t) = 1, \quad g^{\Delta}(t) = 1$$

and

$$g(t) + h\mu(t)g^{\Delta}(t) = t + 1 + h.$$

Hence,

$$f'(g(t) + h\mu(t))g^{\Delta}(t) = f'(t+1+h)$$

$$= -\frac{2(t+1+h)}{(1+(t+1+h)^2)^2}.$$

From Theorem 2.57, we conclude that $f \circ g : \mathbb{T} \to \mathbb{R}$ is delta differentiable and

$$(f \circ g)^{\Delta}(t) = -\int_0^1 \frac{2(t+1+h)}{(1+(t+1+h)^2)^2} \mathrm{d}h$$

$$= -\int_0^1 \frac{\mathrm{d}(t+1+h)^2}{(1+(t+1+h)^2)^2}$$

$$= \frac{1}{1+(t+1+h)^2}\Big|_{h=0}^{h=1}$$

$$= \frac{1}{1+(t+2)^2} - \frac{1}{1+(t+1)^2}$$

$$= \frac{(t+1)^2-(t+2)^2}{(t^2+4t+5)(t^2+2t+3)}$$

$$= \frac{t^2+2t+1-t^2-4t-4}{(t^2+4t+5)(t^2+2t+3)}$$

$$= \frac{-2t-3}{(t^2+4t+5)(t^2+2t+3)}.$$

Example 2.59 Let $\mathbb{T} = 2^{\mathbb{N}_0}$, $f(x) = \sin x$, $g(t) = t^2 + 1$. We have that $f : \mathbb{R} \to \mathbb{R}$ is continuously differentiable, $g : \mathbb{T} \to \mathbb{R}$ is delta differentiable, $\sigma(t) = 2t$, $\mu(t) = t$, and

$$g^{\Delta}(t) = \sigma(t) + t = 2t + t = 3t$$

so that

$$g(t) + h\mu(t)g^{\Delta}(t) = t^2 + 1 + ht(3t) = t^2 + 1 + 3t^2h.$$

Moreover,

$$f'(x) = \cos x,$$

and thus

$$f'(g(t) + h\mu(t)g^{\Delta}(t)) = \cos(t^2 + 1 + 3t^2h).$$

By Theorem 2.57, we conclude that $f \circ g : \mathbb{T} \to \mathbb{R}$ is delta differentiable and

$$\begin{aligned}
(f \circ g)^{\Delta}(t) &= \int_0^1 \cos(t^2 + 1 + 3t^2h)dh(3t) \\
&= \frac{3t}{3t^2} \int_0^1 \cos(t^2 + 1 + 3t^2h)d(t^2 + 1 + 3t^2h) \\
&= \frac{1}{t} \sin(t^2 + 1 + 3t^2h)\Big|_{h=0}^{h=1} \\
&= \frac{1}{t}(\sin(4t^2 + 1) - \sin(t^2 + 1)) \\
&= \frac{2}{t} \sin\frac{3t^2}{2} \cos\left(\frac{5t^2}{2} + 1\right).
\end{aligned}$$

Example 2.60 Let

$$\mathbb{T} = 3^{\mathbb{N}_0}, \quad f(x) = \log(1 + x^2), \quad g(t) = t^3 - 2t.$$

We have that $f : \mathbb{R} \to \mathbb{R}$ is continuously differentiable, $g : \mathbb{T} \to \mathbb{R}$ is delta differentiable, $\sigma(t) = 3t$, $\mu(t) = 2t$, and

$$\begin{aligned}
g^{\Delta}(t) &= (\sigma(t))^2 + t\sigma(t) + t^2 - 2 \\
&= 9t^2 + 3t^2 + t^2 - 2 \\
&= 13t^2 - 2,
\end{aligned}$$

so that

$$
\begin{aligned}
g(t) + h\mu(t)g^{\Delta}(t) &= t^3 - 2t + h(2t)(13t^2 - 2) \\
&= t^3 - 2t + (26t^3 - 4t)h.
\end{aligned}
$$

Moreover,

$$
f'(x) = \frac{2x}{1+x^2},
$$

and therefore

$$
\begin{aligned}
f'(g(t) + h\mu(t)g^{\Delta}(t)) &= \frac{2(t^3 - 2t) + 2(26t^3 - 4t)h}{1 + (t^3 - 2t + (26t^3 - 4t)h)^2} \\
&= 2\frac{t^3 - 2t + (26t^3 - 4t)h}{1 + (t^3 - 2t + (26t^3 - 4t)h)^2}.
\end{aligned}
$$

By Theorem 2.57, we conclude that $f \circ g$ is delta differentiable and

$$
\begin{aligned}
(f \circ g)^{\Delta}(t) &= 2\int_0^1 \frac{t^3 - 2t + (26t^3 - 4t)h}{1 + (t^3 - 2t + (26t^3 - 4t)h)^2} \mathrm{d}h(13t^2 - 2) \\
&= \frac{13t^2 - 2}{13t^3 - 2t}\int_0^1 \frac{t^3 - 2t + (26t^3 - 4t)h}{1 + (t^3 - 2t + (26t^3 - 4t)h)^2} \mathrm{d}(t^3 - 2t + (26t^3 - 4t))h \\
&= \frac{1}{2t}\int_0^1 \frac{\mathrm{d}(t^3 - 2t + (26t^3 - 4t)h)^2}{1 + (t^3 - 2t + (26t^3 - 4t)h)^2} \\
&= \frac{1}{2t}\log(t^3 - 2t + (26t^3 - 4t)h)\Big|_{h=0}^{h=1} \\
&= \frac{1}{2t}\Big(\log(27t^3 - 6t) - \log(t^3 - 2t)\Big) \\
&= \frac{1}{2t}\log\frac{27t^2 - 6}{t^2 - 2}.
\end{aligned}
$$

Exercise 2.61 Let

$$
\mathbb{T} = \mathbb{Z}, \quad f(x) = \cos x, \quad g(t) = t^2 + t.
$$

Using Theorem 2.57, find $(f \circ g)^{\Delta}(t)$.

Solution $-2\sin(t+1)\sin((t+1)^2)$.

Theorem 2.62 (Chain Rule) *Assume* $v : \mathbb{T} \to \mathbb{R}$ *is strictly increasing and* $\tilde{\mathbb{T}} = v(\mathbb{T})$ *is a time scale. Let* $w : \tilde{\mathbb{T}} \to \mathbb{R}$. *If* $v^{\Delta}(t)$ *and* $w^{\tilde{\Delta}}(v(t))$ *exist for* $t \in \mathbb{T}^{\kappa}$, *then*

$$(w \circ v)^{\Delta} = (w^{\tilde{\Delta}} \circ v)v^{\Delta}.$$

Proof Let $\varepsilon \in (0, 1)$ be arbitrarily chosen. We put

$$\varepsilon^* = \frac{\varepsilon}{1 + |v^{\Delta}(t)| + |w^{\tilde{\Delta}}(v(t))|}.$$

Note that $0 < \varepsilon^* < 1$. Since w is differentiable at t, there exists a neighbourhood U_1 of t such that

$$|v(\sigma(t)) - v(t) - (\sigma(t) - s)v^{\Delta}(t)| \le \varepsilon^* |\sigma(t) - s| \quad \text{for all} \quad s \in U_1.$$

Since w is differentiable at $v(t)$, there exists a neighbourhood U_2 of $v(t)$ such that

$$|w(\tilde{\sigma}(v(t))) - w(r) - (\tilde{\sigma}(v(t)) - r)w^{\tilde{\Delta}}(v(t))| \le \varepsilon^* |\tilde{\sigma}(v(t)) - r|$$

for all $r \in U_2$. Let $U = U_1 \cap v^{-1}(U_2)$ and let $s \in U$. Then $s \in U_1$, $v(s) \in U_2$, and

$$|w(v(\sigma(t))) - w(v(s)) - (\sigma(t) - s)w^{\tilde{\Delta}}(v(t))v^{\Delta}(t)|$$

$$= |w(v(\sigma(t))) - w(v(s)) - (\tilde{\sigma}(v(t)) - v(s))w^{\tilde{\Delta}}(v(t))$$

$$+ [(\tilde{\sigma}(v(t)) - v(s)) - (\sigma(t) - s)v^{\Delta}(t)]w^{\tilde{\Delta}}(v(t))|$$

$$\le |w(v(\sigma(t))) - w(v(s)) - (\tilde{\sigma}(v(t)) - v(s))w^{\tilde{\Delta}}(v(t))|$$

$$+ |(\tilde{\sigma}(v(t)) - v(s)) - (\sigma(t) - s)v^{\Delta}(t)||w^{\tilde{\Delta}}(v(t))|$$

$$\le \varepsilon^* |\tilde{\sigma}(v(t)) - v(s)| + \varepsilon^* |\sigma(t) - s||w^{\tilde{\Delta}}(v(t))|$$

$$= \varepsilon^* |\tilde{\sigma}(v(t)) - v(s) - (\sigma(t) - s)v^{\Delta}(t) + (\sigma(t) - s)v^{\Delta}(t)|$$

$$+ \varepsilon^* |\sigma(t) - s||w^{\tilde{\Delta}}(v(t))|$$

$$\le \varepsilon^* |\tilde{\sigma}(v(t)) - v(s) - (\sigma(t) - s)v^{\Delta}(t)| + \varepsilon^* |\sigma(t) - s||v^{\Delta}(t)|$$

$$+ \varepsilon^* |\sigma(t) - s||w^{\tilde{\Delta}}(v(t))|$$

$$\leq \varepsilon^* \left(\varepsilon^* |\sigma(t) - s| + |\sigma(t) - s||v^{\Delta}(t)| + |\sigma(t) - s||w^{\tilde{\Delta}}(v(t))|\right)$$

$$= \varepsilon^*(\varepsilon^* + |v^{\Delta}(t)| + |w^{\tilde{\Delta}}(v(t))|)|\sigma(t) - s|$$

$$\leq \varepsilon^*(1 + |v^{\Delta}(t)| + |w^{\tilde{\Delta}}(v(t))|)|\sigma(t) - s|$$

$$= \varepsilon|\sigma(t) - s|,$$

which completes the proof. □

Example 2.63 Let $\mathbb{T} = \{2^{2n} : n \in \mathbb{N}_0\}$, $v(t) = t^2$, $w(t) = t^2 + 1$. Then $v : \mathbb{T} \to \mathbb{R}$ is strictly increasing and $\tilde{\mathbb{T}} = v(\mathbb{T}) = \{2^{4n} : n \in \mathbb{N}_0\}$ is a time scale. For $t \in \mathbb{T}$, $t = 2^{2n}$, $n \in \mathbb{N}_0$, we have

$$\sigma(t) = \inf\left\{2^{2l} : 2^{2l} > 2^{2n},\ l \in \mathbb{N}_0\right\} = 2^{2n+2} = 4t$$

and

$$v^{\Delta}(t) = \sigma(t) + t = 5t.$$

For $t \in \tilde{\mathbb{T}}$, $t = 2^{4n}$, $n \in \mathbb{N}_0$, we have

$$\tilde{\sigma}(t) = \inf\left\{2^{4l} : 2^{4l} > 2^{4n},\ l \in \mathbb{N}_0\right\} = 2^{4n+4} = 16t.$$

Also, for $t \in \mathbb{T}$, we have

$$(w \circ v)(t) = w(v(t)) = (v(t))^2 + 1 = t^4 + 1$$

and

$$(w \circ v)^{\Delta}(t) = (\sigma(t))^3 + t(\sigma(t))^2 + t^2\sigma(t) + t^3$$

$$= 64t^3 + 16t^3 + 4t^3 + t^3$$

$$= 85t^3.$$

Thus,

$$\left(w^{\tilde{\Delta}} \circ v\right)(t) = \tilde{\sigma}(v(t)) + v(t)$$

$$= 16v(t) + v(t)$$

$$= 17v(t)$$

$$= 17t^2$$

and

$$\left(w^{\tilde{\Delta}} \circ v\right)(t)v^{\Delta}(t) = 17t^2(5t) = 85t^3.$$

Consequently,

$$(w \circ v)^{\Delta}(t) = \left(w^{\tilde{\Delta}} \circ v\right)(t)v^{\Delta}(t), \quad t \in \mathbb{T}^{\kappa}.$$

Example 2.64 Let $\mathbb{T} = \{n + 1 : n \in \mathbb{N}_0\}$, $v(t) = t^2$, $w(t) = t$. Then $v : \mathbb{T} \to \mathbb{R}$ is strictly increasing and $\tilde{\mathbb{T}} = \{(n + 1)^2 : n \in \mathbb{N}_0\}$ is a time scale. For $t \in \mathbb{T}, t = n + 1$, $n \in \mathbb{N}_0$, we have

$$\sigma(t) = \inf\{l + 1 : \ l + 1 > n + 1, \ l \in \mathbb{N}_0\} = n + 2 = t + 1$$

and

$$v^{\Delta}(t) = \sigma(t) + t = t + 1 + t = 2t + 1.$$

For $t \in \tilde{\mathbb{T}}$, $t = (n + 1)^2$, $n \in \mathbb{N}_0$, we have

$$\tilde{\sigma}(t) = \{(l + 1)^2 : \ (l + 1)^2 > (n + 1)^2, \ l \in \mathbb{N}_0\} = (n + 2)^2$$

$$= (n + 1)^2 + 2(n + 1) + 1 = t + 2\sqrt{t} + 1.$$

Hence, for $t \in \mathbb{T}$, we get

$$(w^{\tilde{\Delta}} \circ v)(t) = 1, \quad (w^{\tilde{\Delta}} \circ v)(t)v^{\Delta}(t) = 1(2t + 1) = 2t + 1,$$

and thus

$$(w \circ v)(t) = v(t) = t^2, \quad (w \circ v)^{\Delta}(t) = \sigma(t) + t = 2t + 1.$$

Consequently,

$$(w \circ v)^{\Delta}(t) = \left(w^{\tilde{\Delta}} \circ v\right)(t)v^{\Delta}(t), \quad t \in \mathbb{T}^{\kappa}.$$

Example 2.65 Let $\mathbb{T} = \{2^n : n \in \mathbb{N}_0\}$, $v(t) = t$, $w(t) = t^2$. Then $v : \mathbb{T} \to \mathbb{R}$ is strictly increasing and $v(\mathbb{T}) = \mathbb{T}$. For $t \in \mathbb{T}$, $t = 2^n$, $n \in \mathbb{N}_0$, we have

$$\sigma(t) = \inf\{2^l : 2^l > 2^n, \; l \in \mathbb{N}_0\} = 2^{n+1} = 2t$$

and

$$v^{\Delta}(t) = 1.$$

Moreover,

$$(w \circ v)(t) = w(v(t)) = (v(t))^2 = t^2,$$

and thus

$$(w \circ v)^{\Delta}(t) = \sigma(t) + t = 2t + t = 3t.$$

Therefore,

$$(w^{\Delta} \circ v)(t) = \sigma(v(t)) + v(t) = 2v(t) + v(t) = 3v(t) = 3t,$$

so that

$$(w^{\Delta} \circ v)(t)v^{\Delta}(t) = 3t.$$

Consequently,

$$(w \circ v)^{\Delta}(t) = (w^{\tilde{\Delta}} \circ v)(t)v^{\Delta}(t), \quad t \in \mathbb{T}^{\kappa}.$$

Exercise 2.66 Let $\mathbb{T} = \left\{2^{3n} : n \in \mathbb{N}_0\right\}$, $v(t) = t^2$, $w(t) = t$. Prove

$$(w \circ v)^{\Delta}(t) = (w^{\tilde{\Delta}} \circ v)(t)v^{\Delta}(t), \quad t \in \mathbb{T}^{\kappa}.$$

Theorem 2.67 (Derivative of the Inverse) *Assume $v : \mathbb{T} \to \mathbb{R}$ is strictly increasing and $\tilde{\mathbb{T}} = v(\mathbb{T})$ is a time scale. Then*

$$\left((v^{-1})^{\tilde{\Delta}} \circ v\right)(t) = \frac{1}{v^{\Delta}(t)}$$

for any $t \in \mathbb{T}^{\kappa}$ such that $v^{\Delta}(t) \neq 0$.

Proof Let $w = v^{-1} : \tilde{\mathbb{T}} \to \mathbb{T}$ in Theorem 2.62. □

Example 2.68 Let $\mathbb{T} = \mathbb{N}$ and $v(t) = t^2 + 1$. Then $\sigma(t) = t + 1$, $v : \mathbb{T} \to \mathbb{R}$ is strictly increasing, and

$$v^{\Delta}(t) = \sigma(t) + t = 2t + 1.$$

Hence,

$$\left(\left(v^{-1}\right)^{\tilde{\Delta}} \circ v\right)(t) = \frac{1}{v^{\Delta}(t)} = \frac{1}{2t + 1}.$$

Example 2.69 Let $\mathbb{T} = \{n+3 : n \in \mathbb{N}_0\}$, $v(t) = t^2$. Then $v : \mathbb{T} \to \mathbb{R}$ is strictly increasing, $\sigma(t) = t+1$, and

$$v^{\Delta}(t) = \sigma(t) + t = 2t + 1.$$

Hence,

$$\left(\left(v^{-1}\right)^{\tilde{\Delta}} \circ v\right)(t) = \frac{1}{v^{\Delta}(t)} = \frac{1}{2t+1}.$$

Example 2.70 Let $\mathbb{T} = \left\{2^{n^2} : n \in \mathbb{N}_0\right\}$, $v(t) = t^3$. Then $v : \mathbb{T} \to \mathbb{R}$ is strictly increasing, and for $t \in \mathbb{T}$, $t = 2^{n^2}$, $n \in \mathbb{N}_0$, $n = \left(\log_2 t\right)^{\frac{1}{2}}$, we have

$$\sigma(t) = \inf\left\{2^{l^2} : 2^{l^2} > 2^{n^2},\ l \in \mathbb{N}_0\right\} = 2^{(n+1)^2} = 2^{n^2} 2^{2n+1} = t 2^{2(\log_2 t)^{\frac{1}{2}}+1}.$$

Then

$$v^{\Delta}(t) = (\sigma(t))^2 + t\sigma(t) + t^2 = t^2 2^{4(\log_2 t)^{\frac{1}{2}}+2} + t^2 2^{2(\log_2 t)^{\frac{1}{2}}+1} + t^2.$$

Hence,

$$\left(\left(v^{-1}\right)^{\tilde{\Delta}} \circ v\right)(t) = \frac{1}{t^2 2^{4(\log_2 t)^{\frac{1}{2}}+2} + t^2 2^{2(\log_2 t)^{\frac{1}{2}}+1} + t^2}.$$

Exercise 2.71 Let $\mathbb{T} = \{n+5 : n \in \mathbb{N}_0\}$, $v(t) = t^2 + t$. Find $\left(\left(v^{-1}\right)^{\tilde{\Delta}} \circ v\right)(t)$.

Solution $\frac{1}{2t+2}$.

2.4 One-Sided Derivatives

Definition 2.72 If f is defined on $[t_0, b) \subset \mathbb{T}$, then the *right-hand derivative* of f at t_0 is defined to be

$$f_{+}^{\Delta}(t_0) = \lim_{t \to t_0+} \frac{f(\sigma(t_0)) - f(t)}{\sigma(t_0) - t},$$

while if f is defined on $(a, t_0] \subset \mathbb{T}$, then the left-hand derivative of f at t_0 is defined by

$$f_{-}^{\Delta}(t_0) = \lim_{t \to t_0-} \frac{f(\sigma(t_0)) - f(t)}{\sigma(t_0) - t}.$$

Remark 2.73 f is differentiable at t_0 if and only if $f_{+}^{\Delta}(t_0)$ and $f_{-}^{\Delta}(t_0)$ exist and

$$f^{\Delta}(t_0) = f_{-}^{\Delta}(t_0) = f_{+}^{\Delta}(t_0).$$

Example 2.74 Consider

$$f(t) = \begin{cases} t+1 & \text{for} \quad t \in [-1, 1), \\ t^2 + t & \text{for} \quad t \in \{1, 2, 3\}, \end{cases}$$

where $[-1, 1)$ is the real-valued interval. Note that f is continuous on

$$[-1, 1) \cup \{1, 2, 3\}.$$

Also,

$$\begin{aligned} f_-^{\Delta}(1) &= \lim_{t \to 1-} \frac{f(\sigma(1)) - f(t)}{\sigma(1) - t} \\ &= \lim_{t \to 1-} \frac{f(2) - f(t)}{2 - t} \\ &= \lim_{t \to 1-} \frac{6 - (t+1)}{2 - t} \\ &= \lim_{t \to 1-} \frac{5 - t}{2 - t} \\ &= 4 \end{aligned}$$

and

$$\begin{aligned} f_+^{\Delta}(1) &= \lim_{t \to 1+} \frac{f(\sigma(1)) - f(t)}{\sigma(1) - t} \\ &= \lim_{t \to 1+} \frac{f(2) - f(t)}{2 - t} \\ &= \lim_{t \to 1+} \frac{6 - t^2 - t}{2 - t} \\ &= \lim_{t \to 1+} \frac{(2-t)(t+3)}{2 - t} \\ &= \lim_{t \to 1+} (t+3) \\ &= 4. \end{aligned}$$

Therefore,

$$f_-^\Delta(1) = f_+^\Delta(1).$$

Hence, f is differentiable at $t = 1$.

Example 2.75 Consider

$$f(t) = \begin{cases} t+3 & \text{for} \quad t \in [-2, 2), \\ t^2 + t & \text{for} \quad t \in \{2, 4, 8\}. \end{cases}$$

We have

$$\begin{aligned} f_-^\Delta(2) &= \lim_{t \to 2-} \frac{f(\sigma(2)) - f(t)}{\sigma(2) - t} \\ &= \lim_{t \to 2-} \frac{f(4) - f(t)}{4 - t} \\ &= \lim_{t \to 2-} \frac{20 - (t+3)}{4 - t} \\ &= \lim_{t \to 2-} \frac{17 - t}{4 - t} \\ &= \frac{15}{2} \end{aligned}$$

and

$$\begin{aligned} f_+^\Delta(t) &= \lim_{t \to 2+} \frac{f(\sigma(2)) - f(t)}{\sigma(2) - t} \\ &= \lim_{t \to 2+} \frac{f(4) - f(t)}{4 - t} \\ &= \lim_{t \to 2+} \frac{20 - t^2 - t}{4 - t} \\ &= 7. \end{aligned}$$

Therefore,

$$f_-^\Delta(2) \neq f_+^\Delta(2).$$

Hence, f is not differentiable at $t = 2$.

Example 2.76 Consider

$$f(t) = \begin{cases} 3 & \text{for} \quad t \in [0, 3), \\ \\ 1 & \text{for} \quad t \in \{3, 5, 7, 9\}, \end{cases}$$

where [0, 3) is the real-valued interval. We have

$$\begin{aligned} f_-^{\Delta}(3) &= \lim_{t \to 3-} \frac{f(\sigma(3)) - f(t)}{\sigma(3) - t} \\ &= \lim_{t \to 3-} \frac{f(5) - f(t)}{5 - t} \\ &= \lim_{t \to 3-} \frac{1 - 3}{5 - t} \\ &= \lim_{t \to 3-} \frac{-2}{5 - t} \\ &= -1 \end{aligned}$$

and

$$\begin{aligned} f_+^{\Delta}(3) &= \lim_{t \to 3+} \frac{f(\sigma(3)) - f(t)}{\sigma(3) - t} \\ &= \lim_{t \to 3+} \frac{f(5) - f(t)}{5 - t} \\ &= \lim_{t \to 3+} \frac{1 - 1}{5 - t} \\ &= 0. \end{aligned}$$

Consequently,

$$f_-^{\Delta}(3) \neq f_+^{\Delta}(3).$$

Hence, f is not differentiable at 3.

Exercise 2.77 Consider

$$f(t) = \begin{cases} t^7 + 27 & \text{for} \quad t \in [-3, 0), \\ \\ t^2 + 4t - 5 & \text{for} \quad t \in \{0, 1, 2, 3, 4\}, \end{cases}$$

where $[-3, 0)$ is the real-valued interval. Investigate if f is differentiable at $t = 0$.

Solution No, since it is not continuous at $t = 0$.

2.5 Nabla Derivatives

We define the *backward graininess*

$$\nu(t) = t - \rho(t).$$

If $\mathbb{T}$ has a right-scattered minimum m, then we put $\mathbb{T}_\kappa = \mathbb{T} \setminus \{m\}$. Otherwise, $\mathbb{T}_\kappa = \mathbb{T}$.

Definition 2.78 (*The Nabla Derivative*) A function $f : \mathbb{T} \to \mathbb{R}$ is said to be *nabla differentiable* at $t \in \mathbb{T}_\kappa$ if

1. f is defined in a neighbourhood U of t,
2. f is defined at $\rho(t)$,
3. there exists a unique real number $f^\nabla(t)$, called the nabla derivative of f at t, such that for each $\varepsilon > 0$, there exists a neighbourhood N of t with $N \subseteq U$ and

$$|f(\rho(t)) - f(s) - (\rho(t) - s) f^\nabla(t)| \le \varepsilon |\rho(t) - s| \quad \text{for all} \quad s \in N.$$

Similar to the proofs of Theorems 2.3, 2.10, and 2.18, one can prove the following theorem.

Theorem 2.79 *Let $f, g : \mathbb{T} \to \mathbb{R}$ be functions and let $t \in \mathbb{T}_\kappa$. Then we have the following.*

1. *The nabla derivative is well defined.*
2. *If f is nabla differentiable at t, then f is continuous at t.*
3. *If f is continuous at t and t is left-scattered, then f is nabla differentiable at t with*

$$f^\nabla(t) = \frac{f(t) - f(\rho(t))}{\nu(t)}.$$

4. *If t is left-dense, then f is nabla differentiable at t iff the limit*

$$\lim_{s \to t} \frac{f(t) - f(s)}{t - s}$$

exists as a finite number. In this case,

$$f^\nabla(t) = \lim_{s \to t} \frac{f(t) - f(s)}{t - s}.$$

5. *If f is differentiable at t, then*

$$f(\rho(t)) = f(t) + \nu(t) f^\nabla(t).$$

6. *If f and g are nabla differentiable at t, then*

a. *the sum* $f+g:\mathbb{T}\to\mathbb{R}$ *is nabla differentiable at* t *with*

$$(f+g)^{\nabla}(t)=f^{\nabla}(t)+g^{\nabla}(t).$$

b. *For any constant* α*,* $\alpha f:\mathbb{T}\to\mathbb{R}$ *is nabla differentiable at* t *with*

$$(\alpha f)^{\nabla}(t)=\alpha f^{\nabla}(t).$$

c. *The product* $fg:\mathbb{T}\to\mathbb{R}$ *is nabla differentiable at* t *with*

$$(fg)^{\nabla}(t)=f^{\nabla}(t)g(t)+f(\rho(t))g^{\nabla}(t)=f(t)g^{\nabla}(t)+f^{\nabla}(t)g(\rho(t)).$$

d. *If* $g(t)g(\rho(t))\neq 0$*, then* $\frac{f}{g}:\mathbb{T}\to\mathbb{R}$ *is nabla differentiable at* t *with*

$$\left(\frac{f}{g}\right)^{\nabla}(t)=\frac{f^{\nabla}(t)g(t)-f(t)g^{\nabla}(t)}{g(t)g(\rho(t))}.$$

Definition 2.80 Let $f:\mathbb{T}\to\mathbb{R}$ and let $t\in(\mathbb{T}_{\kappa})_{\kappa}=\mathbb{T}_{\kappa^2}$. We define the second nabla derivative of f at t, provided it exists, by

$$f^{\nabla\nabla}=\left(f^{\nabla}\right)^{\nabla}:\mathbb{T}_{\kappa^2}\to\mathbb{R}.$$

Similarly, we define higher-order nabla derivatives $f^{\nabla^n}:\mathbb{T}_{\kappa^n}\to\mathbb{R}$.

Example 2.81 Let $\mathbb{T}=\mathbb{Z}$ and $f:\mathbb{T}\to\mathbb{R}$ be a function. Then

$$f^{\nabla}(t)=f(t)-f(t-1)\quad\text{for any}\quad t\in\mathbb{T}.$$

Example 2.82 Let $\mathbb{T}=2^{\mathbb{N}_0}$ and $f(t)=t^3+t^2+t+1$, $t\in\mathbb{T}_{\kappa}$. We will find $f^{\nabla}(t)$ for $t\in\mathbb{T}_{\kappa}$. We have that $\mathbb{T}_{\kappa}=\mathbb{T}\setminus\{1\}$ and $\rho(t)=\frac{t}{2}$, $t\in\mathbb{T}_{\kappa}$. Hence,

$$\begin{aligned}
f^{\nabla}(t)&=(\rho(t))^2+t\rho(t)+t^2+\rho(t)+t+1\\
&=\frac{t^2}{4}+\frac{t^2}{2}+t^2+\frac{t}{2}+t+1\\
&=\frac{7}{4}t^2+\frac{3}{2}t+1,\quad t\in\mathbb{T}_{\kappa}.
\end{aligned}$$

Exercise 2.83 Let $\mathbb{T}=2^{\mathbb{N}_0}$. Find $f^{\nabla}(t)$ for $t\in\mathbb{T}_{\kappa}$, where

$$f(t)=\frac{t^2+2t-3}{t-7}.$$

Solution $\frac{t^2-21t-22}{(t-7)(t-14)}$.

2.6 Extreme Values

Assume that f is defined on $D_f \subset \mathbb{T}$. Let $t_0 \in \mathbb{T}$.

Definition 2.84 We say that $f(t_0)$ is a *local maximum value* of f if there exists $\delta > 0$ so that

$$f(t) \leq f(t_0) \quad \text{for all} \quad t \in (t_0 - \delta, t_0 + \delta) \cap D_f$$

and $f(\rho(t_0)), f(\sigma(t_0)) \leq f(t_0)$, or a *local minimum value* of f if there exists $\delta > 0$ such that

$$f(t) \geq f(t_0) \quad \text{for all} \quad t \in (t_0 - \delta, t_0 + \delta) \cap D_f$$

and $f(\rho(t_0)), f(\sigma(t_0)) \geq f(t_0)$. The point t_0 is called a *local extreme point* of f, more specifically, a *local maximum* or *local minimum* point of f.

Theorem 2.85 *Let f be delta and nabla differentiable in a neighbourhood $(t_0 - \delta, t_0 + \delta)$ of t_0. If $f^{\Delta}(t) \leq 0$ in $[t_0, t_0 + \delta)$ and $f^{\nabla}(t) \geq 0$ in $(t_0 - \delta, t_0]$, then t_0 is a local maximum point of f.*

Proof 1. Let t_0 be an isolated point. Then $\rho(t_0) < t_0 < \sigma(t_0)$ and

$$\frac{f(\sigma(t_0)) - f(t_0)}{\sigma(t_0) - t_0} = f^{\Delta}(t_0) \leq 0, \quad \frac{f(t_0) - f(\rho(t_0))}{t_0 - \rho(t_0)} = f^{\nabla}(t_0) \geq 0.$$

Therefore, $f(t_0) \geq f(\sigma(t_0))$ and $f(t_0) \geq f(\rho(t_0))$. Also, there exists $\delta_1 > 0$ such that $f(t) \leq f(t_0)$ for all $t \in (t_0 - \delta_1, t_0 + \delta_1)$. Consequently, t_0 is a local maximum point.

2. Let t_0 be left-dense and right-scattered. Since

$$\frac{f(\sigma(t_0)) - f(t_0)}{\sigma(t_0) - t_0} = f^{\Delta}(t_0) \leq 0,$$

we have that $f(\sigma(t_0)) \leq f(t_0)$.

a. Let $f^{\nabla}(t_0) = 0$. Then there exists $\delta_1 > 0$ such that $f(t) = f(t_0)$ for any $t \in (t_0 - \delta_1, t_0 + \delta_1)$.

b. Let $f^{\nabla}(t_0) > 0$. Then, for every $\varepsilon \in \big(0, f^{\nabla}(t_0)\big)$, there exists $\delta_1 > 0$ such that

$$|f(t_0) - f(t) - (t_0 - t) f^{\nabla}(t_0)| \leq \varepsilon |t_0 - t|$$

for any $t \in (t_0 - \delta_1, t_0 + \delta_1)$ and $[t_0, t_0 + \delta_1) = \{t_0\}$. If $t \in (t_0 - \delta_1, t_0)$, then we get

$$-\varepsilon(t_0 - t) \le f(t_0) - f(t) - (t_0 - t)f^{\nabla}(t_0) \le \varepsilon(t_0 - t),$$

i.e.,

$$\left(f^{\nabla}(t_0) - \varepsilon\right)(t_0 - t) \le f(t_0) - f(t) \le \left(\varepsilon + f^{\nabla}(t_0)\right)(t_0 - t),$$

whereupon

$$f(t) \le f(t_0)$$

for any $t \in (t_0 - \delta_1, t_0 + \delta_1)$, i.e., t_0 is a local maximum point.

3. Let t_0 be left-scattered and right-dense. Since

$$\frac{f(t_0) - f(\rho(t_0))}{t_0 - \rho(t_0)} = f^{\nabla}(t_0) \ge 0,$$

we have that $f(\rho(t_0)) \le f(t_0)$.

a. Let $f^{\Delta}(t_0) = 0$. Then there exists $\delta_1 > 0$ such that $f(t) = f(t_0)$ for any $t \in (t_0 - \delta_1, t_0 + \delta_1)$.
b. Let $f^{\Delta}(t_0) < 0$. Then, for every $\varepsilon \in \left(0, -f^{\Delta}(t_0)\right)$, there exists $\delta_1 > 0$ such that

$$|f(t) - f(t_0) - (t - t_0)f^{\Delta}(t_0)| \le \varepsilon|t_0 - t|$$

for any $t \in (t_0 - \delta_1, t_0 + \delta_1)$ and $(t_0 - \delta_1, t_0] = \{t_0\}$. If $t \in (t_0, t_0 + \delta_1)$, then we get

$$-\varepsilon(t - t_0) \le f(t) - f(t_0) - (t - t_0)f^{\Delta}(t_0) \le \varepsilon(t - t_0),$$

i.e.,

$$\left(f^{\Delta}(t_0) - \varepsilon\right)(t - t_0) \le f(t) - f(t_0) \le \left(\varepsilon + f^{\Delta}(t_0)\right)(t - t_0),$$

whereupon

$$f(t) \le f(t_0)$$

for any $t \in (t_0 - \delta_1, t_0 + \delta_1)$, i.e., t_0 is a local maximum point.

4. Let t_0 be dense.

a. Assume $f^{\Delta}(t_0) = f^{\nabla}(t_0) = 0$. Then there exists $\delta_1 > 0$ such that $f(t) = f(t_0)$ for any $t \in (t_0 - \delta_1, t_0 + \delta_1)$.
b. Assume $f^{\Delta}(t_0) = 0$ and $f^{\nabla}(t_0) > 0$. Then, for any $\varepsilon \in \left(0, f^{\nabla}(t_0)\right)$, there exists $\delta_1 > 0$ such that $f(t) = f(t_0)$ for $t \in [t_0, t_0 + \delta_1)$ and

$$-\varepsilon(t_0 - t) \le f(t_0) - f(t) - (t_0 - t)f^{\nabla}(t_0) \le \varepsilon(t_0 - t)$$

for $t \in (t_0 - \delta_1, t_0]$. Hence, $f(t) \le f(t_0)$ for any $t \in (t_0 - \delta_1, t_0 + \delta_1)$.

c. Assume $f^{\Delta}(t_0) < 0$ and $f^{\nabla}(t_0) = 0$. Then, for every $\varepsilon \in \left(0, -f^{\Delta}(t_0)\right)$, there exists $\delta_1 > 0$ such that $f(t) = f(t_0)$ for any $t \in (t_0 - \delta_1, t_0]$, and for any $t \in [t_0, t_0 + \delta_1)$,

$$\left(f^{\Delta}(t_0) - \varepsilon\right)(t - t_0) \le f(t) - f(t_0) \le \left(\varepsilon + f^{\Delta}(t_0)\right)(t - t_0),$$

whereupon

$$f(t) \le f(t_0)$$

for any $t \in (t_0 - \delta_1, t_0 + \delta_1)$.

d. Assume $f^{\Delta}(t_0) < 0$ and $f^{\nabla}(t_0) > 0$. Then, for every

$$\varepsilon \in \left(0, \min\{-f^{\Delta}(t_0), f^{\nabla}(t_0)\}\right),$$

there exists $\delta_1 > 0$ such that, for any $t \in [t_0, t_0 + \delta_1)$,

$$\left(f^{\Delta}(t_0) - \varepsilon\right)(t - t_0) \le f(t) - f(t_0) \le \left(\varepsilon + f^{\Delta}(t_0)\right)(t - t_0)$$

and

$$-\varepsilon(t_0 - t) \le f(t_0) - f(t) - (t_0 - t) f^{\nabla}(t_0) \le \varepsilon(t_0 - t)$$

for $t \in (t_0 - \delta_1, t_0]$. Hence, $f(t) \le f(t_0)$ for any $t \in (t_0 - \delta_1, t_0 + \delta_1)$. □

As in the proof of Theorem 2.85, one can prove the following theorem.

Theorem 2.86 *Let f be delta and nabla differentiable in a neighbourhood $(t_0 - \delta, t_0 + \delta)$ of t_0. If $f^{\Delta}(t) \ge 0$ in $[t_0, t_0 + \delta)$ and $f^{\nabla}(t) \le 0$ in $(t_0 - \delta, t_0]$, then t_0 is a local minimum point of f.*

Example 2.87 Let $\mathbb{T} = \mathbb{Z}$. Consider the function

$$f(t) = t^2 - 5t + 4.$$

Then

$$f^{\Delta}(t) = \sigma(t) + t - 5 = t + 1 + t - 5 = 2t - 4$$

and

$$f^{\nabla}(t) = \rho(t) + t - 5 = t - 1 + t - 5 = 2t - 6.$$

Hence,

$$f^{\Delta}(t) \le 0 \quad \text{and} \quad f^{\nabla}(t) \ge 0$$

iff

$$2t - 4 \le 0 \quad \text{and} \quad 2t - 6 \ge 0$$

iff

$$t \leq 2 \quad \text{and} \quad t \geq 3.$$

Therefore, f has no local maximum points. Also,

$$f^{\Delta}(t) \geq 0 \quad \text{and} \quad f^{\nabla}(t) \leq 0$$

iff

$$2t - 4 \geq 0 \quad \text{and} \quad 2t - 6 \leq 0$$

iff

$$t \geq 2 \quad \text{and} \quad t \leq 3.$$

Consequently, $t = 2$ and $t = 3$ are local minimum points. We have

$$f(2) = f(3) = -2.$$

Example 2.88 Let $\mathbb{T} = \mathbb{Z}$. We will find the local extreme values of the function

$$f(t) = \frac{t+1}{t^2+1}.$$

Here, $\sigma(t) = t + 1$, $\rho(t) = t - 1$. Also,

$$\begin{aligned} f^{\Delta}(t) &= \frac{t^2 + 1 - (t+1)(\sigma(t) + t)}{(t^2+1)((t+1)^2+1)} \\ &= \frac{t^2 + 1 - (t+1)(t+1+t)}{(t^2+1)(t^2+2t+2)} \\ &= \frac{t^2 + 1 - (t+1)(2t+1)}{(t^2+1)(t^2+2t+2)} \\ &= \frac{t^2 + 1 - (2t^2 + t + 2t + 1)}{(t^2+1)(t^2+2t+2)} \\ &= \frac{t^2 + 1 - 2t^2 - 3t - 1}{(t^2+1)(t^2+2t+2)} \\ &= \frac{-t^2 - 3t}{(t^2+1)(t^2+2t+2)} \\ &= -\frac{t(t+3)}{(t^2+1)(t^2+2t+2)} \end{aligned}$$

and

$$
\begin{aligned}
f^{\nabla}(t) &= \frac{t^2+1-(t+1)(\rho(t)+t)}{(t^2+1)((t-1)^2+1)} \\
&= \frac{t^2+1-(t+1)(2t-1)}{(t^2+1)(t^2-2t+2)} \\
&= \frac{t^2+1-(2t^2-t+2t-1)}{(t^2+1)(t^2-2t+2)} \\
&= \frac{t^2+1-2t^2-t+1}{(t^2+1)(t^2-2t+2)} \\
&= \frac{-t^2-t+2}{(t^2+1)(t^2-2t+2)} \\
&= -\frac{t^2+t-2}{(t^2+1)(t^2-2t+2)} \\
&= -\frac{(t+2)(t-1)}{(t^2+1)(t^2-2t+2)}.
\end{aligned}
$$

Hence,

$$f^{\Delta}(t) \leq 0 \quad \text{and} \quad f^{\nabla}(t) \geq 0$$

iff

$$-\frac{t(t+3)}{(t^2+1)(t^2+2t+2)} \leq 0 \quad \text{and} \quad -\frac{(t+2)(t-1)}{(t^2+1)(t^2-2t+2)} \geq 0$$

iff

$$t(t+3) \geq 0 \quad \text{and} \quad (t-1)(t+2) \leq 0$$

so that

$$t = 0 \quad \text{and} \quad t = 1.$$

Therefore,

$$f_{\max} = f(0) = f(1) = 1.$$

Also,

$$f^{\Delta}(t) \geq 0 \quad \text{and} \quad f^{\nabla}(t) \leq 0$$

iff

$$-\frac{t(t+3)}{(t^2+1)(t^2+2t+2)} \geq 0 \quad \text{and} \quad -\frac{(t+2)(t-1)}{(t^2+1)(t^2-2t+2)} \leq 0$$

iff

$$t(t+3) \leq 0 \quad \text{and} \quad (t-1)(t+2) \geq 0$$

so that

$$t = -2 \quad \text{and} \quad t = -1.$$

Consequently,

$$f_{\min} = f(-2) = \frac{-2+1}{4+1} = -\frac{1}{5}$$

and

$$f_{\min} = f(-1) = 0.$$

Example 2.89 Let $\mathbb{T} = 2^{\mathbb{N}_0}$. We will find the extreme values of the function

$$f(t) = \frac{t^2+2}{t+2} \quad \text{for} \quad t \geq 4.$$

Here, $\sigma(t) = 2t$, $\rho(t) = \frac{1}{2}t$ for all $t \in \mathbb{T}$ and $t \geq 4$. Then, for $t \geq 4$, we have

$$\begin{aligned} f^{\Delta}(t) &= \frac{(\sigma(t)+t)(t+2) - (t^2+2)}{(t+2)(2t+2)} \\ &= \frac{3t(t+2) - (t^2+2)}{2(t+1)(t+2)} \\ &= \frac{3t^2+6t-t^2-2}{2(t+1)(t+2)} \\ &= \frac{t^2+3t-1}{(t+1)(t+2)} \end{aligned}$$

and

$$\begin{aligned} f^{\nabla}(t) &= \frac{(\rho(t)+t)(t+2) - (t^2+2)}{(t+2)\left(\frac{1}{2}t+2\right)} \\ &= \frac{\frac{3}{2}t(t+2) - t^2 - 2}{(t+2)\left(\frac{1}{2}t+2\right)} \\ &= \frac{\frac{3}{2}t^2 + 3t - t^2 - 2}{(t+2)\left(\frac{1}{2}t+2\right)} \\ &= \frac{\frac{1}{2}t^2 + 3t - 2}{(t+2)\left(\frac{1}{2}t+2\right)} \\ &= \frac{t^2+6t-4}{(t+2)(t+4)}. \end{aligned}$$

Note that $f^{\Delta}(t) \geq 0$ and $f^{\nabla}(t) \geq 0$ for all $t \geq 4$. Therefore, the function f has no local extreme values.

Exercise 2.90 Let $\mathbb{T} = \mathbb{Z}$. Find the local extreme values of the function

$$f(t) = t^3 - 3t^2 + 4.$$

Solution $f_{\max} = f(0) = 4$ and $f_{\min} = f(2) = 0$.

Definition 2.91 Suppose that f is Δ-differentiable and ∇-differentiable at t_0. We say that t_0 is a *critical point* of f if

$$f^{\Delta}(t_0) \leq 0 \quad \text{and} \quad f^{\nabla}(t_0) \geq 0$$

or

$$f^{\Delta}(t_0) \geq 0 \quad \text{and} \quad f^{\nabla}(t_0) \leq 0.$$

Exercise 2.92 Let $\mathbb{T} = 2^{\mathbb{N}_0}$. Prove that

$$\frac{t^2+2}{t+3} \geq \frac{3}{4} \quad \text{for all} \quad t \in \mathbb{T}.$$

Solution We have $\sigma(t) = 2t$ for all $t \in \mathbb{T}$ and

$$\begin{aligned} f^{\Delta}(t) &= \frac{(t+\sigma(t))(t+3) - (t^2+2)}{(t+3)(2t+3)} \\ &= \frac{3t(t+3) - t^2 - 2}{(t+3)(2t+3)} \\ &= \frac{3t^2 + 9t - t^2 - 2}{(t+3)(2t+3)} \\ &= \frac{2t^2 + 9t - 2}{(t+3)(2t+3)} \geq 0 \end{aligned}$$

for all $t \in \mathbb{T}$. Consequently, f is increasing in $\mathbb{T}$. Hence

$$f(t) \geq f(1) = \frac{3}{4} \quad \text{for all} \quad t \in \mathbb{T}.$$

Exercise 2.93 Let $\mathbb{T} = 3^{\mathbb{N}_0}$. Find a positive constant a such that

$$1 + a \log t \leq t^2 \quad \text{for all} \quad t \in \mathbb{T}.$$

Solution Let

$$f(t) = t^2 - a \log t - 1, \quad t \in \mathbb{T}.$$

Here, $\sigma(t) = 3t$ for all $t \in \mathbb{T}$ and

$$\begin{aligned} f^{\Delta}(t) &= \sigma(t) + t - a\frac{\log \sigma(t) - \log t}{\sigma(t) - t} \\ &= 3t + t - a\frac{\log(3t) - \log t}{3t - t} \\ &= 4t - a\frac{\log 3}{2t} \quad \text{for all} \quad t \in \mathbb{T}. \end{aligned}$$

Since

$$\frac{\log 3}{2t} \leq \frac{\log 3}{2} \quad \text{for all} \quad t \in \mathbb{T},$$

we conclude that

$$4t - a\frac{\log 3}{2t} \geq 4 - a\frac{\log 3}{2} \quad \text{for all} \quad t \in \mathbb{T}.$$

Hence, if $0 < a < \frac{8}{\log 3}$, then f is increasing in $\mathbb{T}$. From here,

$$f(t) \geq f(1) = 0 \quad \text{for all} \quad t \in \mathbb{T} \quad \text{and for} \quad 0 < a < \frac{8}{\log 3}.$$

2.7 Convex and Concave Functions

Suppose that $f : \mathbb{T} \to \mathbb{R}$.

Definition 2.94 The function f is called *convex* if for all $t_1, t_2 \in \mathbb{T}$ and for all $\lambda \in [0, 1]$, the inequality

$$f(\lambda t_1 + (1 - \lambda)t_2) \leq \lambda f(t_1) + (1 - \lambda) f(t_2)$$

holds.

Definition 2.95 The function f is called *strictly convex* if for all $t_1, t_2 \in \mathbb{T}$ with $t_1 \neq t_2$ and for all $\lambda \in (0, 1)$, the inequality

$$f(\lambda t_1 + (1 - \lambda)t_2) < \lambda f(t_1) + (1 - \lambda) f(t_2)$$

holds.

Definition 2.96 The function f is said to be (*strictly*) *concave* if $-f$ is (strictly) convex.

Theorem 2.97 *Let f be twice delta differentiable on (a, b) and $f^{\Delta\Delta}(t) \geq 0$ for all $t \in (a, b)$. Then f is convex.*

Proof Let $t_1, t_2 \in \mathbb{T}$, $t_1 < t_2$, and $\lambda \in (0, 1)$. Then

$$\begin{aligned}
&\lambda f(t_1) + (1-\lambda) f(t_2) - f(\lambda t_1 + (1-\lambda)t_2) \\
&= \lambda f(t_1) + (1-\lambda) f(t_2) - (1-\lambda+\lambda) f(\lambda t_1 + (1-\lambda)t_2) \\
&= (1-\lambda)(f(t_2) - f(\lambda t_1 + (1-\lambda)t_2)) - \lambda(f(\lambda t_1 + (1-\lambda)t_2) - f(t_1)).
\end{aligned} \tag{2.12}$$

By the mean value theorem (Theorem 2.41), it follows that there exist

$$\xi_1 \in (t_1, \lambda t_1 + (1-\lambda)t_2) \quad \text{and} \quad \xi_2 \in (\lambda t_1 + (1-\lambda)t_2, t_2)$$

so that

$$\begin{aligned}
f(\lambda t_1 + (1-\lambda)t_2) - f(t_1) &\leq f^{\Delta}(\xi_1)(\lambda t_1 + (1-\lambda)t_2 - t_1) \\
&= (1-\lambda) f^{\Delta}(\xi_1)(t_2 - t_1)
\end{aligned}$$

and

$$\begin{aligned}
f(t_2) - f(\lambda t_1 + (1-\lambda)t_2) &\geq f^{\Delta}(\xi_2)(t_2 - \lambda t_1 - (1-\lambda)t_2) \\
&= \lambda f^{\Delta}(\xi_2)(t_2 - t_1).
\end{aligned}$$

By (2.12), we obtain

$$\begin{aligned}
&\lambda f(t_1) + (1-\lambda) f(t_2) - f(\lambda t_1 + (1-\lambda)t_2) \\
&\geq (1-\lambda)\lambda f^{\Delta}(\xi_2)(t_2 - t_1) - \lambda(1-\lambda) f^{\Delta}(\xi_1)(t_2 - t_1) \qquad (2.13) \\
&= \lambda(1-\lambda)(f^{\Delta}(\xi_2) - f^{\Delta}(\xi_1))(t_2 - t_1).
\end{aligned}$$

By the mean value theorem (Theorem 2.41), it follows that there exists $\xi_3 \in (\xi_1, \xi_2)$ so that

$$f^{\Delta}(\xi_2) - f^{\Delta}(\xi_1) \geq f^{\Delta\Delta}(\xi_3)(\xi_2 - \xi_1).$$

From the last inequality and from (2.13), we obtain

$$\lambda f(t_1) + (1-\lambda) f(t_2) - f(\lambda t_1 + (1-\lambda)t_2) \geq \lambda(1-\lambda) f^{\Delta\Delta}(\xi_3)(\xi_2 - \xi_1)(t_2 - t_1)$$

$$\geq 0,$$

which completes the proof.

As in Theorem 2.97, one can prove the following theorem.

Theorem 2.98 *Let f be twice delta differentiable on (a, b) and $f^{\Delta\Delta}(t) \leq 0$ for all $t \in (a, b)$. Then f is concave.*

Example 2.99 Let $\mathbb{T} = \mathbb{Z}$. Consider

$$f(t) = t^3 - 7t^2 + t - 10.$$

Here, $\sigma(t) = t + 1$ and

$$\begin{aligned} f^{\Delta}(t) &= (\sigma(t))^2 + t\sigma(t) + t^2 - 7(\sigma(t) + t) + 1 \\ &= (t+1)^2 + t(t+1) + t^2 - 7(t+1+t) + 1 \\ &= t^2 + 2t + 1 + t^2 + t + t^2 - 14t - 7 + 1 \\ &= 3t^2 - 11t - 5, \\ f^{\Delta\Delta}(t) &= 3(\sigma(t) + t) - 11 \\ &= 3(t+1+t) - 11 \\ &= 6t - 8. \end{aligned}$$

Hence,

$$f^{\Delta\Delta}(t) \geq 0 \quad \text{for} \quad t \geq 2 \quad \text{and} \quad f^{\Delta\Delta}(t) \leq 0 \quad \text{for} \quad t \leq 1.$$

Therefore, f is convex for $t \geq 2$ and concave for $t \leq 1$.

Example 2.100 Let $\mathbb{T} = 2^{\mathbb{N}_0}$. Consider

$$f(t) = t^4 - t^3 - t^2 - t.$$

Here, $\sigma(t) = 2t$ and

$$f^{\Delta}(t) = (\sigma(t))^3 + t(\sigma(t))^2 + t^2\sigma(t) + t^3$$

$$-((\sigma(t))^2 + t\sigma(t) + t^2) - (\sigma(t) + t) - 1$$

$$= 8t^3 + 4t^3 + 2t^3 + t^3 - (4t^2 + 2t^2 + t^2) - (2t + t) - 1$$

$$= 15t^3 - 7t^2 - 3t - 1,$$

$$f^{\Delta\Delta}(t) = 15((\sigma(t))^2 + t\sigma(t) + t^2) - 7(\sigma(t) + t) - 3$$

$$= 15(4t^2 + 2t^2 + t^2) - 7(2t + t) - 3$$

$$= 105t^2 - 21t - 3.$$

Hence, $f^{\Delta\Delta}(t) > 0$ for all $t \in \mathbb{T}$. Therefore, the function f is strictly convex in $\mathbb{T}$.

Example 2.101 Let $\mathbb{T} = 3^{\mathbb{N}_0}$. Consider the function

$$f(t) = \frac{t-3}{t+2}.$$

We have $\sigma(t) = 3t$ and

$$f^{\Delta}(t) = \frac{t+2-(t-3)}{(t+2)(3t+2)}$$

$$= \frac{5}{3t^2 + 2t + 6t + 4}$$

$$= \frac{5}{3t^2 + 8t + 4},$$

$$f^{\Delta\Delta}(t) = -5\frac{3(\sigma(t) + t) + 8}{(3t^2 + 8t + 4)(3(\sigma(t))^2 + 8\sigma(t) + 4)}$$

$$= -5\frac{12t + 8}{(3t^2 + 8t + 4)(27t^2 + 24t + 4)}$$

$$= -20\frac{3t + 2}{(3t^2 + 8t + 4)(27t^2 + 24t + 4)}$$

$$< 0 \quad \text{for all} \quad t \in \mathbb{T}.$$

Therefore, f is a strictly concave function in $\mathbb{T}$.

Exercise 2.102 Let $\mathbb{T} = \mathbb{Z}$. Find the intervals of convexity and concavity of the following functions.

1. $f(t) = t^3 - 6t^2 + 12t + 4$,
2. $f(t) = \frac{1}{t+3}$.

Solution 1. f is convex for $t \geq 1$ and concave for $t \leq 1$.
2. f is convex for $t \geq -4$ and concave for $t \leq -4$.

Exercise 2.103 Let f and g be convex functions. Prove that so are

$$m(t) = \max\{f(t), g(t)\} \quad \text{and} \quad h(t) = f(t) + g(t).$$

2.8 Completely Delta Differentiable Functions

Definition 2.104 A function $f : \mathbb{T} \to \mathbb{R}$ is called *completely delta differentiable* at a point $t^0 \in \mathbb{T}^\kappa$ if there exist constants A_1 and A_2 such that

$$f(t^0) - f(t) = A_1(t^0 - t) + \alpha(t^0 - t) \quad \text{for all} \quad t \in U_\delta(t^0) \tag{2.14}$$

and

$$f(\sigma(t^0)) - f(t) = A_2(\sigma(t^0) - t) + \beta(\sigma(t^0) - t) \quad \text{for all} \quad t \in U_\delta(t^0), \tag{2.15}$$

where $U_\delta(t^0)$ is a δ-neighbourhood of t^0 and

$$\alpha = \alpha(t^0, t) \quad \text{and} \quad \beta = \beta(t^0, t)$$

are equal to zero for $t = t^0$ such that

$$\lim_{t \to t^0} \alpha(t^0, t) = \lim_{t \to t^0} \beta(t^0, t) = 0.$$

Theorem 2.105 *Let a function $f : \mathbb{T} \to \mathbb{R}$ be continuous and have the first-order delta derivative f^Δ in some δ-neighbourhood $U_\delta(t^0)$ of the point $t^0 \in \mathbb{T}^\kappa$. If f^Δ is continuous at the point t^0, then f is completely delta differentiable at t^0.*

Proof Using the definition of delta derivative, we have that

$$f(\sigma(t^0)) - f(t) = f^\Delta(t^0)(\sigma(t^0) - t) + \beta(\sigma(t^0) - t), \tag{2.16}$$

where $\beta = \beta(t^0, t)$ and $\beta \to 0$ as $t \to t^0$, i.e., (2.15) holds. Now, we will prove (2.14).

1. Let $t^0 \in \mathbb{T}$ be isolated. Then (2.14) is satisfied independently of A_1 and α because in this case, $U_\delta(t^0)$ consists of the single point t^0 for sufficiently small $\delta > 0$.

2. Let t^0 be right-dense. In this case, $\sigma(t^0) = t^0$, and (2.16) coincides with (2.14).
3. Let t^0 be left-dense and right-scattered. Then for sufficiently small $\delta > 0$, any point $t \in U_\delta(t^0) \setminus \{t^0\}$ must satisfy $t < t^0$. By Theorem 2.41, we obtain

$$f^\Delta(\xi_1)(t^0 - t) \le f(t^0) - f(t) \le f^\Delta(\xi_2)(t^0 - t),$$

where $\xi_1, \xi_2 \in [t, t^0)$. Since $\xi_1, \xi_2 \to t^0$ as $t \to t^0$ and f^Δ is continuous at t^0, we have

$$\lim_{t \to t^0} \frac{f(t^0) - f(t)}{t^0 - t} = f^\Delta(t^0).$$

Therefore,

$$\frac{f(t^0) - f(t)}{t^0 - t} = f^\Delta(t^0) + \alpha,$$

where $\alpha = \alpha(t^0, t)$ and $\alpha \to 0$ as $t \to t^0$. Thus, (2.14) holds for $A_1 = f^\Delta(t^0)$. □

2.9 Geometric Sense of Differentiability

In this section, we consider the geometric sense of complete delta differentiability in the case of single variable functions on time scales (see also [30]). Let $\mathbb{T}$ be a time scale with the forward jump operator σ and the delta differentiation operator Δ. Consider a real-valued continuous function

$$u = f(t) \quad \text{for} \quad t \in \mathbb{T}. \tag{2.17}$$

Let Γ be the "curve" represented by the function (2.17), that is, the set of points $\{(t, f(t)) : t \in \mathbb{T}\}$ in the xy-plane. Let t^0 be a fixed point in $\mathbb{T}^\kappa$. Then $P_0 = (t^0, f(t^0))$ is a point on Γ.

Definition 2.106 A line $\mathscr{L}_0$ passing through the point P_0 is called the *delta tangent line* to the curve Γ at the point P_0 if

1. $\mathscr{L}_0$ passes also through the point $P_0^\sigma = (\sigma(t^0), f(\sigma(t^0)))$;
2. if P_0 is not an isolated point of the curve Γ, then

$$\lim_{\substack{P \to P_0 \\ P \neq P_0}} \frac{d(P, \mathscr{L}_0)}{d(P, P_0)} = 0, \tag{2.18}$$

where P is the moving point of the curve Γ, $d(P, \mathscr{L}_0)$ is the distance from the point P to the line $\mathscr{L}_0$, and $d(P, P_0^\sigma)$ is the distance from the point P to the point P_0^σ.

Theorem 2.107 *If the function f is completely delta differentiable at the point t^0, then the curve represented by this function has the uniquely determined delta tangent line at the point $P_0 = (t^0, f(t^0))$ specified by the equation*

$$y - f(t^0) = f^\Delta(t^0)(x - t^0), \tag{2.19}$$

where (x, y) is the current point of the line.

Proof Let f be a completely delta differentiable function at a point $t^0 \in \mathbb{T}^\kappa$, Γ be the curve represented by this function, and $\mathcal{L}_0$ be the line described by (2.19). Let us show that $\mathcal{L}_0$ passes also through the point P_0^σ. Indeed, if $\sigma(t^0) = t^0$, then $P_0^\sigma = P_0$ and the statement is true. Let now $\sigma(t^0) > t^0$. Substituting the coordinates $(\sigma(t^0), f(\sigma(t^0)))$ of the point P_0^σ into (2.19), we get

$$f(\sigma(t^0)) - f(t^0) = f^\Delta(t^0)\left[\sigma(t^0) - t^0\right],$$

which is obviously true by virtue of the continuity of f at t^0. Now, we check condition (2.18). Assume that P_0 is not an isolated point of the curve Γ (note that if P_0 is an isolated point of Γ, then from $P \to P_0$ we get $P = P_0$ and the left-hand side of (2.18) becomes meaningless). The variable point $P \in \Gamma$ has the coordinates $(t, f(t))$. As is known from analytic geometry, the distance of the point P from the line $\mathcal{L}_0$ with (2.19) is expressed by the formula

$$d(P, \mathcal{L}_0) = \frac{1}{M}\left|f(t) - f(t^0) - f^\Delta(t^0)\left(t - t^0\right)\right|,$$

where

$$M = \sqrt{1 + \left[f^\Delta(t^0)\right]^2}.$$

Hence, by the differentiability condition (2.14) in which we have $A = f^\Delta(t^0)$ due to the other differentiability condition (2.15),

$$d(P, \mathcal{L}_0) = \frac{1}{M}\left|\alpha(t - t^0)\right| = \frac{1}{M}|\alpha|\left|t - t^0\right|.$$

Next,

$$d(P, P_0) = \sqrt{\left(t - t^0\right)^2 + \left[f(t) - f(t^0)\right]^2} \geq \left|t - t^0\right|.$$

Therefore

$$\frac{d(P, \mathcal{L}_0)}{d(P, P_0)} \leq \frac{1}{M}|\alpha| \to 0 \quad \text{as} \quad P \to P_0.$$

Thus we have proved that the line specified by (2.19) is the delta tangent line to Γ at the point P_0.

Now, we must show that there are no other delta tangent lines to Γ at the point P_0 distinct from $\mathcal{L}_0$. If $P_0 \neq P_0^\sigma$, then the delta tangent line (provided it exists) is unique as it passes through the distinct points P_0 and P_0^σ. Let now $P_0 = P_0^\sigma$ so that P_0 is nonisolated. Suppose that there is a delta tangent line $\mathscr{L}$ to Γ at the point P_0 described by an equation

$$a(x - t^0) - b\left[y - f(t^0)\right] = 0 \quad \text{with} \quad a^2 + b^2 = 1. \tag{2.20}$$

Let $P = (t, f(t))$ be a variable point on Γ. Using (2.20), we have

$$d(P, \mathscr{L}) = \left|a\left(t - t^0\right) - b\left[f(t) - f(t^0)\right]\right|.$$

Hence, by the differentiability condition (2.14) with $A = f^\Delta(t^0)$, the latter being a result of the condition (2.15),

$$d(P, \mathscr{L}) = \left|a - b\left[f^\Delta(t^0) + \alpha\right]\right| \left|t - t^0\right|.$$

Next, by the same differentiability condition,

$$\begin{aligned} d(P, P_0) &= \sqrt{\left(t - t^0\right)^2 + \left[f(t) - f(t^0)\right]^2} \\ &= \sqrt{\left(t - t^0\right)^2 + \left[f^\Delta(t^0) + \alpha\right]^2 (t - t^0)^2} \\ &= \sqrt{1 + \left[f^\Delta(t^0) + \alpha\right]^2} \left|t - t^0\right|. \end{aligned}$$

So, we have

$$\frac{d(P, \mathscr{L})}{d(P, P_0)} = \frac{\left|a - b\left[f^\Delta(t^0) + \alpha\right]\right|}{\sqrt{1 + \left[f^\Delta(t^0) + \alpha\right]^2}}.$$

Passing here to the limit as $t \to t^0$ and taking into account that the left-hand side (by the definition of delta tangent line) and α tend to zero, we obtain

$$a - bf^\Delta(t^0) = 0.$$

We now see that $b \neq 0$ for if otherwise, we would have $a = b = 0$. Hence, the line $\mathscr{L}$ is described by (2.19). □

Remark 2.108 If P_0 is an isolated point of the curve Γ (hence $P_0 \neq P_0^\sigma$), then there exists a delta tangent line at the point P_0 to the curve Γ that coincides with the unique line through the points P_0 and P_0^σ.

Remark 2.109 If P_0 is not an isolated point of the curve Γ and if Γ has a delta tangent line at the point P_0, then the line PP_0, where $P \in \Gamma$ (and $P \neq P_0$), approaches this tangent line as $P \to P_0$. Conversely, if the line PP_0 approaches as $P \to P_0$ some line $\mathscr{L}_0$ passing through the point P_0^σ, then this limiting line is a delta tangent line at P_0. For the proof it is sufficient to note that if φ is the angle between the lines $\mathscr{L}_0$

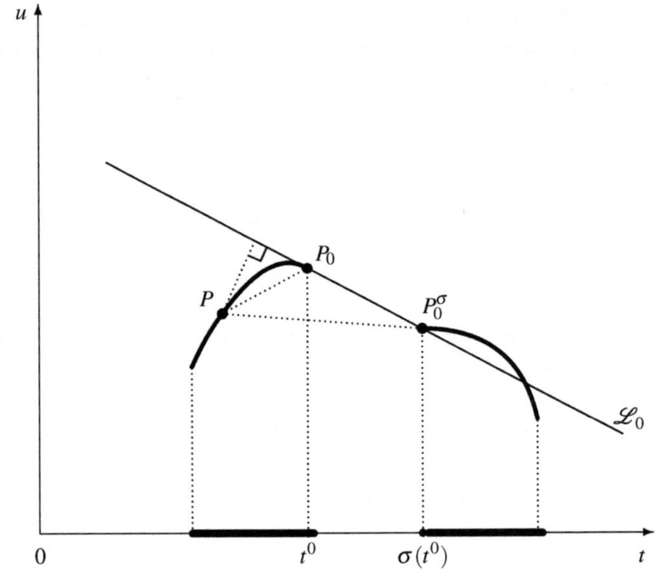

Fig. 2.1 $\mathbb{T}$ consists of two separate real number intervals. Accordingly, the (time scale) curve Γ consists of two arcs of usual curves. In order the curve Γ to have a delta tangent line $\mathscr{L}_0$ at the point P_0, there must exist the usual left-sided tangent line to Γ at P_0 and, moreover, that line must pass through the point P_0^σ

and PP_0, then (see Figure 2.1)

$$\frac{d(P, \mathscr{L}_0)}{d(P, P_0)} = \sin\varphi.$$

Example 2.110 Let $\mathbb{T} = \mathbb{Z}$, $f(t) = t^3 - 3t^2 + 2t$, $t \in \mathbb{T}$. We will find the tangent line of the curve $\{(t, f(t)) : t \in \mathbb{T}\}$ at the point $t^0 = 1$, where $f(t) = t^3 - 3t^2 + 2t$. We have

$$\sigma(t) = t + 1, \quad \mu(t) = 1, \quad t \in \mathbb{T}.$$

Hence,

$$\begin{aligned} f(1) &= 0, \\ f^\Delta(t) &= (\sigma(t))^2 + t\sigma(t) + t^2 - 3(\sigma(t) + t) + 2 \\ &= (t+1)^2 + t(t+1) + t^2 - 3(t+1+t) + 2 \\ &= t^2 + 2t + 1 + t^2 + t + t^2 - 6t - 3 + 2 \end{aligned}$$

$$= 3t^2 - 3t,$$

$$f^{\Delta}(1) = 0.$$

Thus, $y = 0$ is the tangent line of the considered curve at $t^0 = 1$.

Example 2.111 Let $\mathbb{T} = 2^{\mathbb{N}_0}$, $f(t) = t^2 - 7t + 10$, $t \in \mathbb{T}$. We will find the tangent line of the curve $\{(t, f(t)) : t \in \mathbb{T}\}$ at $t^0 = 4$, where $f(t) = t^2 - 7t + 10$. Here,

$$\sigma(t) = 2t, \quad \mu(t) = t, \quad t \in \mathbb{T}.$$

Hence,

$$f(4) = 16 - 28 + 10 = -2,$$

$$f^{\Delta}(t) = \sigma(t) + t - 7 = 3t - 7,$$

$$f^{\Delta}(4) = 12 - 7 = 5.$$

Therefore,

$$y + 2 = 5(x - 4), \quad \text{i.e.,} \quad y = 5x - 22$$

is the tangent line of the considered curve at $t = 4$.

Example 2.112 Let $\mathbb{T} = 3^{\mathbb{N}_0}$, $f(t) = \sqrt{t} - t^3$, $t \in \mathbb{T}$. We will find the tangent line of the curve $\{(t, f(t)) : t \in \mathbb{T}\}$ at the point $t^0 = 1$, where $f(t) = \sqrt{t} - t^3$. Here,

$$\sigma(t) = 3t, \quad \mu(t) = 2t, \quad t \in \mathbb{T}.$$

Hence,

$$f(1) = 0,$$

$$f^{\Delta}(t) = \frac{1}{\sqrt{\sigma(t)} + \sqrt{t}} - \left((\sigma(t))^2 + t\sigma(t) + t^2\right)$$

$$= \frac{1}{\sqrt{3t} + \sqrt{t}} - (9t^2 + 3t^2 + t^2)$$

$$= \frac{1}{(1 + \sqrt{3})\sqrt{t}} - 13t^2,$$

$$f^{\Delta}(1) = \frac{1}{1 + \sqrt{3}} - 13$$

$$= \frac{\sqrt{3}-1}{2} - 13$$

$$= \frac{\sqrt{3}}{2} - \frac{27}{2}.$$

Therefore,

$$y = \left(\frac{\sqrt{3}}{2} - \frac{27}{2} \right)(x-1)$$

is the tangent line of the considered curve at the point $t^0 = 1$.

Exercise 2.113 Let $\mathbb{T} = 2^{\mathbb{N}_0}$, $f(t) = 4t^2 - 7t + 1$, $t \in \mathbb{T}$. Find the tangent line of the curve $\{(t, f(t)) : t \in \mathbb{T}\}$ at the point $t^0 = 2$.

Solution $y = 17x - 31$.

2.10 Advanced Practical Problems

Problem 2.114 Let $\mathbb{T} = \{n^3 : n \in \mathbb{N}_0\}$, $f(t) = t^2 + 2t$, $t \in \mathbb{T}$. Find $f^{\Delta}(t)$, $t \in \mathbb{T}^{\kappa}$.

Solution $2 + t + \left(1 + \sqrt[3]{t}\right)^3$.

Problem 2.115 Let $\mathbb{T} = 2^{\mathbb{N}_0}$. Prove

$$t + t^2 + t^3 = \frac{1}{3} f_2^{\Delta}(t) + \frac{1}{7} f_3^{\Delta}(t) + \frac{1}{15} f_4^{\Delta}(t).$$

Problem 2.116 Let $\mathbb{T} = 2^{\mathbb{N}}$. Find $f^{\Delta}(t)$, where

1. $f(t) = t^2 - 3t + 2$,
2. $f(t) = \frac{t^3 - t^2}{t+1}$,
3. $f(t) = \frac{t-1}{t+1}$.

Solution 1. $3t - 3$,
2. $\frac{6t^3 + 5t^2 - 3t}{(t+1)(2t+1)}$,
3. $\frac{2}{(t+1)(2t+1)}$.

Problem 2.117 Let $\mathbb{T} = \mathbb{Z}$. Find $f^{\Delta\Delta}(t)$, where

$$f(t) = \frac{1}{t+1}.$$

Solution $\frac{2}{(t+1)(t+2)(t+3)}$.

Problem 2.118 Let $\mathbb{T} = \{n+2 : n \in \mathbb{N}_0\}$, $f(t) = t^2 + 2$, $g(t) = t^2$. Find a constant

$$c \in [2, \sigma(2)]$$

such that

$$(f \circ g)^{\Delta}(2) = f'(g(c))g^{\Delta}(2).$$

Solution $c = \sqrt{\frac{13}{2}}$.

Problem 2.119 Let $\mathbb{T} = \mathbb{N}$, $f(t) = e^t$, $g(t) = t^3$. Using Theorem 2.57, find $(f \circ g)^{\Delta}(t)$.

Solution $e^{(t+1)^3} - e^{t^3}$.

Problem 2.120 Let $\mathbb{T} = \left\{2^{4n+2} : n \in \mathbb{N}_0\right\}$, $v(t) = t^3$, $w(t) = t^2 + t$. Prove

$$(w \circ v)^{\Delta}(t) = \left(w^{\tilde{\Delta}} \circ v\right)(t)v^{\Delta}(t), \quad t \in \mathbb{T}^{\kappa}.$$

Problem 2.121 Let $\mathbb{T} = \{n+9 : n \in \mathbb{N}_0\}$, $v(t) = t^2 + 7t + 8$. Find $\left(\left(v^{-1}\right)^{\tilde{\Delta}} \circ v\right)(t)$.

Solution $\frac{1}{2t+8}$.

Problem 2.122 Consider

$$f(t) = \begin{cases} 2 & \text{for } t \in [-3, 3), \\ 4t + 215 & \text{for } t \in \{3, 9, 27\}, \end{cases}$$

where $[-3, 3)$ is the real-valued interval. Investigate whether f is differentiable at $t = 3$.

Solution No.

Problem 2.123 Let $\mathbb{T} = 2^{\mathbb{N}_0} \cup \{0\}$, $f(t) = t^3 - 16t + 1$. Find $\xi_1, \xi_2 \in (0, 4)$ so that

$$f^{\Delta}(\xi_2) \le 0 \le f^{\Delta}(\xi_1).$$

Solution $\xi_1 \in \{2, 3\}$, $\xi_2 = 1$.

Problem 2.124 Let $\mathbb{T} = 3^{\mathbb{N}_0} \cup \{0\}$. Prove that for every $t > 1$, there exist $\xi_1, \xi_2 \in (0, t)$ such that

$$40\xi_1^3 + 4\xi_1 \le t^3 + t \le 40\xi_2^3 + 4\xi_2.$$

Solution Use the function $f(t) = t^4 + t^2$.

Problem 2.125 Let $\mathbb{T} = 2^{\mathbb{N}_0}$. Investigate where the function

$$f(t) = t^4 - 3t^2 - 7$$

is increasing and decreasing.

Solution f is increasing for all $t \in \mathbb{T}$.

Problem 2.126 Let $\mathbb{T} = 2^{\mathbb{N}_0}$. Find the local extreme values of the function

$$f(t) = t^3 + t^2 + t + 1, \quad t \geq 4.$$

Problem 2.127 Let $\mathbb{T} = 2^{\mathbb{N}_0}$. Find the intervals of convexity and concavity of the following functions.

1. $f(t) = t^3 - 2t^2 - 2t - 2$,
2. $f(t) = \frac{1}{t+1}$.

Solution 1. f is convex in $\mathbb{T}$.
2. f is convex in $\mathbb{T}$.

Problem 2.128 Let $\mathbb{T} = 3^{\mathbb{N}_0}$. Find $y^{\Delta}(t)$ in terms of $y(t)$ and $y^{\sigma}(t)$ if

$$y^3 + y^2 t + t^2 - t^4 y = 3.$$

Solution $\frac{40yt^3 - 4t - y^2}{y^2 + yy^{\sigma} + (y^{\sigma})^2 + 3t(y + y^{\sigma}) - 81t^4}$.

2.11 Notes and References

This chapter deals with differential calculus for single-variable functions on time scales. The basic definition of delta differentiation is due to Hilger. Numerous examples on differentiation on various time scales are included. The Leibniz formula for the nth derivative of a product of two functions is given in Theorem 2.32, and it can be found together with its proof in [21, Theorem 1.32]. Mean value results are presented that will be used in the multivariable case. Several versions of the chain rule are included, e.g., Theorem 2.57 is due to Christian Pötzsche [38] and it also appears in Keller [35]. It can be found together with its proof in [21, Theorem 1.90]. The concept of nabla derivative due to Ferhan Atıcı and Gusein Guseinov [3] is briefly discussed. Throughout the book, results are given in terms of delta derivatives, but all results may also be formulated with nabla instead. New sufficient conditions for a local maximum and minimum are given in Theorems 2.85 and 2.86, respectively. Moreover, new sufficient conditions for convexity and concavity of single-variable functions are presented in Theorems 2.97 and 2.98, respectively. In Theorem 2.105, a sufficient condition for complete delta differentiability of a single-variable function is given. The section on the geometric sense of differentiability is extracted from Bohner and Guseinov [8]. Aside from this last section and the new parts that have not been published before, all material in this chapter is taken from Bohner and Peterson [21, 25].

Chapter 3
Integral Calculus of Functions of One Variable

3.1 Regulated, rd-Continuous, and Pre-Differentiable Functions

Definition 3.1 A function $f : \mathbb{T} \to \mathbb{R}$ is called *regulated* provided its right-sided limits exist (finite) at all right-dense points in $\mathbb{T}$ and its left-sided limits exist (finite) at all left-dense points in $\mathbb{T}$.

Example 3.2 Let $\mathbb{T} = \mathbb{N}$ and

$$f(t) = \frac{t^2}{t-1}, \quad g(t) = \frac{t}{t+1}, \quad t \in \mathbb{T}.$$

We note that all points of $\mathbb{T}$ are right-scattered. The points $t \in \mathbb{T}$, $t \neq 1$, are left-scattered. Also, $\lim_{t \to 1-} f(t)$ is not finite and $\lim_{t \to 1-} g(t)$ exists and it is finite. Therefore, the function f is not regulated and the function g is regulated.

Example 3.3 Let $\mathbb{T} = \mathbb{R}$ and

$$f(t) = \begin{cases} \frac{1}{t} & \text{for } t \in \mathbb{R} \setminus \{0\} \\ 0 & \text{for } t = 0. \end{cases}$$

We have that all points of $\mathbb{T}$ are dense and $\lim_{t \to 0-} f(t)$, $\lim_{t \to 0+} f(t)$ are not finite. Therefore, the function f is not regulated.

Exercise 3.4 Let $\mathbb{T} = \mathbb{R}$ and

$$f(t) = \begin{cases} \frac{1}{t-1} & \text{for } t \in \mathbb{R} \setminus \{1\} \\ 11 & \text{for } t = 1. \end{cases}$$

M. Bohner and S.G. Georgiev, *Multivariable Dynamic Calculus on Time Scales*,
DOI 10.1007/978-3-319-47620-9_3

Determine if f is regulated.

Solution The function f is not regulated.

Definition 3.5 A continuous function $f : \mathbb{T} \to \mathbb{R}$ is called *pre-differentiable* with (region of differentiation) D, provided

1. $D \subset \mathbb{T}^{\kappa}$,
2. $\mathbb{T}^{\kappa} \setminus D$ is countable and contains no right-scattered elements of $\mathbb{T}$,
3. f is differentiable at each $t \in D$.

Example 3.6 Let $\mathbb{T} = P_{a,b} = \bigcup_{k=0}^{\infty}[k(a+b), k(a+b)+a]$ for $a > b > 0$. Define $f : \mathbb{T} \to \mathbb{R}$ by

$$f(t) = \begin{cases} 0 & \text{if } t \in \bigcup_{k=0}^{\infty}[k(a+b), k(a+b)+b] \\ t-(a+b)k-b & \text{if } t \in [(a+b)k+b, (a+b)k+a]. \end{cases}$$

Then f is pre-differentiable with $D = \mathbb{T} \setminus \bigcup_{k=0}^{\infty}\{(a+b)k+b\}$.

Example 3.7 Let $\mathbb{T} = \mathbb{R}$ and

$$f(t) = \begin{cases} \frac{1}{t-3} & \text{if } \mathbb{R} \setminus \{3\} \\ 0 & \text{if } t = 3. \end{cases}$$

Since $f : \mathbb{T} \to \mathbb{R}$ is not continuous at $t = 3$, the function f is not pre-differentiable.

Example 3.8 Let $\mathbb{T} = \mathbb{N}_0 \bigcup \left\{1 - \frac{1}{n} : n \in \mathbb{N}\right\}$ and

$$f(t) = \begin{cases} 0 & \text{if } t \in \mathbb{N} \\ t & \text{otherwise.} \end{cases}$$

Since f is not continuous at $t = 1$, it is not pre-differentiable.

Exercise 3.9 Let $\mathbb{T} = \mathbb{R}$ and

$$f(t) = \begin{cases} \frac{1}{t+3} & \text{if } t \in \mathbb{R} \setminus \{-3\} \\ 0 & \text{if } t = -3. \end{cases}$$

Check if $f : \mathbb{T} \to \mathbb{R}$ is pre-differentiable, and if it is, then find the region of differentiation.

Definition 3.10 A function $f : \mathbb{T} \to \mathbb{R}$ is called *rd-continuous* provided it is continuous at right-dense points in $\mathbb{T}$ and its left-sided limits exist (finite) at left-dense points in $\mathbb{T}$. The set of rd-continuous functions $f : \mathbb{T} \to \mathbb{R}$ is denoted by C_{rd} or $\mathrm{C}_{\mathrm{rd}}(\mathbb{T})$ or $\mathrm{C}_{\mathrm{rd}}(\mathbb{T}, \mathbb{R})$. The set of functions $f : \mathbb{T} \to \mathbb{R}$ that are differentiable and whose derivative is rd-continuous is denoted by $\mathrm{C}^1_{\mathrm{rd}}(\mathbb{T})$.

Some results concerning rd-continuous and regulated functions are contained in the following theorem. Since its statements follow directly from the definitions, we leave the proofs to the reader.

Theorem 3.11 *Assume* $f : \mathbb{T} \to \mathbb{R}$.

1. *If f is continuous, then f is rd-continuous.*
2. *If f is rd-continuous, then f is regulated.*
3. *The jump operator σ is rd-continuous.*
4. *If f is regulated or rd-continuous, then so is f^σ.*
5. *Assume f is continuous. If $g : \mathbb{T} \to \mathbb{R}$ is regulated or rd-continuous, then $f \circ g$ has that property.*

Theorem 3.12 *Every regulated function on a compact interval is bounded.*

Proof Assume that $f : [a, b] \to \mathbb{R}$, $[a, b] \subset \mathbb{T}$, is unbounded. Then, for each $n \in \mathbb{N}$, there exists $t_n \in [a, b]$ such that $|f(t_n)| > n$. Because $\{t_n\}_{n\in\mathbb{N}} \subset [a, b]$, there exists a subsequence $\{t_{n_k}\}_{k\in\mathbb{N}} \subset \{t_n\}_{n\in\mathbb{N}}$ such that

$$\lim_{k\to\infty} t_{n_k} = t_0.$$

Since $\mathbb{T}$ is closed, we have that $t_0 \in \mathbb{T}$. Also, t_0 is a right-dense point or a left-dense point. Using that f is regulated, we get

$$\left|\lim_{k\to\infty} f(t_{n_k})\right| \neq \infty,$$

which is a contradiction. □

Theorem 3.13 (Mean Value Theorem) *If $f, g : \mathbb{T} \to \mathbb{R}$ are pre-differentiable with D, then*

$$|f^\Delta(t)| \le |g^\Delta(t)| \quad \textit{for all} \quad t \in D$$

implies

$$|f(s) - f(r)| \le g(s) - g(r) \quad \textit{for all} \quad r, s \in \mathbb{T}, \quad r \le s. \tag{3.1}$$

Proof Let $r, s \in \mathbb{T}$ with $r \le s$. Assume also

$$[r, s) \setminus D = \{t_n : n \in \mathbb{N}\}.$$

We take $\varepsilon > 0$. We consider the statements

$$S(t) : |f(t) - f(r)| \leq g(t) - g(r) + \varepsilon \left(t - r + \sum_{t_n < t} 2^{-n} \right)$$

for $t \in [r, s]$. We will prove, using the induction principle, that $S(t)$ is true for all $t \in [r, s]$.

1. Since $0 \leq \varepsilon \sum_{t_n < t} 2^{-n}$, the statement $S(r)$ is true.
2. Let $t \in [r, s]$ be right-scattered and assume that $S(t)$ is true. Then we have

$$\begin{aligned} |f(\sigma(t)) - f(r)| &= |f(t) + \mu(t) f^{\Delta}(t) - f(r)| \\ &\leq \mu(t)|f^{\Delta}(t)| + |f(t) - f(r)| \\ &\leq \mu(t)|f^{\Delta}(t)| + g(t) - g(r) + \varepsilon \left(t - r + \sum_{t_n < t} 2^{-n} \right) \\ &\leq \mu(t) g^{\Delta}(t) + g(t) - g(r) + \varepsilon \left(t - r + \sum_{t_n < t} 2^{-n} \right) \\ &\leq g(\sigma(t)) - g(r) + \varepsilon \left(t - r + \sum_{t_n < t} 2^{-n} \right) \\ &< g(\sigma(t)) - g(r) + \varepsilon \left(\sigma(t) - r + \sum_{t_n < \sigma(t)} 2^{-n} \right) \end{aligned}$$

   since $t < \sigma(t)$. Hence, $S(\sigma(t))$ is true.
3. Let $t \in [r, s)$ be right-dense.

   a. Let $t \in D$. Then f and g are differentiable at t. Hence, there exists a neighbourhood U of t such that

$$|f(t) - f(\tau) - f^{\Delta}(t)(t - \tau)| \leq \frac{\varepsilon}{2}|t - \tau|$$

   and

$$|g(t) - g(\tau) - g^{\Delta}(t)(t - \tau)| \leq \frac{\varepsilon}{2}|t - \tau|$$

   for all $\tau \in U$. Thus,

$$|f(t) - f(\tau)| \leq \left(|f^{\Delta}(t)| + \frac{\varepsilon}{2} \right) |t - \tau|$$

   for all $\tau \in U$ and

$$g(\tau) - g(t) + g^{\Delta}(t)(t - \tau) \geq -\frac{\varepsilon}{2}|t - \tau|$$

for all $\tau \in U$, i.e.,

$$g(\tau) - g(t) - g^{\Delta}(t)(\tau - t) \geq -\frac{\varepsilon}{2}|t - \tau|$$

for all $\tau \in U$. Hence, for all $\tau \in U \cap (t, \infty)$, we have

$$\begin{aligned}
|f(\tau) - f(r)| &= |f(\tau) - f(t) + f(t) - f(r)| \\
&\leq |f(\tau) - f(t)| + |f(t) - f(r)| \\
&\leq \left(|f^{\Delta}(t)| + \frac{\varepsilon}{2}\right)|t - \tau| + g(t) - g(r) + \varepsilon\left(t - r + \sum_{t_n < t} 2^{-n}\right) \\
&\leq \left(g^{\Delta}(t) + \frac{\varepsilon}{2}\right)|t - \tau| + g(t) - g(r) + \varepsilon\left(t - r + \sum_{t_n < t} 2^{-n}\right) \\
&= g^{\Delta}(t)(\tau - t) + \frac{\varepsilon}{2}(\tau - t) + g(t) - g(r) + \varepsilon\left(t - r + \sum_{t_n < t} 2^{-n}\right) \\
&\leq g(\tau) - g(t) + \frac{\varepsilon}{2}|t - \tau| + \frac{\varepsilon}{2}(\tau - t) + g(t) - g(r) + \varepsilon\left(t - r + \sum_{t_n < t} 2^{-n}\right) \\
&= g(\tau) - g(r) + \varepsilon(\tau - t) + \varepsilon\left(t - r + \sum_{t_n < t} 2^{-n}\right) \\
&= g(\tau) - g(r) + \varepsilon\left(\tau - r + \sum_{t_n < t} 2^{-n}\right).
\end{aligned}$$

Therefore, $S(\tau)$ is true for all $\tau \in U \cap (t, \infty)$.

b. Let $t \notin D$. Then $t = t_m$ for some $m \in \mathbb{N}$. Since f and g are pre-differentiable, they both are continuous. Therefore, there exists a neighbourhood U of t such that

$$|f(\tau) - f(t)| \leq \frac{\varepsilon}{2}2^{-m} \quad \text{for all} \quad \tau \in U$$

and

$$|g(\tau) - g(t)| \leq \frac{\varepsilon}{2} 2^{-m} \quad \text{for all} \quad \tau \in U.$$

Therefore,

$$g(\tau) - g(t) \geq -\frac{\varepsilon}{2} 2^{-m} \quad \text{for all} \quad \tau \in U.$$

Consequently,

$$\begin{aligned}
|f(\tau) - f(r)| &= |f(\tau) - f(t) + f(t) - f(r)| \\
&\leq |f(\tau) - f(t)| + |f(t) - f(r)| \\
&\leq \frac{\varepsilon}{2} 2^{-m} + g(t) - g(r) + \varepsilon \left(t - r + \sum_{t_n < t} 2^{-n} \right) \\
&\leq \frac{\varepsilon}{2} 2^{-m} + g(\tau) + \frac{\varepsilon}{2} 2^{-m} - g(r) + \varepsilon \left(t - r + \sum_{t_n < t} 2^{-n} \right) \\
&= \varepsilon 2^{-m} + g(\tau) - g(r) + \varepsilon \left(t - r + \sum_{t_n < t} 2^{-n} \right) \\
&\leq \varepsilon 2^{-m} + g(\tau) - g(r) + \varepsilon \left(\tau - r + \sum_{t_n < \tau} 2^{-n} \right),
\end{aligned}$$

so $S(\tau)$ is true for all $\tau \in U \cap (t, \infty)$.

4. Let t be left-dense and assume that $S(\tau)$ is true for $\tau \in [r, t)$. Then

$$\begin{aligned}
\lim_{\tau \to t-} |f(\tau) - f(r)| &\leq \lim_{\tau \to t-} \left\{ g(\tau) - g(r) + \varepsilon \left(\tau - r + \sum_{t_n < \tau} 2^{-n} \right) \right\} \\
&\leq \lim_{\tau \to t-} \left\{ g(\tau) - g(r) + \varepsilon \left(\tau - r + \sum_{t_n < t} 2^{-n} \right) \right\}
\end{aligned}$$

implies that $S(t)$ is true since f and g are continuous at t.

Thus, using the induction principle, if follows that $S(t)$ is true for all $t \in [r, s]$. Consequently, (3.1) holds for all $r \leq s$, $r, s \in \mathbb{T}$. □

Theorem 3.14 *Suppose $f : \mathbb{T} \to \mathbb{R}$ is pre-differentiable with D. If U is a compact interval with endpoints $r, s \in \mathbb{T}$, then*

$$|f(s) - f(r)| \leq \left\{ \sup_{t \in U^\kappa \cap D} |f^\Delta(t)| \right\} |s - r|.$$

Proof Without loss of generality, we suppose that $r \leq s$. We set

$$g(t) = \left\{ \sup_{\tau \in U^\kappa \cap D} |f^\Delta(\tau)| \right\} (t - r), \quad t \in \mathbb{T}.$$

Then

$$g^\Delta(t) = \left\{ \sup_{\tau \in U^\kappa \cap D} |f^\Delta(\tau)| \right\} \geq |f^\Delta(t)|$$

for all $t \in D \cap [r, s]^\kappa$. Thus, using Theorem 3.13, we obtain

$$|f(t) - f(r)| \leq g(t) - g(r) \quad \text{for all} \quad t \in [r, s],$$

whereupon

$$|f(s) - f(r)| \leq g(s) - g(r) = g(s) = \left\{ \sup_{t \in U^\kappa \cap D} |f^\Delta(t)| \right\} (s - r).$$

□

Theorem 3.15 *Let f be pre-differentiable with D. If $f^\Delta(t) = 0$ for all $t \in D$, then f is a constant function.*

Proof Let U be a compact interval with endpoints $r, s \in \mathbb{T}$. From Theorem 3.14, it follows that for all $r, s \in \mathbb{T}$,

$$|f(s) - f(r)| \leq \left\{ \sup_{t \in U^\kappa \cap D} |f^\Delta(t)| \right\} |s - r| = 0,$$

i.e., $f(s) = f(r)$. Therefore, f is a constant function. □

Theorem 3.16 *Let f and g be pre-differentiable with D. If $f^\Delta(t) = g^\Delta(t)$ for all $t \in D$, then*

$$g(t) = f(t) + C \quad \textit{for all} \quad t \in \mathbb{T},$$

where C is a constant.

Proof Let $h(t) = f(t) - g(t)$, $t \in \mathbb{T}$. Then

$$h^\Delta(t) = f^\Delta(t) - g^\Delta(t) = 0 \quad \text{for all} \quad t \in D.$$

Thus, using Theorem 3.15, it follows that h is a constant function. □

Theorem 3.17 *Suppose $f_n : \mathbb{T} \to \mathbb{R}$ is pre-differentiable with D for each $n \in \mathbb{N}$. Assume that for each $t \in \mathbb{T}^\kappa$, there exists a compact interval neighbourhood $U(t)$ such that the sequence $\{f_n^\Delta\}_{n\in\mathbb{N}}$ converges uniformly on $U(t) \cap D$.*

(i) If $\{f_n\}_{n\in\mathbb{N}}$ converges at some $t_0 \in U(t)$ for some $t \in \mathbb{T}^\kappa$, then it converges uniformly on $U(t)$.

(ii) If $\{f_n\}_{n\in\mathbb{N}}$ converges at some $t_0 \in \mathbb{T}$, then it converges uniformly on $U(t)$ for all $t \in \mathbb{T}^\kappa$.

(iii) The limit mapping $f = \lim\limits_{n\to\infty} f_n$ is pre-differentiable with D and

$$f^\Delta(t) = \lim_{n\to\infty} f_n^\Delta(t) \quad \textit{for all} \quad t \in D.$$

Proof (i) Since $\{f_n^\Delta\}_{n\in\mathbb{N}}$ converges uniformly on $U(t) \cap D$, there exists $N \in \mathbb{N}$ such that

$$\sup_{s\in U(t)\cap D} |(f_m - f_n)^\Delta(s)|$$

is finite for all $m, n \in \mathbb{N}$. Let $m, n \geq N$ and $r \in U(t)$. Then

$$|f_n(r) - f_m(r)| = |f_n(r) - f_m(r) - (f_n(t_0) - f_m(t_0)) + (f_n(t_0) - f_m(t_0))|$$

$$\leq |f_n(t_0) - f_m(t_0)| + \left\{ \sup_{s\in U(t)\cap D} |(f_n - f_m)^\Delta(s)| \right\} |r - t_0|.$$

Hence, $\{f_n\}_{n\in\mathbb{N}}$ converges uniformly on $U(t)$, i.e., $\{f_n\}_{n\in\mathbb{N}}$ is a locally uniformly convergent sequence.

(ii) Assume $\{f_n(t_0)\}_{n\in\mathbb{N}}$ converges for some $t_0 \in \mathbb{T}$. Consider the statement

$$S(t) : \ \{f_n(t)\}_{n\in\mathbb{N}} \quad \text{converges.}$$

a. Since $\{f_n(t_0)\}$ converges, the statement $S(t_0)$ is true.
b. Let t be right-scattered and assume that $S(t)$ is true. Then

$$f_n(\sigma(t)) = f_n(t) + \mu(t) f_n^\Delta(t)$$

converges by the assumption, i.e., $S(\sigma(t))$ is true.
c. Let t be right-dense and assume that $S(t)$ is true. Then, by (i), $\{f_n\}_{n\in\mathbb{N}}$ converges on $U(t)$, and so $S(r)$ is true for all $r \in U(t) \cap (t, \infty)$.
d. Let t be left-dense and assume that $S(r)$ is true for all $t_0 \leq r < t$. Since $U(t) \cap [t_0, t) \neq \emptyset$, using again part (i), we have that $\{f_n\}_{n\in\mathbb{N}}$ converges on $U(t)$; in particular, $S(t)$ is true.

Consequently, $S(t)$ is true for all $t \in [t_0, \infty)$. Using the dual version of the induction principle for the negative direction, we have that $S(t)$ is also true for all $t \in (-\infty, t_0]$ (we note that the first part of this has already been shown, the second part follows by $f_n(\rho(t)) = f_n(t) - \mu(\rho(t)) f_n^{\Delta}(\rho(t))$, the third part and the fourth part follow again by (i)).

(iii) Let $t \in D$. Without loss of generality, we can assume that $\sigma(t) \in U(t)$. We take $\varepsilon > 0$ arbitrarily. Using (i), there exists $N \in \mathbb{N}$ such that

$$|(f_n - f_m)(r) - (f_n - f_m)(\sigma(t))| \le \left\{ \sup_{s \in U(t) \cap D} |(f_n - f_m)^{\Delta}(s)| \right\} |r - \sigma(t)|$$

for all $r \in U(t)$ and all $m, n \ge N$. Since

$$\{f_n^{\Delta}\}_{n \in \mathbb{N}}$$

converges uniformly on $U(t) \cap D$, there exists $N_1 \ge N$ such that

$$\sup_{s \in U(t) \cap D} |(f_n - f_m)^{\Delta}(s)| \le \frac{\varepsilon}{3} \quad \text{for all} \quad m, n \ge N_1.$$

Hence,

$$|(f_n - f_m)(r) - (f_n - f_m)(\sigma(t))| \le \frac{\varepsilon}{3} |r - \sigma(t)|$$

for all $r \in U(t)$ and for all $m, n \ge N_1$. Now, letting $m \to \infty$, we obtain

$$|(f_n - f)(r) - (f_n - f)(\sigma(t))| \le \frac{\varepsilon}{3} |r - \sigma(t)|$$

for all $r \in U(t)$ and all $n \ge N_1$. Let

$$g = \lim_{n \to \infty} f_n^{\Delta}.$$

Then there exists $M \ge N_1$ such that

$$|f_M^{\Delta}(t) - g(t)| \le \frac{\varepsilon}{3},$$

and since f_M is differentiable at t, there also exists a neighbourhood W of t with

$$|f_M(\sigma(t)) - f_M(r) - f_M^{\Delta}(t)(\sigma(t) - r)| \le \frac{\varepsilon}{3} |\sigma(t) - r|$$

for all $r \in W$. From here, for all $r \in U(t) \cap W$, we get

$$|f(\sigma(t)) - f(r) - g(t)|\sigma(t) - r|| \le |(f_M - f)(\sigma(t)) - (f_M - f)(r)|$$

$$+ |f_M^{\Delta}(t) - g(t)||\sigma(t) - r| + |f_M(\sigma(t)) - f_M(r) - f_M^{\Delta}(t)|\sigma(t) - r||$$

$$\leq \frac{\varepsilon}{3}|\sigma(t) - r| + \frac{\varepsilon}{3}|\sigma(t) - r| + \frac{\varepsilon}{3}|\sigma(t) - r|$$

$$= \varepsilon|\sigma(t) - r|.$$

Consequently, f is differentiable at t with $f^{\Delta}(t) = g(t)$. □

Theorem 3.18 *Let $t_0 \in \mathbb{T}$, $x_0 \in \mathbb{R}$. Assume $f : \mathbb{T}^{\kappa} \to \mathbb{R}$ is regulated. Then there exists exactly one pre-differentiable function f with D satisfying*

$$F^{\Delta}(t) = f(t) \quad \textit{for all} \quad t \in D, \quad F(t_0) = x_0.$$

Proof Let $n \in \mathbb{N}$ and consider the statement

$$S(t): \begin{cases} \text{there exists a pre-differentiable} \quad (F_{nt}, D_{nt}), \\ F_{nt} : [t_0, t] \to \mathbb{R} \quad \text{with} \quad F_{nt}(t_0) = x_0 \quad \text{and} \\ |F_{nt}^{\Delta}(s) - f(s)| \leq \frac{1}{n} \quad \text{for} \quad s \in D_{nt}. \end{cases}$$

1. Let $t = t_0$, $D_{nt_0} = \emptyset$ and $F_{nt_0}(t_0) = x_0$. Then the statement $S(t_0)$ is true.
2. Let t be right-scattered and assume that $S(t)$ is true. Define

$$D_{n\sigma(t)} = D_{nt} \cup \{t\}$$

and $F_{n\sigma(t)}$ on $[t_0, \sigma(t)]$ by

$$F_{n\sigma(t)}(s) = \begin{cases} F_{nt}(s) & \text{if} \quad s \in [t_0, t] \\ F_{nt}(t) + \mu(t)f(t) & \text{if} \quad s = \sigma(t). \end{cases}$$

Then

$$F_{n\sigma(t)}(t_0) = F_{nt}(t_0) = x_0,$$

$$\left|F_{n\sigma(t)}^{\Delta}(s) - f(s)\right| = \left|F_{nt}^{\Delta}(s) - f(s)\right| \leq \frac{1}{n} \quad \text{for} \quad s \in D_{nt},$$

and

$$\left|F_{n\sigma(t)}^{\Delta}(t) - f(t)\right| = \left|\frac{F_{n\sigma(t)}(\sigma(t)) - F_{n\sigma(t)}(t)}{\mu(t)} - f(t)\right|$$

$$= \left|\frac{F_{nt}(t) + \mu(t)f(t) - F_{n\sigma(t)}(t)}{\mu(t)} - f(t)\right|$$

$$= \left| \frac{F_{nt}(t) + \mu(t)f(t) - F_{nt}(t)}{\mu(t)} - f(t) \right|$$

$$= \left| \frac{\mu(t)f(t)}{\mu(t)} - f(t) \right|$$

$$= 0$$

$$\leq \frac{1}{n},$$

and therefore the statement $S(\sigma(t))$ is true.

3. Let t be right-dense and assume that $S(t)$ is true. Since t is right-dense and f is regulated,

$$f(t^+) = \lim_{s \to t, s > t} f(s) \quad \text{exists.}$$

Hence, there exists a neighbourhood U of t with

$$|f(s) - f(t^+)| \leq \frac{1}{n} \quad \text{for all} \quad s \in U \cap (t, \infty). \tag{3.2}$$

Let $r \in U \cap (t, \infty)$. Define

$$D_{nr} = (D_{nt} \setminus \{t\}) \cup [t, r]^\kappa$$

and F_{nr} on $[t_0, r]$ by

$$F_{nr}(s) = \begin{cases} F_{nt}(s) & \text{if} \quad s \in [t_0, t] \\ F_{nt}(t) + f(t^+)(s - t) & \text{if} \quad s \in (t, r]. \end{cases}$$

Then F_{nr} is continuous at t and hence on $[t_0, r]$. Also, F_{nr} is differentiable on $(t, r]^\kappa$ with

$$F_{nr}^\Delta(s) = f(t^+) \quad \text{for all} \quad s \in (t, r]^\kappa.$$

Hence, F_{nr} is pre-differentiable on $[t_0, t)$. Since t is right-dense, we have that F_{nr} is pre-differentiable with D_{nr}. From here and from (3.2), we also have

$$|F_{nr}^\Delta(s) - f(s)| \leq \frac{1}{n} \quad \text{for all} \quad s \in D_{nr}.$$

Therefore, the statement $S(r)$ is true for all $r \in U \cap (t, \infty)$.

4. Now, let t be left-dense and suppose that $S(r)$ is true for $r < t$. Since f is regulated,

$$f(t^-) = \lim_{s \to t, s<t} f(s) \quad \text{exists.} \tag{3.3}$$

Hence, there exists a neighbourhood U of t with

$$|f(s) - f(t^-)| \le \frac{1}{n} \quad \text{for all} \quad s \in U \cap (-\infty, t).$$

Fix some $r \in U \cap (-\infty, t)$ and define

$$D_{nt} = \begin{cases} D_{nr} \cup (r, t) & \text{if } r \text{ is right-dense} \\ \\ D_{nr} \cup [r, t) & \text{if } r \text{ is right-scattered} \end{cases}$$

and F_{nt} on $[t_0, t]$ by

$$F_{nt}(s) = \begin{cases} F_{nr}(s) & \text{if} \quad s \in (t_0, r] \\ \\ F_{nr}(t) + f(t^-)(s - r) & \text{if} \quad s \in (r, t]. \end{cases}$$

We note that F_{nt} is continuous at r and hence in $[t_0, t]$. Since

$$F_{nt}^{\Delta}(s) = f(t^-) \quad \text{for all} \quad s \in (r, t],$$

F_{nt} is pre-differentiable with D_{nt} and

$$|F_{nt}^{\Delta}(s) - f(s)| \le \frac{1}{n} \quad \text{for all} \quad s \in D_{nt}.$$

Hence, the statement $S(t)$ is true.

By the induction principle, it follows that $S(t)$ is true for all $t \ge t_0$, $t \in \mathbb{T}$. Similarly, we can show that $S(t)$ is valid for $t \le t_0$. Hence, F_n is pre-differentiable with D_n, $F_n(t_0) = x_0$, and

$$|F_n^{\Delta}(t) - f(t)| \le \frac{1}{n} \quad \text{for all} \quad t \in D_n.$$

Now, let

$$F = \lim_{n \to \infty} F_n \quad \text{and} \quad D = \bigcap_{n \in \mathbb{N}} D_n.$$

Then $F(t_0) = x_0$, f is pre-differentiable on D, and, using Theorem 3.17,

$$F^{\Delta}(t) = \lim_{n \to \infty} F_n^{\Delta}(t) = f(t) \quad \text{for all} \quad t \in D.$$

This completes the proof. □

3.2 Indefinite Integral

Definition 3.19 Assume $f : \mathbb{T} \to \mathbb{R}$ is a regulated function. Any function F as in Theorem 3.18 is called a *pre-antiderivative* of f. We define the *indefinite integral* of a regulated function f by

$$\int f(t)\Delta t = F(t) + c,$$

where c is an arbitrary constant and F is a pre-antiderivative of f. We define the *Cauchy integral* by

$$\int_s^t f(\tau)\Delta\tau = F(t) - F(s) \quad \text{for all} \quad s, t \in \mathbb{T}.$$

A function $F : \mathbb{T} \to \mathbb{R}$ is called an *antiderivative* of $f : \mathbb{T} \to \mathbb{R}$ provided

$$F^{\Delta}(t) = f(t) \quad \text{holds for all} \quad t \in \mathbb{T}^{\kappa}.$$

Example 3.20 Let $\mathbb{T} = \mathbb{Z}$. Then $\sigma(t) = t + 1$, $t \in \mathbb{T}$. Assume $f(t) = 3t^2 + 5t + 2$, $t \in \mathbb{T}$. Since $g(t) = t^3 + t^2$ satisfies

$$\begin{aligned} g^{\Delta}(t) &= (\sigma(t))^2 + t\sigma(t) + t^2 + \sigma(t) + t \\ &= (t+1)^2 + t(t+1) + t^2 + t + 1 + t \\ &= t^2 + 2t + 1 + t^2 + t + t^2 + 2t + 1 \\ &= 3t^2 + 5t + 2, \end{aligned}$$

we have

$$\int (3t^2 + 5t + 2)\Delta t = t^3 + t^2 + c.$$

Example 3.21 Let $\mathbb{T} = 2^{\mathbb{N}}$ and define $f : \mathbb{T} \to \mathbb{R}$ by $f(t) = \frac{2}{t}\sin\frac{t}{2}\cos\frac{3t}{2}$, $t \in \mathbb{T}$. In this case, we have that $\sigma(t) = 2t$. Since $g(t) = \sin t$ satisfies

$$\begin{aligned} g^{\Delta}(t) &= \frac{\sin\sigma(t) - \sin t}{\sigma(t) - t} \\ &= \frac{\sin(2t) - \sin t}{t} \\ &= \frac{2}{t}\sin\frac{t}{2}\cos\frac{3t}{2}, \end{aligned}$$

we get

$$\int \frac{2}{t} \sin \frac{t}{2} \cos \frac{3t}{2} \Delta t = \sin t + c.$$

Example 3.22 Let $\mathbb{T} = \mathbb{N}_0^2$ and define $f : \mathbb{T} \to \mathbb{R}$ by $f(t) = \frac{1}{1+2\sqrt{t}} \log \frac{(1+\sqrt{t})^2}{t}$, $t \in \mathbb{T}$. Here, $\sigma(t) = (1+\sqrt{t})^2$. Since $g(t) = \log t$ satisfies

$$\begin{aligned} g^{\Delta}(t) &= \frac{\log \sigma(t) - \log t}{\sigma(t) - t} \\ &= \frac{\log(1+\sqrt{t})^2 - \log t}{(1+\sqrt{t})^2 - t} \\ &= \frac{1}{1+2\sqrt{t}} \log \frac{(1+\sqrt{t})^2}{t}, \end{aligned}$$

we get

$$\int \frac{1}{1+2\sqrt{t}} \log \frac{(1+\sqrt{t})^2}{t} \Delta t = \log t + c.$$

Exercise 3.23 Let $\mathbb{T} = \mathbb{N}_0^3$. Prove that

$$\int (2t + 3\sqrt[3]{t^2} + 3\sqrt[3]{t} + 2) \Delta t = t^2 + t + c.$$

Theorem 3.24 *Every rd-continuous function $f : \mathbb{T} \to \mathbb{R}$ has an antiderivative. In particular, if $t_0 \in \mathbb{T}$, then f defined by*

$$F(t) = \int_{t_0}^{t} f(\tau) \Delta \tau \quad \textit{for} \quad t \in \mathbb{T},$$

is an antiderivative of f.

Proof Since f is rd-continuous, it is regulated. Let F be a function guaranteed to exist by Theorem 3.18, together with D, satisfying

$$F^{\Delta}(t) = f(t) \quad \text{for} \quad t \in D.$$

We have that F is pre-differentiable with D. Let $t \in \mathbb{T}^{\kappa} \setminus D$. Then t is right-dense. Since f is rd-continuous, f is continuous at t. Let $\varepsilon > 0$ be arbitrarily chosen. Then there exists a neighbourhood U of t such that

$$|f(s) - f(t)| \le \varepsilon \quad \text{for all} \quad s \in U.$$

We define

$$h(\tau) = F(\tau) - f(t)(\tau - t_0) \quad \text{for} \quad \tau \in \mathbb{T}.$$

Then h is pre-differentiable with D and

$$\begin{aligned} h^{\Delta}(\tau) &= F^{\Delta}(\tau) - f(t) \\ &= f(\tau) - f(t) \quad \text{for all} \quad \tau \in D. \end{aligned}$$

Hence,

$$\begin{aligned} |h^{\Delta}(s)| &= |f(s) - f(t)| \\ &\leq \varepsilon \quad \text{for all} \quad s \in D \cap U. \end{aligned}$$

Therefore,

$$\sup_{s \in D \cap U} |h^{\Delta}(s)| \leq \varepsilon,$$

whereupon

$$\begin{aligned} |F(t) - F(r) - f(t)(t - r)| &= |h(t) + f(t)(t - t_0) - (h(r) \\ &\quad + f(t)(r - t_0)) - f(t)(t - r)| \\ &= |h(t) - h(r)| \\ &\leq \left\{ \sup_{s \in D \cap U} |h^{\Delta}(s)| \right\} |t - r| \\ &\leq \varepsilon |t - r|, \end{aligned}$$

which shows that f is differentiable at t and $F^{\Delta}(t) = f(t)$. □

3.3 The Riemann Delta Integral

Let $a < b$ be points in $\mathbb{T}$ and $[a, b)$ be the half-closed bounded interval in $\mathbb{T}$.

Definition 3.25 A *partition* of $[a, b)$ is any finite ordered subset

$$P = \{t_0, t_1, t_2, \ldots, t_n\} \subset [a, b],$$

where

$$a = t_0 < t_1 < \ldots < t_n = b.$$

The number n depends on the particular partition. The intervals

$$[t_{i-1}, t_i), \quad i \in \{1, \ldots, n\}$$

are called *subintervals* of the partition P.

Let f be a real-valued bounded function on $[a, b)$. We set

$$M = \sup\{f(t) : t \in [a, b)\}, \quad m = \inf\{f(t) : t \in [a, b)\}$$

and

$$M_i = \sup\{f(t) : t \in [t_{i-1}, t_i)\}, \quad m_i = \inf\{f(t) : t \in [t_{i-1}, t_i)\}, \quad i \in \{1, \ldots, n\}.$$

Definition 3.26 The *upper Darboux* Δ*-sum* $U(f, P)$ and the *lower Darboux* Δ*-sum* $L(f, P)$ of f with respect to P are defined by

$$U(f, P) = \sum_{i=1}^{n} M_i(t_i - t_{i-1}), \quad L(f, P) = \sum_{i=1}^{n} m_i(t_i - t_{i-1}),$$

respectively.

Proposition 3.27 *We have*

$$m(b - a) \le L(f, P) \le U(f, P) \le M(b - a). \tag{3.4}$$

Proof We have

$$m_i \ge m \quad \text{and} \quad M_i \le M \quad \text{for all} \quad i \in \{1, \ldots, n\}.$$

Then

$$\begin{aligned} L(f, P) &= \sum_{i=1}^{n} m_i(t_i - t_{i-1}) \\ &\ge m \sum_{i=1}^{n} (t_i - t_{i-1}) \\ &= m(b - a), \end{aligned}$$

$$\begin{aligned} U(f, P) &= \sum_{i=1}^{n} M_i(t_i - t_{i-1}) \\ &\le M \sum_{i=1}^{n}(t_i - t_{i-1}) \\ &= M(b-a), \end{aligned}$$

which completes the proof. □

Definition 3.28 The *upper Darboux Δ-integral* $U(f)$ of f from a to b is defined by

$$U(f) = \inf\{U(f, P) : \ P \quad \text{is a partition of} \quad [a, b)\}$$

and the *lower Darboux Δ-integral* $L(f)$ is defined by

$$L(f) = \sup\{L(f, P) : \ P \quad \text{is a partition of} \quad [a, b)\}.$$

Proposition 3.29 *Let f be a bounded function on $[a, b)$. If P and Q are partitions of $[a, b)$ and Q is a refinement of P, i.e., $P \subset Q$, then*

$$L(f, P) \le L(f, Q) \le U(f, Q) \le U(f, P).$$

Proof Let $P = \{t_0, t_1, t_2, \ldots, t_n\}$. Without loss of generality, we suppose that

$$Q = \{t_0, t_1, \ldots, t_k, t', t_{k+1}, \ldots, t_n\},$$

i.e., $Q \setminus P = \{t'\}$. Define also

$$m_k^1 = \inf\{f(t) : t \in [t_k, t')\}, \quad m_k^2 = \inf\{f(t) : t \in [t', t_{k+1})\}$$

and

$$M_k^1 = \sup\{f(t) : t \in [t_k, t')\}, \quad M_k^2 = \sup\{f(t) : t \in [t', t_{k+1})\}.$$

We have

$$m_k^1 \ge m_{k+1}, \quad m_k^2 \ge m_{k+1}, \quad M_k^1 \le M_{k+1}, \quad M_k^2 \le M_{k+1}.$$

Then

$$\begin{aligned} L(f, Q) &= \sum_{i=1}^{k} m_i(t_i - t_{i-1}) + m_k^1(t' - t_k) + m_k^2(t_{k+1} - t') + \sum_{i=k+2}^{n} m_i(t_i - t_{i-1}) \\ &\ge \sum_{i=1}^{k} m_i(t_i - t_{i-1}) + m_k(t' - t_k) + m_k(t_{k+1} - t') + \sum_{i=k+2}^{n} m_i(t_i - t_{i-1}) \end{aligned}$$

$$
\begin{aligned}
&= \sum_{i=1}^{k} m_i(t_i - t_{i-1}) + m_k(t_{k+1} - t_k) + \sum_{i=k+2}^{n} m_i(t_i - t_{i-1}) \\
&= \sum_{i=1}^{n} m_i(t_i - t_{i-1}) \\
&= L(f, P)
\end{aligned}
$$

and

$$
\begin{aligned}
U(f, Q) &= \sum_{i=1}^{k} M_i(t_i - t_{i-1}) + M_k^1(t' - t_k) + M_k^2(t_{k+1} - t') + \sum_{i=k+2}^{n} M_i(t_i - t_{i-1}) \\
&\le \sum_{i=1}^{k} M_i(t_i - t_{i-1}) + M_k(t' - t_k) + M_k(t_{k+1} - t') + \sum_{i=k+2}^{n} M_i(t_i - t_{i-1}) \\
&= \sum_{i=1}^{k} M_i(t_i - t_{i-1}) + M_k(t_{k+1} - t_k) + \sum_{i=k+2}^{n} M_i(t_i - t_{i-1}) \\
&= \sum_{i=1}^{n} M_i(t_i - t_{i-1}) \\
&= U(f, P),
\end{aligned}
$$

which completes the proof. □

Proposition 3.30 *If f is a bounded function on $[a, b)$ and if P and Q are any two partitions of $[a, b)$, then $L(f, P) \le U(f, Q)$.*

Proof Note that $P \cup Q$ is also a partition of $[a, b)$. Since

$$P \subset P \cup Q \quad \text{and} \quad Q \subset P \cup Q,$$

using Proposition 3.29 twice, we get

$$
\begin{aligned}
L(f, P) &\le L(f, P \cup Q) \\
&\le U(f, P \cup Q) \\
&\le U(f, Q),
\end{aligned}
$$

which completes the proof. □

Theorem 3.31 *If f is a bounded function on $[a, b)$, then $L(f) \leq U(f)$.*

Proof Let P and Q be arbitrarily chosen partitions of $[a, b)$. From Proposition 3.30, it follows that

$$L(f, P) \leq U(f, Q) \leq \inf_R U(f, R) = U(f),$$

and therefore

$$L(f) = \sup_R L(f, R) \leq U(f),$$

which completes the proof.

Definition 3.32 We say that f is *Δ-integrable* from a to b (or on $[a, b)$) provided $L(f) = U(f)$. We write $\int_a^b f(t)\Delta t$ for this common value. We call this integral the *Darboux Δ-integral.*

Theorem 3.33 *If $L(f, P) = U(f, P)$ for some partition P of $[a, b)$, then the function f is Δ-integrable from a to b and*

$$\int_a^b f(t)\Delta t = L(f, P) = U(f, P).$$

Proof From Theorem 3.31, it follows that

$$L(f, P) \leq L(f) \leq U(f) \leq U(f, P) = L(f, P),$$

which completes the proof. □

Theorem 3.34 *A bounded function f on $[a, b)$ is Δ-integrable if and only if for each $\varepsilon > 0$, there exists a partition P of $[a, b)$ such that*

$$U(f, P) - L(f, P) < \varepsilon.$$

Proof 1. Let f be Δ-integrable. Then $L(f) = U(f)$. Let $\varepsilon > 0$ be arbitrarily chosen. Then there exist partitions P and Q of $[a, b)$ such that

$$L(f) - L(f, P) < \frac{\varepsilon}{2} \quad \text{and} \quad U(f, Q) - U(f) < \frac{\varepsilon}{2}.$$

a. Assume that P is a refinement of Q. Then

$$U(f, P) \leq U(f, Q)$$

and

$$U(f, P) - L(f, P) \leq U(f, Q) - L(f, P)$$

$$\begin{aligned} &= U(f, Q) - U(f) + L(f) - L(f, P) \\ &< \frac{\varepsilon}{2} + \frac{\varepsilon}{2} \\ &= \varepsilon. \end{aligned}$$

b. Assume that Q is a refinement of P. Then

$$L(f, Q) \geq L(f, P)$$

and

$$\begin{aligned} U(f, Q) - L(f, Q) &\leq U(f, Q) - L(f, P) \\ &= U(f, Q) - U(f) + L(f) - L(f, P) \\ &< \frac{\varepsilon}{2} + \frac{\varepsilon}{2} \\ &= \varepsilon. \end{aligned}$$

2. Suppose for each $\varepsilon > 0$, there exists a partition P so that

$$U(f, P) - L(f, P) < \varepsilon.$$

Hence, using that

$$U(f) \leq U(f, P) \quad \text{and} \quad -L(f) \leq -L(f, P),$$

we get

$$U(f) - L(f) < \varepsilon$$

for each $\varepsilon > 0$. Consequently, $U(f) = L(f)$, which completes the proof. □

Proposition 3.35 *For every $\delta > 0$, there exists at least one partition*

$$P = \{a = t_0 < t_1 < \ldots < t_n = b\}$$

of $[a, b)$ such that, for each $i \in \{1, \ldots, n\}$, either

$$t_i - t_{i-1} \leq \delta$$

or

$$t_i - t_{i-1} > \delta \quad \text{and} \quad \rho(t_i) = t_{i-1}.$$

Proof Let $\delta > 0$ be arbitrarily chosen.

1. If $b - a \leq \delta$, then, for every partition $P = \{t_0 = a < t_1 < \ldots < t_n = b\}$, we have $t_i - t_{i-1} \leq \delta$ for all $i \in \{1, \ldots, n\}$.
2. Let $b - a > \delta$. We set $a = t_0$.
   a. Assume that t_0 is right-scattered. Let $t_1 = \sigma(t_0)$. Hence, $t_1 - t_0 \leq \delta$ or $t_1 - t_0 > \delta$ and $\rho(t_1) = t_0$.
   b. Assume that t_0 is right-dense. Then there exists $t_1 \in (a, b]$ such that $t_1 - t_0 \leq \delta$.

   If $t_1 = b$, then $P = \{t_0 = a < t_1 = b\}$ is the desired partition. Otherwise, we consider the interval $[t_1, b)$. □

Definition 3.36 For given $\delta > 0$, we denote by $\mathscr{P}_\delta([a, b))$ the set of all partitions

$$P = \{a = t_0 < t_1 < \ldots < t_n = b\}$$

that possess the property indicated in Proposition 3.35.

Theorem 3.37 *A bounded function f on $[a, b)$ is Δ-integrable if and only if for each $\varepsilon > 0$, there exists $\delta > 0$ such that $P \in \mathscr{P}_\delta([a, b))$ implies*

$$U(f, P) - L(f, P) < \varepsilon. \tag{3.5}$$

Proof 1. Let f be Δ-integrable on $[a, b)$. Let also $\varepsilon > 0$ be arbitrarily chosen. Then there exists a partition P_1 of $[a, b)$ such that

$$U(f, P_1) - L(f, P_1) < \varepsilon. \tag{3.6}$$

If $P_1 \in \mathscr{P}_\delta([a, b))$ for some $\delta > 0$, then the assertion follows. Let $P_1 \notin \mathscr{P}_\delta([a, b))$ for any $\delta > 0$. Then there exist $\delta > 0$ and $P \in \mathscr{P}_\delta([a, b))$ so that P is a refinement of P_1. Hence,

$$U(f, P) \leq U(f, P_1) \quad \text{and} \quad L(f, P_1) \geq L(f, P).$$

From here and from (3.6), we get (3.5).

2. Suppose that for each $\varepsilon > 0$, there exists $\delta > 0$ so that $P \in \mathscr{P}_\delta([a, b))$ implies (3.5). Thus, using Theorem 3.34, it follows that f is Δ-integrable on $[a, b)$. □

Definition 3.38 Let f be a bounded function on $[a, b)$. Assume

$$P = \{a = t_0 < t_1 < \ldots < t_n = b\}$$

is a partition of $[a, b)$. In each interval $[t_{i-1}, t_i)$, $i \in \{1, \ldots, b\}$, choose an arbitrary point ξ_i and form the sum

$$S = \sum_{i=1}^{n} f(\xi_i)(t_i - t_{i-1}).$$

We call S a *Riemann Δ-sum* of f corresponding to the partition P. We say that f is *Riemann Δ-integrable* from a to b (or on $[a, b)$) if there exists a number I with the following property. For each $\varepsilon > 0$, there exists $\delta > 0$ such that $|S - I| < \varepsilon$ for every Riemann Δ-sum of f corresponding to a partition $P \in \mathscr{P}_\delta([a, b))$, independent of the way in which we choose $\xi_i \in [t_{i-1}, t_i)$, $i \in \{1, \ldots, n\}$. The number I is called the *Riemann Δ-integral* of f from a to b.

Exercise 3.39 Prove that the number I is unique.

Theorem 3.40 *A bounded function f on $[a, b)$ is Riemann Δ-integrable if and only if it is Darboux Δ-integrable, in which case the values of the integrals coincide.*

Proof 1. Suppose that f is Darboux Δ-integrable on $[a, b)$. Then

$$U(f) = L(f) = \int_a^b f(t)\Delta t.$$

Let $\varepsilon > 0$ be arbitrarily chosen. From Theorem 3.37, it follows that there exists $\delta > 0$ such that $P \in \mathscr{P}_\delta([a, b))$ satisfies

$$U(f, P) - L(f, P) < \varepsilon.$$

Note that

$$\begin{aligned} U(f, P) &< \varepsilon + L(f, P) \\ &\le \varepsilon + L(f) \\ &= \varepsilon + \int_a^b f(t)\Delta t, \end{aligned}$$

$$\begin{aligned} L(f, P) &\ge U(f, P) - \varepsilon \\ &\ge U(f) - \varepsilon \\ &= \int_a^b f(t)\Delta t - \varepsilon. \end{aligned}$$

Since

$$L(f, P) \leq S \leq U(f, P),$$

we have that

$$\begin{aligned} 0 &\geq S - U(f, P) \\ &> S - \varepsilon - \int_a^b f(t)\Delta t, \end{aligned}$$

i.e.,

$$S - \int_a^b f(t)\Delta t \leq \varepsilon. \tag{3.7}$$

Also,

$$\begin{aligned} 0 &\leq S - L(f, P) \\ &\leq S - \int_a^b f(t)\Delta t + \varepsilon, \end{aligned}$$

i.e.,

$$S - \int_a^b f(t)\Delta t \geq -\varepsilon.$$

From the last inequality and from (3.7), we get that

$$\left| S - \int_a^b f(t)\Delta t \right| \leq \varepsilon,$$

which proves that f is Riemann Δ-integrable and

$$I = \int_a^b f(t)\Delta t.$$

2. Assume that f is Riemann Δ-integrable in the sense of Definition 3.38 and let $\varepsilon > 0$. Let $\delta > 0$ and I be as given in Definition 3.38. We take a partition

$$P = \{a = t_0 < t_1 < \ldots < t_n = b\}$$

of $[a, b)$ such that $P \in \mathscr{P}_\delta([a, b))$. For each $i \in \{1, \ldots, n\}$, we choose $\xi_i \in [t_{i-1}, t_i)$ so that $f(\xi_i) < m_i + \varepsilon$, where $m_i = \inf\{f(t) : t \in [t_{i-1}, t_i)\}$. Hence,

$$\begin{aligned} S &= \sum_{i=1}^{n} f(\xi_i)(t_i - t_{i-1}) \\ &< \sum_{i=1}^{n} (m_i + \varepsilon)(t_i - t_{i-1}) \\ &= \sum_{i=1}^{n} m_i(t_i - t_{i-1}) + \varepsilon \sum_{i=1}^{n} (t_i - t_{i-1}) \\ &= L(f, P) + \varepsilon(b - a), \end{aligned}$$

i.e.,

$$L(f, P) > S - \varepsilon(b - a).$$

Also, we have

$$|S - I| < \varepsilon, \quad \text{i.e.,} \quad -\varepsilon + I < S < I + \varepsilon.$$

Hence,

$$\begin{aligned} L(f) &\geq L(f, P) \\ &> S - \varepsilon(b - a) \\ &> I - \varepsilon - \varepsilon(b - a). \end{aligned}$$

Because $\varepsilon > 0$ was arbitrarily chosen, we conclude that

$$L(f) \geq I. \tag{3.8}$$

Let $\eta_i \in [t_{i-1}, t_i)$ be so that

$$f(\eta_i) > M_i - \varepsilon,$$

where

$$M_i = \sup\{f(t) : t \in [t_{i-1}, t_i)\}, \quad i \in \{1, \ldots, n\}.$$

Hence,

$$\begin{aligned} S &= \sum_{i=1}^{n} f(\eta_i)(t_i - t_{i-1}) \\ &> \sum_{i=1}^{n} (M_i - \varepsilon)(t_i - t_{i-1}) \end{aligned}$$

$$= \sum_{i=1}^{n} M_i(t_i - t_{i-1}) - \varepsilon \sum_{i=1}^{n}(t_i - t_{i-1})$$

$$= U(f, P) - \varepsilon(b - a),$$

i.e.,

$$U(f, P) < S + \varepsilon(b - a).$$

From here,

$$U(f) \le U(f, P)$$

$$< S + \varepsilon(b - a)$$

$$< I + \varepsilon + \varepsilon(b - a).$$

Since $\varepsilon > 0$ was arbitrarily chosen, we conclude that

$$U(f) \le I.$$

From the last inequality and from (3.8), we obtain

$$I \le L(f) \le U(f) \le I,$$

i.e.,

$$L(f) = U(f) = I.$$

This shows that f is Darboux Δ-integrable and $\int_a^b f(t)\Delta t = I$. □

Remark 3.41 In our definition of $\int_a^b f(t)\Delta t$, we assumed that $a < b$. We remove this restriction with the following definitions.

$$\int_a^a f(t)\Delta t = 0, \quad \int_a^b f(t)\Delta t = -\int_b^a f(t)\Delta t \quad \text{if} \quad a > b.$$

Theorem 3.42 *Let $a, b \in \mathbb{T}$. Then every constant function*

$$f(t) = c, \quad t \in \mathbb{T},$$

is Δ-integrable from a to b and

$$\int_a^b c\Delta t = c(b - a).$$

Proof Without loss of generality, we assume that $a < b$. For any partition

$$P = \{a = t_0 < t_1 < \ldots < t_n = b\},$$

we have

$$U(f, P) = L(f, P) = c(b - a).$$

Therefore,

$$U(f) = L(f) = c(b - a),$$

which completes the proof. □

Theorem 3.43 *Let t be an arbitrary point in $\mathbb{T}$. Every function f defined on $\mathbb{T}$ is Δ-integrable from t to $\sigma(t)$ and*

$$\int_t^{\sigma(t)} f(s)\Delta s = \mu(t) f(t).$$

Proof 1. Let $\sigma(t) = t$. Then the assertion is valid.

2. Let $\sigma(t) > t$. Then a single partition of $[t, \sigma(t))$ is

$$P = \{t_0 = t < t_1 = \sigma(t)\}.$$

Since $[t, \sigma(t)) = \{t\}$, we have that

$$U(f, P) = L(f, P) = (\sigma(t) - t) f(t).$$

Therefore,

$$U(f) = L(f) = (\sigma(t) - t) f(t),$$

which completes the proof. □

Example 3.44 Let $\mathbb{T} = \mathbb{Z}$. Then $\sigma(t) = t + 1$, $\mu(t) = 1$, and

$$\int_t^{t+1} (\tau^3 + \tau^2 + \tau + 1)\Delta\tau = t^3 + t^2 + t + 1.$$

Example 3.45 Let $\mathbb{T} = \mathbb{N}_0^4$. Then $\sigma(t) = (1 + \sqrt[4]{t})^4$, $\mu(t) = (1 + \sqrt[4]{t})^4 - t$, and

$$\int_t^{(1+\sqrt[4]{t})^4} \tau^2 \Delta\tau = \left((1 + \sqrt[4]{t})^4 - t\right) t^2.$$

Example 3.46 Let $\mathbb{T} = 3^{\mathbb{N}_0}$. Then $\sigma(t) = 3t$, $\mu(t) = 2t$, and

$$\int_t^{3t} \sin\tau \Delta\tau = 2t \sin t.$$

Exercise 3.47 Let $\mathbb{T} = 3^{\mathbb{N}_0}$. Prove that

$$\int_t^{3t} (\tau^2 - 3\tau + 4\sin\tau)\Delta\tau = 2t(t^2 - 3t + 4\sin t).$$

3.4 Basic Properties of the Riemann Integral

Theorem 3.48 *Let f be Δ-integrable on $[a, b)$ and let M and m be its supremum and infimum on $[a, b)$, respectively. Assume $\phi : \mathbb{R} \to \mathbb{R}$ is a function defined on $[m, M]$ such that*

$$|\phi(x) - \phi(y)| \leq B|x - y|$$

for some positive constant B and for all $x, y \in [m, M]$. Then the composite function $h = \phi \circ f$ is Δ-integrable on $[a, b)$.

Proof Let $\varepsilon > 0$ be arbitrarily chosen. Since f is Δ-integrable on $[a, b)$, using Theorem 3.34, there exists a partition

$$P = \{a = t_0 < t_1 < \ldots < t_n = b\}$$

of $[a, b)$ such that

$$U(f, P) - L(f, P) < \frac{\varepsilon}{B}.$$

Define

$$m_i = \inf_{t \in [t_{i-1}, t_i)} f(t), \quad M_i = \sup_{t \in [t_{i-1}, t_i)} f(t)$$

and

$$m_i^* = \inf_{t \in [t_{i-1}, t_i)} h(t), \quad M_i^* = \sup_{t \in [t_{i-1}, t_i)} h(t).$$

Then, for every $s, \tau \in [t_{i-1}, t_i)$, we have

$$\begin{aligned} h(s) - h(\tau) \leq & |h(s) - h(\tau)| \\ = & |\phi(f(s)) - \phi(f(\tau))| \\ \leq & B|f(s) - f(\tau)| \\ \leq & B(M_i - m_i). \end{aligned} \tag{3.9}$$

There exist sequences $\{s_k\}_{k \in \mathbb{N}}$ and $\{\tau_k\}_{k \in \mathbb{N}}$ of points of $[t_{i-1}, t_i)$ such that

$$h(s_k) \to M_i^* \quad \text{and} \quad h(\tau_k) \to m_i^*$$

as $k \to \infty$. Thus, using (3.9), we obtain

$$h(s_k) - h(\tau_k) \leq B(M_i - m_i),$$

whereupon

$$M_i^* - m_i^* \leq B(M_i - m_i).$$

From here,

$$\begin{aligned}
U(h, P) - L(h, P) &= \sum_{i=1}^{n} (M_i^* - m_i^*)(t_i - t_{i-1}) \\
&\leq \sum_{i=1}^{n} B(M_i - m_i)(t_i - t_{i-1}) \\
&= B\left(\sum_{i=1}^{n} M_i(t_i - t_{i-1}) - \sum_{i=1}^{n} m_i(t_i - t_{i-1})\right) \\
&= B(U(f, P) - L(f, P)) \\
&< B\frac{\varepsilon}{B} \\
&= \varepsilon.
\end{aligned}$$

Thus, using Theorem 3.34, it follows that h is Δ-integrable on $[a, b)$. □

Theorem 3.49 *Let f be Δ-integrable on $[a, b)$ and let M and m be its supremum and infimum on $[a, b)$, respectively. Assume that $\phi : \mathbb{R} \to \mathbb{R}$ is a continuous function on $[m, M]$. Then the composite function $h = \phi \circ f$ is Δ-integrable on $[a, b)$.*

Proof Let $\varepsilon > 0$ be arbitrarily chosen. We take a partition

$$P = \{a = t_0 < t_1 < \ldots < t_n = b\}$$

of $[a, b)$ so that

$$\sup_{t \in [t_{i-1}, t_i)} h(t) - \inf_{t \in [t_{i-1}, t_i)} h(t) < \frac{\varepsilon}{b - a}, \quad i \in \{1, \ldots, n\}.$$

Then

$$
\begin{aligned}
U(h,P) - L(h,P) &= \sum_{i=1}^{n} \left(\sup_{t \in [t_{i-1},t_i)} h(t) - \inf_{t \in [t_{i-1},t_i)} h(t) \right) (t_i - t_{i-1}) \\
&< \frac{\varepsilon}{b-a} \sum_{i=1}^{n} (t_i - t_{i-1}) \\
&= \varepsilon .
\end{aligned}
$$

From here and from Theorem 3.34, it follows that h is Δ-integrable on $[a,b)$. □

Corollary 3.50 *If f is Δ-integrable on $[a,b)$, then, for an arbitrary positive number α, the function $|f|^{\alpha}$ is Δ-integrable on $[a,b)$.*

Proof We take the function $\phi(x) = |x|^{\alpha}$ and apply Theorem 3.49. □

Theorem 3.51 *Let f be a bounded function that is Δ-integrable on $[a,b)$. Then f is Δ-integrable on every subinterval $[c,d)$ of the interval $[a,b)$.*

Proof Let $\varepsilon > 0$ be arbitrarily chosen. Since f is Δ-integrable on $[a,b)$, using Theorem 3.34, there exists a partition P of $[a,b)$ such that

$$U(f,P) - L(f,P) < \varepsilon.$$

We set $P' = P \cup \{c\} \cup \{d\}$. Then P' is a refinement of P and

$$U(f,P') - L(f,P') < \varepsilon.$$

Let $P'' = P' \cap [c,d]$. Then P'' is a partition of $[c,d)$ and $P'' \subset P'$. Hence,

$$U(f,P'') - L(f,P'') \le U(f,P') - L(f,P') < \varepsilon.$$

From here and from Theorem 3.34, it follows that f is Δ-integrable on $[c,d)$. □

Theorem 3.52 *Let f and g be Δ-integrable functions on $[a,b)$ and $\alpha \in \mathbb{R}$. Then*

1. *αf is Δ-integrable on $[a,b)$ and*

$$\int_a^b (\alpha f)(t)\Delta t = \alpha \int_a^b f(t)\Delta t,$$

2. *$f+g$ is Δ-integrable on $[a,b)$ and*

$$\int_a^b (f+g)(t)\Delta t = \int_a^b f(t)\Delta t + \int_a^b g(t)\Delta t.$$

Proof 1. a. Let $\alpha > 0$. Suppose $\varepsilon > 0$ is arbitrarily chosen. Since f is Δ-integrable on $[a, b)$, there exists a partition P of $[a, b)$ such that

$$U(f, P) - L(f, P) < \frac{\varepsilon}{\alpha}.$$

Note that

$$U(\alpha f, P) = \alpha U(f, P), \quad L(\alpha f, P) = \alpha L(f, P)$$

and thus

$$U(\alpha f) = \alpha U(f), \quad L(\alpha f) = \alpha L(f).$$

Hence,

$$\begin{aligned} U(\alpha f, P) - L(\alpha f, P) &= \alpha U(f, P) - \alpha L(f, P) \\ &< \alpha \frac{\varepsilon}{\alpha} \\ &= \varepsilon. \end{aligned}$$

Therefore, αf is Δ-integrable on $[a, b)$. Also,

$$U(\alpha f) = \alpha U(f) = \alpha L(f) = L(\alpha f),$$

i.e.,

$$\int_a^b (\alpha f)(t)\Delta t = \alpha \int_a^b f(t)\Delta t.$$

b. Let $\alpha = -1$. Then

$$U(-f, P) = -L(f, P), \quad L(-f, P) = -U(f, P)$$

and thus

$$U(-f) = -L(f), \quad L(-f) = -U(f).$$

Therefore,

$$U(-f) = -L(f) = -U(f) = L(-f).$$

Consequently, $-f$ is Δ-integrable on $[a, b)$ and

$$\int_a^b (-f)(t)\Delta t = -\int_a^b f(t)\Delta t.$$

c. Let $\alpha < 0$. Then, by the parts a and b,

$$\begin{aligned}\int_a^b (\alpha f)(t)\Delta t &= \int_a^b (-(-\alpha))f(t)\Delta t\\ &= -\int_a^b (-\alpha)f(t)\Delta t\\ &= -(-\alpha)\int_a^b f(t)\Delta t\\ &= \alpha\int_a^b f(t)\Delta t,\end{aligned}$$

which completes the proof.

2. Let $\varepsilon > 0$ be arbitrarily chosen. Since f and g are Δ-integrable on $[a, b)$, there exist partitions P_1 and P_2 of $[a, b)$ so that

$$U(f, P_1) - L(f, P_1) < \frac{\varepsilon}{2} \quad \text{and} \quad U(g, P_2) - L(g, P_2) < \frac{\varepsilon}{2}.$$

We set $P = P_1 \cup P_2$. Then P is a refinement of P_1 and P_2. Hence,

$$U(f, P) - L(f, P) \leq U(f, P_1) - L(f, P_1) < \frac{\varepsilon}{2},$$

$$U(g, P) - L(g, P) \leq U(g, P_2) - L(g, P_2) < \frac{\varepsilon}{2},$$

and

$$\begin{aligned}U(f+g, P) - L(f+g, P) &\leq U(f, P) + U(g, P) - L(f, P) - L(g, P)\\ &= U(f, P) - L(f, P) + U(g, P) - L(g, P)\\ &< \frac{\varepsilon}{2} + \frac{\varepsilon}{2}\\ &= \varepsilon.\end{aligned}$$

Thus, using Theorem 3.34, we conclude that $f + g$ is Δ-integrable on $[a, b)$. Also,

$$\int_a^b (f+g)(t)\Delta t = U(f+g)$$

$$
\begin{aligned}
&\le U(f+g,P)\\
&\le U(f,P)+U(g,P)\\
&< \frac{\varepsilon}{2}+L(f,P)+\frac{\varepsilon}{2}+L(g,P)\\
&= \varepsilon+L(f,P)+L(g,P)\\
&\le \varepsilon+L(f)+L(g)\\
&= \varepsilon+\int_a^b f(t)\Delta t+\int_a^b g(t)\Delta t
\end{aligned}
$$

and

$$
\begin{aligned}
\int_a^b (f+g)(t)\Delta t &= L(f+g)\\
&\ge L(f+g,P)\\
&\ge L(f,P)+L(g,P)\\
&> U(f,P)-\frac{\varepsilon}{2}+U(g,P)-\frac{\varepsilon}{2}\\
&= U(f,P)+U(g,P)-\varepsilon\\
&\ge U(f)+U(g)-\varepsilon\\
&= \int_a^b f(t)\Delta t+\int_a^b g(t)\Delta t-\varepsilon,
\end{aligned}
$$

i.e.,

$$
\begin{aligned}
-\varepsilon+\int_a^b f(t)\Delta t+\int_a^b g(t)\Delta t &< \int_a^b (f+g)(t)\Delta t\\
&< \varepsilon+\int_a^b f(t)\Delta t+\int_a^b g(t)\Delta t.
\end{aligned}
$$

Because $\varepsilon > 0$ was arbitrarily chosen, we obtain that

$$\int_a^b (f+g)(t)\Delta t = \int_a^b f(t)\Delta t + \int_a^b g(t)\Delta t.$$

□

Theorem 3.53 *Let f and g be Δ-integrable on $[a, b)$. Then fg is Δ-integrable on $[a, b)$.*

Proof Let $\phi(x) = x^2$. Then $\phi : \mathbb{R} \to \mathbb{R}$ is continuous and satisfies the Lipschitz condition on any finite interval $[m, M]$. We observe that $(f(t))^2 = \phi(f(t))$. Thus, using Theorem 3.48, it follows that f^2 is Δ-integrable on $[a, b)$. Because f and g are Δ-integrable on $[a, b)$, using Theorem 3.52, we get that $f + g$, $-g$, $f - g$ are Δ-integrable functions on $[a, b)$. From here, $(f + g)^2$ and $(f - g)^2$ are Δ-integrable functions on $[a, b)$. Thus, using Theorem 3.52, we conclude that

$$\frac{1}{4}(f+g)^2, \quad -\frac{1}{4}(f-g)^2, \quad \text{and} \quad \frac{1}{4}(f+g)^2 - \frac{1}{4}(f-g)^2$$

are Δ-integrable on $[a, b)$, which completes the proof. □

Theorem 3.54 *Let f be a function defined on $[a, b)$ and let $c \in \mathbb{T}$ with $a < c < b$. If f is Δ-integrable from a to c and from c to b, then f is Δ-integrable on $[a, b)$ and*

$$\int_a^b f(t)\Delta t = \int_a^c f(t)\Delta t + \int_c^b f(t)\Delta t. \tag{3.10}$$

Proof Let $\varepsilon > 0$ be arbitrarily chosen. Since f is Δ-integrable on $[a, c)$, there exists a partition P_1 of $[a, c)$ so that

$$U_a^c(f, P_1) - L_a^c(f, P_1) < \frac{\varepsilon}{2}.$$

Because f is Δ-integrable on $[c, b)$, there exists a partition P_2 of $[c, b)$ such that

$$U_c^b(f, P_2) - L_c^b(f, P_2) < \frac{\varepsilon}{2}.$$

Let $P = P_1 \cup P_2$. Then

$$\begin{aligned} U_a^b(f, P) - L_a^b(f, P) &= U_a^c(f, P_1) + U_c^b(f, P_2) - L_a^c(f, P_1) - L_c^b(f, P_2) \\ &= U_a^c(f, P_1) - L_a^c(f, P_1) + U_c^b(f, P_2) - L_c^b(f, P_2) \\ &< \frac{\varepsilon}{2} + \frac{\varepsilon}{2} \\ &= \varepsilon. \end{aligned}$$

Consequently, f is Δ-integrable on $[a, b)$. Also,

$$\begin{aligned}\int_a^b f(t)\Delta t &\leq U_a^b(f, P)\\ &= U_a^c(f, P_1) + U_c^b(f, P_2)\\ &< L_a^c(f, P_1) + L_c^b(f, P_2) + \varepsilon\\ &\leq \int_a^c f(t)\Delta t + \int_c^b f(t)\Delta t + \varepsilon\end{aligned}$$

and

$$\begin{aligned}\int_a^b f(t)\Delta t &\geq L_a^b(f, P)\\ &= L_a^c(f, P_1) + L_c^b(f, P_2)\\ &> U_a^c(f, P_1) + U_c^b(f, P_2) - \varepsilon\\ &\geq \int_a^c f(t)\Delta t + \int_c^b f(t)\Delta t - \varepsilon,\end{aligned}$$

i.e.,

$$\begin{aligned}\int_a^c f(t)\Delta t + \int_c^b f(t)\Delta t - \varepsilon < \int_a^b f(t)\Delta t\\ < \int_a^c f(t)\Delta t + \int_c^b f(t)\Delta t + \varepsilon.\end{aligned}$$

Since $\varepsilon > 0$ was arbitrarily chosen, we obtain (3.10). □

Theorem 3.55 *If f and g are Δ-integrable on $[a, b)$ and $f(t) \leq g(t)$ for all $t \in [a, b)$, then*

$$\int_a^b f(t)\Delta t \leq \int_a^b g(t)\Delta t.$$

Proof Since f and g are Δ-integrable on $[a, b)$, we have that $h = g - f$ is Δ-integrable on $[a, b)$. Because $h(t) \geq 0$ for all $t \in [a, b)$, we conclude that $L(h, P) \geq 0$ for all partitions P of $[a, b)$. Therefore,

$$L(h) = \int_a^b h(t)\Delta t \geq 0,$$

i.e.,

$$\int_a^b (g - f)(t)\Delta t \geq 0.$$

Now, applying Theorem 3.52, we obtain

$$\int_a^b g(t)\Delta t - \int_a^b f(t)\Delta t \geq 0,$$

which completes the proof. □

Theorem 3.56 *If f is Δ-integrable on $[a, b)$, then $|f|$ is Δ-integrable on $[a, b)$ and*

$$\left|\int_a^b f(t)\Delta t\right| \leq \int_a^b |f(t)|\Delta t.$$

Proof Let $\phi(x) = |x|$. Then

$$\phi : \mathbb{R} \to \mathbb{R}$$

satisfies the Lipschitz condition on every finite interval $[m, M]$. Hence, using Theorem 3.48, we conclude that $|f|$ is Δ-integrable on $[a, b)$. Since

$$-|f(t)| \leq f(t) \leq |f(t)| \quad \text{for any} \quad t \in [a, b),$$

using Theorem 3.55, we obtain

$$-\int_a^b |f(t)|\Delta t \leq \int_a^b f(t)\Delta t \leq \int_a^b |f(t)|\Delta t,$$

which completes the proof. □

Theorem 3.57 *If f and g are Δ-integrable on $[a, b)$, then*

$$\left|\int_a^b f(t)g(t)\Delta t\right| \leq \int_a^b |f(t)g(t)|\Delta t \leq \left(\sup_{\tau\in[a,b)} |f(\tau)|\right)\int_a^b |g(t)|\Delta t.$$

Proof Since f and g are Δ-integrable on $[a, b)$, using Theorem 3.53, we have that $|fg|$ is Δ-integrable on $[a, b)$. By Theorem 3.56, we get

$$\left|\int_a^b f(t)g(t)\Delta t\right| \leq \int_a^b |f(t)g(t)|\Delta t.$$

Because

$$|f(t)g(t)| \le \left(\sup_{\tau\in[a,b)} |f(\tau)|\right)|g(t)| \quad \text{for any} \quad t \in [a,b),$$

using Theorem 3.55, we obtain

$$\begin{aligned}\int_a^b |f(t)g(t)|\Delta t &\le \int_a^b \left(\sup_{\tau\in[a,b)} |f(\tau)|\right)|g(t)|\Delta t\\ &= \left(\sup_{\tau\in[a,b)} |f(\tau)|\right)\int_a^b |g(t)|\Delta t,\end{aligned}$$

which completes the proof. □

Theorem 3.58 *Let $\{f_n\}_{n\in\mathbb{N}}$ be a sequence of Δ-integrable functions on $[a,b)$ and suppose that $f_n \to f$ uniformly on $[a,b)$ for a function f defined on $[a,b)$. Then f is Δ-integrable on $[a,b)$ and*

$$\int_a^b f(t)\Delta t = \lim_{n\to\infty} \int_a^b f_n(t)\Delta t.$$

Proof Let $\varepsilon > 0$ be arbitrarily chosen. Since $f_n \to f$ uniformly on $[a,b)$, there exists $n_0 \in \mathbb{N}$ so that

$$|f_n(t) - f(t)| < \varepsilon \quad \text{for all} \quad t \in [a,b) \quad \text{and} \quad n \ge n_0.$$

Because f_n, $n \ge n_0$, are Δ-integrable on $[a,b)$, there exists a partition P of $[a,b)$ so that

$$U(f_n, P) - L(f_n, P) < \varepsilon, \quad n \ge n_0.$$

Note that

$$U(f_n - f, P) < \varepsilon(b-a), \quad L(f_n - f, P) > -\varepsilon(b-a), \quad n \ge n_0.$$

Therefore,

$$\begin{aligned}U(f,P) - L(f,P) &= U(f - f_n + f_n, P) - L(f - f_n + f_n, P)\\ &\le U(f - f_n, P) + U(f_n, P) - L(f - f_n, P) - L(f_n, P)\\ &= U(f - f_n, P) - L(f - f_n, P) + U(f_n, P) - L(f_n, P)\\ &< \varepsilon(b-a) + \varepsilon(b-a) + \varepsilon, \quad n \ge n_0.\end{aligned}$$

Hence, using Theorem 3.34, it follows that f is Δ-integrable on $[a, b)$. From here, $f_n - f, |f_n - f|, n \geq n_0$, are Δ-integrable on $[a, b)$. Then

$$\begin{aligned}\left|\int_a^b (f_n(t) - f(t))\Delta t\right| &= \left|\int_a^b f_n(t)\Delta t - \int_a^b f(t)\Delta t\right| \\ &\leq \int_a^b |f_n(t) - f(t)|\Delta t \\ &< \varepsilon \int_a^b \Delta t \\ &= \varepsilon(b-a),\end{aligned}$$

which completes the proof. □

Theorem 3.59 *Suppose that* $\sum\limits_{k=1}^{\infty} g_k$ *is a series of* Δ*-integrable functions* g_k *on* $[a, b)$ *that converges uniformly to* g *on* $[a, b)$. *Then* g *is* Δ*-integrable and*

$$\int_a^b g(t)\Delta t = \sum_{k=1}^{\infty} \int_a^b g_k(t)\Delta t.$$

Proof Since $\sum\limits_{k=1}^{\infty} g_k$ is uniformly convergent to g on $[a, b)$, the sequence

$$\left\{S_n = \sum_{k=1}^{n} g_k\right\}_{k\in\mathbb{N}}$$

is uniformly convergent to g on $[a, b)$. Hence, using Theorem 3.58, it follows that g is Δ-integrable on $[a, b)$ and

$$\lim_{n\to\infty} \int_a^b S_n(t)\Delta t = \int_a^b g(t)\Delta t,$$

whereupon

$$\lim_{n\to\infty} \int_a^b \sum_{k=1}^{n} g_k(t)\Delta t = \int_a^b g(t)\Delta t$$

and

$$\lim_{n\to\infty} \sum_{k=1}^{n} \int_a^b g_k(t)\Delta t = \int_a^b g(t)\Delta t,$$

which completes the proof. □

Theorem 3.60 *Let f be Δ-integrable on $[a, b)$. If f has a Δ-antiderivative*

$$F : [a, b] \to \mathbb{R},$$

then

$$\int_a^b f(t)\Delta t = F(b) - F(a).$$

Proof Since f is a Δ-antiderivative of f on $[a, b]$, we have that

$$F^{\Delta}(t) = f(t) \quad \text{for all} \quad t \in [a, b].$$

Let $\varepsilon > 0$ be arbitrarily chosen. Because f is Δ-integrable on $[a, b)$, there exists a partition $P = \{a = t_0 < t_1 < \ldots < t_n = b\}$ of $[a, b)$ so that

$$U(f, P) - L(f, P) < \varepsilon. \tag{3.11}$$

For every $i \in \{1, \ldots, n\}$, there exist $\xi_i, \eta_i \in (t_{i-1}, t_i)$ so that

$$F^{\Delta}(\eta_i)(t_i - t_{i-1}) \le F(t_i) - F(t_{i-1}) \le F^{\Delta}(\xi_i)(t_i - t_{i-1}),$$

whereupon

$$f(\eta_i)(t_i - t_{i-1}) \le F(t_i) - F(t_{i-1}) \le f(\xi_i)(t_i - t_{i-1}).$$

Hence,

$$m_i(t_i - t_{i-1}) \le F(t_i) - F(t_{i-1}) \le M_i(t_i - t_{i-1}),$$

where

$$m_i = \inf_{t \in [t_{i-1}, t_i)} f(t), \quad M_i = \sup_{t \in [t_{i-1}, t_i)} f(t),$$

and

$$\sum_{i=1}^{n} m_i(t_i - t_{i-1}) \le \sum_{i=1}^{n} (F(t_i) - F(t_{i-1})) \le \sum_{i=1}^{n} M_i(t_i - t_{i-1}).$$

Therefore,

$$L(f, P) \le F(b) - F(a) \le U(f, P).$$

Since

$$L(f, P_1) \le \int_a^b f(t)\Delta t \le U(f, P_1)$$

for all partitions P_1 of $[a, b)$, (3.11) yields

$$\left| \int_a^b f(t)\Delta t - (F(b) - F(a)) \right| < \varepsilon,$$

which completes the proof. □

Items 5 and 6 of the following theorem are called *integration by parts* rules.

Theorem 3.61 *If $a, b, c \in \mathbb{T}$, $\alpha \in \mathbb{R}$, and $f, g \in \mathrm{C_{rd}}(\mathbb{T})$, then*

1. $\int_a^b (f+g)(t)\Delta t = \int_a^b f(t)\Delta t + \int_a^b g(t)\Delta t$,
2. $\int_a^b (\alpha f)(t)\Delta t = \alpha \int_a^b f(t)\Delta t$,
3. $\int_a^b f(t)\Delta t = -\int_b^a f(t)\Delta t$,
4. $\int_a^b f(t)\Delta t = \int_a^c f(t)\Delta t + \int_c^b f(t)\Delta t$,
5. $\int_a^b f(\sigma(t)) g^\Delta(t)\Delta t = (fg)(b) - (fg)(a) - \int_a^b f^\Delta(t) g(t)\Delta t$,
6. $\int_a^b f(t) g^\Delta(t)\Delta t = (fg)(b) - (fg)(a) - \int_a^b f^\Delta(t) g(\sigma(t))\Delta t$,
7. $\int_a^a f(t)\Delta t = 0$,
8. *if $|f(t)| \le g(t)$ on $[a, b]$, then*

$$\left| \int_a^b f(t)\Delta t \right| \le \int_a^b g(t)\Delta t,$$

9. *if $f(t) \ge 0$ for all $a \le t < b$, then $\int_a^b f(t)\Delta t \ge 0$.*

Proof Since $f, g \in \mathrm{C_{rd}}(\mathbb{T})$, they possess antiderivatives F and G. We have

$$F^\Delta(t) = f(t) \quad \text{and} \quad G^\Delta(t) = g(t) \quad \text{for all} \quad t \in \mathbb{T}^\kappa.$$

1. For all $t \in \mathbb{T}^\kappa$, we have

$$(F+G)^\Delta(t) = F^\Delta(t) + G^\Delta(t) = f(t) + g(t) = (f+g)(t).$$

Hence,

$$\begin{aligned}\int_a^b (f+g)(t)\Delta t &= (F+G)(b) - (F+G)(a)\\ &= F(b) - F(a) + G(b) - G(a)\\ &= \int_a^b f(t)\Delta t + \int_a^b g(t)\Delta t.\end{aligned}$$

2. Since

$$(\alpha F)^\Delta(t) = \alpha F^\Delta(t) = \alpha f(t) = (\alpha f)(t) \quad \text{for all} \quad t \in \mathbb{T}^\kappa,$$

we get

$$\int_a^b (\alpha f)(t)\Delta t = (\alpha F)(b) - (\alpha F)(a)$$

$$= \alpha(F(b) - F(a))$$

$$= \alpha \int_a^b f(t)\Delta t.$$

3. We have

$$\int_a^b f(t)\Delta t = F(b) - F(a)$$

$$= -(F(a) - F(b))$$

$$= -\int_b^a f(t)\Delta t.$$

4. We have

$$\int_a^b f(t)\Delta t = F(b) - F(a)$$

$$= F(c) - F(a) + F(b) - F(c)$$

$$= \int_a^c f(t)\Delta t + \int_c^b f(t)\Delta t.$$

5. For all $t \in \mathbb{T}^\kappa$, we have

$$(fg)^\Delta(t) = f^\Delta(t)g(t) + f(\sigma(t))g^\Delta(t),$$

i.e.,

$$f(\sigma(t))g^\Delta(t) = (fg)^\Delta(t) - f^\Delta(t)g(t).$$

Hence, by using items 1 and 2, we get

$$\int_a^b f(\sigma(t))g^\Delta(t)\Delta t = \int_a^b \left((fg)^\Delta(t) - f^\Delta(t)g(t)\right) \Delta t$$

$$= \int_a^b (fg)^\Delta(t)\Delta t - \int_a^b f^\Delta(t)g(t)\Delta t$$

$$= (fg)(b) - (fg)(a) - \int_a^b f^\Delta(t)g(t)\Delta t.$$

6. For all $t \in \mathbb{T}^\kappa$, we have

$$(fg)^\Delta(t) = f(t)g^\Delta(t) + f^\Delta(t)g(\sigma(t)),$$

i.e.,

$$f(t)g^\Delta(t) = (fg)^\Delta(t) - f^\Delta(t)g(\sigma(t)).$$

Hence, by using items 1 and 2, we find

$$\int_a^b f(t)g^\Delta(t)\Delta t = \int_a^b (fg)^\Delta(t)\Delta t - \int_a^b f^\Delta(t)g(\sigma(t))\Delta t$$

$$= (fg)(b) - (fg)(a) - \int_a^b f^\Delta(t)g(\sigma(t))\Delta t.$$

7. We have

$$\int_a^a f(t)\Delta t = F(a) - F(a) = 0.$$

8. We note that

$$|F^\Delta(t)| \le G^\Delta(t) \quad \text{on} \quad [a, b].$$

Thus, employing Theorem 3.13, we get

$$|F(b) - F(a)| \le G(b) - G(a),$$

i.e.,

$$\left|\int_a^b f(t)\Delta t\right| \le \int_a^b g(t)\Delta t.$$

9. This property follows directly from the property 8. □

Exercise 3.62 Let $a, b \in \mathbb{T}$ and $f \in C_{\mathrm{rd}}(\mathbb{T})$.

1. Let $\mathbb{T} = \mathbb{R}$. Prove

$$\int_a^b f(t)\Delta t = \int_a^b f(t)\mathrm{d}t,$$

where the integral on the right-hand side is the usual Riemann integral.

2. Prove that, if $[a, b]$ consists only of isolated points, then

$$\int_a^b f(t)\Delta t = \begin{cases} \sum\limits_{t\in[a,b)} \mu(t) f(t) & \text{if } a < b \\ 0 & \text{if } a = b \\ -\sum\limits_{t\in[b,a)} \mu(t) f(t) & \text{if } a > b. \end{cases}$$

Example 3.63 Let $\mathbb{T} = \mathbb{Z}$. We will compute

$$I = \int_{-2}^{3} (t^2 + t + 1)\Delta t.$$

1. First way (using Theorem 3.60). Let

$$f(t) = t^2 + t + 1 \quad \text{and} \quad F(t) = \frac{1}{3}t^3 + \frac{2}{3}t.$$

   Then

$$F^{\Delta}(t) = \frac{1}{3}\left((t+1)^2 + t(t+1) + t^2\right) + \frac{2}{3} = t^2 + t + 1 = f(t).$$

   Therefore,

$$I = \int_{-2}^{3} f(t)\Delta t = F(3) - F(2) = \left(\frac{27}{3} + 2\right) - \left(-\frac{8}{3} - \frac{4}{3}\right) = 11 + 4 = 15.$$

2. Second way (using Theorem 3.43 and item 4 of Theorem 3.61). Since all points of $\mathbb{T}$ are isolated and $\mu(t) = 1$, we have that

$$\begin{aligned} I &= \sum_{t=-2}^{2} (t^2 + t + 1) \\ &= (4 - 2 + 1) + (1 - 1 + 1) + 1 + (1 + 1 + 1) + (4 + 2 + 1) \\ &= 3 + 1 + 1 + 3 + 7 \\ &= 15. \end{aligned}$$

Example 3.64 Let $\mathbb{T} = 2^{\mathbb{N}_0}$. We will compute

$$I = \int_1^4 \left(\frac{\sin\frac{t}{2}\sin\frac{3t}{2}}{t} + t^2 \right)\Delta t.$$

1. First way (using Theorem 3.60). Note that $\sigma(t) = 2t$ and $\mu(t) = t$. Let

$$f(t) = \frac{\sin\frac{t}{2}\sin\frac{3t}{2}}{t} \quad \text{and} \quad F(t) = \frac{1}{7}t^3 - \frac{1}{2}\cos t.$$

Then

$$\begin{aligned} F^{\Delta}(t) &= \frac{1}{7}\left((2t)^2 + t(2t) + t^2\right) - \frac{1}{2}\frac{\cos(2t) - \cos t}{t} \\ &= t^2 + \frac{\sin\frac{t}{2}\sin\frac{3t}{2}}{t} \\ &= f(t). \end{aligned}$$

Therefore,

$$\begin{aligned} I &= \int_1^4 f(t)\Delta t \\ &= F(4) - F(1) \\ &= \frac{64}{7} - \frac{1}{2}\cos 4 - \frac{1}{7} + \frac{1}{2}\cos 1 \\ &= 9 - \frac{1}{2}(\cos 4 - \cos 1) \\ &= 9 + \sin\frac{3}{2}\sin\frac{5}{2}. \end{aligned}$$

2. Second way (using Theorem 3.43 and item 4 of Theorem 3.61). Since all points of $\mathbb{T}$ are isolated, we obtain

$$\begin{aligned} I &= \sum_{t\in\{1,2\}} \mu(t)\left(\frac{\sin\frac{t}{2}\sin\frac{3t}{2}}{t} + t^2\right) \\ &= \sin\frac{1}{2}\sin\frac{3}{2} + 1 + 2\left(\frac{\sin 1\sin 3}{2} + 4\right) \\ &= \sin\frac{1}{2}\sin\frac{3}{2} + \sin 1\sin 3 + 9 \\ &= \frac{1}{2}\left(\cos 1 - \cos 2 + \cos 2 - \cos 4\right) + 9 \end{aligned}$$

$$= \frac{1}{2}(\cos 1 - \cos 4) + 9$$

$$= -\sin\frac{-3}{2}\sin\frac{5}{2} + 9$$

$$= \sin\frac{3}{2}\sin\frac{5}{2} + 9.$$

Example 3.65 Let $\mathbb{T} = [-1, 0] \cup 3^{\mathbb{N}_0}$, where $[-1, 0]$ is the real-valued interval. Define

$$f(t) = \begin{cases} \frac{1}{(t+2)^3} - \frac{1}{8} & \text{for } t \in [-1, 0) \\ 0 & \text{for } t = 0 \\ t^2 - t & \text{for } t \in 3^{\mathbb{N}_0}. \end{cases}$$

We will compute

$$I = \int_{-1}^{3} f(t)\Delta t.$$

We have

$$\begin{aligned} I &= \int_{-1}^{0} f(t)\Delta t + \int_{0}^{1} f(t)\Delta t + \int_{1}^{3} f(t)\Delta t \\ &= \int_{-1}^{0}\left(\frac{1}{(t+2)^3} - \frac{1}{8}\right)dt + \int_{1}^{3}(t^2 - t)\Delta t \\ &= -\frac{1}{2(t+2)^2}\Big|_{t=-1}^{t=0} - \frac{1}{8} + 2t(t^2 - t)\Big|_{t=1} \\ &= -\frac{1}{4} + \frac{1}{2} \\ &= \frac{1}{4}. \end{aligned}$$

Exercise 3.66 Let $\mathbb{T} = 2\mathbb{Z}$. Compute

$$\int_{-2}^{2}(t^2 + t)\Delta t.$$

Solution 4.

Theorem 3.67 *Let f be a function which is Δ-integrable from a to b. Let*

$$F(t) = \int_a^t f(s)\Delta s, \quad t \in [a, b].$$

Then F is continuous on $[a, b]$. Further, let $t_0 \in [a, b)$ and suppose that f is continuous at t_0 if t_0 is right-dense. Then F is Δ-differentiable at t_0 and $F^\Delta(t_0) = f(t_0)$.

Proof Let $B > 0$ be such that

$$|f(t)| \leq B \quad \text{for all} \quad t \in [a, b].$$

Let $\varepsilon > 0$ be arbitrarily chosen. Take $t_1, t_2 \in [a, b]$ such that $|t_1 - t_2| < \frac{\varepsilon}{B}$. Then

$$\begin{aligned}
|F(t_1) - F(t_2)| &= \left| \int_a^{t_1} f(s)\Delta s - \int_a^{t_2} f(s)\Delta s \right| \\
&= \left| \int_a^{t_1} f(s)\Delta s + \int_{t_1}^{a} f(s)\Delta s - \int_{t_1}^{a} f(s)\Delta s - \int_a^{t_2} f(s)\Delta s \right| \\
&= \left| \int_{t_1}^{t_2} f(s)\Delta s \right| \\
&\leq \left| \int_{t_1}^{t_2} |f(s)|\Delta s \right| \\
&\leq B|t_2 - t_1| \\
&< \varepsilon.
\end{aligned}$$

Therefore, F is continuous on $[a, b]$.

1. Let t_0 be right-scattered. Since f is continuous on $[a, b]$, it is Δ-differentiable at t_0. Hence,

$$\begin{aligned}
F^\Delta(t_0) &= \frac{F(\sigma(t_0)) - F(t_0)}{\sigma(t_0) - t_0} \\
&= \frac{1}{\sigma(t_0) - t_0} \left[\int_a^{\sigma(t_0)} f(s)\Delta s - \int_a^{t_0} f(s)\Delta s \right] \\
&= \frac{1}{\sigma(t_0) - t_0} \int_{t_0}^{\sigma(t_0)} f(s)\Delta s \\
&= \frac{1}{\sigma(t_0) - t_0} (\sigma(t_0) - t_0) f(t_0) \\
&= f(t_0).
\end{aligned}$$

2. Let t_0 be right-dense and assume that f is continuous at t_0. Then

$$F^{\Delta}(t_0) = \lim_{t\to t_0} \frac{F(t)-F(t_0)}{t-t_0}$$

$$= \lim_{t\to t_0} \frac{1}{t-t_0}\left[\int_a^t f(s)\Delta s - \int_a^{t_0} f(s)\Delta s\right] \tag{3.12}$$

$$= \lim_{t\to t_0} \frac{1}{t-t_0}\int_{t_0}^t f(s)\Delta s.$$

Let $\varepsilon > 0$ be arbitrarily chosen. Since f is continuous at t_0, there exists $\delta > 0$ such that $s \in [a, b)$ and $|s - t_0| < \delta$ imply $|f(s) - f(t_0)| < \varepsilon$. Then

$$\begin{aligned}\left|\frac{1}{t-t_0}\int_{t_0}^t f(s)\Delta s - f(t_0)\right| &= \left|\frac{1}{t-t_0}\int_{t_0}^t (f(s)-f(t_0))\Delta s\right| \\ &\le \frac{1}{|t-t_0|}\left|\int_{t_0}^t |f(s)-f(t_0)\Delta s\right| \\ &< \frac{\varepsilon}{|t-t_0|}\left|\int_{t_0}^t \Delta s\right| \\ &= \varepsilon\end{aligned}$$

for all $t \in [a, b]$ such that

$$|t-t_0| < \delta$$

and

$$t \neq t_0.$$

Thus, using (3.12), we get

$$F^{\Delta}(t_0) = f(t_0),$$

which completes the proof. □

Exercise 3.68 Let f be a Δ-integrable function on $[a, b)$. Assume that f has a pre-antiderivative $F : [a, b] \to \mathbb{R}$. Prove

$$\int_a^b f(t)\Delta t = F(b) - F(a).$$

Theorem 3.69 (Mean Value Theorem) *Let f and g be defined on $[a, b]$. Suppose f and g are Δ-integrable on $[a, b)$ and*

$$m \leq f(t) \leq M, \quad g(t) \geq 0$$

for all $t \in [a, b]$. *Then there exists* $\lambda \in [m, M]$ *such that*

$$\int_a^b f(t)g(t)\Delta t = \lambda \int_a^b g(t)\Delta t.$$

Proof Since $g(t) \geq 0$ and $m \leq f(t) \leq M$ for all $t \in [a, b]$, we have

$$mg(t) \leq f(t)g(t) \leq Mg(t) \quad \text{for all} \quad t \in [a, b].$$

Hence, using Theorem 3.55, we obtain

$$\int_a^b mg(t)\Delta t \leq \int_a^b f(t)g(t)\Delta t \leq \int_a^b Mg(t)\Delta t,$$

i.e.,

$$m \int_a^b g(t)\Delta t \leq \int_a^b f(t)g(t)\Delta t \leq M \int_a^b g(t)\Delta t. \tag{3.13}$$

1. If $\int_a^b g(t)\Delta t = 0$, then the assertion is valid.
2. If $\int_a^b g(t)\Delta t \neq 0$, then $\int_a^b g(t)\Delta t > 0$, and, using (3.13), we obtain

$$m \leq \frac{\int_a^b f(t)g(t)\Delta t}{\int_a^b g(t)\Delta t} \leq M.$$

Then there exists $\lambda \in [m, M]$ such that

$$\lambda = \frac{\int_a^b f(t)g(t)\Delta t}{\int_a^b g(t)\Delta t},$$

which completes the proof. □

Lemma 3.70 (Abel's Lemma) *Suppose that the numbers* p_i, $1 \leq i \leq n$, *satisfy the inequalities*

$$p_1 \geq p_2 \geq \ldots \geq p_n \geq 0$$

and the numbers $S_k = \sum_{i=1}^{k} q_i$, $1 \leq k \leq n$, *satisfy the inequalities*

$$m \leq S_k \leq M \quad \textit{for all} \quad k \in \{1, \ldots, n\},$$

where q_i, $i \in \{1, \ldots, n\}$, m *and* M *are some numbers. Then*

$$mp_1 \le \sum_{i=1}^{n} p_i q_i \le M p_1.$$

Proof Let $l \in \{1, \dots, n\}$ and $j_r, r \in \{1, \dots, n\}$, be chosen so that

$$j_1, \dots, j_l \in \{1, \dots, n\} \quad \text{and} \quad q_{j_i} \ge 0 \quad \text{for} \quad i \in \{1, \dots, l\}$$

and

$$j_{l+1}, \dots, j_n \in \{1, \dots, n\} \quad \text{and} \quad q_{j_i} \le 0 \quad \text{for} \quad i \in \{l+1, \dots, n\}.$$

Then

$$\sum_{i=1}^{n} p_i q_i = \sum_{i=1}^{l} p_{j_i} q_{j_i} + \sum_{i=l+1}^{n} p_{j_i} q_{j_i}.$$

We have

$$\sum_{i=1}^{l} p_{j_i} q_{j_i} \le p_1 \sum_{i=1}^{l} q_{j_i} \quad \text{and} \quad \sum_{i=l+1}^{n} p_{j_i} q_{j_i} \le p_n \sum_{i=l+1}^{n} q_{j_i}.$$

Then

$$\begin{aligned}
\sum_{i=1}^{n} p_i q_i &\le p_1 \sum_{i=1}^{l} q_{j_i} + p_n \sum_{i=l+1}^{n} q_{j_i} \\
&= (p_1 - p_n) \sum_{i=1}^{l} q_{j_i} + p_n \sum_{i=1}^{n} q_i \\
&\le M(p_1 - p_n) + M p_n \\
&= M p_1.
\end{aligned}$$

Also,

$$\sum_{i=1}^{l} p_{j_i} q_{j_i} \ge p_n \sum_{i=1}^{l} q_{j_i} \quad \text{and} \quad \sum_{i=l+1}^{n} p_{j_i} q_{j_i} \ge p_1 \sum_{i=l+1}^{n} q_{j_i}.$$

Hence,

$$\begin{aligned}
\sum_{i=1}^{n} p_i q_i &\ge p_n \sum_{i=1}^{l} q_{j_i} + p_1 \sum_{i=l+1}^{n} q_{j_i} \\
&= p_n \sum_{i=1}^{n} q_i + (p_1 - p_n) \sum_{i=l+1}^{n} q_{j_i}
\end{aligned}$$

$$\geq mp_n + (p_1 - p_n)m$$

$$= mp_1,$$

which completes the proof. □

Theorem 3.71 (Second Mean Value Theorem) *Let f be a bounded function that is integrable on $[a, b]$. Let m_F and M_F be the infimum and supremum, respectively, of the function $F(t) = \int_a^t f(s)\Delta s$ on $[a, b]$. Then*

1. *if a function g is nonincreasing with $g(t) \geq 0$ on $[a, b]$, then there exists a number Λ such that*
$$m_F \leq \Lambda \leq M_F \quad \text{and} \quad \int_a^b f(t)g(t)\Delta t = g(a)\Lambda,$$
2. *if g is any monotone function on $[a, b]$, then there exists some number Λ such that $m_F \leq \Lambda \leq M_F$ and*
$$\int_a^b f(t)g(t)\Delta t = (g(a) - g(b))\Lambda + g(b)\int_a^b f(t)\Delta t.$$

Proof 1. Let $\varepsilon > 0$ be arbitrarily chosen. Note that the functions f and fg are integrable on $[a, b]$. Then there exists a partition

$$P = \{a = t_0 < t_1 < \ldots < t_n = b\} \in \mathscr{P}([a, b])$$

such that

$$\sum_{i=1}^{n} (M_i - m_i)(t_i - t_{i-1}) < \varepsilon$$

and

$$\left| \sum_{i=1}^{n} f(t_{i-1})g(t_{i-1})(t_i - t_{i-1}) - \int_a^b f(t)g(t)\Delta t \right| < \varepsilon, \tag{3.14}$$

where m_i and M_i are the infimum and supremum, respectively, of f on $[t_{i-1}, t_i)$. We have

$$\sum_{i=1}^{n} m_i g(t_{i-1})(t_i - t_{i-1}) \leq \sum_{i=1}^{n} f(t_{i-1})g(t_{i-1})(t_i - t_{i-1})$$

$$\leq \sum_{i=1}^{n} M_i g(t_{i-1})(t_i - t_{i-1}).$$

From Theorem 3.69, it follows that there exist numbers Λ_i, $1 \leq i \leq n$, such that

$$m_i \le \Lambda_i \le M_i \quad \text{and} \quad \int_{t_{i-1}}^{t_i} f(t)\Delta t = \Lambda_i(t_i - t_{i-1}).$$

Consider the numbers

$$S_k = \sum_{i=1}^{k} \Lambda_i(t_i - t_{i-1}) = \int_a^{t_k} f(t)\Delta t, \quad 1 \le k \le n.$$

We have that

$$m_F \le S_k \le M_F, \quad 1 \le k \le n.$$

Let

$$p_i = g(t_{i-1}) \quad \text{and} \quad q_i = \Lambda_i(t_i - t_{i-1}), \quad 1 \le i \le n.$$

Since g is nonincreasing and $g(t) \ge 0$, we have that

$$p_1 \ge p_2 \ge \ldots \ge p_n \ge 0.$$

Therefore, using Lemma 3.70, we get

$$m_F g(a) \le \sum_{i=1}^{n} g(t_{i-1})\Lambda_i(t_i - t_{i-1}) \le M_F g(a). \tag{3.15}$$

On the other hand, we have

$$\begin{aligned}\sum_{i=1}^{n} m_i g(t_{i-1})(t_i - t_{i-1}) &\le \sum_{i=1}^{n} g(t_{i-1})\Lambda_i(t_i - t_{i-1}) \\ &\le \sum_{i=1}^{n} M_i g(t_{i-1})(t_i - t_{i-1}).\end{aligned}$$

Therefore,

$$\begin{aligned}\left|\sum_{i=1}^{n} g(t_{i-1})(f(t_{i-1}) - \Lambda_i)(t_i - t_{i-1})\right| &\le \sum_{i=1}^{n}(M_i - m_i)g(t_{i-1})(t_i - t_{i-1}) \\ &\le g(a)\sum_{i=1}^{n}(M_i - m_i)(t_i - t_{i-1}) \\ &\le g(a)\varepsilon.\end{aligned}$$

From here and from (3.14), we get

$$\begin{aligned}
&\left|\int_a^b f(t)g(t)\Delta t - \sum_{i=1}^n g(t_{i-1})\Lambda_i(t_i - t_{i-1})\right|\\
&= \Bigg|\int_a^b f(t)g(t)\Delta t - \sum_{i=1}^n f(t_{i-1})g(t_{i-1})(t_i - t_{i-1})\\
&\quad + \sum_{i=1}^n g(t_{i-1})(f(t_{i-1}) - \Lambda_i)(t_i - t_{i-1})\Bigg|\\
&\le \left|\int_a^b f(t)g(t)\Delta t - \sum_{i=1}^n f(t_{i-1})g(t_{i-1})(t_i - t_{i-1})\right|\\
&\quad + \left|\sum_{i=1}^n g(t_{i-1})(f(t_{i-1}) - \Lambda_i)(t_i - t_{i-1})\right|\\
&< \varepsilon + g(a)\varepsilon,
\end{aligned}$$

whereupon

$$\begin{aligned}
-\varepsilon - g(a)\varepsilon + \sum_{i=1}^n g(t_{i-1})\Lambda_i(t_i - t_{i-1}) &< \int_a^b f(t)g(t)\Delta t\\
&< \varepsilon + g(a)\varepsilon + \sum_{i=1}^n g(t_{i-1})\Lambda_i(t_i - t_{i-1}).
\end{aligned}$$

Thus, using (3.15), we get

$$-\varepsilon - g(a)\varepsilon + m_F g(a) < \int_a^b f(t)g(t)\Delta t < \varepsilon + \varepsilon g(a) + M_F g(a).$$

Since $\varepsilon > 0$ was arbitrarily chosen, we obtain

$$m_F g(a) \le \int_a^b f(t)g(t)\Delta t \le M_F g(a).$$

If $g(a) = 0$, then we get that $\int_a^b f(t)g(t)\Delta t = 0$, which completes the proof. Suppose $g(a) > 0$. Then

$$m_F \leq \frac{\int_a^b f(t)g(t)\Delta t}{g(a)} \leq M_F.$$

From here, there exists $\Lambda \in [m_F, M_F]$ such that

$$\Lambda = \frac{\int_a^b f(t)g(t)\Delta t}{g(a)},$$

which completes the proof.

2. Let g be an arbitrary nonincreasing function on $[a, b]$ and define $h(t) = g(t) - g(b)$, $t \in [a, b]$. We have that $h(t) \geq 0$ and h is nonincreasing on $[a, b]$. Then there exists $\Lambda \in [m_F, M_F]$ such that

$$\int_a^b f(t)h(t)\Delta t = h(a)\Lambda,$$

whereupon

$$\int_a^b f(t)(g(t) - g(b))\Delta t = (g(a) - g(b))\Lambda,$$

i.e.,

$$\int_a^b f(t)g(t)\Delta t = (g(a) - g(b))\Lambda + g(b)\int_a^b f(t)\Delta t.$$

The proof is complete. □

As above, one can prove the following theorem.

Theorem 3.72 (Second Mean Value Theorem) *Let f be a bounded function that is integrable on $[a, b]$. Let m_F and M_F be the infimum and supremum, respectively, of the function $F(t) = \int_t^b f(s)\Delta s$ on $[a, b]$. Then*

1. *if a function g is nondecreasing with $g(t) \geq 0$ on $[a, b]$, then there exists a number Λ such that*

$$m_F \leq \Lambda \leq M_F \quad \text{and} \quad \int_a^b f(t)g(t)\Delta t = g(b)\Lambda,$$

2. *if g is any monotone function on $[a, b]$, then there exists some number Λ such that $m_F \leq \Lambda \leq M_F$ and*

$$\int_a^b f(t)g(t)\Delta t = (g(b) - g(a))\Lambda + g(a)\int_a^b f(t)\Delta t.$$

3.5 Some Elementary Functions

3.5.1 Hilger's Complex Plane

Definition 3.73 Let $h > 0$.

1. The *Hilger complex numbers* are defined by

$$\mathbb{C}_h = \left\{ z \in \mathbb{C} : \, z \neq -\frac{1}{h} \right\}.$$

2. The *Hilger real axis* is defined as

$$\mathbb{R}_h = \left\{ z \in \mathbb{C} : \, z > -\frac{1}{h} \right\}.$$

3. The *Hilger alternative axis* is defined as

$$A_h = \left\{ z \in \mathbb{C} : \, z < -\frac{1}{h} \right\}.$$

4. The *Hilger imaginary circle* is defined by

$$\mathbb{I}_h = \left\{ z \in \mathbb{C} : \, \left| z + \frac{1}{h} \right| = \frac{1}{h} \right\}.$$

For $h = 0$, we set

$$\mathbb{C}_0 = \mathbb{C}, \quad \mathbb{R}_0 = \mathbb{R}, \quad \mathbb{A}_0 = \emptyset, \quad \mathbb{I}_0 = i\mathbb{R}.$$

Definition 3.74 Let $h > 0$ and $z \in \mathbb{C}_h$. We define the *Hilger real part* of z by

$$\mathrm{Re}_h(z) = \frac{|zh + 1| - 1}{h}$$

and the *Hilger imaginary part* of z by

$$\mathrm{Im}_h(z) = \frac{\mathrm{Arg}(zh + 1)}{h},$$

where $\mathrm{Arg}(z)$ denotes the principal argument of z, i.e.,

$$-\pi < \mathrm{Arg}(z) \leq \pi.$$

We note that

$$-\frac{1}{h} < \mathrm{Re}_h(z) < \infty \quad \text{and} \quad -\frac{\pi}{h} < \mathrm{Im}_h(z) < \frac{\pi}{h}.$$

In particular, $\mathrm{Re}_h(z) \in \mathbb{R}_h$.

Definition 3.75 Let $-\frac{\pi}{h} < w \le \frac{\pi}{h}$. We define the *Hilger purely imaginary number* $\mathring{\iota}$ by

$$\mathring{\iota} w = \frac{e^{iwh} - 1}{h}.$$

Theorem 3.76 *Let $z \in \mathbb{C}_h$. Then $\mathring{\iota}\,\mathrm{Im}_h(z) \in \mathbb{I}_h$.*

Proof We have

$$\mathring{\iota}\,\mathrm{Im}_h(z) = \frac{e^{iw\,\mathrm{Im}_h(z)} - 1}{h}$$

and

$$\begin{aligned} \left|\mathring{\iota}\,\mathrm{Im}_h(z) + \frac{1}{h}\right| &= \left|\frac{e^{iw\,\mathrm{Im}_h(z)} - 1}{h} + \frac{1}{h}\right| \\ &= \frac{\left|e^{iw\,\mathrm{Im}_h(z)}\right|}{h} \\ &= \frac{1}{h}, \end{aligned}$$

completing the proof. □

Theorem 3.77 *We have*

$$\lim_{h\to 0}[\mathrm{Re}_h(z) + \mathring{\iota}\,\mathrm{Im}_h(z)] = \mathrm{Re}(z) + i\,\mathrm{Im}(z).$$

Proof We have

$$z = \mathrm{Re}(z) + \mathrm{Im}(z),$$

$$\begin{aligned} zh + 1 &= (\mathrm{Re}(z) + i\,\mathrm{Im}(z))h + 1 \\ &= h\,\mathrm{Re}(z) + 1 + ih\,\mathrm{Im}(z), \end{aligned}$$

$$\mathrm{Arg}(zh+1) = \arcsin\frac{h\,\mathrm{Im}(z)}{\sqrt{(h\,\mathrm{Re}(z)+1)^2 + h^2\,\mathrm{Im}^2(z)}},$$

$$\mathrm{Im}_h(z) = \frac{\mathrm{Arg}(zh+1)}{h}$$

$$
\begin{aligned}
&= \frac{1}{h} \arcsin \frac{h \operatorname{Im}(z)}{\sqrt{(h \operatorname{Re}(z)+1)^2 + h^2 \operatorname{Im}^2(z)}}, \\
|zh+1| &= \sqrt{(h \operatorname{Re}(z)+1)^2 + h^2 \operatorname{Im}^2(z)}, \\
\operatorname{Re}_h(z) &= \frac{|zh+1|-1}{h} \\
&= \frac{\sqrt{(h \operatorname{Re}(z)+1)^2 + h^2 \operatorname{Im}^2(z)} - 1}{h}.
\end{aligned}
$$

Hence,

$$
\begin{aligned}
\lim_{h\to 0} \operatorname{Re}_h(z) &= \lim_{h\to 0} \frac{\sqrt{(h \operatorname{Re}(z)+1)^2 + h^2 \operatorname{Im}^2(z)} - 1}{h} \\
&= \lim_{h\to 0} \frac{(h \operatorname{Re}(z)+1) \operatorname{Re}(z) + h \operatorname{Im}^2(z)}{\sqrt{(h \operatorname{Re}(z)+1)^2 + h^2 \operatorname{Im}^2(z)}} \\
&= \operatorname{Re}(z)
\end{aligned}
$$

and

$$
\lim_{h\to 0} \operatorname{Im}_h(z) = \lim_{h\to 0} \frac{1}{h} \arcsin \frac{h \operatorname{Im}(z)}{\sqrt{(h \operatorname{Re}(z)+1)^2 + h^2 \operatorname{Im}^2(z)}}
$$

$$
= \lim_{h\to 0} \frac{1}{\sqrt{1 - \frac{h^2 \operatorname{Im}^2(z)}{(h \operatorname{Re}(z)+1)^2 + h^2 \operatorname{Im}^2(z)}}}
$$

$$
\times \frac{\operatorname{Im}(z)\sqrt{(h \operatorname{Re}(z)+1)^2 + h^2 \operatorname{Im}^2(z)} - h \operatorname{Im}(z) \frac{(h \operatorname{Re}(z)+1) \operatorname{Re}(z) + h \operatorname{Im}^2(z)}{\sqrt{(h \operatorname{Re}(z)+1)^2 + h^2 \operatorname{Im}^2(z)}}}{(h \operatorname{Re}(z)+1)^2 + h^2 \operatorname{Im}^2(z)}
$$

$$
= \operatorname{Im}(z),
$$

which completes the proof. □

Theorem 3.78 *Let* $-\frac{\pi}{h} < w \le \frac{\pi}{h}$. *Then*

$$
|\mathring{\iota} w|^2 = \frac{4}{h^2} \sin^2 \frac{wh}{2}.
$$

Proof We have

$$\mathring{\iota} w = \frac{e^{iwh} - 1}{h} = \frac{\cos(wh) - 1 + i\sin(wh)}{h}.$$

Hence,

$$\begin{aligned}
|\mathring{\iota} w|^2 &= \frac{(\cos(wh) - 1)^2}{h^2} + \frac{\sin^2(wh)}{h^2} \\
&= \frac{\cos^2(wh) - 2\cos(wh) + 1 + \sin^2(wh)}{h^2} \\
&= \frac{2(1 - \cos(wh))}{h^2} \\
&= \frac{4}{h^2}\sin^2\frac{wh}{2},
\end{aligned}$$

completing the proof. □

Definition 3.79 The *circle plus addition* $\oplus$ on $\mathbb{C}_h$ is defined by

$$z \oplus w = z + w + zwh.$$

Theorem 3.80 $(\mathbb{C}_h, \oplus)$ *is an Abelian group.*

Proof Let $z, w \in \mathbb{C}_h$. Then $z, w \in \mathbb{C}$ and $z, w \neq -\frac{1}{h}$. Therefore, $z \oplus w \in \mathbb{C}$. Since

$$\begin{aligned}
h(z \oplus w) + 1 &= h(z + w + zwh) + 1 \\
&= 1 + hz + hw + zwh^2 \\
&= 1 + hz + hw(1 + hz) \\
&= (1 + hw)(1 + hz) \\
&\neq 0,
\end{aligned}$$

we conclude that $z \oplus w \in \mathbb{C}_h$. Also,

$$0 \oplus z = z \oplus 0 = z,$$

i.e., 0 is the additive identity for $\oplus$. For $z \in \mathbb{C}_h$, we have

$$z \oplus \left(-\frac{z}{1+zh}\right) = z - \frac{z}{1+zh} - z\frac{z}{1+zh}h$$

$$= \frac{z^2h}{1+zh} - \frac{z^2h}{1+zh}$$

$$= 0,$$

i.e., the additive inverse of z under the addition $\oplus$ is $-\frac{z}{1+zh}$. We note that

$$-\frac{z}{1+zh} \in \mathbb{C}$$

and

$$1 - \frac{zh}{1+zh} = \frac{1}{1+zh} \neq 0,$$

i.e., $-\frac{z}{1+zh} \neq -\frac{1}{h}$. Therefore, $-\frac{z}{1+zh} \in \mathbb{C}_h$. For $z, w, v \in \mathbb{C}_h$, we have

$$(z \oplus w) \oplus v = (z + w + zwh) \oplus v$$

$$= z + w + zwh + v + (z + w + zwh)vh$$

$$= z + w + zwh + v + zvh + wvh + zwvh^2$$

and

$$z \oplus (w \oplus v) = z + (w \oplus v) + z(w \oplus v)h$$

$$= z + w + v + wvh + z(w + v + wvh)h$$

$$= z + w + v + wvh + zwh + zvh + zwvh^2.$$

Consequently,

$$z \oplus (w \oplus v) = (z \oplus w) \oplus v,$$

i.e., the associative law holds in $(\mathbb{C}_h, \oplus)$. For $z, w \in \mathbb{C}_h$, we have

$$z \oplus w = z + w + zwh$$

$$= w + z + wzh$$

$$= w \oplus z,$$

which completes the proof. □

Example 3.81 Let $z \in \mathbb{C}_h$ and $w \in \mathbb{C}$ be such that $z + w \in \mathbb{C}_h$. We will simplify the expression

$$A = z \oplus \frac{w}{1 + hz}.$$

We have

$$\begin{aligned} A &= z + \frac{w}{1+hz} + \frac{zw}{1+hz}h \\ &= z + \frac{(1+hz)w}{1+hz} \\ &= z + w. \end{aligned}$$

Theorem 3.82 *For $z \in \mathbb{C}_h$, we have*

$$z = \mathrm{Re}_h(z) \oplus \mathring{\iota}\, \mathrm{Im}_h(z).$$

Proof We have

$$\begin{aligned} \mathrm{Re}_h(z) \oplus \mathring{\iota}\, \mathrm{Im}_h(z) &= \frac{|zh+1|-1}{h} \oplus \mathring{\iota} \frac{\mathrm{Arg}(zh+1)}{h} \\ &= \frac{|zh+1|-1}{h} \oplus \frac{e^{i\,\mathrm{Arg}(zh+1)}-1}{h} \\ &= \frac{|zh+1|-1}{h} + \frac{e^{i\,\mathrm{Arg}(zh+1)}-1}{h} \\ &\quad + \frac{|zh+1|-1}{h}\frac{e^{i\,\mathrm{Arg}(zh+1)}-1}{h}h \\ &= \frac{1}{h}\Big(|zh+1| - 1 + e^{i\,\mathrm{Arg}(zh+1)} - 1 \\ &\quad + |zh+1|e^{i\,\mathrm{Arg}(zh+1)} \\ &\quad - |zh+1| - e^{i\,\mathrm{Arg}(zh+1)} + 1\Big) \\ &= \frac{1}{h}\left(|zh+1|e^{i\,\mathrm{Arg}(zh+1)} - 1\right) \\ &= \frac{1}{h}(zh + 1 - 1) \end{aligned}$$

$$= z,$$

completing the proof. □

Definition 3.83 Let $n \in \mathbb{N}$ and $z \in \mathbb{C}_h$. We define the *circle dot multiplication* $\odot$ by

$$n \odot z = z \oplus z \oplus \cdots \oplus z.$$

Theorem 3.84 *Let $n \in \mathbb{N}$ and $z \in \mathbb{C}_h$. Then*

$$n \odot z = \frac{(zh+1)^n - 1}{h}. \tag{3.16}$$

Proof 1. Let $n = 2$. Then

$$\begin{aligned}
2 \odot z &= z \oplus z \\
&= z + z + z^2 h \\
&= 2z + zh \\
&= \frac{1}{h}(z^2h^2 + 2zh) \\
&= \frac{1}{h}(z^2h^2 + 2zh + 1 - 1) \\
&= \frac{(zh+1)^2 - 1}{h}.
\end{aligned}$$

2. Assume

$$n \odot z = \frac{(zh+1)^n - 1}{h}$$

for some $n \in \mathbb{N}$.

3. We will prove that

$$(n+1) \odot z = \frac{(zh+1)^{n+1}-1}{h}.$$

Indeed,

$$\begin{aligned}
(n+1) \odot z &= (n \odot z) \oplus z \\
&= \frac{(zh+1)^n - 1}{h} \oplus z \\
&= \frac{(zh+1)^n - 1}{h} + z + \frac{(zh+1)^n - 1}{h} zh \\
&= \frac{(zh+1)^n - 1 + zh + zh(zh+1)^n - zh}{h} \\
&= \frac{(zh+1)^{n+1} - 1}{h}.
\end{aligned}$$

Hence, we conclude that (3.16) holds for all $n \in \mathbb{N}$. □

Definition 3.85 Let $z \in \mathbb{C}_h$. We define the *circle minus* $\ominus$ of z as

$$\ominus z = \frac{-z}{1+zh}.$$

Theorem 3.86 *Let $z \in \mathbb{C}_h$. Then $\ominus z$ is the additive inverse of z under the operation $\oplus$, i.e.,*

$$\ominus(\ominus z) = z.$$

Proof We have

$$\begin{aligned}
\ominus(\ominus z) &= -\frac{\ominus z}{1 + (\ominus z)h} \\
&= -\frac{\frac{-z}{1+zh}}{1 + \frac{-z}{1+zh} h} \\
&= \frac{\frac{z}{1+zh}}{\frac{1+zh-zh}{1+zh}} \\
&= z,
\end{aligned}$$

completing the proof. □

Definition 3.87 Let $z, w \in \mathbb{C}_h$. We define the *circle minus subtraction* by

$$z \ominus w = z \oplus (\ominus w).$$

Remark 3.88 For $z, w \in \mathbb{C}_h$, we have

$$\begin{aligned} z \ominus w &= z \oplus (\ominus w) \\ &= z + (\ominus w) + z(\ominus w)h \\ &= z - \frac{w}{1+wh} - \frac{zwh}{1+wh} \\ &= \frac{z + zwh - w - zwh}{1+wh} \\ &= \frac{z-w}{1+wh}, \end{aligned}$$

i.e.,

$$z \ominus w = \frac{z-w}{1+wh}. \tag{3.17}$$

Theorem 3.89 *Let $z \in \mathbb{C}_h$. Then $\overline{z} = \ominus z$ iff $z \in \mathbb{I}_h$.*

Proof By (3.17), we have

$$\overline{z} = \ominus z = -\frac{z}{1+zh}$$

iff

$$\overline{z} + \overline{z}zh = -z$$

iff

$$2\,\mathrm{Re}(z) + |z|^2 h = 0.$$

Also, $z \in \mathbb{I}_h$ iff

$$\left| z + \frac{1}{h} \right| = \frac{1}{h}$$

iff

$$\begin{aligned} \frac{1}{h^2} &= \left| z + \frac{1}{h} \right|^2 \\ &= \left(\mathrm{Re}(z) + \frac{1}{h} \right)^2 + \mathrm{Im}^2(z) \end{aligned}$$

$$= \mathrm{Re}^2(z) + \frac{2}{h}\mathrm{Re}(z) + \frac{1}{h^2} + \mathrm{Im}^2(z)$$

iff

$$|z|^2 + \frac{2}{h}\mathrm{Re}(z) = 0$$

iff

$$2\,\mathrm{Re}(z) + |z|^2 h = 0,$$

which completes the proof. □

Theorem 3.90 *Let* $-\frac{\pi}{h} < w \leq \frac{\pi}{h}$*. Then*

$$\ominus(\overset{\circ}{\iota}w) = \overline{\overset{\circ}{\iota}w}$$

Proof We have

$$\begin{aligned}\ominus(\overset{\circ}{\iota}w) &= -\frac{\overset{\circ}{\iota}w}{1+\overset{\circ}{\iota}wh}\\ &= -\frac{\frac{e^{iwh}-1}{h}}{1+\frac{e^{iwh}-1}{h}h}\\ &= -\frac{e^{iwh}-1}{he^{iwh}}\\ &= \frac{e^{-iwh}-1}{h}\\ &= \overline{\overset{\circ}{\iota}w},\end{aligned}$$

completing the proof. □

Definition 3.91 Let $z \in \mathbb{C}_h$. The *circle square*② of z is defined by

$$z^{②} = (-z)(\ominus z).$$

Remark 3.92 We have

$$z^{②} = -z\frac{-z}{1+zh} = \frac{z^2}{1+zh}.$$

Theorem 3.93 *For* $z \in \mathbb{C}_h$*, we have*

$$(\ominus z)^{②} = z^{②}.$$

Proof We have

$$
\begin{aligned}
(\ominus z)^{⃝2} &= -(\ominus z)(\ominus(\ominus z)) \\
&= \frac{z}{1+zh} z \\
&= \frac{z^2}{1+zh} \\
&= z^{⃝2},
\end{aligned}
$$

completing the proof. □

Theorem 3.94 *For $z \in \mathbb{C}_h$, we have*

$$1 + zh = \frac{z^2}{z^{⃝2}}.$$

Proof We have

$$\frac{z^2}{z^{⃝2}} = \frac{z^2}{\frac{z^2}{1+zh}} = 1 + zh,$$

completing the proof. □

Theorem 3.95 *For $z \in \mathbb{C}_h$, we have*

$$z + (\ominus z) = z^{⃝2} h.$$

Proof We have

$$z^{⃝2} h = \frac{z^2}{1+zh} h$$

and

$$z + (\ominus z) = z - \frac{z}{1+zh} = \frac{z^2 h}{1+zh},$$

which completes the proof. □

Theorem 3.96 *For $z \in \mathbb{C}_h$, we have*

$$z \oplus z^{⃝2} = z + z^2.$$

Proof We have

$$z \oplus z^{⃝2} = z + z^{⃝2} + z z^{⃝2} h$$

$$
\begin{aligned}
&= z + \frac{z^2}{1+zh} + \frac{z^3 h}{1+zh} \\
&= z + \frac{z^2(1+zh)}{1+zh} \\
&= z + z^2,
\end{aligned}
$$

completing the proof. □

Theorem 3.97 *Let* $-\frac{\pi}{h} < w \leq \frac{\pi}{h}$. *Then*

$$
-(\mathring{\iota} w)^{②} = \frac{4}{h^2} \sin^2\left(\frac{wh}{2}\right).
$$

Proof We have

$$
\begin{aligned}
-(\mathring{\iota} w)^{②} &= -(\mathring{\iota} w)(\ominus \mathring{\iota} w) \\
&= (\mathring{\iota} w)\overline{\mathring{\iota} w} \\
&= |\mathring{\iota} w|^2 \\
&= \frac{4}{h^2} \sin^2\left(\frac{wh}{2}\right),
\end{aligned}
$$

completing the proof. □

Exercise 3.98 Let $z \in \mathbb{C}_h$. Prove that

$$
z^{②} \in \mathbb{R} \quad \text{iff} \quad z \in \mathbb{R}_h \cup \mathbb{A}_h \cup \mathbb{I}_h.
$$

3.5.2 Exponential Function

For $h > 0$, we define the strip

$$
\mathbb{Z}_h = \left\{ z \in \mathbb{C} : -\frac{\pi}{h} < \operatorname{Im}(z) \leq \frac{\pi}{h} \right\}.
$$

For $h = 0$, we set $\mathbb{Z}_0 = \mathbb{C}$.

Definition 3.99 For $h > 0$, we define the *cylinder transformation* $\xi_h : \mathbb{C}_h \to \mathbb{Z}_h$ by

$$\xi_h(z) = \frac{1}{h} \operatorname{Log}(1 + zh),$$

where Log is the principal logarithm function. Moreover, we define $\xi_0(z) = z$ for all $z \in \mathbb{C}$.

Remark 3.100 We note that

$$\xi_h^{-1}(z) = \frac{e^{zh} - 1}{h}$$

for $z \in \mathbb{Z}_h$.

Definition 3.101 We say that a function $f : \mathbb{T} \to \mathbb{R}$ is *regressive* provided

$$1 + \mu(t)p(t) \neq 0 \quad \text{for all} \quad t \in \mathbb{T}^\kappa$$

holds. The set of all regressive and rd-continuous functions $f : \mathbb{T} \to \mathbb{R}$ is denoted by $\mathscr{R}$ or $\mathscr{R}(\mathbb{T})$ or $\mathscr{R}(\mathbb{T}, \mathbb{R})$.

Definition 3.102 In $\mathscr{R}$, we define the *circle plus addition* by

$$f \oplus g = f + g + \mu f g, \quad f, g \in \mathscr{R}.$$

Exercise 3.103 Prove that $(\mathscr{R}, \oplus)$ is an Abelian group.

Definition 3.104 The group $(\mathscr{R}, \oplus)$ is called the *regressive group.*

Definition 3.105 For $f \in \mathscr{R}$, we define the *circle minus* by

$$\ominus f = -\frac{f}{1 + \mu f}.$$

Exercise 3.106 Let $f \in \mathscr{R}$. Prove that $(\ominus f) \in \mathscr{R}$.

Definition 3.107 We define the *circle minus subtraction* $\ominus$ on $\mathscr{R}$ by

$$f \ominus g = f \oplus (\ominus g), \quad f, g \in \mathscr{R}.$$

Remark 3.108 For $f, g \in \mathscr{R}$, we have

$$\begin{aligned} f \ominus g &= f \oplus (\ominus g) \\ &= f \oplus \left(-\frac{g}{1 + \mu g} \right) \end{aligned}$$

$$= f - \frac{g}{1+\mu g} - \frac{\mu f g}{1+\mu g}$$

$$= \frac{f-g}{1+\mu g}.$$

Theorem 3.109 *Let $f, g \in \mathscr{R}$. Then*

1. $f \ominus f = 0$,
2. $\ominus(\ominus f) = f$,
3. $f \ominus g \in \mathscr{R}$,
4. $\ominus(f \ominus g) = g \ominus f$,
5. $\ominus(f \oplus g) = (\ominus f) \oplus (\ominus g)$,
6. $f \oplus \frac{g}{1+\mu f} = f + g$.

Proof 1. We have

$$\begin{aligned} f \ominus f &= f \oplus (\ominus f) \\ &= f \oplus \left(-\frac{f}{1+\mu f}\right) \\ &= f - \frac{f}{1+\mu f} - \frac{\mu f^2}{1+\mu f} \\ &= \frac{f + \mu f^2 - f - \mu f^2}{1+\mu f} \\ &= 0. \end{aligned}$$

2. We have

$$\begin{aligned} \ominus(\ominus f) &= \ominus\left(-\frac{f}{1+\mu f}\right) \\ &= \frac{\frac{f}{1+\mu f}}{1 - \frac{\mu f}{1+\mu f}} \\ &= f. \end{aligned}$$

3. We have

$$\begin{aligned} 1 + \mu(f \ominus g) &= 1 + \frac{\mu f - \mu g}{1+\mu g} \\ &= \frac{1+\mu f}{1+\mu g} \neq 0. \end{aligned}$$

We note that $\frac{f-g}{1+\mu g}$ is rd-continuous. Therefore, $f \ominus g \in \mathscr{R}$.

4. We have

$$\begin{aligned}
\ominus(f \ominus g) &= \ominus\left(\frac{f-g}{1+\mu g}\right)\\
&= -\frac{\frac{f-g}{1+\mu g}}{1+\mu\frac{f-g}{1+\mu g}}\\
&= -\frac{f-g}{1+\mu f}\\
&= \frac{g-f}{1+\mu f}\\
&= g \ominus f.
\end{aligned}$$

5. We have

$$\begin{aligned}
\ominus(f \oplus g) &= \ominus(f+g+\mu f g)\\
&= -\frac{f+g+\mu f g}{1+\mu f+\mu g+\mu^2 f g}\\
&= -\frac{f+g+\mu f g}{(1+\mu f)(1+\mu g)}.
\end{aligned}$$

Since

$$\ominus f = -\frac{f}{1+\mu f} \quad \text{and} \quad \ominus g = -\frac{g}{1+\mu g},$$

we also have

$$\begin{aligned}
(\ominus f) \oplus (\ominus g) &= \ominus f + (\ominus g) + \mu(\ominus f)(\ominus g)\\
&= -\frac{f}{1+\mu f} - \frac{g}{1+\mu g} + \frac{\mu f g}{(1+\mu f)(1+\mu g)}\\
&= \frac{-f(1+\mu g) - g(1+\mu f) + \mu f g}{(1+\mu f)(1+\mu g)}\\
&= -\frac{f+g+\mu f g}{(1+\mu f)(1+\mu g)}.
\end{aligned}$$

6. We have

$$
\begin{aligned}
f \oplus \frac{g}{1+\mu f} &= f + \frac{g}{1+\mu f} + \frac{\mu f g}{1+\mu f} \\
&= f + g.
\end{aligned}
$$

The proof is complete. □

Definition 3.110 If $f \in \mathscr{R}$, then we define the *generalized exponential function* by

$$e_f(t,s) = e^{\int_s^t \xi_{\mu(\tau)}(f(\tau))\Delta\tau} \quad \text{for} \quad s,t \in \mathbb{T}.$$

Remark 3.111 In fact, using the definition for the cylinder transformation, we have

$$e_f(t,s) = e^{\int_s^t \frac{1}{\mu(\tau)} \operatorname{Log}(1+\mu(\tau)f(\tau))\Delta\tau} \quad \text{for} \quad s,t \in \mathbb{T}.$$

Theorem 3.112 (Semigroup Property) *If $f \in \mathscr{R}$, then*

$$e_f(t,r)e_f(r,s) = e_f(t,s) \quad \textit{for all} \quad t,r,s \in \mathbb{T}.$$

Proof We have

$$
\begin{aligned}
e_f(t,r)e_f(r,s) &= e^{\int_r^t \xi_{\mu(\tau)}(f(\tau))\Delta\tau} e^{\int_s^r \xi_{\mu(\tau)}(f(\tau))\Delta\tau} \\
&= e^{\int_r^t \xi_{\mu(\tau)}(f(\tau))\Delta\tau + \int_s^r \xi_{\mu(\tau)}(f(\tau))\Delta\tau} \\
&= e^{\int_s^t \xi_{\mu(\tau)}(f(\tau))\Delta\tau} \\
&= e_f(t,s),
\end{aligned}
$$

completing the proof. □

Exercise 3.113 Let $f \in \mathscr{R}$. Prove that

$$e_0(t,s) = 1 \quad \text{and} \quad e_f(t,t) = 1.$$

Theorem 3.114 *Let $f \in \mathscr{R}$ and fix $t_0 \in \mathbb{T}$. Then*

$$e_f^{\Delta}(t,t_0) = f(t)e_f(t,t_0).$$

Proof 1. If $\sigma(t) > t$, then

$$
\begin{aligned}
e_f^{\Delta}(t,t_0) &= \frac{e_f(\sigma(t),t_0)-e_f(t,t_0)}{\mu(t)}\\
&= \frac{e^{\int_{t_0}^{\sigma(t)}\xi_{\mu(\tau)}(f(\tau))\Delta\tau}-e^{\int_{t_0}^{t}\xi_{\mu(\tau)}(f(\tau))\Delta\tau}}{\mu(t)}\\
&= \frac{e^{\int_{t_0}^{t}\xi_{\mu(\tau)}(f(\tau))\Delta\tau+\int_{t}^{\sigma(t)}\xi_{\mu(\tau)}(f(\tau))\Delta\tau}-e^{\int_{t_0}^{t}\xi_{\mu(\tau)}(f(\tau))\Delta\tau}}{\mu(t)}\\
&= \frac{e^{\int_{t_0}^{\sigma(t)}\xi_{\mu(\tau)}(f(\tau))\Delta\tau}-1}{\mu(t)}e^{\int_{t_0}^{t}\xi_{\mu(\tau)}(f(\tau))\Delta\tau}\\
&= \frac{e^{\mu(t)\xi_{\mu(t)}(f(t))}-1}{\mu(t)}e_f(t,t_0)\\
&= f(t)e_f(t,t_0).
\end{aligned}
$$

2. If $\sigma(t)=t$, then

$$
\begin{aligned}
&|e_f(t,t_0)-e_f(s,t_0)-f(t)e_f(t,t_0)(t-s)|\\
&= |e_f(t,t_0)-e_f(t,t_0)e_f(s,t)-f(t)e_f(t,t_0)(t-s)|\\
&= |e_f(t,t_0)||1-e_f(s,t)-f(t)(t-s)|\\
&= |e_f(t,t_0)|\Big|1-\int_s^t\xi_{\mu(\tau)}(f(\tau))\Delta\tau-e_f(s,t)\\
&\quad+\int_s^t\xi_{\mu(\tau)}(f(\tau))\Delta\tau-f(t)(t-s)\Big|\\
&\le |e_f(t,t_0)|\bigg(\Big|1-\int_s^t\xi_{\mu(\tau)}(f(\tau))\Delta\tau-e_f(s,t)\Big|\\
&\quad+\Big|\int_s^t\xi_{\mu(\tau)}(f(\tau))\Delta\tau-f(t)(t-s)\Big|\bigg)\\
&\le |e_f(t,t_0)|\bigg(\Big|1-\int_s^t\xi_{\mu(\tau)}(f(\tau))\Delta\tau-e_f(s,t)\Big|
\end{aligned}
$$

$$+\left|\int_s^t (\xi_{\mu(\tau)}(f(\tau)) - \xi_0(f(t)))\Delta\tau\right|\Bigg),$$

i.e.,

$$\begin{aligned}
&|e_f(t,t_0) - e_f(s,t_0) - f(t)e_f(t,t_0)(t-s)| \\
&\le |e_f(t,t_0)|\left|1 - \int_s^t \xi_{\mu(\tau)}(f(\tau))\Delta\tau - e_f(s,t)\right| \\
&\quad + |e_f(t,t_0)|\left|\int_s^t (\xi_{\mu(\tau)}(f(\tau)) - \xi_0(f(t)))\Delta\tau\right|.
\end{aligned} \tag{3.18}$$

Since $\sigma(t) = t$ and $f \in C_{rd}$, we get

$$\lim_{r\to t} \xi_{\mu(r)}(f(r)) = \xi_0(f(t)).$$

Therefore, there exists a neighbourhood U_1 of t such that

$$|\xi_{\mu(\tau)}(f(\tau)) - \xi_0(f(t))| < \frac{\varepsilon}{3|e_f(t,t_0)|} \quad \text{for all} \quad \tau \in U_1.$$

Let $s \in U_1$. Then

$$|e_f(t,t_0)|\left|\int_s^t (\xi_{\mu(\tau)}(f(\tau)) - \xi_0(f(t)))\Delta\tau\right| \le \frac{\varepsilon}{3}|t-s|. \tag{3.19}$$

Also, using that

$$\lim_{z\to 0} \frac{1 - z - e^{-z}}{z} = 0,$$

we conclude that there exists a neighbourhood U_2 of t so that if $s \in U_2$ and $s < t$, then we have

$$\left|\frac{1 - \int_s^t \xi_{\mu(\tau)}(f(\tau))\Delta\tau - e_f(s,t)}{\int_s^t \xi_{\mu(\tau)}(f(\tau))\Delta\tau}\right| < \varepsilon^*,$$

where

$$\varepsilon^* = \min\left\{1, \frac{\varepsilon}{1 + 3|f(t)||e_f(t,t_0)|}\right\}.$$

Let $s \in U = U_1 \cap U_2$, $s \ne t$. Then

$$|e_f(t,t_0)|\left|1 - \int_s^t \xi_{\mu(\tau)}(f(\tau))\Delta\tau - e_f(s,t)\right|$$

$$= |e_f(t,t_0)| \frac{\left|1 - \int_s^t \xi_{\mu(\tau)}(f(\tau))\Delta\tau - e_f(s,t)\right|}{\left|\int_s^t \xi_{\mu(\tau)}(f(\tau))\Delta\tau\right|} \left|\int_s^t \xi_{\mu(\tau)}(f(\tau))\Delta\tau\right|$$

$$\le |e_f(t,t_0)|\varepsilon^* \left|\int_s^t \xi_{\mu(\tau)}(f(\tau))\Delta\tau\right|$$

$$\le |e_f(t,t_0)|\varepsilon^* \left\{\left|\int_s^t (\xi_{\mu(\tau)}(f(\tau)) - \xi_0(f(t)))\Delta\tau\right| + |f(t)||t-s|\right\}$$

$$\le |e_f(t,t_0)| \left|\int_s^t (\xi_{\mu(\tau)}(f(\tau)) - \xi_0(f(t)))\Delta\tau\right| + |e_f(t,t_0)|\varepsilon^*|f(t)||t-s|$$

$$\le \frac{\varepsilon}{3}|t-s| + \frac{\varepsilon}{3}|t-s|$$

$$= \frac{2\varepsilon}{3}|t-s|.$$

From the last inequality and from (3.18) and (3.19), we conclude that

$$|e_f(t,t_0) - e_f(s,t_0) - f(t)e_f(t,t_0)(t-s)| \le \frac{2\varepsilon}{3}|t-s| + \frac{\varepsilon}{3}|t-s| = \varepsilon|t-s|,$$

which completes the proof. □

Corollary 3.115 *Let $f \in \mathscr{R}$ and fix $t_0 \in \mathbb{T}$. Then $e_f(\cdot, t_0)$ is a solution to the Cauchy problem*

$$y^\Delta(t) = f(t)y(t), \quad y(t_0) = 1. \tag{3.20}$$

Theorem 3.116 *Let $f \in \mathscr{R}$ and fix $t_0 \in \mathbb{T}$. Then $e_f(\cdot, t_0)$ is the unique solution of the problem* (3.20).

Proof Let y be any solution of the problem (3.20). Then

$$\left(\frac{y}{e_f(\cdot,t_0)}\right)^\Delta (t) = \frac{y^\Delta(t)e_f(t,t_0) - y(t)e_f^\Delta(t,t_0)}{e_f(\sigma(t),t_0)e_f(t,t_0)}$$

$$= \frac{f(t)y(t)e_f(t,t_0) - y(t)f(t)e_f(t,t_0)}{e_f(\sigma(t),t_0)e_f(t,t_0)}$$

$$= 0.$$

Consequently, $y = ce_f(\cdot, t_0)$, where c is a constant. Thus,

$$1 = y(t_0) = ce_f(t_0, t_0) = c.$$

Therefore, $y = e_f(\cdot, t_0)$. □

Theorem 3.117 *If $f \in \mathscr{R}$, then*

$$e_f(\sigma(t), s) = (1 + \mu(t) f(t)) e_f(t, s).$$

Proof Using the simple useful formula, we have

$$\begin{aligned} e_f(\sigma(t), s) - e_f(t, s) &= \mu(t) e_f^{\Delta}(t, s) \\ &= \mu(t) f(t) e_f(t, s), \end{aligned}$$

which completes the proof. □

Theorem 3.118 *If $f \in \mathscr{R}$, then*

$$e_f(t, s) = \frac{1}{e_f(s, t)} = e_{\ominus f}(s, t).$$

Proof We have

$$\begin{aligned} e_f(t, s) &= e^{\int_s^t \xi_{\mu(\tau)}(f(\tau))\Delta\tau} \\ &= e^{-\int_t^s \xi_{\mu(\tau)}(f(\tau))\Delta\tau} \\ &= \frac{1}{e^{\int_t^s \xi_{\mu(\tau)}(f(\tau))\Delta\tau}} \\ &= \frac{1}{e_f(s, t)}. \end{aligned}$$

Now, we fix $t_0 \in \mathbb{T}$ and consider the problem

$$y^{\Delta}(t) = (\ominus f)(t) y(t), \quad y(t_0) = 1.$$

Its solution is $e_{\ominus f}(t, s)$. Also, using the quotient rule, we have

$$\begin{aligned} \left(\frac{1}{e_f(\cdot, s)}\right)^{\Delta}(t) &= -\frac{e_f^{\Delta}(t, s)}{e_f(\sigma(t), s) e_f(t, s)} \\ &= -\frac{f(t) e_f(t, s)}{(1 + \mu(t) f(t)) e_f(t, s) e_f(t, s)} \end{aligned}$$

$$= -\frac{f(t)}{(1+\mu(t)f(t))e_f(t,s)}$$

$$= (\ominus f)(t)\frac{1}{e_f(t,s)}.$$

Therefore,

$$\frac{1}{e_f(t,s)} = e_{\ominus f}(t,s),$$

completing the proof. □

Theorem 3.119 *If $f, g \in \mathscr{R}$, then*

$$e_f(t,s)e_g(t,s) = e_{f\oplus g}(t,s).$$

Proof We have

$$e_f(t,s)e_g(t,s) = e^{\int_s^t \xi_{\mu(\tau)}(f(\tau))\Delta\tau} e^{\int_s^t \xi_{\mu(\tau)}(g(\tau))\Delta\tau}$$

$$= e^{\int_s^t (\xi_{\mu(\tau)}(f(\tau))+\xi_{\mu(\tau)}(g(\tau)))\Delta\tau}$$

$$= e^{\int_s^t \frac{1}{\mu(\tau)}(\mathrm{Log}(1+\mu(\tau)f(\tau))+\mathrm{Log}(1+\mu(\tau)g(\tau)))\Delta\tau}$$

$$= e^{\int_s^t \frac{1}{\mu(\tau)}\mathrm{Log}((1+\mu(\tau)f(\tau))(1+\mu(\tau)g(\tau)))\Delta\tau}$$

$$= e^{\int_s^t \frac{1}{\mu(\tau)}\mathrm{Log}(1+\mu(\tau)(f(\tau)+g(\tau)+\mu(\tau)f(\tau)g(\tau)))\Delta\tau}$$

$$= e^{\int_s^t \xi_{\mu(\tau)}((f\oplus g)(\tau))\Delta\tau}$$

$$= e_{f\oplus g}(t,s),$$

completing the proof. □

Theorem 3.120 *If $f, g \in \mathscr{R}$, then*

$$\frac{e_f(t,s)}{e_g(t,s)} = e_{f\ominus g}(t,s).$$

Proof We have

$$\frac{e_f(t,s)}{e_g(t,s)} = e_f(t,s)e_{\ominus g}(t,s)$$

$$= e_{f \oplus (\ominus g)}(t, s)$$

$$= e_{f \ominus g}(t, s),$$

completing the proof. □

Theorem 3.121 *If $f \in \mathscr{R}$, then*

$$e_f(t, \sigma(s)) e_f(s, r) = \frac{1}{1 + \mu(s) f(s)} e_f(t, r).$$

Proof We have

$$\begin{aligned}
e_f(t, \sigma(s)) e_f(s, r) &= e^{\int_{\sigma(s)}^t \xi_{\mu(\tau)}(f(\tau))\Delta\tau} e^{\int_r^s \xi_{\mu(\tau)}(f(\tau))\Delta\tau} \\
&= e^{\int_{\sigma(s)}^s \xi_{\mu(\tau)}(f(\tau))\Delta\tau + \int_s^t \xi_{\mu(\tau)}(f(\tau))\Delta\tau + \int_r^s \xi_{\mu(\tau)}(f(\tau))\Delta\tau} \\
&= e^{-\xi_{\mu(s)}(f(s))\mu(s) + \int_r^t \xi_{\mu(\tau)}(f(\tau))\Delta\tau} \\
&= e^{-\operatorname{Log}(1 + f(s)\mu(s))} e^{\int_r^t \xi_{\mu(\tau)}(f(\tau))\Delta\tau} \\
&= \frac{1}{1 + \mu(s) f(s)} e_f(t, r),
\end{aligned}$$

completing the proof. □

Theorem 3.122 *If $f, g \in \mathscr{R}$, then*

$$e^{\Delta}_{f \ominus g}(t, t_0) = \frac{(f(t) - g(t)) e_f(t, t_0)}{e_g(\sigma(t), t_0)}.$$

Proof We have

$$\begin{aligned}
e^{\Delta}_{f \ominus g}(t, t_0) &= \left(\frac{e_f(\cdot, t_0)}{e_g(\cdot, t_0)} \right)^{\Delta} (t) \\
&= \frac{e^{\Delta}_f(t, t_0) e_g(t, t_0) - e_f(t, t_0) e^{\Delta}_g(t, t_0)}{e_g(t, t_0) e_g(\sigma(t), t_0)} \\
&= \frac{f(t) e_f(t, t_0) e_g(t, t_0) - g(t) e_f(t, t_0) e_g(t, t_0)}{e_g(t, t_0) e_g(\sigma(t), t_0)} \\
&= \frac{(f(t) - g(t)) e_f(t, t_0)}{e_g(\sigma(t), t_0)},
\end{aligned}$$

completing the proof. □

Theorem 3.123 *Let $f \in \mathscr{R}$ and $a, b, c \in \mathbb{T}$. Then*

$$(e_f(c,\cdot))^{\Delta}(t) = -f(t)e_f(c,\sigma(t))$$

and

$$\int_a^b f(t)e_f(c,\sigma(t))\Delta t = e_f(c,a) - e_f(c,b).$$

Proof We have

$$\begin{aligned} f(t)e_f(c,\sigma(t)) &= f(t)e_{\ominus f}(\sigma(t),c) \\ &= f(t)(1+\mu(t)\ominus f(t))e_{\ominus f}(t,c) \\ &= f(t)\left(1 - \frac{\mu(t)f(t)}{1+\mu(t)f(t)}\right)e_{\ominus f}(t,c) \\ &= \frac{f(t)}{1+\mu(t)f(t)}e_{\ominus f}(t,c) \\ &= -(\ominus f)(t)e_{\ominus f}(t,c) \\ &= -e^{\Delta}_{\ominus f}(t,c). \end{aligned}$$

Hence,

$$\begin{aligned} \int_a^b f(t)e_f(c,\sigma(t))\Delta t &= -\int_a^b e^{\Delta}_{\ominus f}(t,c)\Delta t \\ &= e_{\ominus f}(a,c) - e_{\ominus f}(b,c) \\ &= e_f(c,a) - e_f(c,b), \end{aligned}$$

which completes the proof. □

Exercise 3.124 Assume $1+\mu(t)\frac{2}{t} \neq 0$ and $1+\mu(t)\frac{5}{t} \neq 0$ for all $t \in \mathbb{T} \cap (0,\infty)$. Let $t_0 \in \mathbb{T} \cap (0,\infty)$. Evaluate the integral

$$I = \int_{t_0}^{t} \frac{e_f(s,t_0)}{s e_g^{\sigma}(s,t_0)}\Delta s, \quad \text{where} \quad f(t) = \frac{5}{t},\ g(t) = \frac{2}{t}.$$

Solution By Theorem 3.122,

$$e^{\Delta}_{f\ominus g}(t,t_0) = (f(t)-g(t))\frac{e_f(t,t_0)}{e_g(\sigma(t),t_0)} = 3\frac{e_f(t,t_0)}{e_g(\sigma(t),t_0)}$$

so that

$$I = \frac{1}{3}\int_{t_0}^{t} e^{\Delta}_{f\ominus g}(s,t_0)\Delta s = \frac{1}{3}\left(e_{f\ominus g}(t,t_0)-1\right) = \frac{1}{3}e_h(t,t_0) - \frac{1}{3}$$

with

$$h(t) = (f\ominus g)(t) = \frac{\frac{3}{t}}{1+\mu(t)\frac{2}{t}} = \frac{3}{t+2\mu(t)}.$$

Exercise 3.125 Let $\alpha \in \mathbb{R}$. Suppose that the exponentials

$$e_f(t,t_0) \quad \text{and} \quad e_g(t,t_0) \quad \text{with} \quad f(t) = \frac{\alpha^2}{t} - \frac{(\alpha-1)^2}{\sigma(t)}, \; g(t) = \frac{\alpha}{t}$$

exist for all $t \in \mathbb{T}\cap(0,\infty)$. Prove that

1. $\frac{e_f(t,t_0)}{e_g(t,t_0)} = e_{\frac{\alpha-1}{\sigma}}(t,t_0)$,
2. $\frac{e_{\frac{\alpha-1}{\sigma}}(t,t_0)}{e_g(t,t_0)} = e_{-\frac{1}{\sigma}}(t,t_0) = \frac{t_0}{t}$.

for all $t, t_0 \in \mathbb{T}\cap(0,\infty)$.

Solution 1. We have

$$\begin{aligned}(f\ominus g)(t) &= \frac{f(t)-g(t)}{1+\mu(t)g(t)}\\ &= \frac{\frac{\alpha^2}{t} - \frac{(\alpha-1)^2}{\sigma(t)} - \frac{\alpha}{t}}{1+\mu(t)\frac{\alpha}{t}}\\ &= \frac{1}{\sigma(t)}\cdot\frac{\alpha(\alpha-1)\sigma(t)-(\alpha-1)^2 t}{t+\alpha\mu(t)}\\ &= \frac{\alpha-1}{\sigma(t)}\cdot\frac{\alpha\sigma(t)-(\alpha-1)t}{t+\alpha\mu(t)}\\ &= \frac{\alpha-1}{\sigma(t)}.\end{aligned}$$

Hence,

$$\frac{e_f(t,t_0)}{e_g(t,t_0)} = e_{f\ominus g}(t,t_0) = e_{\frac{\alpha-1}{\sigma}}(t,t_0).$$

2. We have

$$\left(\frac{\alpha-1}{\sigma}\ominus g\right)(t)=\frac{\frac{\alpha-1}{\sigma(t)}-\frac{\alpha}{t}}{1+\mu(t)\frac{\alpha}{t}}$$

$$=\frac{1}{\sigma(t)}\cdot\frac{(\alpha-1)t-\alpha\sigma(t)}{t+\alpha\mu(t)}$$

$$=-\frac{1}{\sigma(t)}.$$

Hence,

$$\frac{e_{\frac{\alpha-1}{\sigma}}(t,t_0)}{e_g(t,t_0)}=e_{-\frac{1}{\sigma}}(t,t_0).$$

Now we show that

$$e_{-\frac{1}{\sigma}}(t,t_0)=\frac{t_0}{t}. \tag{3.21}$$

Define $p(t)=\frac{t_0}{t}$. Then $p(t_0)=1$ and

$$p^{\Delta}(t)=-\frac{t_0}{t\sigma(t)}=-\frac{1}{\sigma(t)}p(t),$$

By Theorem 3.116, (3.21) holds.

Exercise 3.126 Let $\mathbb{T}=\mathbb{Z}$. Prove that

$$e_\alpha(t,t_0)=\prod_{s\in[t_0,t)}(1+\alpha(s)).$$

Exercise 3.127 Let $\mathbb{T}=h\mathbb{Z}$, $h>0$. Prove that

$$e_\alpha(t,t_0)=\prod_{s\in[t_0,t)}(1+h\alpha(s)).$$

3.5.3 Hyperbolic Functions

In this and the following subsection, we often "skip" the arguments, e.g., we write e_f for $e_f(t,s)$.

Definition 3.128 Let $f\in \mathrm{C_{rd}}$ and $-\mu f^2\in\mathscr{R}$. Define the *hyperbolic functions* $\cosh_f$ and $\sinh_f$ by

$$\cosh_f=\frac{e_f+e_{-f}}{2}\quad\text{and}\quad\sinh_f=\frac{e_f-e_{-f}}{2}.$$

Theorem 3.129 *Let $f \in \mathrm{C}_{\mathrm{rd}}$ and $-\mu f^2 \in \mathscr{R}$. Then*

$$\cosh_f^{\Delta} = f \sinh_f, \quad \sinh_f^{\Delta} = f \cosh_f \quad \textit{and} \quad \cosh_f^2 - \sinh_f^2 = e_{-\mu f^2}.$$

Proof We have

$$\begin{aligned}
\cosh_f^{\Delta} &= \left(\frac{e_f + e_{-f}}{2}\right)^{\Delta} \\
&= \frac{e_f^{\Delta} + e_{-f}^{\Delta}}{2} \\
&= \frac{f e_f - f e_{-f}}{2} \\
&= f \frac{e_f - e_{-f}}{2} \\
&= f \sinh_f,
\end{aligned}$$

$$\begin{aligned}
\sinh_f^{\Delta} &= \left(\frac{e_f - e_{-f}}{2}\right)^{2} \\
&= \frac{e_f^{\Delta} - e_{-f}^{\Delta}}{2} \\
&= \frac{f e_f + f e_{-f}}{2} \\
&= f \frac{e_f + e_{-f}}{2} \\
&= f \cosh_f,
\end{aligned}$$

and

$$\begin{aligned}
\cosh_f^2 - \sinh_f^2 &= \left(\frac{e_f + e_{-f}}{2}\right)^2 - \left(\frac{e_f - e_{-f}}{2}\right)^2 \\
&= \frac{e_f^2 + 2e_f e_{-f} + e_{-f}^2 - e_f^2 + 2e_f e_{-f} - e_{-f}^2}{4} \\
&= e_f e_{-f} \\
&= e_{f \oplus (-f)}
\end{aligned}$$

$$= e_{-\mu f^2}.$$

The proof is complete. □

Example 3.130 Let $\mathbb{T} = \mathbb{Z}$. We compute $\cosh_{\frac{1}{2}}(t, t_0)$ and $\sinh_{\frac{1}{2}}(t, t_0)$ for $t, t_0 \in \mathbb{T}$. Here, $\mu(t) = 1$. Let $\alpha = \frac{1}{2}$. Then

$$1 - \alpha^2(\mu(t))^2 = 1 - \frac{1}{4} = \frac{3}{4} \neq 0.$$

Therefore, $\cosh_{\frac{1}{2}}(t, t_0)$ and $\sinh_{\frac{1}{2}}(t, t_0)$ are well defined. We have

$$e_{\frac{1}{2}}(t, t_0) = \left(\frac{3}{2}\right)^{t-t_0} \quad \text{and} \quad e_{-\frac{1}{2}}(t, t_0) = \left(\frac{1}{2}\right)^{t-t_0}.$$

Hence,

$$\begin{aligned}
\cosh_{\frac{1}{2}}(t, t_0) &= \frac{e_{\frac{1}{2}}(t, t_0) + e_{-\frac{1}{2}}(t, t_0)}{2} \\
&= \frac{\frac{3^{t-t_0}}{2^{t-t_0}} + \frac{1}{2^{t-t_0}}}{2} \\
&= \frac{3^{t-t_0} + 1}{2^{1+t-t_0}}, \\
\sinh_{\frac{1}{2}}(t, t_0) &= \frac{e_{\frac{1}{2}}(t, t_0) - e_{-\frac{1}{2}}(t, t_0)}{2} \\
&= \frac{3^{t-t_0} - 1}{2^{1+t-t_0}}.
\end{aligned}$$

Example 3.131 Let $\mathbb{T} = 2^{\mathbb{N}_0}$. We compute $\cosh_2(t, t_0)$ and $\sinh_2(t, t_0)$ for $t, t_0 \in \mathbb{T}$. Here, $\mu(t) = t$. Let $\alpha = 2$. Then

$$1 - \alpha^2(\mu(t))^2 = 1 - 4t^2 \neq 0 \quad \text{for any} \quad t \in \mathbb{T}.$$

Therefore, $\cosh_2(t, t_0)$ and $\sinh_2(t, t_0)$ are well defined. We have

$$e_2(t, t_0) = \prod_{s \in [t_0, t)} (1 + 2s), \quad e_{-2}(t, t_0) = \prod_{s \in [t_0, t)} (1 - 2s).$$

Hence,

$$\cosh_2(t, t_0) = \frac{e_2(t, t_0) + e_{-2}(t, t_0)}{2}$$

$$= \frac{1}{2}\left(\prod_{s\in[t_0,t)} (1+2s) + \prod_{s\in[t_0,t)} (1-2s)\right),$$

$$\sinh_2(t, t_0) = \frac{e_2(t, t_0) - e_{-2}(t, t_0)}{2}$$

$$= \frac{1}{2}\left(\prod_{s\in[t_0,t)} (1+2s) - \prod_{s\in[t_0,t)} (1-2s)\right).$$

Example 3.132 Let $\mathbb{T} = \mathbb{N}_0^2$. We compute $\cosh_2(t, t_0)$ and $\sinh_2(t, t_0)$ for $t, t_0 \in \mathbb{T}$. Here,

$$\sigma(t) = (1+\sqrt{t})^2, \quad \mu(t) = 1 + 2\sqrt{t}.$$

Let $\alpha = 2$. Thus,

$$1 - \alpha^2(\mu(t))^2 = 1 - 4(1+2\sqrt{t})^2 \neq 0 \quad \text{for all} \quad t \in \mathbb{T}.$$

Therefore, $\cosh_2(t, t_0)$ and $\sinh_2(t, t_0)$ are well defined. Note that

$$e_2(t, t_0) = \prod_{s\in[t_0,t)} \big(1 + 2(1+2\sqrt{s})\big)$$

$$= \prod_{s\in[t_0,t)} (3+4\sqrt{s}),$$

$$e_{-2}(t, t_0) = \prod_{s\in[t_0,t)} \big(1 - 2(1+2\sqrt{s})\big)$$

$$= \prod_{s\in[t_0,t)} (-1-4\sqrt{s}).$$

Hence,

$$\cosh_2(t, t_0) = \frac{1}{2}(e_2(t, t_0) + e_{-2}(t, t_0))$$

$$= \frac{1}{2}\left(\prod_{s\in[t_0,t)} (3+4\sqrt{s}) + \prod_{s\in[t_0,t)} (-1-4\sqrt{s})\right),$$

$$\sinh_2(t, t_0) = \frac{1}{2}(e_2(t, t_0) - e_{-2}(t, t_0))$$

$$= \frac{1}{2}\left(\prod_{s\in[t_0,t)}(3+4\sqrt{s}) - \prod_{s\in[t_0,t)}(-1-4\sqrt{s})\right).$$

Exercise 3.133 Let $\mathbb{T} = 3^{\mathbb{N}_0}$. Compute $\cosh_2(t, t_0)$ and $\sinh_2(t, t_0)$ for $t, t_0 \in \mathbb{T}$.

Solution

$$\cosh_2(t,t_0) = \frac{1}{2}\left(\prod_{s\in[t_0,t)}(1+4s) + \prod_{s\in[t_0,t)}(1-4s)\right),$$

$$\sinh_2(t,t_0) = \frac{1}{2}\left(\prod_{s\in[t_0,t)}(1+4s) - \prod_{s\in[t_0,t)}(1-4s)\right).$$

Theorem 3.134 *Let $f \in \mathrm{C_{rd}}$ and $-\mu f^2 \in \mathcal{R}$. Then*

$$(\cosh_f(c,\cdot))^{\Delta}(t) = -f(t)\sinh_f(c,\sigma(t)).$$

Proof Using Theorem 3.123, we have

$$\begin{aligned}(\cosh_f(c,\cdot))^{\Delta}(t) &= \left(\frac{e_f(c,\cdot)+e_{-f}(c,\cdot)}{2}\right)^{\Delta}(t)\\ &= \frac{-f(t)e_f(c,\sigma(t)) + f(t)e_{-f}(c,\sigma(t))}{2}\\ &= -f(t)\frac{e_f(c,\sigma(t)) - e_{-f}(c,\sigma(t))}{2}\\ &= -f(t)\sinh_f(c,\sigma(t)),\end{aligned}$$

completing the proof. □

Exercise 3.135 Let $f \in \mathrm{C_{rd}}$ and $-\mu f^2 \in \mathcal{R}$. Prove that

$$(\sinh_f(c,\cdot))^{\Delta}(t) = -f(t)\cosh_f(c,\sigma(t)).$$

Theorem 3.136 *Let $f \in \mathrm{C_{rd}}$ and $-\mu f^2 \in \mathcal{R}$. Then*

$$\int_a^b f(t)\sinh_f(c,\sigma(t))\Delta t = \cosh_f(c,a) - \cosh_f(c,b).$$

Proof By Theorem 3.134, we have

$$f(t)\sinh_f(c,\sigma(t)) = -\left(\cosh_f(c,\cdot)\right)^{\Delta}(t).$$

Hence,

$$\int_a^b f(t)\sinh_f(c,\sigma(t)) = -\int_a^b \left(\cosh_f(c,\cdot)\right)^{\Delta}(t)\Delta t$$

$$= \cosh_f(c,a) - \cosh_f(c,b),$$

completing the proof. □

Exercise 3.137 Let $f \in \mathrm{C}_{\mathrm{rd}}$ and $-\mu f^2 \in \mathscr{R}$. Prove that

$$\int_a^b f(t)\cosh_f(c,\sigma(t))\Delta t = \sinh_f(c,a) - \sinh_f(c,b).$$

Exercise 3.138 Let $f \in \mathrm{C}_{\mathrm{rd}}$ and $-\mu f^2 \in \mathscr{R}$. Simplify

1. $A = \frac{\cosh_f(s,t_0)\cosh_f(t,t_0)-\sinh_f(s,t_0)\sinh_f(t,t_0)}{\cosh_f^2(s,t_0)-\sinh_f^2(s,t_0)}$,
2. $B = \frac{\cosh_f(s,t_0)\sinh_f(t,t_0)-\sinh_f(s,t_0)\cosh_f(t,t_0)}{\cosh_f^2(s,t_0)-\sinh_f^2(s,t_0)}$.

Solution 1. We have

$$\cosh_f(s,t_0)\cosh_f(t,t_0) - \sinh_f(s,t_0)\sinh_f(t,t_0)$$

$$= \frac{e_f(s,t_0)+e_{-f}(s,t_0)}{2}\frac{e_f(t,t_0)+e_{-f}(t,t_0)}{2}$$

$$- \frac{e_f(s,t_0)-e_{-f}(s,t_0)}{2}\frac{e_f(t,t_0)-e_{-f}(t,t_0)}{2}$$

$$= \frac{1}{4}\Big(e_f(s,t_0)e_f(t,t_0) + e_f(s,t_0)e_{-f}(t,t_0)$$

$$+ e_{-f}(s,t_0)e_f(t,t_0) + e_{-f}(s,t_0)e_{-f}(t,t_0)$$

$$- e_f(s,t_0)e_f(t,t_0) + e_f(s,t_0)e_{-f}(t,t_0)$$

$$+ e_{-f}(s,t_0)e_f(t,t_0) - e_{-f}(s,t_0)e_{-f}(t,t_0)\Big)$$

$$= \frac{e_f(s,t_0)e_{-f}(t,t_0) + e_{-f}(s,t_0)e_f(t,t_0)}{2}.$$

Also,

$$
\begin{aligned}
\cosh_f^2(s,t_0)-\sinh_f^2(s,t_0) &= \left(\frac{e_f(s,t_0)+e_{-f}(s,t_0)}{2}\right)^2\\
&\quad -\left(\frac{e_f(s,t_0)-e_{-f}(s,t_0)}{2}\right)^2\\
&= \frac{e_f^2(s,t_0)+2e_f(s,t_0)e_{-f}(s,t_0)+e_{-f}^2(s,t_0)}{4}\\
&\quad -\frac{e_f^2(s,t_0)-2e_f(s,t_0)e_{-f}(s,t_0)+e_{-f}^2(s,t_0)}{4}\\
&= e_f(s,t_0)e_{-f}(s,t_0).
\end{aligned}
$$

Therefore,

$$
\begin{aligned}
A &= \frac{1}{2}\frac{e_f(s,t_0)e_{-f}(t,t_0)+e_{-f}(s,t_0)e_f(t,t_0)}{e_f(s,t_0)e_{-f}(s,t_0)}\\
&= \frac{1}{2}\left(\frac{e_{-f}(t,t_0)}{e_{-f}(s,t_0)}+\frac{e_f(t,t_0)}{e_f(s,t_0)}\right)\\
&= \frac{1}{2}(e_f(t,s)+e_{-f}(t,s))\\
&= \cosh_f(t,s).
\end{aligned}
$$

2. We have

$$
\begin{aligned}
&\cosh_f(s,t_0)\sinh_f(t,t_0)-\sinh_f(s,t_0)\cosh_f(t,t_0)\\
&= \frac{e_f(s,t_0)+e_{-f}(s,t_0)}{2}\,\frac{e_f(t,t_0)-e_{-f}(t,t_0)}{2}\\
&\quad -\frac{e_f(s,t_0)-e_{-f}(s,t_0)}{2}\,\frac{e_f(t,t_0)+e_{-f}(t,t_0)}{2}\\
&= \frac{1}{4}\Big(e_f(s,t_0)e_f(t,t_0)-e_f(s,t_0)e_{-f}(t,t_0)\\
&\quad +e_{-f}(s,t_0)e_f(t,t_0)-e_{-f}(s,t_0)e_{-f}(t,t_0)\Big)
\end{aligned}
$$

$$-\frac{1}{4}\Big(e_f(s,t_0)e_f(t,t_0)+e_f(s,t_0)e_{-f}(t,t_0)$$

$$-e_{-f}(s,t_0)e_f(t,t_0)-e_{-f}(s,t_0)e_{-f}(t,t_0)\Big)$$

$$=\frac{-e_f(s,t_0)e_{-f}(t,t_0)+e_{-f}(s,t_0)e_f(t,t_0)}{2}.$$

Hence,

$$\begin{aligned}B&=\frac{1}{2}\frac{-e_f(s,t_0)e_{-f}(t,t_0)+e_{-f}(s,t_0)e_f(t,t_0)}{e_f(s,t_0)e_{-f}(s,t_0)}\\&=\frac{1}{2}\left(-\frac{e_{-f}(t,t_0)}{e_{-f}(s,t_0)}+\frac{e_f(t,t_0)}{e_f(s,t_0)}\right)\\&=\frac{1}{2}(e_f(t,s)-e_{-f}(t,s))\\&=\sinh_f(t,s).\end{aligned}$$

Definition 3.139 Let $f, g \in \mathrm{C}_{\mathrm{rd}}$ and suppose that

$$2f+\mu(f^2-g^2)=0. \tag{3.22}$$

We define the *hyperbolic functions* ch_{fg} and sh_{fg} by

$$\mathrm{ch}_{fg}=\frac{e_{f+g}+e_{f-g}}{2}\quad\text{and}\quad\mathrm{sh}_{fg}=\frac{e_{f+g}-e_{f-g}}{2}.$$

Theorem 3.140 *If $f, g \in \mathrm{C}_{\mathrm{rd}}$ satisfy* (3.22), *then*

$$\mathrm{ch}_{fg}^{\Delta}=f\,\mathrm{ch}_{fg}+g\,\mathrm{sh}_{fg},\quad \mathrm{sh}_{fg}^{\Delta}=g\,\mathrm{ch}_{fg}+f\,\mathrm{sh}_{fg}$$

and

$$\mathrm{ch}_{fg}^2-\mathrm{sh}_{fg}^2=1.$$

Proof We have

$$\begin{aligned}\mathrm{ch}_{fg}^{\Delta}&=\left(\frac{e_{f+g}+e_{f-g}}{2}\right)^{\Delta}\\&=\frac{e_{f+g}^{\Delta}+e_{f-g}^{\Delta}}{2}\end{aligned}$$

$$= \frac{(f+g)e_{f+g} + (f-g)e_{f-g}}{2}$$

$$= f\frac{e_{f+g} + e_{f-g}}{2} + g\frac{e_{f+g} - e_{f-g}}{2}$$

$$= f\,\mathrm{ch}_{fg} + g\,\mathrm{sh}_{fg}$$

and

$$\mathrm{sh}_{fg}^{\Delta} = \left(\frac{e_{f+g} - e_{f-g}}{2}\right)^{\Delta}$$

$$= \frac{e_{f+g}^{\Delta} - pe_{f-g}^{\Delta}}{2}$$

$$= \frac{(f+g)e_{f+g} - (f-g)e_{f-g}}{2}$$

$$= f\frac{e_{f+g} - e_{f-g}}{2} + g\frac{e_{f+g} + e_{f-g}}{2}$$

$$= f\,\mathrm{sh}_{fg} + g\,\mathrm{ch}_{fg}\,.$$

Next,

$$\mathrm{ch}_{fg}^2 - \mathrm{sh}_{fg}^2 = \left(\frac{e_{f+g} + e_{f-g}}{2}\right)^2 - \left(\frac{e_{f+g} - e_{f-g}}{2}\right)^2$$

$$= \frac{e_{f+g}^2 + 2e_{f+g}e_{f-g} + e_{f-g}^2}{4}$$

$$- \frac{e_{f+g}^2 - 2e_{f+g}e_{f-g} + e_{f-g}^2}{4}$$

$$= e_{f+g}e_{f-g}$$

$$= e_{(f+g)\oplus(f-g)}.$$

Note that

$$(f+g)\oplus(f-g) = f+g+f-g+\mu(f+g)(f-g)$$

$$= 2f + \mu(f^2 - g^2)$$

$$= 0.$$

Therefore,

$$\mathrm{ch}_{fg}^2 - \mathrm{sh}_{fg}^2 = 1,$$

completing the proof. □

Exercise 3.141 Let $f, g \in \mathrm{C_{rd}}$ and satisfy (3.22). Prove that

$$e_{f+g} = \mathrm{ch}_{fg} + \mathrm{sh}_{fg}$$

and

$$e_{f-g} = \mathrm{ch}_{fg} - \mathrm{sh}_{fg}\,.$$

Exercise 3.142 Let $f, g \in \mathrm{C_{rd}}$ satisfy (3.22). Prove that

1. $\mathrm{sh}_{fg}(t,s) = -\mathrm{sh}_{fg}(s,t)$,
2. $\mathrm{sh}_{fg}(t,s) = \mathrm{sh}_{fg}(t,r)\,\mathrm{ch}_{fg}(r,s) - \mathrm{ch}_{fg}(t,r)\,\mathrm{sh}_{fg}(s,r)$,
3. $\mathrm{sh}_{fg}(t,r) = \mathrm{sh}_{fg}(t,s)\,\mathrm{ch}_{fg}(s,r) + \mathrm{ch}_{fg}(t,s)\,\mathrm{sh}_{fg}(s,r)$.

Solution 1. We have

$$\begin{aligned}
\mathrm{sh}_{fg}(t,s) &= \frac{e_{f+g}(t,s) - e_{f-g}(t,s)}{2} \\
&= \frac{\frac{1}{e_{f+g}(s,t)} - \frac{1}{e_{f-g}(s,t)}}{2} \\
&= \frac{e_{f-g}(s,t) - e_{f+g}(s,t)}{2e_{f+g}(s,t)e_{f-g}(s,t)} \\
&= -\frac{e_{f+g}(s,t) - e_{f-g}(s,t)}{2} \\
&= -\mathrm{sh}_{fg}(s,t).
\end{aligned}$$

2. We have

$$\begin{aligned}
&\mathrm{sh}_{fg}(t,r)\,\mathrm{ch}_{fg}(r,s) - \mathrm{ch}_{fg}(t,r)\,\mathrm{sh}_{fg}(s,r) \\
&= \mathrm{sh}_{fg}(t,r)\,\mathrm{ch}_{fg}(r,s) + \mathrm{ch}_{fg}(t,r)\,\mathrm{sh}_{fg}(r,s) \\
&= \frac{e_{f+g}(t,r) - e_{f-g}(t,r)}{2}\,\frac{e_{f+g}(r,s) + e_{f-g}(r,s)}{2} \\
&\quad + \frac{e_{f+g}(t,r) + e_{f-g}(t,r)}{2}\,\frac{e_{f+g}(r,s) + e_{f-g}(r,s)}{2}
\end{aligned}$$

$$= \frac{1}{4}\Big(e_{f+g}(t,r)e_{f+g}(r,s) + e_{f+g}(t,r)e_{f-g}(r,s)$$

$$- e_{f-g}(t,r)e_{f+g}(r,s) - e_{f-g}(t,r)e_{f-g}(r,s)\Big)$$

$$+ \frac{1}{4}\Big(e_{f+g}(t,r)e_{f+g}(r,s) - e_{f+g}(t,r)e_{f-g}(r,s)$$

$$- e_{f-g}(t,r)e_{f+g}(r,s) - e_{f-g}(t,r)e_{f-g}(r,s)\Big)$$

$$= \frac{e_{f+g}(t,s) - e_{f-g}(t,s)}{2}$$

$$= \operatorname{sh}_{fg}(t,s).$$

3. We have

$$\operatorname{sh}_{fg}(t,s)\operatorname{ch}_{fg}(s,r) + \operatorname{ch}_{fg}(t,s)\operatorname{sh}_{fg}(s,r)$$

$$= \frac{e_{f+g}(t,s) - e_{f-g}(t,s)}{2}\,\frac{e_{f+g}(s,r) + e_{f-g}(s,r)}{2}$$

$$+ \frac{e_{f+g}(t,s) + e_{f-g}(t,s)}{2}\,\frac{e_{f+g}(s,r) - e_{f-g}(s,r)}{2}$$

$$= \frac{1}{4}\Big(e_{f+g}(t,s)e_{f+g}(s,r) + e_{f+g}(t,s)e_{f-g}(s,r)$$

$$- e_{f-g}(t,s)e_{f+g}(s,r) - e_{f-g}(t,s)e_{f-g}(s,r)\Big)$$

$$+ \frac{1}{4}\Big(e_{f+g}(t,s)e_{f+g}(s,r) - e_{f+g}(t,s)e_{f-g}(s,r)$$

$$+ e_{f-g}(t,s)e_{f+g}(s,r) - e_{f-g}(t,s)e_{f-g}(s,r)\Big)$$

$$= \frac{e_{f+g}(t,r) - e_{f-g}(t,r)}{2}$$

$$= \operatorname{sh}_{fg}(t,r).$$

Exercise 3.143 Let $f, g \in \mathrm{C_{rd}}$ satisfy (3.22). Prove that

1. $\operatorname{ch}_{fg}(t,s) = \operatorname{ch}_{fg}(s,t)$,
2. $\operatorname{ch}_{fg}(t,s) = \operatorname{ch}_{fg}(t,r)\operatorname{ch}_{fg}(s,r) - \operatorname{sh}_{fg}(t,r)\operatorname{sh}_{fg}(s,r)$,
3. $\operatorname{ch}_{fg}(t,r) = \operatorname{ch}_{fg}(t,s)\operatorname{ch}_{fg}(s,r) + \operatorname{sh}_{fg}(t,s)\operatorname{sh}_{fg}(s,r)$.

3.5.4 Trigonometric Functions

Definition 3.144 Let $f \in \mathrm{C_{rd}}$ and $\mu f^2 \in \mathscr{R}$. Define the *trigonometric functions* $\cos_f$ and $\sin_f$ by

$$\cos_f = \frac{e_{if} + e_{-if}}{2} \quad \text{and} \quad \sin_f = \frac{e_{if} - e_{-if}}{2i}.$$

Theorem 3.145 *Let $f \in \mathrm{C_{rd}}$ and $\mu f^2 \in \mathscr{R}$. Then*

$$\cos_f^{\Delta} = -f\sin_f \quad \textit{and} \quad \sin_f^{\Delta} = f\cos_f$$

and

$$\cos_f^2 + \sin_f^2 = e_{\mu f^2}.$$

Proof We have

$$\begin{aligned}
\cos_f^{\Delta} &= \left(\frac{e_{if} + e_{-if}}{2}\right)^{\Delta} \\
&= \frac{e_{if}^{\Delta} + e_{-if}^{\Delta}}{2} \\
&= \frac{if e_{if} - if e_{-if}}{2} \\
&= -f\frac{e_{if} - e_{-if}}{2i} \\
&= -f\sin_f,
\end{aligned}$$

$$\begin{aligned}
\sin_f^{\Delta} &= \left(\frac{e_{if} - e_{-if}}{2i}\right)^{\Delta} \\
&= \frac{e_{if}^{\Delta} - e_{-if}^{\Delta}}{2i} \\
&= \frac{if e_{if} + if e_{-if}}{2i} \\
&= f \frac{e_{if} + e_{-if}}{2} \\
&= f \cos_f,
\end{aligned}$$

$$\begin{aligned}
\cos_f^2 + \sin_f^2 &= \left(\frac{e_{if} + e_{-if}}{2}\right)^2 + \left(\frac{e_{if} - e_{-if}}{2}\right)^2 \\
&= \frac{e_{if}^2 + 2e_{if}e_{-if} + e_{-if}^2}{4} - \frac{e_{if} - 2e_{if}e_{-if} + e_{-if}^2}{4} \\
&= e_{if}e_{-if} \\
&= e_{(if)\oplus(-if)}.
\end{aligned}$$

Since

$$(if) \oplus (-if) = (if) + (-if) + \mu(if)(-if) = \mu f^2,$$

we get the desired result. □

Example 3.146 Let $f \in \mathrm{C_{rd}}$ and $\mu f^2 \in \mathscr{R}$. We simplify

$$A = \frac{\cos_f(s, t_0)\cos_f(t, t_0) + \sin_f(s, t_0)\sin_f(t, t_0)}{\cos_f^2(s, t_0) + \sin_f^2(s, t_0)}.$$

We have

$$\begin{aligned}
&\cos_f(s, t_0)\cos_f(t, t_0) + \sin_f(s, t_0)\sin_f(t, t_0) \\
&= \frac{e_{if}(s, t_0) + e_{-if}(s, t_0)}{2} \frac{e_{if}(t, t_0) + e_{-if}(t, t_0)}{2} \\
&\quad + \frac{e_{if}(s, t_0) - e_{-if}(s, t_0)}{2i} \frac{e_{if}(t, t_0) - e_{-if}(t, t_0)}{2i}
\end{aligned}$$

$$= \frac{1}{4}\Big(e_{if}(s,t_0)e_{if}(t,t_0) + e_{-if}(s,t_0)e_{if}(t,t_0)$$

$$+ e_{if}(s,t_0)e_{-if}(t,t_0) + e_{-if}(s,t_0)e_{-if}(t,t_0)\Big)$$

$$- \frac{1}{4}\Big(e_{if}(s,t_0)e_{if}(t,t_0) - e_{if}(s,t_0)e_{-if}(t,t_0)$$

$$- e_{-if}(s,t_0)e_{if}(t,t_0) + e_{-if}(s,t_0)e_{-if}(t,t_0)\Big)$$

$$= \frac{e_{-if}(s,t_0)e_{if}(t,t_0) + e_{if}(s,t_0)e_{-if}(t,t_0)}{2},$$

$$\cos_f^2(s,t_0) + \sin_f^2(s,t_0)$$

$$= \left(\frac{e_{if}(s,t_0) + e_{-if}(s,t_0)}{2}\right)^2 + \left(\frac{e_{if}(s,t_0) - e_{-if}(s,t_0)}{2i}\right)^2$$

$$= \frac{e_{if}^2(s,t_0) + 2e_{if}(s,t_0)e_{-if}(s,t_0) + e_{-if}^2(s,t_0)}{4}$$

$$- \frac{e_{if}^2(s,t_0) - 2e_{if}(s,t_0)e_{-if}(s,t_0) + e_{-if}^2(s,t_0)}{4}$$

$$= e_{if}(s,t_0)e_{-if}(s,t_0).$$

Hence,

$$A = \frac{1}{2}\frac{e_{-if}(s,t_0)e_{if}(t,t_0) + e_{if}(s,t_0)e_{-if}(t,t_0)}{e_{if}(s,t_0)e_{-if}(s,t_0)}$$

$$= \frac{1}{2}\left(\frac{e_{if}(t,t_0)}{e_{if}(s,t_0)} + \frac{e_{-if}(t,t_0)}{e_{-if}(s,t_0)}\right)$$

$$= \frac{1}{2}(e_{if}(t,s) + e_{-if}(t,s))$$

$$= \cos_f(t,s).$$

Exercise 3.147 Let $f \in \mathrm{C}_{\mathrm{rd}}$ and $\mu f^2 \in \mathscr{R}$. Simplify

$$B = \frac{\cos_f(s, t_0) \sin_f(t, t_0) - \sin_f(s, t_0) \cos_f(t, t_0)}{\cos_f^2(s, t_0) + \sin_f^2(s, t_0)}.$$

Solution $B = \sin_f(t, s)$.

Definition 3.148 Let $f, g \in \mathrm{C}_{\mathrm{rd}}$ and assume

$$2f + \mu(f^2 + g^2) = 0. \tag{3.23}$$

Define the *trigonometric functions* c_{fg} and s_{fg} by

$$\mathrm{c}_{fg} = \frac{e_{f+ig} + e_{f-ig}}{2} \quad \text{and} \quad \mathrm{s}_{fg} = \frac{e_{f+ig} - e_{f-ig}}{2i}.$$

Theorem 3.149 *Let $f, g \in \mathrm{C}_{\mathrm{rd}}$ satisfy* (3.23). *Then*

$$\mathrm{c}_{fg}^{\Delta} = f\,\mathrm{c}_{fg} - g\,\mathrm{s}_{fg} \quad \textit{and} \quad \mathrm{s}_{fg}^{\Delta} = g\,\mathrm{c}_{fg} + f\,\mathrm{s}_{fg}$$

and

$$\mathrm{c}_{fg}^2 + \mathrm{s}_{fg}^2 = 1.$$

Proof We have

$$\begin{aligned}
\mathrm{c}_{fg}^{\Delta} &= \left(\frac{e_{f+ig} + e_{f-ig}}{2}\right)^{\Delta} \\
&= \frac{e_{f+ig}^{\Delta} + e_{f-ig}^{\Delta}}{2} \\
&= \frac{(f+ig)e_{f+ig} + (f-ig)e_{f-ig}}{2} \\
&= f\frac{e_{f+ig} + e_{f-ig}}{2} + ig\frac{e_{f+ig} - e_{f-ig}}{2} \\
&= f\,\mathrm{c}_{fg} - g\frac{e_{f+ig} - e_{f-ig}}{2i} \\
&= f\,\mathrm{c}_{fg} - g\,\mathrm{s}_{fg},
\end{aligned}$$

$$\mathrm{s}_{fg}^{\Delta} = \left(\frac{e_{f+ig} - e_{f-ig}}{2i}\right)^{\Delta}$$

$$
\begin{aligned}
&= \frac{e^{\Delta}_{f+ig} - e^{\Delta}_{f-ig}}{2i} \\
&= \frac{(f+ig)e_{f+ig} - (f-ig)e_{f-ig}}{2i} \\
&= f\frac{e_{f+ig} - e_{f-ig}}{2i} + g\frac{e_{f+ig} + e_{f-ig}}{2} \\
&= f\,\mathrm{s}_{fg} + g\,\mathrm{c}_{fg},
\end{aligned}
$$

and

$$
\begin{aligned}
\mathrm{c}^2_{fg} + \mathrm{s}^2_{fg} &= \left(\frac{e_{f+ig} + e_{f-ig}}{2}\right)^2 + \left(\frac{e_{f+ig} - e_{f-ig}}{2i}\right)^2 \\
&= \frac{e^2_{f+ig} + 2e_{f+ig}e_{f-ig} + e^2_{f-ig}}{4} \\
&\quad - \frac{e^2_{f+ig} - 2e_{f+ig}e_{f-ig} + e^2_{f-ig}}{4} \\
&= e_{(f+ig)\oplus(f-ig)}.
\end{aligned}
$$

Since

$$
\begin{aligned}
(f+ig)\oplus(f-ig) &= f + ig + f - ig + \mu(f+ig)(f-ig) \\
&= 2f + \mu(f^2+g^2) \\
&= 0,
\end{aligned}
$$

we get

$$
\mathrm{c}^2_{fg} + \mathrm{s}^2_{fg} = 1,
$$

completing the proof. □

Exercise 3.150 Let $f, g \in \mathrm{C_{rd}}$ satisfy (3.23). Prove

1. $\mathrm{c}_{fg}(s,t) = \mathrm{c}_{fg}(t,s)$,
2. $\mathrm{c}_{fg}(t,s) = \mathrm{c}_{fg}(t,r)\,\mathrm{c}_{fg}(s,r) + \mathrm{s}_{fg}(t,r)\,\mathrm{s}_{fg}(s,r)$,
3. $\mathrm{c}_{fg}(t,r) = \mathrm{c}_{fg}(t,s)\,\mathrm{c}_{fg}(s,r) - \mathrm{s}_{fg}(t,s)\,\mathrm{s}_{fg}(s,r)$.

Solution 1. We have

$$
\mathrm{c}_{fg}(t,s) = \frac{e_{f+ig}(t,s) + e_{f-ig}(t,s)}{2}
$$

$$= \frac{\frac{1}{e_{f+ig}(s,t)} + \frac{1}{e_{f-ig}(s,t)}}{2}$$

$$= \frac{e_{f+ig}(s,t) + e_{f-ig}(s,t)}{2e_{f+ig}(s,t)e_{f-ig}(s,t)}$$

$$= \frac{e_{f+ig}(s,t) + e_{f-ig}(s,t)}{2}$$

$$= \mathrm{c}_{fg}(s,t).$$

2. We have

$$\mathrm{c}_{fg}(t,r)\,\mathrm{c}_{fg}(s,r) = \frac{e_{f+ig}(t,r) + e_{f-ig}(t,r)}{2}\,\frac{e_{f+ig}(s,r) + e_{f-ig}(s,r)}{2}$$

$$= \frac{1}{4}\Big(e_{f+ig}(t,r)e_{f+ig}(s,r) + e_{f+ig}(t,r)e_{f-ig}(s,r)$$

$$+e_{f-ig}(t,r)e_{f+ig}(s,r) + e_{f-ig}(t,r)e_{f-ig}(s,r)\Big),$$

$$\mathrm{s}_{fg}(t,r)\,\mathrm{s}_{fg}(s,r) = \frac{e_{f+ig}(t,r) - e_{f-ig}(t,r)}{2i}\,\frac{e_{f+ig}(s,r) - e_{f-ig}(s,r)}{2i}$$

$$= -\frac{1}{4}\Big(e_{f+ig}(t,r)e_{f+ig}(s,r) - e_{f+ig}(t,r)e_{f-ig}(s,r)$$

$$-e_{f-ig}(t,r)e_{f+ig}(s,r) + e_{f-ig}(t,r)e_{f-ig}(s,r)\Big).$$

Hence,

$$\mathrm{c}_{fg}(t,r)\,\mathrm{c}_{fg}(s,r) + \mathrm{s}_{fg}(t,r)\,\mathrm{s}_{fg}(s,r)$$

$$= \frac{1}{2}\left(e_{f+ig}(t,r)e_{f-ig}(s,r) + e_{f-ig}(t,r)e_{f+ig}(s,r)\right)$$

$$= \frac{1}{2}\left(e_{f+ig}(t,r)e_{f+ig}(r,s) + e_{f-ig}(t,r)e_{f-ig}(r,s)\right)$$

$$= \frac{1}{2}(e_{f+ig}(t,s) + e_{f-ig}(t,s))$$

$$= \mathrm{c}_{fg}(t,s).$$

3. We have

$$\begin{aligned}
\mathrm{c}_{fg}(t,s)\,\mathrm{c}_{fg}(s,r) &= \frac{e_{f+ig}(t,s)+e_{f-ig}(t,s)}{2}\,\frac{e_{f+ig}(s,r)+e_{f-ig}(s,r)}{2} \\
&= \frac{1}{4}\Big(e_{f+ig}(t,r)+e_{f-ig}(t,r)+e_{f+ig}(t,s)e_{f-ig}(s,r) \\
&\quad + e_{f-ig}(t,s)e_{f+ig}(s,r)\Big), \\
\mathrm{s}_{fg}(t,s)\,\mathrm{s}_{fg}(s,r) &= \frac{e_{f+ig}(t,s)-e_{f-ig}(t,s)}{2i}\,\frac{e_{f+ig}(s,r)-e_{f-ig}(s,r)}{2i} \\
&= -\frac{1}{4}\Big(e_{f+ig}(t,r)+e_{f-ig}(t,r)-e_{f+ig}(t,s)e_{f-ig}(s,r) \\
&\quad - e_{f-ig}(t,s)e_{f+ig}(s,r)\Big).
\end{aligned}$$

Hence,

$$\begin{aligned}
\mathrm{c}_{fg}(t,s)\,\mathrm{c}_{fg}(s,r)-\mathrm{s}_{fg}(t,s)\,\mathrm{s}_{fg}(s,r) &= \frac{1}{2}(e_{f+ig}(t,r)+e_{f-ig}(t,r)) \\
&= \mathrm{c}_{fg}(t,r).
\end{aligned}$$

Exercise 3.151 Let $f, g \in \mathrm{C_{rd}}$ satisfy (3.23). Prove

1. $\mathrm{s}_{fg}(s,t) = -\,\mathrm{s}_{fg}(t,s)$,
2. $\mathrm{s}_{fg}(t,s) = \mathrm{s}_{fg}(t,r)\,\mathrm{c}_{fg}(s,r) - \mathrm{c}_{fg}(t,r)\,\mathrm{s}_{fg}(s,r)$,
3. $\mathrm{s}_{fg}(t,r) = \mathrm{s}_{fg}(t,s)\,\mathrm{c}_{fg}(s,r) + \mathrm{c}_{fg}(t,s)\,\mathrm{s}_{fg}(s,r)$.

3.6 Taylor's Formula

Definition 3.152 Let $s, t \in \mathbb{T}$. We define the time scales *monomials* recursively by

$$g_0(t,s) = h_0(t,s) = 1$$

and

$$g_{k+1}(t,s) = \int_s^t g_k(\sigma(\tau),s)\Delta\tau, \quad h_{k+1}(t,s) = \int_s^t h_k(\tau,s)\Delta\tau, \quad k \in \mathbb{N}_0.$$

Remark 3.153 We have

$$\begin{aligned} g_1(t,s) &= \int_s^t g_0(\sigma(\tau),s)\Delta\tau \\ &= \int_s^t \Delta\tau \\ &= t-s, \\ h_1(t,s) &= \int_s^t h_0(\tau,s)\Delta\tau \\ &= \int_s^t \Delta\tau \\ &= t-s, \\ g_2(t,s) &= \int_s^t g_1(\sigma(\tau),s)\Delta\tau \\ &= \int_s^t (\sigma(\tau)-s)\Delta\tau, \\ h_2(t,s) &= \int_s^t h_1(\tau,s)\Delta\tau \\ &= \int_s^t (\tau-s)\Delta\tau, \end{aligned}$$

and so on. Moreover,

$$g_k^{\Delta}(t,s) = g_{k-1}(\sigma(t),s), \quad h_k^{\Delta}(t,s) = h_{k-1}(t,s), \quad k\in\mathbb{N}.$$

Lemma 3.154 *Let $n\in\mathbb{N}$. If f is n times differentiable and p_k, $0\le k\le n-1$, are differentiable at some $t\in\mathbb{T}$ with*

$$p_{k+1}^{\Delta}(t) = p_k^{\sigma}(t) \ \text{ for all } \ 0\le k\le n-2, \quad n\in\mathbb{N}\setminus\{1\},$$

then

$$\left(\sum_{k=0}^{n-1}(-1)^k f^{\Delta^k} p_k\right)^{\Delta}(t) = (-1)^{n-1} f^{\Delta^n}(t)p_{n-1}^{\sigma}(t) + f(t)p_0^{\Delta}(t).$$

Proof We have

$$\left(\sum_{k=0}^{n-1}(-1)^k f^{\Delta^k} p_k\right)^{\Delta}(t) = \sum_{k=0}^{n-1}(-1)^k \left(f^{\Delta^k} p_k\right)^{\Delta}(t)$$

$$= \sum_{k=0}^{n-1}(-1)^k \left(f^{\Delta^{k+1}}(t)p_k^{\sigma}(t) + f^{\Delta^k}(t)p_k^{\Delta}(t)\right)$$

$$= \sum_{k=0}^{n-1}(-1)^k f^{\Delta^{k+1}}(t)p_k^{\sigma}(t) + \sum_{k=0}^{n-1}(-1)^k f^{\Delta^k}(t)p_k^{\Delta}(t)$$

$$= \sum_{k=0}^{n-2}(-1)^k f^{\Delta^{k+1}}(t)p_k^{\sigma}(t) + (-1)^{n-1} f^{\Delta^n}(t)p_{n-1}^{\sigma}(t)$$

$$+ \sum_{k=1}^{n-1}(-1)^k f^{\Delta^k}(t)p_k^{\Delta}(t) + f^{\Delta^0}(t)p_0^{\Delta}(t)$$

$$= \sum_{k=0}^{n-2}(-1)^k f^{\Delta^{k+1}}(t)p_{k+1}^{\Delta}(t) + (-1)^{n-1} f^{\Delta^n}(t)p_{n-1}^{\sigma}(t)$$

$$+ \sum_{k=0}^{n-2}(-1)^{k+1} f^{\Delta^{k+1}}(t)p_{k+1}^{\Delta}(t) + f(t)p_0^{\Delta}(t)$$

$$= (-1)^{n-1} f^{\Delta^n}(t)p_{n-1}^{\sigma}(t) + f(t)p_0^{\Delta}(t),$$

completing the proof. □

Lemma 3.155 *The functions g_n satisfy for all $t \in \mathbb{T}$ the relationship*

$$g_n(\rho^k(t), t) = 0 \quad \textit{for all} \quad n \in \mathbb{N} \quad \textit{and all} \quad 0 \le k \le n-1.$$

Proof Let $n \in \mathbb{N}$ be arbitrarily chosen. Then

$$g_n(\rho^0(t), t) = g_n(t, t)$$

$$= \int_t^t g_{n-1}(\sigma(\tau), t)\Delta\tau$$

$$= 0.$$

Assume that

$$g_{n-1}(\rho^k(t), t) = 0 \quad \text{and} \quad g_n(\rho^k(t), t) = 0$$

for some $0 \le k < n - 1$. We will prove that

$$g_n(\rho^{k+1}(t), t) = 0.$$

1. Let $\rho^k(t)$ be left-dense. Then

$$\rho^{k+1}(t) = \rho(\rho^k(t)) = \rho^k(t).$$

Consequently, using the induction assumption, we have

$$g_n(\rho^{k+1}(t), t) = g_n(\rho^k(t), t) = 0.$$

2. Let $\rho^k(t)$ be left-scattered. Then

$$\rho(\rho^k(t)) < \rho^k(t),$$

and there is no $s \in \mathbb{T}$ such that $\rho^{k+1}(t) < s < \rho^k(t)$. Hence,

$$\sigma(\rho^{k+1}(t)) = \rho^k(t).$$

Therefore

$$g_n(\sigma(\rho^{k+1}(t)), t) = g_n(\rho^{k+1}(t), t) + \mu(\rho^{k+1}(t)) g_n^{\Delta}(\rho^{k+1}(t), t),$$

i.e.,

$$g_n(\rho^k(t), t) = g_n(\rho^{k+1}(t), t) + \mu(\rho^{k+1}(t)) g_n^{\Delta}(\rho^{k+1}(t), t),$$

whereupon

$$\begin{aligned} g_n(\rho^{k+1}(t), t) &= g_n(\rho^k(t), t) - \mu(\rho^{k+1}(t)) g_n^{\Delta}(\rho^{k+1}(t), t) \\ &= g_n(\rho^k(t), t) - \mu(\rho^{k+1}(t)) g_{n-1}(\sigma(\rho^{k+1}(t)), t) \\ &= g_n(\rho^k(t), t) - \mu(\rho^{k+1}(t)) g_{n-1}(\rho^k(t), t) \\ &= 0. \end{aligned}$$

The proof is complete. □

Lemma 3.156 *Let $n \in \mathbb{N}$ and suppose that f is $n-1$ times differentiable at $\rho^{n-1}(t)$. Then*

$$f(t) = \sum_{k=0}^{n-1} (-1)^k f^{\Delta^k}(\rho^{n-1}(t)) g_k(\rho^{n-1}(t), t).$$

Proof 1. Since,

$$\begin{aligned}\sum_{k=0}^{0} (-1)^k f^{\Delta^k}(\rho^0(t)) g_k(\rho^0(t), t) &= (-1)^0 f^{\Delta^0}(t) g_0(t, t)\\ &= f(t),\end{aligned}$$

the claim holds for $n = 1$.

2. Assume that

$$f(t) = \sum_{k=0}^{m-1} (-1)^k f^{\Delta^k}(\rho^{m-1}(t)) g_k(\rho^{m-1}(t), t)$$

holds for some $m \in \mathbb{N}$. We will prove that

$$f(t) = \sum_{k=0}^{m} (-1)^k f^{\Delta^k}(\rho^m(t)) g_k(\rho^m(t), t).$$

a. Let $\rho^{m-1}(t)$ be left-dense. Then

$$\rho^m(t) = \rho(\rho^{m-1}(t)) = \rho^{m-1}(t).$$

Thus, using the induction assumption, we obtain

$$\begin{aligned}&\sum_{k=0}^{m} (-1)^k f^{\Delta^k}(\rho^m(t)) g_k(\rho^m(t), t)\\ &= \sum_{k=0}^{m-1} (-1)^k f^{\Delta}[k(\rho^m(t)) g_k(\rho^m(t), t) + (-1)^m f^{\Delta^m}(\rho^m(t)) g_m(\rho^m(t), t)\\ &= \sum_{k=0}^{m-1} (-1)^k f^{\Delta^k}(\rho^{m-1}(t)) g_k(\rho^{m-1}(t), t)\\ &\quad + (-1)^m f^{\Delta^m}(\rho^{m-1}(t)) g_m(\rho^{m-1}(t), t)\end{aligned}$$

$$= \sum_{k=0}^{m-1} (-1)^k f^{\Delta^k}(\rho^{m-1}(t)) g_k(\rho^{m-1}(t), t)$$

$$= f(t),$$

where we have applied Lemma 3.155 for the third equality.

b. Let $\rho^{m-1}(t)$ be left-scattered. Then

$$\rho^m(t) = \rho(\rho^{m-1}(t)) < \rho^{m-1}(t),$$

and there is no $s \in \mathbb{T}$ such that

$$\rho^m(t) < s < \rho^{m-1}(t).$$

Also,

$$\sigma(\rho^m(t)) = \rho^{m-1}(t).$$

Hence,

$$g_k(\sigma(\rho^m(t)), t) = g_k(\rho^{m-1}(t), t).$$

Therefore,

$$\begin{aligned} g_k(\rho^{m-1}(t), t) &= g_k(\sigma(\rho^m(t)), t) \\ &= g_k(\rho^m(t), t) + \mu(\rho^m(t)) g_k^{\Delta}(\rho^m(t), t) \\ &= g_k(\rho^m(t), t) + \mu(\rho^m(t)) g_{k-1}(\sigma(\rho^m(t)), t) \\ &= g_k(\rho^m(t), t) + \mu(\rho^m(t)) g_{k-1}(\rho^{m-1}(t), t), \end{aligned}$$

whereupon

$$g_k(\rho^m(t), t) = g_k(\rho^{m-1}(t), t) - \mu(\rho^m(t)) g_{k-1}(\rho^{m-1}(t), t).$$

Consequently,

$$\sum_{k=0}^{m} (-1)^k f^{\Delta^k}(\rho^m(t)) g_k(\rho^m(t), t)$$

$$= f(\rho^m(t)) + \sum_{k=1}^{m} (-1)^k f^{\Delta^k}(\rho^m(t)) g_k(\rho^m(t), t)$$

$$= f(\rho^m(t)) + \sum_{k=1}^{m} (-1)^k f^{\Delta^k}(\rho^m(t)) g_k(\rho^{m-1}(t), t)$$

$$+ \sum_{k=1}^{m} (-1)^{k-1} f^{\Delta^k}(\rho^m(t)) \mu(\rho^m(t)) g_{k-1}(\rho^{m-1}(t), t)$$

$$= f(\rho^m(t)) + \sum_{k=1}^{m-1} (-1)^k f^{\Delta^k}(\rho^m(t)) g_k(\rho^{m-1}(t), t)$$

$$+ (-1)^m f^{\Delta^m}(\rho^m(t)) g_m(\rho^{m-1}(t), t)$$

$$+ \sum_{k=0}^{m-1} (-1)^k f^{\Delta^{k+1}}(\rho^m(t)) \mu(\rho^m(t)) g_k(\rho^{m-1}(t), t)$$

$$= \sum_{k=0}^{m-1} (-1)^k f^{\Delta^k}(\rho^m(t)) g_k(\rho^{m-1}(t), t)$$

$$+ \sum_{k=0}^{m-1} (-1)^k \mu(\rho^m(t)) f^{\Delta^{k+1}}(\rho^m(t)) g_k(\rho^{m-1}(t), t)$$

$$= \sum_{k=0}^{m-1} (-1)^k \left(f^{\Delta^k}(\rho^m(t)) + \mu(\rho^m(t)) (f^{\Delta^k})^{\Delta}(\rho^m(t)) \right) g_k(\rho^{m-1}(t), t)$$

$$= \sum_{k=0}^{m-1} (-1)^k f^{\Delta^k}(\sigma(\rho^m(t))) g_k(\rho^{m-1}(t), t)$$

$$= \sum_{k=0}^{m-1} (-1)^k f^{\Delta^k}(\rho^{m-1}(t)) g_k(\rho^{m-1}(t), t)$$

$$= f(t).$$

This completes the proof. □

Theorem 3.157 (Taylor's Formula) *Let $n \in \mathbb{N}$. Suppose that f is n times differentiable on $\mathbb{T}^{\kappa^n}$. Let $\alpha \in \mathbb{T}^{\kappa^{n-1}}$ and $t \in \mathbb{T}$. Then*

$$f(t) = \sum_{k=0}^{n-1} (-1)^k g_k(\alpha, t) f^{\Delta^k}(\alpha)$$

$$+ \int_\alpha^{\rho^{n-1}(t)} (-1)^{n-1} g_{n-1}(\sigma(\tau), t) f^{\Delta^n}(\tau) \Delta\tau.$$

Proof We note that, applying Lemma 3.154 for $p_k = g_k$, we have

$$\left(\sum_{k=0}^{n-1} (-1)^k g_k(\cdot, t) f^{\Delta^k} \right)^{\Delta} (\tau)$$

$$= (-1)^{n-1} f^{\Delta^n}(\tau) g_{n-1}(\sigma(\tau), t) + f(\tau) g_0^{\Delta}(\tau, t)$$

$$= (-1)^{n-1} f^{\Delta^n}(\tau) g_{n-1}(\sigma(\tau), t) \quad \text{for all} \quad \tau \in \mathbb{T}^{\kappa^n}.$$

We integrate the last relation from α to $\rho^{n-1}(t)$ to obtain

$$\int_\alpha^{\rho^{n-1}(t)} \left(\sum_{k=0}^{n-1} (-1)^k g_k(\cdot, t) f^{\Delta^k} \right)^{\Delta} (\tau) \Delta\tau$$

$$= \int_\alpha^{\rho^{n-1}(t)} (-1)^{n-1} f^{\Delta^n}(\tau) g_{n-1}(\sigma(\tau), t) \Delta\tau,$$

i.e.,

$$\sum_{k=0}^{n-1} (-1)^k g_k(\rho^{n-1}(t), t) f^{\Delta^k}(\rho^{n-1}(t)) - \sum_{k=0}^{n-1} (-1)^k g_k(\alpha, t) f^{\Delta^k}(\alpha)$$

$$= \int_{\alpha}^{\rho^{n-1}(t)} (-1)^{n-1} f^{\Delta^n}(\tau) g_{n-1}(\sigma(\tau), t) \Delta\tau.$$

Hence, applying Lemma 3.156, we get

$$f(t) - \sum_{k=0}^{n-1} (-1)^k g_k(\alpha, t) f^{\Delta^k}(\alpha) = \int_{\alpha}^{\rho^{n-1}(t)} (-1)^{n-1} f^{\Delta^n}(\tau) g_{n-1}(\sigma(\tau), t) \Delta\tau,$$

completing the proof. □

Theorem 3.158 *The functions g_n and h_n satisfy the relationship*

$$h_n(t, s) = (-1)^n g_n(s, t)$$

for all $t \in \mathbb{T}$ and all $s \in \mathbb{T}^{\kappa^n}$.

Proof Let $t \in \mathbb{T}$ and $s \in \mathbb{T}^{\kappa^n}$ be arbitrarily chosen. We apply Theorem 3.157 for $\alpha = s$ and $f(\tau) = h_n(\tau, s)$. We observe that

$$f^{\Delta^k}(\tau) = h_{n-k}(\tau, s), \quad 0 \le k \le n.$$

Hence,

$$f^{\Delta^k}(s) = h_{n-k}(s, s) = 0, \quad 0 \le k \le n-1$$

and

$$f^{\Delta^n}(s) = h_0(s, s) = 1, \quad f^{\Delta^{n+1}}(\tau) = 0.$$

From here, using Taylor's formula, we get

$$\begin{aligned} f(t) &= h_n(t, s) \\ &= \sum_{k=0}^{n} (-1)^k g_k(\alpha, t) f^{\Delta^k}(\alpha) + \int_{\alpha}^{\rho^n(t)} (-1)^n g_n(\sigma(\tau), t) f^{\Delta^{n+1}}(\tau) \Delta\tau \\ &= \sum_{k=0}^{n} (-1)^k g_k(s, t) f^{\Delta^k}(s) + \int_{s}^{\rho^n(t)} (-1)^n g_n(\sigma(\tau), t) f^{\Delta^{n+1}}(\tau) \Delta\tau \\ &= \sum_{k=0}^{n-1} (-1)^k g_k(s, t) f^{\Delta^k}(s) + (-1)^n g_n(s, t) f^{\Delta^n}(s) \\ &= (-1)^n g_n(s, t) f^{\Delta^n}(s) \\ &= (-1)^n g_n(s, t), \end{aligned}$$

i.e.,

$$h_n(t,s) = (-1)^n g_n(s,t),$$

completing the proof. □

From Theorem 3.157 and Theorem 3.158, the following result is clear.

Theorem 3.159 (Taylor's Formula) *Let $n \in \mathbb{N}$. Suppose that f is n times differentiable on $\mathbb{T}^{\kappa^n}$. Let $\alpha \in \mathbb{T}^{\kappa^{n-1}}$ and $t \in \mathbb{T}$. Then*

$$f(t) = \sum_{k=0}^{n-1} h_k(t,\alpha) f^{\Delta^k}(\alpha) + \int_\alpha^{\rho^{n-1}(t)} h_{n-1}(t,\sigma(\tau)) f^{\Delta^n}(\tau)\Delta\tau.$$

3.7 L'Hôpital's Rule

Definition 3.160 Let $\overline{\mathbb{T}} = \mathbb{T} \cup \{\sup \mathbb{T}\} \cup \{\inf \mathbb{T}\}$. If $\infty \in \overline{\mathbb{T}}$, then ∞ is called left-dense. If $-\infty \in \overline{\mathbb{T}}$, then $-\infty$ is called right-dense.

Remark 3.161 For any left-dense point $t_0 \in \overline{\mathbb{T}}$ and for any $\varepsilon > 0$, we set

$$L_\varepsilon(t_0) = \{t \in \mathbb{T} :\ 0 < t_0 - t < \varepsilon\}.$$

If $t_0 \in \overline{\mathbb{T}}$ is left-dense, then $L_\varepsilon(t_0)$ is nonempty. If $\infty \in \overline{\mathbb{T}}$, then

$$L_\varepsilon(\infty) = \left\{t \in \mathbb{T} :\ t > \frac{1}{\varepsilon}\right\}.$$

For a right-dense point $t_1 \in \overline{\mathbb{T}}$, we define

$$R_\varepsilon(t_1) = \{t \in \mathbb{T} :\ 0 < t - t_1 < \varepsilon\}.$$

For every right-dense point $t_1 \in \overline{\mathbb{T}}$, the set $R_\varepsilon(t_1)$ is nonempty. If $-\infty \in \overline{\mathbb{T}}$, then

$$R_\varepsilon(-\infty) = \left\{t \in \mathbb{T} :\ t < -\frac{1}{\varepsilon}\right\}.$$

Definition 3.162 Let $h : \mathbb{T} \to \mathbb{R}$.

1. Let $t_0 \in \overline{\mathbb{T}}$ be left-dense. We define

$$\liminf_{t\to t_0-} h(t) = \lim_{\varepsilon\to 0+} \inf_{t\in L_\varepsilon(t_0)} h(t), \quad \limsup_{t\to t_0-} h(t) = \lim_{\varepsilon\to 0+} \sup_{t\in L_\varepsilon(t_0)} h(t).$$

2. Let $t_1 \in \overline{\mathbb{T}}$ be right-dense. We define

$$\liminf_{t\to t_1+} h(t) = \lim_{\varepsilon\to 0+} \inf_{t\in R_\varepsilon(t_1)} h(t), \quad \limsup_{t\to t_1+} h(t) = \lim_{\varepsilon\to 0+} \sup_{t\in R_\varepsilon(t_1)} h(t).$$

Theorem 3.163 (L'Hôpital's Rule) *Let f and g be differentiable on $\mathbb{T}$ and*

$$\lim_{t\to t_0-} f(t) = \lim_{t\to t_0-} g(t) = 0 \quad \textit{for some left-dense} \quad t_0 \in \overline{\mathbb{T}}. \tag{3.24}$$

Suppose there exists $\varepsilon > 0$ such that

$$g(t) > 0, \quad g^\Delta(t) < 0 \quad \textit{for all} \quad t \in L_\varepsilon(t_0). \tag{3.25}$$

Then

$$\liminf_{t\to t_0-} \frac{f^\Delta(t)}{g^\Delta(t)} \le \liminf_{t\to t_0-} \frac{f(t)}{g(t)} \le \limsup_{t\to t_0-} \frac{f(t)}{g(t)} \le \limsup_{t\to t_0-} \frac{f^\Delta(t)}{g^\Delta(t)}.$$

Proof Let $\delta \in (0, \varepsilon]$. We set

$$\alpha = \inf_{\tau\in L_\delta(t_0)} \frac{f^\Delta(\tau)}{g^\Delta(\tau)}, \quad \beta = \sup_{\tau\in L_\delta(t_0)} \frac{f^\Delta(\tau)}{g^\Delta(\tau)}.$$

Then, using (3.25),

$$\alpha g^\Delta(\tau) \ge f^\Delta(\tau) \ge \beta g^\Delta(\tau) \quad \text{for any} \quad \tau \in L_\delta(t_0).$$

Hence,

$$\int_s^t \alpha g^\Delta(\tau)\Delta\tau \ge \int_s^t f^\Delta(\tau)\Delta\tau \ge \int_s^t \beta g^\Delta(\tau)\Delta\tau$$

for all $s, t \in L_\delta(t_0)$, $s < t$. Therefore,

$$\alpha(g(t) - g(s)) \ge f(t) - f(s) \ge \beta(g(t) - g(s))$$

for all $s, t \in L_\delta(t_0)$, $s < t$. From here,

$$-\alpha g(s) \ge -f(s) \ge -\beta g(s) \quad \text{for all} \quad s \in L_\delta(t_0)$$

as $t \to t_0-$, whereupon

$$\alpha \le \frac{f(s)}{g(s)} \le \beta \quad \text{for all} \quad s \in L_\delta(t_0).$$

Letting $\delta \to 0+$, we get the desired result. □

Exercise 3.164 Let f and g be differentiable on $\mathbb{T}$ and suppose

$$\lim_{t\to t_0+} f(t) = \lim_{t\to t_0+} g(t) = 0 \quad \text{for some right-dense} \quad t_0 \in \overline{\mathbb{T}}.$$

Assume there exists $\varepsilon > 0$ such that

$$g(t) > 0, \quad g^{\Delta}(t) < 0 \quad \text{for all} \quad t \in R_{\varepsilon}(t_0).$$

Prove that then

$$\liminf_{t\to t_0+} \frac{f^{\Delta}(t)}{g^{\Delta}(t)} \le \liminf_{t\to t_0+} \frac{f(t)}{g(t)} \le \limsup_{t\to t_0+} \frac{f(t)}{g(t)} \le \limsup_{t\to t_0+} \frac{f^{\Delta}(t)}{g^{\Delta}(t)}$$

holds.

Example 3.165 Let $\mathbb{T} = [-1, 1) \cup 2^{\mathbb{N}_0}$, where $[-1, 1)$ is the real-valued interval. We compute

$$l = \lim_{t\to 1-} \frac{t^3 - 2t^2 - t + 2}{t^3 - 7t + 6}.$$

Note that $t = 1$ is left-dense and $f^{\Delta}(t) = f'(t)$ for all $t \in [-1, 1)$. Then

$$\begin{aligned} l &= \lim_{t\to 1-} \frac{t^2(t-2) - (t-2)}{t^3 - t - 6t + 6} \\ &= \lim_{t\to 1-} \frac{(t^2-1)(t-2)}{t(t-1)(t+1) - 6(t-1)} \\ &= \lim_{t\to 1-} \frac{(t-1)(t+1)(t-2)}{(t-1)(t^2+t-6)} \\ &= \lim_{t\to 1-} \frac{(t+1)(t-2)}{t^2+t-6} \\ &= \frac{-2}{-4} \\ &= \frac{1}{2}. \end{aligned}$$

Example 3.166 Let $\mathbb{T} = [-3, 0) \cup \mathbb{N}_0$, where $[-3, 0)$ is the real-valued interval. We find

$$l = \lim_{t\to 0-} \frac{t\cos t - \sin t}{t}.$$

Note that $t = 0$ is left-dense and $f^{\Delta}(t) = f'(t)$ for $t \in [-3, 0)$. Hence,

$$l = \lim_{t\to 0-} \frac{(t\cos t - \sin t)'}{t'}$$

$$= \lim_{t\to 0-} (\cos t - t\sin t - \cos t)$$

$$= \lim_{t\to 0-} (-t\sin t)$$

$$= 0.$$

Example 3.167 Let $\mathbb{T} = [-3, 1) \cup 2^{\mathbb{N}_0}$, where $[-3, 1)$ is the real-valued interval. We compute

$$l = \lim_{t\to 0} \frac{\tan t - \sin t}{t - \sin t}.$$

Here, $t = 0$ is left-dense and $f^{\Delta}(t) = f'(t)$ for all $t \in [-3, 1)$. Then

$$l = \lim_{t\to 0} \frac{(\tan t - \sin t)'}{(t - \sin t)'}$$

$$= \lim_{t\to 0} \frac{\frac{1}{\cos^2 t} - \cos t}{1 - \cos t}$$

$$= \lim_{t\to 0} \frac{1 - \cos^3 t}{\cos^2 t(1 - \cos t)}$$

$$= \lim_{t\to 0} \frac{(1 - \cos t)(1 + \cos t + \cos^2 t)}{\cos^2 t(1 - \cos t)}$$

$$= \lim_{t\to 0} \frac{1 + \cos t + \cos^2 t}{\cos^2 t}$$

$$= 3.$$

Exercise 3.168 Let $\mathbb{T} = [-3, 3) \cup \{3, 9, 27, \ldots\}$, where $[-3, 3)$ is the real valued interval. Compute

1. $\lim_{t\to 0} \frac{\tan t - \sin t}{t - \sin t}$,
2. $\lim_{t\to \frac{\pi}{2}} \frac{\tan t}{\tan(5t)}$,
3. $\lim_{t\to 0} \frac{\log(\sin(10t))}{\log(\sin t)}$,
4. $\lim_{t\to 0}(1 - \cos t)\cot t$,
5. $\lim_{t\to 3-} \left(\frac{1}{t-3} - \frac{5}{t^2-t-6}\right)$,
6. $\lim_{t\to \frac{\pi}{2}} \left(\frac{t}{\cot t} - \frac{\pi}{2\cos t}\right)$.

Solution 1. 3,
2. 5,
3. 1,
4. 0,
5. $\frac{1}{5}$,
6. -1.

Theorem 3.169 (L'Hôpital's Rule) *Let f and g be differentiable on $\mathbb{T}$ and*

$$\lim_{t \to t_0-} g(t) = \infty \quad \textit{for some left-dense point} \quad t_0 \in \overline{\mathbb{T}}. \tag{3.26}$$

Assume there exists $\varepsilon > 0$ such that

$$g(t) > 0, \quad g^{\Delta}(t) > 0 \quad \textit{for all} \quad t \in L_{\varepsilon}(t_0). \tag{3.27}$$

Then

$$\lim_{t \to t_0-} \frac{f^{\Delta}(t)}{g^{\Delta}(t)} = r \in \overline{\mathbb{R}}$$

implies

$$\lim_{t \to t_0-} \frac{f(t)}{g(t)} = r.$$

Proof 1. Let $r \in \mathbb{R}$ and $c > 0$. Then there exists $\delta \in (0, \varepsilon]$ such that

$$\left| \frac{f^{\Delta}(\tau)}{g^{\Delta}(\tau)} - r \right| \leq c \quad \text{for all} \quad \tau \in L_{\delta}(t_0).$$

Hence, using (3.27), we obtain

$$-cg^{\Delta}(\tau) \leq f^{\Delta}(\tau) - rg^{\Delta}(\tau) \leq cg^{\Delta}(\tau) \quad \text{for all} \quad \tau \in L_{\delta}(t_0).$$

From here,

$$-\int_s^t cg^{\Delta}(\tau)\Delta\tau \leq \int_s^t (f^{\Delta}(\tau) - rg^{\Delta}(\tau))\Delta\tau \leq \int_s^t cg^{\Delta}(\tau)\Delta\tau$$

for all $s, t \in L_{\delta}(t_0)$, $s < t$. Therefore,

$$-c(g(t) - g(s)) \leq f(t) - f(s) - r(g(t) - g(s)) \leq c(g(t) - g(s),$$

i.e.,

$$(r - c)(g(t) - g(s)) \leq f(t) - f(s) \leq (c + r)(g(t) - g(s))$$

for all $s, t \in L_{\delta}(t_0)$, $s < t$. Hence,

$$(r-c)\left(1-\frac{g(s)}{g(t)}\right)\le\frac{f(t)}{g(t)}-\frac{f(s)}{g(t)}\le(c+r)\left(1-\frac{g(s)}{g(t)}\right)$$

for all $s, t \in L_\delta(t_0)$, $s < t$. Letting $t \to t_0-$ and using (3.26), we find

$$r-c\le\liminf_{t\to t_0-}\frac{f(t)}{g(t)}\le\limsup_{t\to t_0-}\frac{f(t)}{g(t)}\le c+r.$$

Now, we let $c \to 0+$, and then $\lim_{t\to t_0-}\frac{f(t)}{g(t)}$ exists and equals r.

2. Let $r = \infty$ and $c > 0$. Then there exists $\delta \in (0, \varepsilon]$ such that

$$\frac{f^\Delta(\tau)}{g^\Delta(\tau)}\ge\frac{1}{c}\quad\text{for all}\quad \tau\in L_\delta(t_0),$$

and hence

$$f^\Delta(\tau)\ge\frac{1}{c}g^\Delta(\tau)\quad\text{for all}\quad \tau\in L_\delta(t_0).$$

From here,

$$\int_s^t f^\Delta(\tau)\Delta\tau\ge\int_s^t\frac{1}{c}g^\Delta(\tau)\Delta\tau\quad\text{for all}\quad s,t\in L_\delta(t_0),\quad s<t,$$

whereupon

$$f(t)-f(s)\ge\frac{1}{c}(g(t)-g(s))\quad\text{for all}\quad s,t\in L_\delta(t_0),\quad s<t.$$

Therefore,

$$\frac{f(t)}{g(t)}-\frac{f(s)}{g(t)}\ge\frac{1}{c}\left(1-\frac{g(s)}{g(t)}\right)\quad\text{for all}\quad s,t\in L_\delta(t_0),\quad s<t.$$

Then letting $t \to t_0-$, we find

$$\lim_{t\to t_0-}\frac{f(t)}{g(t)}\ge\frac{1}{c}.$$

Now, letting $c \to 0+$, we obtain

$$\lim_{t\to t_0-}\frac{f(t)}{g(t)}=\infty=r.$$

3. The case $r = -\infty$ is left to the reader.

This completes the proof. □

Exercise 3.170 Let f and g be differentiable on $\mathbb{T}$ and suppose that

$$\lim_{t\to t_0+} g(t) = \infty \quad \text{for some right-dense point} \quad t_0 \in \overline{\mathbb{T}}.$$

Assume that there exists $\varepsilon > 0$ such that

$$g(t) > 0, \quad g^{\Delta}(t) > 0 \quad \text{for all} \quad t \in \mathbb{R}_{\varepsilon}(t_0).$$

Prove that

$$\lim_{t\to t_0+} \frac{f^{\Delta}(t)}{g^{\Delta}(t)} = r \in \overline{\mathbb{R}}$$

implies

$$\lim_{t\to t_0+} \frac{f(t)}{g(t)} = r.$$

3.8 Improper Integrals of the First Kind

Throughout this section, we assume that $\mathbb{T}$ is unbounded above.

Definition 3.171 Suppose that the real-valued function f is defined on $[a, \infty)$ and is integrable from a to any point $A \in \mathbb{T}$, $A \geq a$. Assume that the integral

$$F(A) = \int_a^A f(t)\Delta t$$

approaches a finite limit as $A \to \infty$. Then we call that limit the *improper integral of the first kind* from a to ∞ and write

$$\int_a^{\infty} f(t)\Delta t = \lim_{A\to\infty} \left\{ \int_a^A f(t)\Delta t \right\}. \tag{3.28}$$

In such a case, we say that the improper integral

$$\int_a^{\infty} f(t)\Delta t \tag{3.29}$$

exists or that it is *convergent*. If the limit (3.28) does not exist, then the improper integral (3.29) is said to be not existent or *divergent*.

Example 3.172 Let $\mathbb{T} = \mathbb{Z}$. We consider

$$I = \int_1^{\infty} \frac{3t^2 + 3t + 1}{t^3(t+1)^3} \Delta t.$$

Here, $\sigma(t) = t + 1$. With $F(t) = \frac{1}{t^3}$, we have

$$\begin{aligned} F^{\Delta}(t) &= -\frac{(\sigma(t))^2 + t\sigma(t) + t^2}{t^3(\sigma(t))^3} \\ &= -\frac{(t+1)^2 3 + t(t+1) + t^2}{t^3(t+1)^3} \\ &= -\frac{t^2 + 2t + 1 + t^2 + t + t^2}{t^3(t+1)^3} \\ &= -\frac{3t^2 + 3t + 1}{t^3(t+1)^3}. \end{aligned}$$

Therefore,

$$\begin{aligned} I &= \lim_{A\to\infty} \int_1^A \frac{3t^2 + 3t + 1}{t^3(t+1)^3} \Delta t \\ &= -\lim_{A\to\infty} \int_1^A F^{\Delta}(t)\Delta t \\ &= -\lim_{A\to\infty} (F(A) - F(1)) \\ &= -\lim_{A\to\infty} \left(\frac{1}{A^3} - 1\right) \\ &= 1. \end{aligned}$$

Example 3.173 Let $\mathbb{T} = 2^{\mathbb{N}_0}$. We consider

$$I = \int_1^{\infty} \frac{1}{t^3} \Delta t.$$

Here, $\sigma(t) = 2t$. With $F(t) = \frac{1}{t^2}$, we have

$$\begin{aligned} F^{\Delta}(t) &= -\frac{\sigma(t) + t}{t^2(\sigma(t))^2} \\ &= -\frac{2t + t}{4t^4} \\ &= -\frac{3}{4t^3}. \end{aligned}$$

Therefore,

$$\begin{aligned}
I &= \lim_{A\to\infty} \int_1^A \frac{1}{t^2}\Delta t \\
&= -\frac{4}{3}\lim_{A\to\infty}\int_1^A F^{\Delta}(t)\Delta t \\
&= -\frac{4}{3}\lim_{A\to\infty}(F(A)-F(1)) \\
&= -\frac{4}{3}\lim_{A\to\infty}\left(\frac{1}{A^2}-1\right) \\
&= \frac{4}{3}.
\end{aligned}$$

Example 3.174 Let $\mathbb{T} = 3^{\mathbb{N}_0}$. We consider

$$I = \int_1^{\infty} \frac{1}{\sqrt{t}}\Delta t.$$

Here, $\sigma(t) = 3t$. With $F(t) = \sqrt{t}$, we have

$$\begin{aligned}
F^{\Delta}(t) &= \frac{\sqrt{\sigma(t)}-\sqrt{t}}{\sigma(t)-t} \\
&= \frac{\sqrt{\sigma(t)}-\sqrt{t}}{(\sqrt{\sigma(t)}-\sqrt{t})(\sqrt{\sigma(t)}+\sqrt{t})} \\
&= \frac{1}{\sqrt{\sigma(t)}+\sqrt{t}} \\
&= \frac{1}{(1+\sqrt{3})\sqrt{t}}.
\end{aligned}$$

Therefore,

$$\begin{aligned}
I &= \lim_{A\to\infty} \int_1^A \frac{1}{\sqrt{t}}\Delta t \\
&= (1+\sqrt{3})\lim_{A\to\infty}\int_1^A F^{\Delta}(t)\Delta t \\
&= (1+\sqrt{3})\lim_{A\to\infty}(F(A)-F(1))
\end{aligned}$$

$$= (1+\sqrt{3}) \lim_{A\to\infty} (\sqrt{A}-1)$$

$$= \infty.$$

Consequently, this improper integral diverges to ∞.

Exercise 3.175 Let $\mathbb{T} = 2\mathbb{Z}$. Investigate the following integrals for convergence and divergence.

1. $\int_1^\infty \frac{2t+2}{(t^2+1)(t^2+4t+5)} \Delta t$,
2. $\int_2^\infty \frac{1}{4t^2+12t+5} \Delta t$,
3. $\int_0^\infty \frac{1}{t^2+10t+24} \Delta t$.

Solution 1. $\frac{1}{2}$,
2. $\frac{1}{10}$,
3. $\frac{1}{4}$.

Theorem 3.176 (Cauchy's Criterion) *For the existence of the integral* (3.28), *it is necessary and sufficient that for any given* $\varepsilon > 0$, *there exists* $A_0 > a$ *such that*

$$\left| \int_{A_1}^{A_2} f(t)\Delta t \right| < \varepsilon \tag{3.30}$$

for any $A_1, A_2 \in \mathbb{T}$ *satisfying the inequalities* $A_1 > A_0$ *and* $A_2 > A_0$.

Proof The convergence of the integral (3.28) is equivalent to the existence of the limit $\lim_{A\to\infty} F(A)$. Using Cauchy's criterion for the existence of the limit of a function, it follows that the existence of the integral (3.28) is equivalent to the condition (3.30). □

Example 3.177 Let $\mathbb{T} = 3\mathbb{Z}$. We prove that the integral

$$\int_1^\infty \frac{1}{t^2+11t+28} \Delta t$$

is convergent. Here, $\sigma(t) = t+3$. With $F(t) = \frac{1}{t+4}$, we have

$$\begin{aligned} F^\Delta(t) &= -\frac{1}{(t+4)(\sigma(t)+4)} \\ &= -\frac{1}{(t+4)(t+7)} \\ &= -\frac{1}{t^2+11t+28}. \end{aligned}$$

Let $\varepsilon > 0$ be arbitrarily chosen. We take

$$A > \max\left\{\frac{2-4\varepsilon}{\varepsilon}, 1\right\}.$$

Then, for any $A_1, A_2 > A$, we have

$$\begin{aligned}
\left|\int_{A_1}^{A_2} \frac{1}{t^2+11t+28}\Delta t\right| &= \left|-\int_{A_1}^{A_2} F^{\Delta}(t)\Delta t\right| \\
&= |F(A_2) - F(A_1)| \\
&= \left|\frac{1}{A_2+4} - \frac{1}{A_1+4}\right| \\
&\leq \frac{1}{A_2+4} + \frac{1}{A_1+4} \\
&< \frac{2}{A+4} \\
&< \varepsilon.
\end{aligned}$$

Therefore, employing the Cauchy criterion, we conclude that the considered integral is convergent.

Example 3.178 Let $\mathbb{T} = 2^{\mathbb{N}_0}$. We investigate the integral

$$\int_1^{\infty} \frac{3t+7}{2(t+2)(t+3)(t+4)(2t+3)}\Delta t$$

for convergence. Here, $\sigma(t) = 2t$. With $F(t) = \frac{1}{(t+3)(t+4)}$, we have

$$\begin{aligned}
F^{\Delta}(t) &= -\frac{t+4+\sigma(t)+3}{(t+3)(t+4)(\sigma(t)+3)(\sigma(t)+4)} \\
&= -\frac{t+4+2t+3}{2(t+2)(t+3)(t+4)(2t+3)} \\
&= -\frac{3t+7}{2(t+2)(t+3)(t+4)(2t+3)}.
\end{aligned}$$

Let $\varepsilon > 0$ be arbitrarily chosen. We take $A > 1$ such that

$$A^2 + 7A + 21 - \frac{2}{\varepsilon} > 0.$$

Then, for any $A_1, A_2 > A$, we have

$$\begin{aligned}\left|\int_{A_1}^{A_2} \frac{3t+7}{2(t+2)(t+3)(t+4)(2t+3)}\Delta t\right| &= \left|-\int_{A_1}^{A_2} F^{\Delta}(t)\Delta t\right| \\ &= |F(A_2)-F(A_1)| \\ &= \left|\frac{1}{(A_2+3)(A_2+4)}-\frac{1}{(A_1+3)(A_1+4)}\right| \\ &\le \frac{1}{(A_2+3)(A_2+4)}+\frac{1}{(A_1+3)(A_1+4)} \\ &< \frac{2}{(A+3)(A+4)} \\ &< \varepsilon.\end{aligned}$$

Thus, using the Cauchy criterion, it follows that the considered integral is convergent.

Example 3.179 Let $\mathbb{T} = 2^{\mathbb{N}_0}$. We investigate the integral

$$\int_1^{\infty} \frac{2\sin t - \sin(2t)}{2t^2}\Delta t$$

for convergence. Here, $\sigma(t) = 2t$. With $F(t) = \frac{\sin t}{t}$, we have

$$\begin{aligned}F^{\Delta}(t) &= \frac{t\frac{\sin(2t)-\sin t}{2t-t} - \sin t}{2t^2} \\ &= \frac{\sin(2t) - 2\sin t}{2t^2}.\end{aligned}$$

Let $\varepsilon > 0$ be arbitrarily chosen. We take $A > \frac{2}{\varepsilon}$. Then, for any $A_1, A_2 > A$, we have

$$\begin{aligned}\left|\int_{A_1}^{A_2} \frac{2\sin t - t\sin(2t)}{2t^2}\Delta t\right| &= \left|-\int_{A_1}^{A_2} F^{\Delta}(t)\Delta t\right| \\ &= |F(A_2) - F(A_1)|\end{aligned}$$

$$
\begin{aligned}
&= \left| \frac{\sin A_2}{A_2} - \frac{\sin A_1}{A_1} \right| \\
&\le \frac{|\sin A_2|}{A_2} + \frac{|\sin A_1|}{A_1} \\
&\le \frac{1}{A_2} + \frac{1}{A_1} \\
&< \frac{2}{A} \\
&< \varepsilon.
\end{aligned}
$$

Thus, using the Cauchy criterion, we conclude that the considered integral is convergent.

Exercise 3.180 Let $\mathbb{T} = 3\mathbb{Z}$. Using Cauchy's criterion, prove that the integral

$$\int_1^\infty \frac{1}{(2t+1)(2t+7)} \Delta t$$

is convergent.

Definition 3.181 An integral of type (3.29) is said to be *absolutely convergent* provided the integral

$$\int_a^\infty |f(t)| \Delta t \tag{3.31}$$

of the modulus of the function f is convergent. If an integral of type (3.29) is convergent but not absolutely convergent, then it is called *conditionally convergent*.

Theorem 3.182 *If the integral* (3.29) *is absolutely convergent, then it is convergent.*

Proof Let (3.29) be absolutely convergent. Then the integral (3.31) is convergent. Suppose that $\varepsilon > 0$ is arbitrarily chosen. Hence, employing the Cauchy criterion, it follows that there exists $A > a$ such that for any $A_1, A_2 > A$, we have

$$\left| \int_{A_1}^{A_2} |f(t)| \Delta t \right| < \varepsilon.$$

From here, for any $A_1, A_2 > A$, we have

$$\left| \int_{A_1}^{A_2} f(t) \Delta t \right| \le \left| \int_{A_1}^{A_2} |f(t)| \Delta t \right| < \varepsilon,$$

which completes the proof. □

Theorem 3.183 *An integral* (3.29) *with* $f(t) \geq 0$ *for all* $t \geq a$ *is convergent if and only if there exists a constant* $M > 0$ *such that*

$$\int_a^A f(t)\Delta t \leq M \quad \textit{whenever} \quad A \geq a.$$

Proof 1. Let $F(A) = \int_a^A f(t)\Delta t \leq M$ whenever $A \geq a$. Then

$$\int_a^\infty f(t)\Delta t = \lim_{A\to\infty} F(A) \leq M.$$

Therefore, the integral (3.29) is convergent.

2. Let the integral (3.29) be convergent. Assume that $F(A)$, $A \geq a$, is unbounded. Then

$$\int_a^\infty f(t)\Delta t = \lim_{A\to\infty} F(A) = \infty,$$

which is a contradiction. □

Theorem 3.184 *Let the inequalities* $0 \leq f(t) \leq g(t)$ *be satisfied for all* $t \in [a, \infty)$. *Then the convergence of the improper integral*

$$\int_a^\infty g(t)\Delta t \tag{3.32}$$

implies the convergence of the improper integral (3.29), *while the divergence of the improper integral* (3.29) *implies the divergence of the improper integral* (3.32).

Proof Since $0 \leq f(t) \leq g(t)$ for any $t \in [a, \infty)$, we get

$$0 \leq \int_a^A f(t)\Delta t \leq \int_a^A g(t)\Delta t \quad \text{for any} \quad A \in [a, \infty),$$

which completes the proof. □

Example 3.185 Let $\mathbb{T} = \mathbb{Z}$. Consider the integral

$$I = \int_1^\infty (t^6 + 7t^3 + 100) \log \frac{t+1}{t} \Delta t.$$

Here, $\sigma(t) = t + 1$ and

$$(t^6 + 7t^3 + 100) \log \frac{t+1}{t} \geq \log \frac{t+1}{t} \quad \text{for any} \quad t \in [1, \infty)$$

and

$$I \geq \int_1^\infty \log\frac{t+1}{t}\Delta t.$$

Note that, with $F(t) = \log t$, we have

$$\begin{aligned} F^\Delta(t) &= \frac{\log\sigma(t) - \log t}{\sigma(t) - t} \\ &= \log(t+1) - \log t \\ &= \log\frac{t+1}{t}. \end{aligned}$$

Therefore,

$$\int_1^\infty \log\frac{t+1}{t}\Delta t = \lim_{A\to\infty}\int_1^A F^\Delta(t)\Delta t = \lim_{A\to\infty}(F(A) - F(1)) = \lim_{A\to\infty}\log A = \infty.$$

Hence, the improper integral I is divergent.

Example 3.186 Let $\mathbb{T} = 2^{\mathbb{N}_0}$. Consider the integral

$$I = \int_1^\infty \frac{1}{t^3(t^2+5)(t^2+7t+1)}\Delta t.$$

Here, $\sigma(t) = 2t$ and

$$\frac{1}{t^3(t^2+5)(t^2+7t+1)} \leq \frac{1}{t^3} \quad \text{for any} \quad t \in [1, \infty).$$

Also, with $F(t) = \frac{1}{t^2}$, we have

$$F^\Delta(t) = -\frac{\sigma(t)+t}{t^2(\sigma(t))^2} = -\frac{3}{4t^3}.$$

Hence,

$$\begin{aligned} \int_1^\infty \frac{1}{t^3}\Delta t &= \lim_{A\to\infty}\int_1^A \frac{1}{t^3}\Delta t \\ &= -\frac{4}{3}\lim_{A\to\infty}\int_1^A F^\Delta(t)\Delta t \\ &= -\frac{4}{3}\lim_{A\to\infty}(F(A) - F(1)) \end{aligned}$$

$$= \frac{4}{3}.$$

Therefore, the integral I is convergent.

Example 3.187 Let $\mathbb{T} = 3^{\mathbb{N}_0}$. Consider the integral

$$I = \int_1^\infty \frac{\sin^3 t + e_1(t,1) + \cos^3 t + 5}{t} \Delta t.$$

Here, $\sigma(t) = 3t$ and

$$\frac{\sin^3 t + e_1(t,1) + \cos^3 t + 5}{t} \geq \frac{1}{t} \quad \text{for any} \quad t \in [1, \infty),$$

whereupon

$$I \geq \int_1^\infty \frac{1}{t} \Delta t.$$

Also, with $F(t) = \log t$, we have

$$\begin{aligned} F^\Delta(t) &= \frac{\log(\sigma(t)) - \log t}{\sigma(t) - t} \\ &= \frac{\log(3t) - \log t}{2t} \\ &= \frac{\log 3}{2t}. \end{aligned}$$

Hence,

$$\begin{aligned} \int_1^\infty \frac{1}{t} \Delta t &= \frac{2}{\log 3} \lim_{A \to \infty} \int_1^A F^\Delta(t) \Delta t \\ &= \frac{2}{\log 3} \lim_{A \to \infty} (F(A) - F(1)) \\ &= \frac{2}{\log 3} \lim_{A \to \infty} \log A \\ &= \infty. \end{aligned}$$

Therefore, the considered integral is divergent.

Exercise 3.188 Investigate the following integrals for convergence and divergence.

1. $\int_1^\infty (t^7 + 11t^6 + 12t^5 + 13t + 4) \log \frac{t+1}{t} \Delta t$, $\mathbb{T} = \mathbb{Z}$,

2. $\int_1^\infty (t^4 + 11t^2 + 100) \log \frac{t+3}{t} \Delta t$, $\mathbb{T} = 3\mathbb{Z}$,
3. $\int_1^\infty \frac{1}{t^2}(t^4 + 5t^3 + 5t^2 + 5t + 5)\Delta t$, $\mathbb{T} = 2^{\mathbb{N}_0}$,
4. $\int_1^\infty \frac{1}{t^2(t^3+11t^2+12t+13)(t+1)} \Delta t$, $\mathbb{T} = 3^{\mathbb{N}_0}$,
5. $\int_1^\infty \frac{1}{(t+1)^2(t+3)^3(t^4+10t+10)} \Delta t$, $\mathbb{T} = \mathbb{Z}$,
6. $\int_1^\infty t^2(t^6 + 5t + 1)\Delta t$, $\mathbb{T} = 3^{\mathbb{N}_0}$.

Solution 1. Divergent,
2. divergent,
3. divergent,
4. convergent,
5. convergent,
6. divergent.

Theorem 3.189 *Let $|f(t)| \leq g(t)$ for all $t \in \mathbb{T}$ with $t \geq a$. Then the convergence of the integral $\int_a^\infty g(t)\Delta t$ implies the convergence of the integral $\int_a^\infty f(t)\Delta t$.*

Proof Since $\int_a^\infty g(t)\Delta t$ is convergent, using Theorem 3.184, we have that the integral $\int_a^\infty |f(t)|\Delta t$ is convergent. Therefore, the integral $\int_a^\infty f(t)\Delta t$ is absolutely convergent. From here and from Theorem 3.182, it follows that the integral $\int_a^\infty f(t)\Delta t$ is convergent. □

Theorem 3.190 (Comparison Criterion) *Let $\int_a^\infty f(t)\Delta t$ and $\int_a^\infty g(t)\Delta t$ be improper integrals of the first kind with positive integrands. Suppose that the limit*

$$\lim_{t\to\infty} \frac{f(t)}{g(t)} = L \tag{3.33}$$

exists (finite) and is not zero. Then the integrals are simultaneously convergent or divergent.

Proof Let $\varepsilon \in (0, L)$ be arbitrarily chosen. From (3.33), it follows that there exists $A_0 > a$ such that

$$L - \varepsilon \leq \frac{f(t)}{g(t)} \leq L + \varepsilon \quad \text{for any} \quad t \geq A_0,$$

from where

$$(L - \varepsilon)g(t) \leq f(t) \leq (L + \varepsilon)g(t) \quad \text{for any} \quad t \geq A_0.$$

Hence,

$$(L - \varepsilon)\int_{A_0}^\infty g(t)\Delta t \leq \int_{A_0}^\infty f(t)\Delta t \leq (L + \varepsilon)\int_{A_0}^\infty g(t)\Delta t. \tag{3.34}$$

1. Let $\int_a^\infty g(t)\Delta t$ be convergent. Then $\int_{A_0}^\infty g(t)\Delta t$ is convergent. Hence,
$$(L+\varepsilon)\int_{A_0}^\infty g(t)\Delta t$$
is convergent. From here and from Theorem 3.184, using (3.34), we obtain that $\int_{A_0}^\infty f(t)\Delta t$ is convergent. Therefore, $\int_a^\infty f(t)\Delta t$ is convergent.
2. Let $\int_a^\infty f(t)\Delta t$ be convergent. Then $\int_{A_0}^\infty f(t)\Delta t$ is convergent. From here and from Theorem 3.184, using (3.34), we obtain that $\int_{A_0}^\infty g(t)\Delta t$ is convergent. Therefore, $\int_a^\infty g(t)\Delta t$ is convergent.
3. Let $\int_a^\infty f(t)\Delta t$ be divergent. Hence, $\int_{A_0}^\infty f(t)\Delta t$ is divergent. Thus, using (3.34), it follows that $\int_{A_0}^\infty g(t)\Delta t$ is divergent. Therefore, $\int_a^\infty g(t)\Delta t$ is divergent.
4. Let $\int_a^\infty g(t)\Delta t$ be divergent. Hence, $\int_{A_0}^\infty g(t)\Delta t$ is divergent. Thus, using (3.34), it follows that $\int_{A_0}^\infty f(t)\Delta t$ is divergent. Therefore, $\int_a^\infty f(t)\Delta t$ is divergent. □

Example 3.191 Let $\mathbb{T}=\mathbb{Z}$. Consider the integral
$$I=\int_1^\infty \frac{t^4}{(t^2+11t+30)(t^4+t^3+t^2+1)}\Delta t.$$
Define
$$f(t)=\frac{t^4}{(t^2+11t+30)(t^4+t^3+t^2+1)},\quad g(t)=\frac{1}{t^2+11t+30}$$
and
$$J=\int_1^\infty \frac{1}{t^2+11t+30}\Delta t.$$
We have
$$\begin{aligned}\lim_{t\to\infty}\frac{f(t)}{g(t)}&=\lim_{t\to\infty}\frac{\frac{t^4}{(t^2+11t+30)(t^4+t^3+t^2+1)}}{\frac{1}{t^2+11t+30}}\\&=\lim_{t\to\infty}\frac{t^4}{t^4+t^3+t^2+1}\\&=1.\end{aligned}$$
Thus, using Theorem 3.190, it follows that the integrals I and J are simultaneously convergent or divergent. Note that, with $F(t)=\frac{1}{t+5}$, we have
$$F^\Delta(t)=-\frac{1}{(t+5)(\sigma(t)+5)}$$

$$= -\frac{1}{(t+5)(t+6)}$$

$$= -\frac{1}{t^2+11t+30}.$$

Therefore,

$$J = -\lim_{A\to\infty}\int_1^A F^{\Delta}(t)\Delta t = -\lim_{A\to\infty}(F(A)-F(1)) = \frac{1}{6}.$$

Consequently, the integral I is convergent.

Example 3.192 Let $\mathbb{T} = 2^{\mathbb{N}_0}$. Consider the integral

$$I = \int_1^{\infty}(t^2+2t+3)\Delta t.$$

Let

$$f(t) = t^2+2t+3, \quad g(t) = t^2, \quad J = \int_1^{\infty} g(t)\Delta t.$$

We have that

$$\lim_{t\to\infty}\frac{f(t)}{g(t)} = \lim_{t\to\infty}\frac{t^2+2t+3}{t^2}$$

$$= 1.$$

Thus, employing Theorem 3.190, it follows that the integrals I and J are simultaneously convergent or divergent. Since, with $F(t) = t^3$, we have

$$J = \lim_{A\to\infty}\int_1^A t^2\Delta t = \frac{1}{7}\lim_{A\to\infty}\int_1^A F^{\Delta}(t)\Delta t = \frac{1}{7}\lim_{A\to\infty}(F(A)-F(1)) = \infty,$$

we conclude that I is divergent.

Example 3.193 Let $\mathbb{T} = 2^{\mathbb{N}_0}$. Consider the integral

$$I = \int_1^{\infty}\frac{e^{-t}-e^{-2t}}{t^2}(t+1)\Delta t.$$

We set

$$f(t) = \frac{e^{-t}-e^{-2t}}{t^2}(t+1), \quad g(t) = \frac{e^{-t}-e^{-2t}}{t}$$

and

$$J = \int_1^{\infty} \frac{e^{-t} - e^{-2t}}{t} \Delta t.$$

Then

$$\begin{aligned}
\lim_{t \to \infty} \frac{f(t)}{g(t)} &= \lim_{t \to \infty} \frac{\frac{e^{-t} - e^{-2t}}{t^2}(t+1)}{\frac{e^{-t} - e^{-2t}}{t}} \\
&= \lim_{t \to \infty} \frac{t+1}{t} \\
&= 1.
\end{aligned}$$

By using Theorem 3.190, we conclude that the integrals I and J are simultaneously convergent or divergent. Note that, with $F(t) = e^{-t}$, we have

$$F^{\Delta}(t) = \frac{e^{-\sigma(t)} - e^{-t}}{\sigma(t) - t} = \frac{e^{-2t} - e^{-t}}{t}.$$

Hence,

$$J = - \lim_{A \to \infty} \int_1^A F^{\Delta}(t) \Delta t = - \lim_{A \to \infty} (F(A) - F(1)) = \frac{1}{e}.$$

Consequently, the integral I is convergent.

Exercise 3.194 Let $\mathbb{T} = 2^{\mathbb{N}_0}$. Investigate the integral

$$\int_2^{\infty} \frac{1}{t+1} \operatorname{arctanh} \frac{t}{1 - 2t^2} \Delta t$$

for convergence and divergence.

Solution Convergent. Hint. Use

$$f(t) = \frac{1}{t+1} \operatorname{arctanh} \frac{t}{1 - 2t^2}, \quad g(t) = \frac{1}{t} \operatorname{arctanh} \frac{t}{1 - 2t^2}$$

and show that, with $F(t) = \operatorname{arctanh} t$, we have $F^{\Delta} = g$.

Theorem 3.195 *Let f be integrable from a to any point $A \in \mathbb{T}$, $A > a$. Suppose that the integral*

$$F(A) = \int_a^A f(t) \Delta t$$

is bounded for any $A \geq a$. Suppose that g is monotone on $[a, \infty)$ and $\lim_{t \to \infty} g(t) = 0$. Then the improper integral of the first kind of the form

$$\int_a^\infty f(t)g(t)\Delta t \tag{3.35}$$

is convergent.

Proof Let $A_1, A_2 \in \mathbb{T}$, $A_2 > A_1 \geq a$. By the mean value theorem, Theorem 3.71, it follows that there exists Λ between $\inf\limits_{A\in[A_1,A_2]} F(A)$ and $\sup\limits_{A\in[A_1,A_2]} F(A)$ such that

$$\int_{A_1}^{A_2} f(t)g(t)\Delta t = (g(A_1) - g(A_2))\Lambda + g(A_2)\int_{A_1}^{A_2} f(t)\Delta t. \tag{3.36}$$

Let $M > 0$ be a constant such that

$$|F(A)| \leq M \quad \text{on} \quad [a, \infty).$$

From (3.36), we get

$$\int_{A_1}^{A_2} f(t)g(t)\Delta t = (g(A_1) - g(A_2))\Lambda + g(A_2)(F(A_2) - F(A_1))$$

and

$$\begin{aligned}\left|\int_{A_1}^{A_2} f(t)g(t)\Delta t\right| &= |(g(A_1) - g(A_2))\Lambda + g(A_2)(F(A_2) - F(A_1))|\\ &\leq |g(A_1)||\Lambda| + |g(A_2)||\Lambda| + |g(A_2)|(|F(A_2)| + |F(A_1)|)\\ &\leq M|g(A_1)| + 3M|g(A_2)|\\ &= M(|g(A_1)| + 3|g(A_2)|).\end{aligned} \tag{3.37}$$

Let $\varepsilon > 0$ be arbitrarily chosen. Since $g(t) \to 0$ as $t \to \infty$, there exists $A_3 > a$ such that

$$|g(A)| < \frac{\varepsilon}{4M} \quad \text{for any} \quad A > A_3.$$

Hence, using (3.37), for $A_1, A_2 > A_3$, we get

$$\left|\int_{A_1}^{A_2} f(t)g(t)\Delta t\right| < M\left(\frac{\varepsilon}{4M} + \frac{3\varepsilon}{4M}\right) = \varepsilon.$$

From here and from Cauchy's criterion, Theorem 3.72, it follows that the integral (3.35) is convergent. □

Example 3.196 Let $\mathbb{T} = \mathbb{Z}$. Consider the integral

$$I = \int_1^\infty \frac{t \sin t}{(t^2 + 3t + 2)(t^2 + 1)} \Delta t.$$

Let

$$f(t) = \frac{\sin t}{t^2 + 3t + 2}, \quad g(t) = \frac{t}{t^2 + 1}.$$

We have

$$\begin{aligned} g^\Delta(t) &= \frac{t^2 + 1 - t(t + \sigma(t))}{(t^2 + 1)((\sigma(t))^2 + 1)} \\ &= \frac{t^2 + 1 - t(t + t + 1)}{(t^2 + 1)((t + 1)^2 + 1)} \\ &= \frac{t^2 + 1 - t(2t + 1)}{(t^2 + 1)(t^2 + 2t + 2)} \\ &= \frac{t^2 + 1 - 2t^2 - t}{(t^2 + 1)(t^2 + 2t + 2)} \\ &= \frac{1 - t - t^2}{(t^2 + 1)(t^2 + 2t + 2)}. \end{aligned}$$

Therefore, the function g is monotone on $[1, \infty)$. Moreover,

$$\lim_{t \to \infty} g(t) = \lim_{t \to \infty} \frac{t}{t^2 + 1} = 0$$

and

$$\begin{aligned} \left| \int_1^\infty f(t) \Delta t \right| &\leq \int_1^\infty |f(t)| \Delta t \\ &= \int_1^\infty \frac{|\sin t|}{(t + 1)(t + 2)} \Delta t \\ &\leq \int_1^\infty \frac{1}{(t + 1)(t + 2)} \Delta t \\ &= -\frac{1}{t + 1} \Big|_{t=1}^{t=\infty} \\ &= \frac{1}{2}. \end{aligned}$$

Thus, using Theorem 3.195, it follows that the integral I is convergent.

Example 3.197 Let $\mathbb{T} = 2^{\mathbb{N}_0}$. Consider the integral

$$I = \int_1^\infty \frac{\sin^2 t + \cos t + 3}{t^2(t^2+1)} \Delta t.$$

Let

$$f(t) = \frac{\sin^2 t + \cos t + 3}{t^2}, \quad g(t) = \frac{1}{t^2+1}.$$

Then

$$\begin{aligned} g^\Delta(t) &= -\frac{t+\sigma(t)}{(t^2+1)((\sigma(t))^2+1)} \\ &= -\frac{2t+t}{(t^2+1)(4t^2+1)} \\ &= -\frac{3t}{(t^2+1)(4t^2+1)}. \end{aligned}$$

Therefore, the function g is monotone on $[1, \infty)$. Moreover,

$$\lim_{t\to\infty} g(t) = \lim_{t\to\infty} \frac{1}{t^2+1} = 0$$

and

$$\begin{aligned} \left| \int_1^\infty f(t)\Delta t \right| &\le \int_1^\infty |f(t)|\Delta t \\ &= \int_1^\infty \frac{|\sin^2 t + \cos t + 3|}{t^2} \Delta t \\ &\le \int_1^\infty \frac{\sin^2 t + |\cos t| + 3}{t^2} \Delta t \\ &\le 5 \int_1^\infty \frac{1}{t^2} \Delta t \\ &= -10 \frac{1}{t} \Big|_{t=1}^{t=\infty} \\ &= 10. \end{aligned}$$

Therefore, using Theorem 3.195, it follows that the integral I is convergent.

Example 3.198 Let $\mathbb{T} = 3^{\mathbb{N}_0}$. Consider the integral

$$I = \int_1^\infty \frac{1}{t(t^{10} + t^{11} + t^{12} + 1)} \Delta t.$$

We set

$$f(t) = \frac{1}{t^{10} + t^{11} + t^{12} + 1}, \quad g(t) = \frac{1}{t}.$$

Then

$$g^\Delta(t) = -\frac{1}{3t^2}.$$

Therefore, the function g is a monotonic function on $[1, \infty)$. Also,

$$\lim_{t\to\infty} g(t) = \lim_{t\to\infty} \frac{1}{t} = 0$$

and

$$\begin{aligned} \left| \int_1^\infty f(t)\Delta t \right| &\le \int_1^\infty |f(t)|\Delta t \\ &\le \int_1^\infty \frac{1}{t^{10} + t^{11} + t^{12} + 1} \Delta t \\ &\le \int_1^\infty \frac{1}{t^2} \Delta t \\ &= -3\frac{1}{t}\Big|_{t=1}^{t=\infty} \\ &= 3. \end{aligned}$$

Hence, using Theorem 3.195, it follows that the integral I is convergent.

Exercise 3.199 Let $\mathbb{T} = \mathbb{Z}$. Using Theorem 3.195, prove that the integral

$$\int_1^\infty \frac{\sin t - 2\cos t + 10}{(t^2 + 1)(t^2 - 3t + 5)} \Delta t$$

is convergent.

Exercise 3.200 Let $\mathbb{T}$ be a time scale of the form

$$\mathbb{T} = \{t_k : k \in \mathbb{N}_0\} \text{ with } 0 < t_0 < t_1 < \dots \text{ and } \lim_{k\to\infty} t_k = \infty. \tag{3.38}$$

Suppose that $f : [t_0, \infty) \to \mathbb{R}$ is nonincreasing with $\int_{t_0}^{\infty} f(t)\Delta t < \infty$. Assume that $g : \mathbb{T} \to \mathbb{R}_+$ satisfies

$$g(t_k) \le K f(t_{k+1}) \quad \text{for all} \quad k \in \mathbb{N}_0,$$

where $K > 0$ is a constant. Prove that $\int_{t_0}^{\infty} g(t)\Delta t < \infty$.

Solution We have

$$\begin{aligned}
\int_{t_0}^{\infty} g(t)\Delta t &= \sum_{k=0}^{\infty} g(t_k)\mu(t_k) \\
&\le K \sum_{k=0}^{\infty} f(t_{k+1})\mu(t_k) \\
&\le K \sum_{k=0}^{\infty} f(t_k)\mu(t_k) \\
&= K \int_{t_0}^{\infty} f(t)\Delta t \\
&< \infty.
\end{aligned}$$

Exercise 3.201 Let $\mathbb{T} = 2^{\mathbb{N}_0}$. Prove that the integral

$$\int_1^{\infty} \frac{1}{t^p}\Delta t$$

is divergent for $p \in [0, 1]$ and convergent for $p > 1$.

Solution Let $t_k = 2^k$, $k \in \mathbb{N}_0$. Then

$$t_{k+1} = 2^{k+1}, \quad \mu(t_k) = t_{k+1} - t_k = 2^{k+1} - 2^k$$

and

$$\begin{aligned}
\int_1^{\infty} \frac{1}{t^p}\Delta t &= \sum_{k=0}^{\infty} \frac{1}{t_k^p}\mu(t_k) \\
&= \sum_{k=0}^{\infty} \frac{1}{2^{kp}}(2^{k+1} - 2^k) \\
&= 2\sum_{k=0}^{\infty} \frac{1}{2^{k(p-1)}} - \sum_{k=0}^{\infty} \frac{1}{2^{k(p-1)}}
\end{aligned}$$

$$= \sum_{k=0}^{\infty} \frac{1}{2^{k(p-1)}} \begin{cases} = \infty & \text{if } \ p \in [0, 1] \\ < \infty & \text{if } \ p > 1. \end{cases}$$

Theorem 3.202 *Let* $\mathbb{T}$ *be a time scale that satisfies* (3.38). *If* $f : [t_0, \infty) \to \mathbb{R}$ *is nonincreasing, then*

$$\int_{t_0}^{\infty} f(t)\nabla t \leq \int_{t_0}^{\infty} f(t)\mathrm{d}t \leq \int_{t_0}^{\infty} f(t)\Delta t,$$

where the first and last integrals are taken over $\mathbb{T}$, *while the middle integral is taken over the interval* $[t_0, \infty)$ *of* $\mathbb{R}$.

Proof Since $\mathbb{T}$ is a time scale that satisfies (3.38), we have

$$\int_{t_0}^{\infty} f(t)\nabla t = \sum_{k=0}^{\infty} f(t_{k+1})(t_{k+1} - t_k) \quad \text{and} \quad \int_{t_0}^{\infty} f(t)\Delta t = \sum_{k=0}^{\infty} f(t_k)(t_{k+1} - t_k).$$

Because the function f is nonincreasing on $[t_0, \infty)$, we get

$$f(t_{k+1})(t_{k+1} - t_k) \leq \int_{t_k}^{t_{k+1}} f(t)\mathrm{d}t \leq f(t_k)(t_{k+1} - t_k)$$

for all $k \in \mathbb{N}_0$. Hence,

$$\sum_{k=0}^{\infty} f(t_{k+1})(t_{k+1} - t_k) \leq \sum_{k=0}^{\infty} \int_{t_k}^{t_{k+1}} f(t)\mathrm{d}t \leq \sum_{k=0}^{\infty} f(t_k)(t_{k+1} - t_k),$$

which completes the proof. □

Corollary 3.203 *Let* $\mathbb{T}$ *be a time scale that satisfies* (3.38) *and assume that* $f : [t_0, \infty) \to \mathbb{R}_+$ *is nonincreasing.*

1. *If* $\int_{t_0}^{\infty} f(t)\mathrm{d}t = \infty$, *then* $\int_{t_0}^{\infty} f(t)\Delta t = \infty$.
2. *If* $\int_{t_0}^{\infty} f(t)\mathrm{d}t < \infty$, *then* $\int_{t_0}^{\infty} f(t)\nabla t < \infty$.

Theorem 3.204 *Let* $\mathbb{T}$ *be a time scale that satisfies* (3.38). *Then*

$$\int_{t_0}^{\infty} \frac{1}{t^p}\Delta t = \infty \quad \textit{if} \quad 0 \leq p \leq 1. \tag{3.39}$$

Proof Let $f(t) = \frac{1}{t^p}$. Then f is nonincreasing on $[t_0, \infty)$ and

$$\int_{t_0}^{\infty} \frac{1}{t^p}\mathrm{d}t = \infty.$$

Thus, employing Corollary 3.203, we get (3.39). □

Theorem 3.205 *Let* $\mathbb{T}$ *be a time scale satisfying* (3.38). *Then*

$$\int_{t_0}^{\infty} \frac{\nabla t}{t^p} < \infty \quad \text{if} \quad p > 1. \tag{3.40}$$

Proof Let $f(t) = \frac{1}{t^p}$. Then f is nonincreasing on $[t_0, \infty)$. Since

$$\int_{t_0}^{\infty} \frac{1}{t^p} \mathrm{d}t < \infty \quad \text{for} \quad p > 1,$$

using Corollary 3.203, we get (3.40). □

Theorem 3.206 *Let* $\mathbb{T}$ *be a time scale satisfying* (3.38). *Then*

$$\int_{t_0}^{\infty} \frac{\nabla t}{t^p} = \infty \quad \text{if} \quad 0 \le p \le 1. \tag{3.41}$$

Proof 1. Firstly, we prove that

$$\int_{t_0}^{\infty} \frac{\nabla t}{t} = \infty. \tag{3.42}$$

We have

$$\int_{t_0}^{\infty} \frac{\nabla t}{t} = \sum_{k=0}^{\infty} \frac{t_{k+1} - t_k}{t_{k+1}}.$$

Assume that

$$\sum_{k=0}^{\infty} \frac{t_{k+1} - t_k}{t_{k+1}} < \infty.$$

Then

$$\lim_{k \to \infty} \frac{t_{k+1} - t_k}{t_{k+1}} = 0,$$

whereupon

$$\lim_{k \to \infty} \frac{t_{k+1}}{t_k} = 1.$$

Since $t_k < t_{k+1}$ for any $k \in \mathbb{N}_0$, there exists $N \in \mathbb{N}$ such that

$$\frac{t_{k+1}}{t_k} < 2 \quad \text{for any} \quad k > N.$$

We note that by Theorem 3.204, we have

$$\sum_{k=0}^{\infty} \frac{t_{k+1} - t_k}{t_k} = \infty.$$

On the other hand, we have

$$\begin{aligned}
\sum_{k=0}^{\infty} \frac{t_{k+1} - t_k}{t_k} &= \sum_{k=0}^{N-1} \frac{t_{k+1} - t_k}{t_k} + \sum_{k=N}^{\infty} \frac{t_{k+1} - t_k}{t_k} \\
&= \sum_{k=0}^{N-1} \frac{t_{k+1} - t_k}{t_k} + \sum_{k=N}^{\infty} \frac{t_{k+1} - t_k}{t_{k+1}} \frac{t_{k+1}}{t_k} \\
&< \sum_{k=0}^{N-1} \frac{t_{k+1} - t_k}{t_k} + 2 \sum_{k=N}^{\infty} \frac{t_{k+1} - t_k}{t_{k+1}} \\
&< \infty,
\end{aligned}$$

which is a contradiction. Therefore

$$\sum_{k=0}^{\infty} \frac{t_{k+1} - t_k}{t_{k+1}} = \infty,$$

from where (3.42) follows.

2. Let $p \in [0, 1)$. Since $\lim_{k \to \infty} t_k = \infty$, there exists $t_l \in \mathbb{T}$ so that $t_l > 1$. Hence,

$$\int_{t_0}^{\infty} \frac{1}{t^p} \nabla t = \int_{t_0}^{t_l} \frac{1}{t^p} \nabla t + \int_{t_l}^{\infty} \frac{1}{t^p} \nabla t. \tag{3.43}$$

Because $t_l > 1$ and $p \in [0, 1)$, we have

$$\int_{t_l}^{\infty} \frac{1}{t} \nabla t = \infty \quad \text{and} \quad \int_{t_l}^{\infty} \frac{1}{t^p} \nabla t \geq \int_{t_l}^{\infty} \frac{1}{t} \nabla t.$$

Thus, using (3.43), we get the desired result (3.41).
This completes the proof. □

3.9 Improper Integrals of the Second Kind

In the following situation, the ordinary Riemann integral of f on $[a, b]$ cannot exist since a Riemann integrable function from a to b must be bounded on $[a, b)$.

Definition 3.207 Let $\mathbb{T}$ be a time scale, $a, b \in \mathbb{T}$ with $a < b$. Suppose that b is left-dense. Assume that the function f is defined in the interval $[a, b)$. Suppose that f is integrable on any interval $[a, c]$ with $c < b$ and is unbounded on $[a, b)$. The formal expression

$$\int_a^b f(t)\Delta t \tag{3.44}$$

is called the *improper integral of the second kind*. We say that the integral (3.44) is improper at $t = b$. We also say that f has a singularity at $t = b$. If the left-sided limit

$$\lim_{c \to b-} \int_a^c f(t)\Delta t \tag{3.45}$$

exists as a finite number, then the improper integral (3.44) is said to exist or be *convergent*. In such a case, we call this limit the *value* of the improper integral (3.44) and write

$$\int_a^b f(t)\Delta t = \lim_{c \to b-} \int_a^c f(t)\Delta t.$$

If the limit (3.45) does not exist, then the integral (3.44) is said to be not existent or *divergent*.

Example 3.208 Let $\mathbb{T} = [0, 1] \cup 2^{\mathbb{N}}$, where $[0, 1]$ is the real-valued interval. Define

$$f(t) = \begin{cases} \sqrt{1 - t^2} & \text{for } t \in [0, 1] \\ t^4 & \text{for } t \in 2^{\mathbb{N}}. \end{cases}$$

Consider the integral

$$I = \int_0^8 \frac{1}{f(t)}\Delta t.$$

We have

$$\begin{aligned} I &= \int_0^1 \frac{1}{f(t)}\Delta t + \int_2^8 \frac{1}{f(t)}\Delta t \\ &= \int_0^1 \frac{1}{\sqrt{1-t^2}}dt + \int_2^8 \frac{1}{t^4}\Delta t \\ &= \lim_{c \to 1-} \int_0^c \frac{1}{\sqrt{1-t^2}}dt + \frac{1}{t^4}\mu(t)\Big|_{t=2} + \frac{1}{t^4}\mu(t)\Big|_{t=4} \\ &= \lim_{c \to 1-} \arcsin t\Big|_{t=0}^{t=c} + \frac{2}{16} + \frac{4}{256} \end{aligned}$$

$$= \frac{\pi}{2} + \frac{9}{64}.$$

Therefore, the considered integral is convergent.

Example 3.209 Let $\mathbb{T} = \{-4, -2\} \cup [0, 1]$, where $[0, 1]$ is the real-valued interval. Consider the integral

$$I = \int_{-4}^{1} \frac{\Delta t}{\sqrt{1-t}}.$$

We have

$$\begin{aligned} I &= \int_{-4}^{-2} \frac{\Delta t}{\sqrt{1-t}} + \int_{-2}^{0} \frac{\Delta t}{\sqrt{1-t}} + \int_{0}^{1} \frac{dt}{\sqrt{1-t}} \\ &= \frac{1}{\sqrt{1-t}}\mu(t)\Big|_{t=-4} + \frac{1}{\sqrt{1-t}}\mu(t)\Big|_{t=-2} + \lim_{c \to 1-} \int_0^c \frac{dt}{\sqrt{1-t}} \\ &= \frac{2}{\sqrt{5}} + \frac{2}{\sqrt{3}} - 2 \lim_{c \to 1-} \sqrt{1-t}\Big|_{t=0}^{t=c} \\ &= \frac{2}{\sqrt{5}} + \frac{2}{\sqrt{3}} + 2. \end{aligned}$$

Therefore, the considered integral is convergent.

Example 3.210 Let $\mathbb{T} = \{-1, 0\} \cup [1, 2]$, where $[1, 2]$ is the real-valued interval. Consider the integral

$$I = \int_{-1}^{2} \frac{t^3}{\sqrt{4-t^2}} \Delta t.$$

We have

$$\begin{aligned} I &= \int_{-1}^{0} \frac{t^3}{\sqrt{4-t^2}} \Delta t + \int_{0}^{1} \frac{t^3}{\sqrt{4-t^2}} \Delta t + \int_{1}^{2} \frac{t^3}{\sqrt{4-t^2}} dt \\ &= \frac{t^3\mu(t)}{\sqrt{4-t^2}}\Big|_{t=-1} + \frac{t^3\mu(t)}{\sqrt{4-t^2}}\Big|_{t=0} - \lim_{c \to 2-} \int_1^c t^2 d\sqrt{4-t^2} \\ &= -\frac{1}{\sqrt{3}} - \lim_{c \to 2-} t^2\sqrt{4-t^3}\Big|_{t=1}^{t=c} + 2 \lim_{c \to 2-} \int_1^c t\sqrt{4-t^2} dt \\ &= -\frac{1}{\sqrt{3}} + \sqrt{3} - \lim_{c \to 2-} \int_1^c \sqrt{4-t^2} d(4-t^2) \\ &= \frac{2\sqrt{3}}{3} - \lim_{c \to 2-} \frac{(4-t^2)^{\frac{3}{2}}}{\frac{3}{2}}\Big|_{t=1}^{t=c} \end{aligned}$$

$$= \frac{2\sqrt{3}}{3} + 2\sqrt{3}$$

$$= \frac{8\sqrt{3}}{3}.$$

Therefore, the considered integral is convergent.

Exercise 3.211 Investigate the following integrals for convergence and divergence.

1. $\int_{-3}^{1} \frac{\Delta t}{(1-t)(2t-1)}$, $\mathbb{T} = \{-3, -2, -1, 0\} \cup \left[\frac{1}{2}, 1\right]$, where $\left[\frac{1}{2}, 1\right]$ is the real-valued interval.
2. $\int_{-3}^{10} \frac{2t}{(t^2-1)^2} \Delta t$, $\mathbb{T} = [-3, 3] \cup \{4, 7, 10\}$, where $[-3, 3]$ is the real-valued interval.
3. $\int_{-3}^{2} \frac{\Delta t}{t\sqrt{3t^2-2t-1}}$, $\mathbb{T} = \{-3, -2, -1, 0\} \cup [1, 2]$, where $[1, 2]$ is the real-valued interval.
4. $\int_{-3}^{3} \frac{\Delta t}{(t-7)\sqrt{t^2-3}}$, $\mathbb{T} = \{-3, -2, -1\} \cup [\sqrt{3}, 3]$, where $[\sqrt{3}, 3]$ is the real-valued interval.
5. $\int_{-7}^{1} \frac{\Delta t}{\sqrt[3]{t(1-t)}}$, $\mathbb{T} = \{-7, -4, -1\} \cup [0, 1]$, where $[0, 1]$ is the real-valued interval.
6. $\int_{-\frac{1}{2}}^{1} \frac{\Delta t}{(10-t)\sqrt{1-t^2}}$, $\mathbb{T} = \left\{-\frac{1}{2}, -\frac{1}{4}, 0\right\} \cup \left[\frac{1}{2}, 1\right]$, where $\left[\frac{1}{2}, 1\right]$ is the real-valued interval.

Solution 1. Divergent,
2. divergent,
3. convergent,
4. convergent,
5. convergent,
6. convergent.

Remark 3.212 All theorems in Section 3.8 have exact analogues for improper integrals of the second kind.

1. For the existence of the integral (3.44), it is necessary and sufficient that for any given $\varepsilon > 0$, there exists $b_0 < b$ such that

$$\left| \int_{c_1}^{c_2} f(t)\Delta t \right| < \varepsilon$$

for any $c_1, c_2 \in \mathbb{T}$ satisfying the inequalities $b_0 < c_1 < b$ and $b_0 < c_2 < b$.
2. Suppose that $f(t) \geq 0$. Then, for any $c \in [a, b]$,

$$F(c) = \int_a^c f(t)\Delta t$$

does not decrease as c increases, and the integral (3.44) is convergent if and only if f is bounded, in which case the value of the integral is $\lim_{c \to b-} F(c)$.

3. Let the limit

$$\lim_{t \to b-} \frac{f(t)}{g(t)} = L$$

exist (finite) and suppose it is not zero. Then the integrals $\int_a^b f(t)\Delta t$ and $\int_a^b g(t)\Delta t$ are simultaneously convergent or divergent.

Similar definitions are made and entirely similar results are obtained for integrals of the second kind that are improper at the lower limit of integration.

Example 3.213 Let $\mathbb{T}$ be an arbitrary time scale, $a, b \in \mathbb{T}$ with $a < b$, and suppose that b is left-dense. Let $p \geq 1$. We prove that the integral

$$\int_a^b \frac{\Delta t}{(b-t)^p} \tag{3.46}$$

is divergent.

1. Let $p = 1$. Let us choose points $t_n \in \mathbb{T}$ for $n \in \mathbb{N}_0$ such that

$$a = t_0 < t_1 < \ldots < b \quad \text{and} \quad \lim_{n \to \infty} t_n = b. \tag{3.47}$$

We set

$$\tau_n = \frac{1}{b - t_n} \quad \text{for any} \quad n \in \mathbb{N}_0. \tag{3.48}$$

Then $\lim\limits_{n \to \infty} \tau_n = \infty$, $t_n = b - \frac{1}{\tau_n}$, and

$$\begin{aligned} t_{n+1} - t_n &= \frac{1}{\tau_n} - \frac{1}{\tau_{n+1}} \\ &= \frac{\tau_{n+1} - \tau_n}{\tau_n \tau_{n+1}} \quad \text{for all} \quad n \in \mathbb{N}_0. \end{aligned}$$

Hence,

$$\begin{aligned} \int_a^b \frac{\Delta t}{b-t} &= \sum_{n=0}^{\infty} \int_{t_n}^{t_{n+1}} \frac{\Delta t}{b-t} \\ &\geq \sum_{n=0}^{\infty} \frac{1}{b-t_n} \int_{t_n}^{t_{n+1}} \Delta t \\ &= \sum_{n=0}^{\infty} \frac{t_{n+1} - t_n}{b - t_n} \end{aligned}$$

$$= \sum_{n=0}^{\infty} \tau_n \frac{\tau_{n+1} - \tau_n}{\tau_n \tau_{n+1}}$$

$$= \sum_{n=0}^{\infty} \frac{\tau_{n+1} - \tau_n}{\tau_{n+1}}$$

$$= \infty.$$

2. Let $p > 1$. There exists $d \in [a, b)$ such that

$$0 < b - t < 1 \quad \text{for} \quad t \in [d, b).$$

Then

$$(b-t)^p < b - t \quad \text{for} \quad t \in [d, b).$$

Hence,

$$\int_a^b \frac{\Delta t}{(b-t)^p} = \int_a^d \frac{\Delta t}{(b-t)^p} + \int_d^b \frac{\Delta t}{(b-t)^p}$$

$$> \int_a^d \frac{\Delta t}{(b-t)^p} + \int_d^b \frac{\Delta t}{b-t}$$

$$= \infty.$$

Example 3.214 Let $\mathbb{T}$ be a time scale satisfying (3.47). Let $p < 1$ and suppose that for some $\alpha \in \left[1, \frac{1}{p}\right)$,

$$\frac{1}{b - t_{k+1}} = \mathrm{O}\left(\frac{1}{(b-t_k)^\alpha}\right) \quad \text{as} \quad k \to \infty.$$

We prove that the improper integral (3.46) is convergent. Let τ_n be defined by (3.48). Then

$$\tau_{k+1} = \mathrm{O}(\tau_k^\alpha) \quad \text{as} \quad k \to \infty.$$

Hence, $\tau_{k+1} \le K \tau_k^\alpha$ for all $k \in \mathbb{N}_0$, where $K > 0$ is a constant. Then

$$\int_a^b \frac{\Delta t}{(b-t)^p} = \sum_{k=0}^{\infty} \int_{t_k}^{t_{k+1}} \frac{\Delta t}{(b-t)^p}$$

$$\le \sum_{k=0}^{\infty} \frac{1}{(b - t_{k+1})^p} \int_{t_k}^{t_{k+1}} \Delta t$$

$$
\begin{aligned}
&= \sum_{k=0}^{\infty} \frac{t_{k+1} - t_k}{(b - t_{k+1})^p} \\
&= \sum_{k=0}^{\infty} \frac{\tau_{k+1} - \tau_k}{\tau_k \tau_{k+1}^{1-p}} \\
&\leq K^{\frac{1}{\alpha}} \sum_{k=0}^{\infty} \frac{\tau_{k+1} - \tau_k}{\tau_{k+1}^{\frac{1}{\alpha}+1-p}} \\
&\leq K^{\frac{1}{\alpha}} \sum_{k=0}^{\infty} \int_{\tau_k}^{\tau_{k+1}} \frac{\mathrm{d}t}{t^{\frac{1}{\alpha}+1-p}} \\
&= K^{\frac{1}{\alpha}} \int_{\tau_0}^{\infty} \frac{\mathrm{d}t}{t^{\frac{1}{\alpha}+1-p}} \\
&< \infty.
\end{aligned}
$$

Example 3.215 Let $\mathbb{T}$ be a time scale satisfying (3.47). We consider the integral

$$I = \int_a^b \frac{\Delta t}{(t^4 + t^2 + 1)(b - t)^{\frac{1}{2}}}.$$

We have

$$I \leq \int_a^b \frac{\Delta t}{(b - t)^{\frac{1}{2}}} < \infty.$$

Example 3.216 Let $\mathbb{T}$ be a time scale satisfying (3.47). Assume $a = 0$, $b = 2$, $\frac{1}{2}, 1 \in \mathbb{T}$. We consider the integral

$$I = \int_0^2 \frac{t^{\alpha-1}}{|1 - t|} \Delta t.$$

We have

$$I = \int_0^{\frac{1}{2}} \frac{t^{\alpha-1}}{1 - t} \Delta t + \int_{\frac{1}{2}}^1 \frac{t^{\alpha-1}}{1 - t} \Delta t + \int_1^2 \frac{t^{\alpha-1}}{t - 1} \Delta t.$$

Since $\int_{\frac{1}{2}}^1 \frac{t^{\alpha-1}}{1-t} \Delta t$ is divergent for all $\alpha \in \mathbb{R}$, we conclude that the integral I is divergent.

Example 3.217 Let $\mathbb{T}$ be a time scale satisfying (3.47). Assume $a = 0$, $b = 1$, $\frac{1}{2} \in \mathbb{T}$. Consider the integral

$$I = \int_0^1 t^{\alpha-1}(1 - t)^{\beta-1} \Delta t.$$

We have

$$
\begin{aligned}
I &= \int_0^{\frac{1}{2}} t^{\alpha-1}(1-t)^{\beta-1}\Delta t + \int_{\frac{1}{2}}^1 t^{\alpha-1}(1-t)^{\beta-1}\Delta t \\
&= I_1 + I_2.
\end{aligned}
$$

Note that I_1 is convergent for $\alpha > 0$ and divergent for $\alpha \le 0$. Also, I_2 is convergent for $\beta > 0$ and divergent for $\beta \le 0$. Therefore, I is convergent for $\alpha > 0$ and $\beta > 0$, and I is divergent for $\alpha \le 0$ or $\beta \le 0$.

Exercise 3.218 Let $\mathbb{T}$ be a time scale satisfying (3.47). Investigate the integral

$$
\int_a^b (t-a)^\alpha (b-t)^\beta \Delta t
$$

for convergence and divergence.

Solution Convergent for $\alpha > -1$ and $\beta > -1$, divergent for $\alpha \le -1$ or $\beta \le -1$.

3.10 Advanced Practical Problems

Problem 3.219 Let $\mathbb{T} = \mathbb{R}$ and

$$
f(t) = \begin{cases} \frac{10}{t-2} & \text{for } \ t \in \mathbb{R} \setminus \{2\} \\ 1 & \text{for } \ t = 2. \end{cases}
$$

Determine if f is regulated.

Solution No.

Problem 3.220 Let $\mathbb{T} = \mathbb{R}$ and

$$
f(t) = \begin{cases} \frac{1}{t-5} & \text{if } \ t \in \mathbb{R} \setminus \{5\} \\ 0 & \text{if } \ t = 5. \end{cases}
$$

Check if $f : \mathbb{T} \to \mathbb{R}$ is pre-differentiable, and if it is, then find the region of differentiation.

Solution No.

Problem 3.221 Let $\mathbb{T} = 3^{\mathbb{N}}$. Prove that

$$
-\int \frac{1}{t} \sin t \sin(2t)\Delta t = \cos t + c.
$$

Problem 3.222 Let $\mathbb{T} = 2^{\mathbb{N}_0}$. Prove that

$$\int_t^{2t} (\tau^4 - 3\tau^2 + \tau)\Delta\tau = t^5 - 3t^3 + t^2.$$

Problem 3.223 Let $\mathbb{T} = \mathbb{Z}$. Compute

$$\int_{-1}^{2} (t^3 - t)\Delta t.$$

Solution 0.

Problem 3.224 Let $\mathbb{T} = q^{\mathbb{N}_0}$, $q > 1$. Prove that

$$e_\alpha(t, t_0) = \prod_{s \in [t_0, t)} (1 + (q-1)s\alpha(s)).$$

Problem 3.225 Let $\mathbb{T} = q^{\mathbb{N}_0}$, where $0 < q < 1$. Prove that

$$e_\alpha(t, t_0) = \prod_{s \in [t_0, t)} \left(1 + \frac{1-q}{q}\alpha(s)s\right).$$

Problem 3.226 Let $\mathbb{T} = \mathbb{N}_0^k$, where $k \in \mathbb{N}$. Prove that

$$e_\alpha(t, t_0) = \prod_{s \in [t_0, t)} \left(1 + \left((1 + \sqrt[k]{s})^k - s\right)\alpha(s)\right).$$

Problem 3.227 Let $\mathbb{T} = 3\mathbb{Z}$. Compute $\cosh_2(t, t_0)$ and $\sinh_2(t, t_0)$ for $t, t_0 \in \mathbb{T}, t \geq t_0$.

Solution

$$\cosh_2(t, t_0) = \frac{1}{2}\left(7^{t-t_0} + (-5)^{t-t_0}\right), \quad \sinh_2(t, t_0) = \frac{1}{2}\left(7^{t-t_0} - (-5)^{t-t_0}\right).$$

Problem 3.228 Let $\mathbb{T} = [-2, 2) \cup \{7, 10, 13, \ldots\}$, where $[-2, 2]$ is the real number interval. Compute

1. $\lim\limits_{t \to 1} \left(\frac{1}{2(1-\sqrt{t})} - \frac{1}{3(1-\sqrt[3]{t})}\right)$,
2. $\lim\limits_{t \to \frac{\pi}{2}} \left(\frac{t}{\cot t} - \frac{\pi}{2\cos t}\right)$,
3. $\lim\limits_{t \to 0} t^t$.

Solution 1. $\frac{1}{12}$,
2. -1,
3. 1.

Problem 3.229 Investigate the following integrals for convergence and divergence.

1. $\int_1^\infty \frac{\sin\frac{t}{2}\cos\frac{3t}{2}}{t}\Delta t,\ \mathbb{T}=2^{\mathbb{N}_0}$,
2. $\int_1^\infty \log\frac{t+1}{t}\Delta t,\ \mathbb{T}=\mathbb{Z}$,
3. $\int_1^\infty \frac{e^{3t}-e^t}{2t}\Delta t,\ \mathbb{T}=3^{\mathbb{N}_0}$,
4. $\int_1^\infty \frac{1}{\sqrt{t}(1+\sqrt{t})(2+\sqrt{t})}\Delta t,\ \mathbb{T}=4^{\mathbb{N}_0}$.

Solution 1. Divergent,
2. divergent,
3. divergent,
4. 2.

Problem 3.230 Let $\mathbb{T}=2^{\mathbb{N}_0}$. Using Cauchy's criterion, prove that the integral

$$\int_2^\infty \frac{t}{(t^2-1)(4t^2-1)}\Delta t$$

is convergent.

Problem 3.231 Investigate the following integrals for convergence and divergence.

1. $\int_1^\infty \frac{t^3+2t^2+3t+10}{t^7+t^4+11t+20}\Delta t,\ \mathbb{T}=\mathbb{Z}$,
2. $\int_1^\infty \cos\left(t+\frac{1}{2}\right)\Delta t,\ \mathbb{T}=\mathbb{Z}$,
3. $\int_2^\infty \frac{1}{t}(t^2+t+1)\Delta t,\ \mathbb{T}=2^{\mathbb{N}_0}$,
4. $\int_1^\infty \frac{\sin t\sin(2t)}{2t(t^2+1)(t^4+10)}\Delta t,\ \mathbb{T}=3^{\mathbb{N}_0}$,
5. $\int_1^\infty (\sin^2 t+2\sin^3 t-\cos^4 t+e_1(t,1)+10)(t^2+t)\Delta t,\ \mathbb{T}=2^{\mathbb{N}_0}$,
6. $\int_2^\infty \frac{1}{t^2+10t+20}\Delta t,\ \mathbb{T}=4^{\mathbb{N}_0}$.

Solution 1. Convergent,
2. divergent,
3. divergent,
4. convergent,
5. divergent,
6. convergent.

Problem 3.232 Let $\mathbb{T}=3^{\mathbb{N}_0}$. Investigate the integral

$$\int_1^\infty \frac{1}{2t+3}\operatorname{arctanh}\frac{2t}{1-3t^2}\Delta t$$

for convergence and divergence.

Solution Convergent.

Problem 3.233 Let $\mathbb{T} = 2^{\mathbb{N}_0}$. Using Theorem 3.195, prove that the integral

$$\int_1^\infty \frac{\sin^2 t - 4\cos^2 t + 10}{(t^4 + 2t^2 + 11)(t^2 + 3t + 15)} \Delta t$$

is convergent.

Problem 3.234 Investigate the following integrals for convergence and divergence.

1. $\int_{-1}^7 \frac{e^{\frac{1}{t}}}{t^3} \Delta t$, $\mathbb{T} = [-1, 0] \cup \{1, 4, 7\}$, where $[-1, 0]$ is the real-valued interval,
2. $\int_0^4 \frac{\Delta t}{t \log^2 t}$, $\mathbb{T} = \left[0, \frac{1}{2}\right] \cup \{2, 3, 4\}$, where $\left[0, \frac{1}{2}\right]$ is the real-valued interval.
3. $\int_0^5 \frac{\Delta t}{t \log^2 t}$, $\mathbb{T} = [0, 1] \cup \{2, 3, 4, 5\}$, where $[0, 1]$ is the real-valued interval,
4. $\int_{-3}^1 \frac{\Delta t}{t \log |t|}$, $\mathbb{T} = \{-3, -2, -1\} \cup [0, 1]$, where $[0, 1]$ is the real-valued interval,
5. $\int_{-4}^1 \frac{\log |t|}{\sqrt{|t|}} \Delta t$, $\mathbb{T} = \{-4, -1\} \cup [0, 1]$, where $[0, 1]$ is the real-valued interval,
6. $\int_{-3}^\pi \cot t \Delta t$, $\mathbb{T} = \{-3, -2, -1\} \cup [0, \pi]$, where $[0, \pi]$ is the real-valued interval.

Solution 1. Convergent,
2. convergent,
3. divergent,
4. divergent,
5. convergent,
6. divergent.

Problem 3.235 Let $\mathbb{T}$ be a time scale satisfying (3.47). Investigate the following integrals for convergence and divergence.

1. $\int_a^b \frac{(t-a)^{2\alpha+3}}{t^4+2t^2+7} \Delta t$,
2. $\int_a^b e_{t^2}(t, a)(b - t)^{\beta+2} \Delta t$,
3. $\int_a^b (t - a)^{\alpha+1}(b - t)^{2\beta+7} \Delta t$.

Solution 1. Convergent for $\alpha > -2$, divergent for $\alpha \le -2$,
2. convergent for $\beta > -3$, divergent for $\beta \le -3$,
3. convergent for $\alpha > -2$ and $\beta > -4$, divergent for $\alpha \le -2$ or $\beta \le -4$.

3.11 Notes and References

In this chapter, the concepts of integral calculus for single-variable functions on time scales are introduced, starting with the two crucial notions of rd-continuity and regularity. These are classes of functions that possess an antiderivative and a pre-antiderivative, respectively. Corresponding existence theorems are presented. A weak form of an integral, the Cauchy integral, is defined in terms of antiderivatives. The concept of the Riemann integral on time scales was investigated in [39], where only the Darboux definition of the integral was introduced. In [7, 27], the Riemann definition of the integral on time scales was given and the equivalence of the Darboux and Riemann definitions of the integral was proved. The main properties of the Riemann integral were presented in [7, 27–29]. In this chapter, trigonometric and hyperbolic functions are introduced, and they can be used for solving various dynamic equations. Trigonometric and hyperbolic functions have been defined in [21–23, 33]. Many of the results for trigonometric and hyperbolic functions are contained in [37]. The time scales versions of the L'Hôpital rules are taken from [1, 21]. Many of the results connected with polynomials are contained in [1, 21]. The functions g_k and h_k are time scales "substitutes" for the usual monomials t^k. There are two of them, and this is the reason why we have two versions of Taylor's theorem, one using the functions g_k and the other one using the functions h_k as coefficients. Some versions of mean value theorems for integrals are taken from [6, 7, 20, 21, 27–29]. The definitions for improper integrals of first and second kind and their properties are taken from [6, 25]. We refer to [10, 12, 13, 15, 16, 18, 19, 24] for some related results that are not given in this chapter. All results presented in this chapter are taken from Bohner and Peterson [21, 25].

Chapter 4
Sequences and Series of Functions

4.1 Uniform Convergence of Sequences of Functions

Suppose that $f_n : \mathbb{T} \to \mathbb{R}$, $n \in \mathbb{N}$, $S \subset \mathbb{T}$.

Definition 4.1 We say that the sequence $\{f_n\}_{n\in\mathbb{N}}$ *converges pointwise*, i.e., at each point, to a function f defined on S if

$$\lim_{n\to\infty} f_n(t) = f(t) \quad \text{for all} \quad t \in S.$$

We often write $\lim_{n\to\infty} f_n = f$ pointwise on S or $f_n \to f$ pointwise on S.

Example 4.2 Let $\mathbb{T} = \left\{\frac{1}{n} : n \in \mathbb{N}\right\} \cup \{0\}$. Consider

$$f_n(t) = t^n, \quad t \in [0, 1].$$

We have

$$\lim_{n\to\infty} f_n(t) = \lim_{n\to\infty} t^n = \begin{cases} 0 & \text{for } t \in [0, 1) \\ 1 & \text{for } t = 1. \end{cases}$$

If

$$f(t) = \begin{cases} 0 & \text{for } t \in [0, 1) \\ 1 & \text{for } t = 1, \end{cases}$$

then $f_n \to f$ pointwise on $[0, 1]$.

Example 4.3 Let $\mathbb{T} = 2^{\mathbb{N}_0}$. If

$$f_n(t) = t\left(1 + e^{-nt}\right),$$

M. Bohner and S.G. Georgiev, *Multivariable Dynamic Calculus on Time Scales*,
DOI 10.1007/978-3-319-47620-9_4

then

$$\lim_{n\to\infty} f_n(t) = \lim_{n\to\infty} t\left(1+e^{-nt}\right) = t.$$

If we set $f(t) = t$, then $f_n \to f$ pointwise on $\mathbb{T}$.

Example 4.4 If $f_n(t) = \frac{n+1}{n+t^2}$, then

$$\lim_{n\to\infty} f_n(t) = \lim_{n\to\infty} \frac{n+1}{n+t^2} = 1.$$

If $f(t) = 1$, $t \in \mathbb{T}$, then $f_n \to f$ pointwise on $\mathbb{T}$.

Exercise 4.5 Let

$$f_n(t) = \frac{nt^2}{n+t}, \quad f(t) = t^2.$$

Prove that $f_n \to f$ pointwise on $\mathbb{T}$.

Definition 4.6 We say that the sequence $\{f_n\}_{n\in\mathbb{N}}$ *converges uniformly* on S to a function f defined on S if for every $\varepsilon > 0$, there exists $N \in \mathbb{N}$ such that

$$|f_n(t) - f(t)| < \varepsilon \quad \text{for all} \quad n > N \quad \text{and all} \quad t \in S.$$

Example 4.7 Let $\mathbb{T} = \left\{\frac{1}{n} : n \in \mathbb{N}\right\} \cup \{0\}$. Consider

$$f_n(t) = t^n, \quad 0 \le t \le a, \quad 0 < a < 1.$$

Note that

$$\lim_{n\to\infty} f_n(t) = \lim_{n\to\infty} t^n = 0 \quad \text{for all} \quad t \in \mathbb{T} \setminus \{1\}.$$

Let $\varepsilon > 0$ be arbitrarily chosen. We choose $N \in \mathbb{N}$ such that

$$N > \frac{\log \varepsilon}{\log a}.$$

Hence, $a^N < \varepsilon$. Then, for every $n > N$, we have

$$a^n \le a^N < \varepsilon$$

and

$$|t^n - 0| = t^n \le a^n < \varepsilon.$$

Therefore, $f_n \to 0$ uniformly on $[0, a]$.

Example 4.8 Let $\mathbb{T} = 2^{\mathbb{N}_0}$. Consider

$$f_n(t) = \sqrt{t^2 + \frac{1}{n^2}}.$$

If we take $f(t) = t$ and let $\varepsilon > 0$ be arbitrarily chosen, then

$$\begin{aligned}|f_n(t) - f(t)| &= \left|\sqrt{t^2 + \frac{1}{n^2}} - t\right| \\ &= \frac{\left|\left(\sqrt{t^2 + \frac{1}{n^2}} - t\right)\left(\sqrt{t^2 + \frac{1}{n^2}} + t\right)\right|}{\sqrt{t^2 + \frac{1}{n^2}} + t} \\ &= \frac{t^2 + \frac{1}{n^2} - t^2}{\sqrt{t^2 + \frac{1}{n^2}} + t} \\ &= \frac{1}{n^2\left(\sqrt{t^2 + \frac{1}{n^2}} + t\right)} \\ &\le \frac{1}{n^2 \frac{1}{n}} \\ &= \frac{1}{n}.\end{aligned}$$

If we take $N = \frac{1}{\varepsilon}$, then, for every $n > N$, we have

$$\frac{1}{n} < \frac{1}{N} = \varepsilon \quad \text{and} \quad |f_n(t) - f(t)| < \varepsilon \quad \text{for any} \quad t \in \mathbb{T}.$$

Hence, $\{f_n\}_{n \in \mathbb{N}}$ is uniformly convergent to f on $\mathbb{T}$.

Example 4.9 Let $\mathbb{T} = \left\{\frac{1}{n} : n \in \mathbb{N}\right\} \cup \{0\}$. Consider

$$f_n(t) = \frac{nt}{2 + n^3 t^3}, \quad t \in \mathbb{T}.$$

If $f(t) = 0$, then $f_n \to f$ pointwise on $\mathbb{T}$. Assume that $\{f_n\}_{n \in \mathbb{N}}$ is uniformly convergent to f on $\mathbb{T}$. We take $0 < \varepsilon < \frac{1}{3}$. Then there exists $N = N(\varepsilon)$ such that for every $n > N$, we have

$$|f_n(t)| < \varepsilon \quad \text{for any} \quad t \in \mathbb{T}.$$

In particular, when $n > N$ and $t = \frac{1}{n}$, we get

$$\left| f_n \left(\frac{1}{n} \right) \right| < \varepsilon,$$

which is a contradiction because $f_n\left(\frac{1}{n}\right) = \frac{1}{3}$. Therefore, $\{f_n\}_{n\in\mathbb{N}}$ is not uniformly convergent to f on $\mathbb{T}$.

Exercise 4.10 Let $\mathbb{T} = \left\{\frac{1}{n} : n \in \mathbb{N}\right\} \cup \{0\}$. Check if the following sequences are uniformly convergent on $\mathbb{T}$.

1. $f_n(t) = e^{-(t-3n)^2}$,
2. $f_n(t) = \frac{t}{4+2n^2t^2}$,
3. $f_n(t) = \frac{1}{2+3nt}$.

Solution 1. uniformly convergent to 0 on $\mathbb{T}$,
2. uniformly convergent to 0 on $\mathbb{T}$,
3. not uniformly convergent to 0 on $\mathbb{T}$.

Theorem 4.11 *If $\{f_n\}_{n\in\mathbb{N}}$ converges pointwise to f on $D \subset \mathbb{T}$, then $\{f_n\}_{n\in\mathbb{N}}$ converges uniformly to f on D if and only if*

$$\lim_{n\to\infty} \sup_{t\in D} |f_n(t) - f(t)| = 0. \tag{4.1}$$

Proof 1. Suppose that $\{f_n\}_{n\in\mathbb{N}}$ converges uniformly to f on D. Then, for every $\varepsilon > 0$, there exists $N = N(\varepsilon)$ so that $n > N$ implies

$$|f_n(t) - f(t)| < \varepsilon \quad \text{for any} \quad t \in D.$$

Hence,

$$\sup_{t\in D} |f_n(t) - f(t)| < \varepsilon \quad \text{for any} \quad n > N.$$

2. Suppose that $\{f_n\}_{n\in\mathbb{N}}$ converges pointwise to f on D and (4.1) holds.Then, for every $\varepsilon > 0$, there exists $N = N(\varepsilon)$ so that $n > N$ implies

$$\sup_{t\in D} |f_n(t) - f(t)| < \varepsilon.$$

Hence, for any $n > N$, we have

$$|f_n(t) - f(t)| < \varepsilon \quad \text{for any} \quad t \in D.$$

The proof is complete. □

Example 4.12 Let $\mathbb{T} = 2^{\mathbb{N}_0}$, $f_n(t) = \frac{1}{n+t^2}$ and $f(t) = 0$, $t \in \mathbb{T}$. We have that $f_n \to f$ pointwise on $\mathbb{T}$. Also,

$$\sup_{t\in\mathbb{T}} |f_n(t) - f(t)| = \sup_{t\in\mathbb{T}} \frac{1}{n+t^2} = \frac{1}{n+1} \to 0 \quad \text{as} \quad n \to \infty.$$

Hence, using Theorem 4.11, it follows that $\{f_n\}_{n\in\mathbb{N}}$ converges uniformly to f on $\mathbb{T}$.

Example 4.13 Let $\mathbb{T} = \mathbb{N}$, $f_n(t) = \frac{nt}{nt+1}$, $f(t) = 1$, $t \in \mathbb{T}$. We have that $f_n \to 1$ pointwise on $\mathbb{T}$. Also,

$$\begin{aligned} |f_n(t) - f(t)| &= \left|\frac{nt}{nt+1} - 1\right| \\ &= \left|-\frac{1}{nt+1}\right| \\ &= \frac{1}{nt+1}, \end{aligned}$$

$$\begin{aligned} \sup_{t\in\mathbb{T}} |f_n(t) - f(t)| &= \sup_{t\in\mathbb{T}} \frac{1}{nt+1} \\ &= \frac{1}{n+1} \to 0 \quad \text{as} \quad n \to \infty. \end{aligned}$$

Hence, using Theorem 4.11, it follows that $\{f_n\}_{n\in\mathbb{N}}$ converges uniformly to f on $\mathbb{T}$.

Example 4.14 Let $\mathbb{T} = \mathbb{Z}$. We will investigate for uniform convergence of the sequence $\left\{f_n(t) = \frac{n+1}{n+t^2}\right\}_{n\in\mathbb{N}}$ on $D_1 = (-4, 4)$ and $D_2 = [1, \infty)$. Let $f(t) = 1$. Note that $f_n \to f$ pointwise on $\mathbb{Z}$. Moreover,

$$|f_n(t) - f(t)| = \left|\frac{n+1}{n+t^2} - 1\right| = \left|\frac{1-t^2}{n+t^2}\right|.$$

1. If $t \in D_1$, then

$$\begin{aligned} |f_n(t) - f(t)| &= \frac{1-t^2}{n+t^2} =: g(t), \\ g^{\Delta}(t) &= \frac{-(\sigma(t)+t)(n+t^2) - (1-t^2)(\sigma(t)+t)}{\left(n+(t+1)^2\right)\left(n+t^2\right)} \\ &= -\frac{(n+1)(2t+1)}{\left(n+(t+1)^2\right)(n+t^2)} \\ &\leq 0 \quad \text{if} \quad t \geq 0, \end{aligned}$$

$$g^{\nabla}(t) = \frac{-(\rho(t)+t)(n+t^2) - (1-t^2)(\rho(t)+t)}{\left(n+(t-1)^2\right)\left(n+t^2\right)}$$

$$= -\frac{(n+1)(2t-1)}{\left(n+(t-1)^2\right)(n+t^2)}$$

$$\geq 0 \ \text{ if } \ t \leq 0.$$

Therefore, the function $\frac{1-t^2}{n+t^2}$ has a maximum at $t = 0$. Note that

$$\left.\frac{1-t^2}{n+t^2}\right|_{t=0} = \frac{1}{n} \to 0 \quad \text{as} \quad n \to \infty.$$

Hence, using Theorem 4.11, it follows that $\{f_n\}_{n\in\mathbb{N}}$ converges uniformly to f on D_1.

2. If $t \in D_2$, then

$$|f_n(t) - f(t)| = \frac{t^2-1}{n+t^2}.$$

Assume that $\{f_n\}_{n\in\mathbb{N}}$ converges uniformly to f on D_2. Then, for $0 < \varepsilon < \frac{1}{2}$, there exists $N = N(\varepsilon)$ so that $n > N$ implies

$$\frac{t^2-1}{n+t^2} < \varepsilon \quad \text{for any} \quad t \in D_2.$$

We take $t = n + 2 \in D_2$. Then

$$\frac{(n+2)^2-1}{n+(n+2)^2} < \frac{1}{2},$$

so

$$\frac{n^2+4n+3}{2n^2+4n+4} < \frac{1}{2},$$

so

$$n^2+4n+3 < n^2+2n+2,$$

so

$$2n+1 < 0,$$

which is a contradiction. Therefore, $\{f_n\}_{n\in\mathbb{N}}$ does not converge uniformly to f on D_2.

Exercise 4.15 Let

$$\mathbb{T} = \{0\} \cup \left\{\frac{1}{n} : n \in \mathbb{N}\right\} \cup 2^{\mathbb{N}_0}.$$

Using Theorem 4.11, investigate the sequence

$$\left\{ f_n(t) = \frac{n+t}{nt+1} \right\}_{n\in\mathbb{N}}$$

for uniform convergence on $D_1 = [0, 1]$ and $D_2 = 2^{\mathbb{N}_0}$.

Solution The sequence is not uniformly convergent on D_1, and it is uniformly convergent on D_2.

Theorem 4.16 *If the function sequence $\{f_n\}_{n\in\mathbb{N}}$ is pointwise convergent to f on $D \subset \mathbb{T}$, then $\{f_n\}_{n\in\mathbb{N}}$ is uniformly convergent to f on D if and only if for an arbitrary sequence $\{t_n\}_{n\in\mathbb{N}}$, $t_n \in D$, we have*

$$\lim_{n\to\infty} (f_n(t_n) - f(t_n)) = 0. \tag{4.2}$$

Proof 1. Necessity. Suppose that $\{f_n\}_{n\in\mathbb{N}}$ is uniformly convergent on D. Then, using Theorem 4.11, we have

$$\sup_{t\in D} |f_n(t) - f(t)| \to 0 \quad \text{as} \quad n \to \infty.$$

Hence, for any sequence $\{t_n\}_{n\in\mathbb{N}}$, $t_n \in D$, we have

$$|f_n(t_n) - f(t_n)| \le \sup_{t\in D} |f_n(t) - f(t)| \to 0 \quad \text{as} \quad n \to \infty,$$

i.e., (4.2) holds.

2. Sufficiency. Assume that for any sequence $\{t_n\}_{n\in\mathbb{N}}$, $t_n \in D$, (4.2) holds and the sequence $\{f_n\}_{n\in\mathbb{N}}$ does not converge uniformly to f on D. Hence, there exists $\varepsilon_0 > 0$ such that for any $N > 0$, there exist $n > N$ and $t \in D$ so that

$$|f_n(t) - f(t)| \ge \varepsilon_0.$$

For $N_1 = 1$, there exist $n_1 > 1$ and $t_{n_1} \in D$ such that

$$|f_{n_1}(t_{n_1}) - f(t_{n_1})| \ge \varepsilon_0.$$

For $N_2 = n_1$, there exist $n_2 > n_1$ and $t_{n_2} \in D$ such that

$$|f_{n_2}(t_{n_2}) - f(t_{n_2})| \ge \varepsilon_0,$$

and so on. Thus, we get a sequence $\{t_{n_k}\}_{k\in\mathbb{N}}$, $t_{n_k} \in D$, such that

$$|f_{n_k}(t_{n_k}) - f(t_{n_k})| \ge \varepsilon_0,$$

which leads to a contradiction due to (4.2).

The proof is complete. □

Example 4.17 Consider $f_n(t) = nt(1-t)^n$ on $D = \left\{\frac{1}{n} : n \in \mathbb{N}\right\} \cup \{0\}$. We have that $\{f_n\}_{n\in\mathbb{N}}$ is pointwise convergent to 0 on D. Suppose that $\{f_n\}_{n\in\mathbb{N}}$ is uniformly convergent to 0 on D. Then, applying Theorem 4.16 for $t_n = \frac{1}{n}$, we have

$$f_n\left(\frac{1}{n}\right) = \left(1 - \frac{1}{n}\right)^n \not\to 0,$$

which is a contradiction. Therefore, $\{f_n\}_{n\in\mathbb{N}}$ is not uniformly convergent to 0 on D.

Example 4.18 Consider $f_n(t) = \frac{1}{1+nt}$ on $D = \left\{\frac{1}{n} : n \in \mathbb{N}\right\} \cup \{0\}$. We have that $\{f_n\}_{n\in\mathbb{N}}$ is pointwise convergent to 0 on D. Assume that $\{f_n\}_{n\in\mathbb{N}}$ is uniformly convergent to 0 on D. Then, using Theorem 4.16 for $t_n = \frac{1}{n}$, we have

$$f_n\left(\frac{1}{n}\right) = \frac{1}{2} \not\to 0,$$

which is a contradiction. Therefore, $\{f_n\}_{n\in\mathbb{N}}$ is not uniformly convergent to 0 on D.

Example 4.19 Consider $f_n(t) = 1 - (1-t^2)^n$ on $D = \left\{\frac{1}{n} : n \in \mathbb{N}\right\} \cup \{0\}$. We have that $\{f_n\}_{n\in\mathbb{N}}$ is pointwise convergent to 1 on D. Assume that $\{f_n\}_{n\in\mathbb{N}}$ is uniformly convergent to 1 on D. Then, using Theorem 4.16 for $t_n = \frac{1}{n}$, we have

$$f_n\left(\frac{1}{n}\right) - 1 = -\left(1 - \frac{1}{n^2}\right)^n \not\to 0,$$

which is a contradiction. Therefore, $\{f_n\}_{n\in\mathbb{N}}$ is not uniformly convergent to 1 on D.

Exercise 4.20 Consider $f_n(t) = \frac{nt+2}{3+4n^2t^2}$ on $D = \left\{\frac{1}{n} : n \in \mathbb{N}\right\} \cup \{0\}$. Using Theorem 4.16, prove that $\{f_n\}_{n\in\mathbb{N}}$ is not uniformly convergent to 0 on D.

Theorem 4.21 *Let $D \subset \mathbb{T}$. If $f_n : D \to \mathbb{R}$, $n \in \mathbb{N}$, are rd-continuous and $\{f_n\}_{n\in\mathbb{N}}$ is uniformly convergent to $f : D \to \mathbb{R}$ on D, then f is rd-continuous on D and*

$$\int_a^b f(t)\Delta t = \lim_{n\to\infty} \int_a^b f_n(t)\Delta t$$

for every $[a, b] \subset D$.

Proof Since $\{f_n\}_{n\in\mathbb{N}}$ is uniformly convergent to f on D, for any given $\varepsilon > 0$, there exists $M \in \mathbb{N}$ such that

$$|f_M(t) - f(t)| < \frac{\varepsilon}{3} \quad \text{for all} \quad t \in D.$$

First, assume $t_0 \in D$ is left-dense. Because f_M is rd-continuous on D, there exists $\delta > 0$ such that

$$|f_M(t') - f_M(t'')| < \frac{\varepsilon}{3} \quad \text{for any} \quad t', t'' \in (t_0 - \delta, t_0).$$

If $t_n \to t_0^-$, $n \to \infty$, $n \in \mathbb{N}$, then there exists $N \in \mathbb{N}$ such that $m, n > N$ imply $t_m, t_n \in (t_0 - \delta, t_0)$ and

$$|f_M(t_n) - f_M(t_m)| < \frac{\varepsilon}{3}. \tag{4.3}$$

Hence, for $m, n > N$, we have

$$\begin{aligned} |f(t_n) - f(t_m)| =& |f_M(t_n) - f(t_n) - f_M(t_n) - f_M(t_m) + f_M(t_m) - f(t_m)| \\ \leq& |f_M(t_n) - f(t_n)| + |f_M(t_n) - f_M(t_m)| \\ &+ |f_M(t_m) - f(t_m)| \\ <& \frac{\varepsilon}{3} + \frac{\varepsilon}{3} + \frac{\varepsilon}{3} \\ =& \varepsilon. \end{aligned} \tag{4.4}$$

Hence, the left-sided limit of f in t_0 exists and is finite. Second, assume that $t_0 \in D$ is right-dense. Then f_n is continuous in t_0. Thus, there exists $N \in \mathbb{N}$ such that $m, n > N$ imply $t_m, t_n \in (t_0 - \delta, t_0 + \delta)$ and (4.3) holds. Therefore, (4.4) holds and f is continuous in t_0. Thus, f is rd-continuous on D. Hence, f is integrable on every $[a, b] \subset D$. For every $n > M$, we have

$$\begin{aligned} \left| \int_a^b f_n(t)\Delta t - \int_a^b f(t)\Delta t \right| &= \left| \int_a^b (f_n(t) - f(t))\Delta t \right| \\ &\leq \int_a^b |f_n(t) - f(t)|\Delta t \\ &< \frac{\varepsilon}{3}(b - a), \end{aligned}$$

which completes the proof. □

Theorem 4.22 *Suppose that the function sequence*

$$\{f_n\}_{n \in \mathbb{N}}, \quad f_n : [a, b] \to \mathbb{R}, \quad n \in \mathbb{N},$$

satisfies the following conditions.

1. f_n, $n \in \mathbb{N}$, *is differentiable on* $[a, b]$, *and its derivative* f_n^{Δ} *is rd-continuous on* $[a, b]$,

2. *f_n converges pointwise to f on $[a, b]$,*
3. *$\{f_n^\Delta\}_{n\in\mathbb{N}}$ is uniformly convergent to g on $[a, b]$.*

Then f is differentiable on $[a, b]$ and

$$f^\Delta(t) = g(t) \text{ for any } t \in [a, b].$$

Proof By Theorem 4.21, we have that g is rd-continuous on $[a, b]$. Therefore, g is integrable on $[a, b]$. Hence, using Theorem 4.21, we get

$$\begin{aligned}\int_a^t g(s)\Delta s &= \lim_{n\to\infty} \int_a^t f_n^\Delta(s)\Delta s \\ &= \lim_{n\to\infty} (f_n(t) - f_n(a)) \\ &= f(t) - f(a) \quad \text{for any} \quad t \in [a, b].\end{aligned}$$

The left-hand side of the above formula is differentiable, so the right-hand side is also differentiable, and this leads to

$$f^\Delta(t) = g(t) \quad \text{for all} \quad t \in [a, b],$$

completing the proof. □

Theorem 4.23 (Dini Theorem) *Assume that the function sequence $\{f_n\}_{n\in\mathbb{N}}$, $f_n : [a, b] \to \mathbb{R}$, converges pointwise to the function f on $[a, b]$. If the conditions*

1. *f_n, $n \in \mathbb{N}$, are rd-continuous on $[a, b]$,*
2. *f is rd-continuous on $[a, b]$,*
3. *for any given $t \in [a, b]$, $\{f_n(t)\}_{n\in\mathbb{N}}$ is monotone with respect to $n \in \mathbb{N}$*

hold, then $\{f_n\}_{n\in\mathbb{N}}$ is uniformly convergent to f on $[a, b]$.

Proof Suppose that the sequence $\{f_n\}_{n\in\mathbb{N}}$ is not uniformly convergent to f on $[a, b]$. Then there exists $\varepsilon_0 > 0$ such that for any given $N \in \mathbb{N}$, there exist $n > N, t \in [a, b]$, implying

$$|f_n(t) - f(t)| \geq \varepsilon_0. \tag{4.5}$$

For $N = 1$, there exist $n_1 > 1$, $t_1 \in [a, b]$, such that

$$|f_{n_1}(t_1) - f(t_1)| \geq \varepsilon_0.$$

For $N = n_1$, there exist $n_2 > n_1$, $t_2 \in [a, b]$, such that

$$|f_{n_2}(t_2) - f(t_2)| \geq \varepsilon_0,$$

and so on. For $N = n_k$ there exist $n_{k+1} > n_k$, $t_{k+1} \in [a, b]$, such that

$$|f_{n_{k+1}}(t_{k+1}) - f(t_{k+1})| \geq \varepsilon_0.$$

Hence, we obtain a point sequence $\{t_k\}_{k\in\mathbb{N}}$, $t_k \in [a, b]$. This sequence has a convergent subsequence. Let $\{t_{k_l}\}_{l\in\mathbb{N}}$ be a convergent subsequence of the sequence $\{t_k\}_{k\in\mathbb{N}}$ and $t_{k_l} \to \xi$ as $l \to \infty$. We have that $\xi \in [a, b]$. Because $f_n(\xi) \to f(\xi)$ as $n \to \infty$, for the above ε_0, there exists $N \in \mathbb{N}$ such that

$$|f_N(\xi) - f(\xi)| < \frac{\varepsilon_0}{2}.$$

Suppose that ξ is left-dense and right-scattered. Then, for the above sequence $\{t_{k_l}\}_{l\in\mathbb{N}}$, $t_{k_l} \leq \xi$ and $t_{k_l} \to \xi$ as $l \to \infty$. Thus, for $\varepsilon_0 > 0$, as above, there exists $L \in \mathbb{N}$ such that $l > L$ implies

$$|\xi - t_{k_l}| < \varepsilon_0.$$

Since f_N and f are rd-continuous on $[a, b]$, we have that

$$\left|\left(f_N(t_{k_l}) - f(t_{k_l})\right) - \left(f_N(\xi) - f(\xi)\right)\right| < \frac{\varepsilon_0}{2}.$$

Hence,

$$\begin{aligned}\left|f_N(t_{k_l}) - f(t_{k_l})\right| &< \frac{\varepsilon_0}{2} + |f_N(\xi) - f(\xi)| \\ &< \frac{\varepsilon_0}{2} + \frac{\varepsilon_0}{2} \\ &= \varepsilon_0.\end{aligned}$$

Suppose that ξ is not left-dense and right-scattered. Then ξ is a point of continuity of f_N and f. Because $t_{k_l} \to \xi$, $l \to \infty$, there exists $L_1 \in \mathbb{N}$ such that $l > L_1$ implies

$$|f_N(t_{k_l}) - f(t_{k_l})| < \varepsilon_0.$$

By using the monotonicity condition, we get

$$|f_n(t_{k_l}) - f(t_{k_l})| \leq |f_N(t_{k_l}) - f(t_{k_l})| < \varepsilon_0$$

with $n > N$, $l > \max\{L, L_1\}$. So when n is sufficiently large, $n_l > N$ and $l > \max\{L, L_1\}$ are satisfied. Thus,

$$|f_{n_{k_l}(t_{k_l})} - f(t_{k_l})| < \varepsilon_0,$$

which contradicts to (4.5). This completes the proof. □

4.2 Uniform Convergence of Series of Functions

Let $f_n : \mathbb{T} \to \mathbb{R}$, $n \in \mathbb{N}$. We consider the infinite series

$$\sum_{n=1}^{\infty} f_n(t). \tag{4.6}$$

Definition 4.24 If the numerical series $\sum_{n=1}^{\infty} f_n(t_0)$, $t_0 \in \mathbb{T}$, is convergent, then t_0 is called a *point of convergence* of the function series (4.6). The set D of all points of convergence of the series (4.6) is called its *domain of convergence*. If a domain of convergence of a function series is not empty, then the function series is called *pointwise convergent* on its domain of convergence. We define the *sum function* $F = \sum_{n=1}^{\infty} f_n$ on D.

Definition 4.25 The function

$$F_n(t) = f_1(t) + f_2(t) + \cdots + f_n(t), \quad n \in \mathbb{N},$$

is called the *partial sum* or, more precisely, the nth partial sum of the function series (4.6).

Definition 4.26 If the partial sum sequence $\{F_n\}_{n\in\mathbb{N}}$ of the function series (4.6) is uniformly convergent to F on D, then we term that the function series (4.6) is *uniformly convergent* to F on D.

Example 4.27 Let

$$\mathbb{T} = (-1, 0] \cup \left\{\frac{1}{n} : n \in \mathbb{N} \setminus \{1\}\right\},$$

where $(-1, 0]$ is the real-valued interval. Consider the series

$$\sum_{l=1}^{\infty} t^{l-1}. \tag{4.7}$$

Note that the series (4.7) is pointwise convergent on $\mathbb{T}$ to $F(t) = \frac{1}{1-t}$. Moreover,

$$F_n(t) = \sum_{l=1}^{n} t^{l-1} = \frac{1-t^n}{1-t}$$

and

$$F(t) - F_n(t) = \frac{1}{1-t} - \frac{1-t^n}{1-t} = \frac{t^n}{1-t}.$$

Assume that the series (4.7) is uniformly convergent to F on $\mathbb{T}$. Take $\varepsilon > 0$ arbitrarily. Then there exists $N \in \mathbb{N}$ such that $n > N$ implies

$$\left|\frac{t^n}{1-t}\right| < \varepsilon \quad \text{for all} \quad t \in \mathbb{T}.$$

If $t \in [0, 1)$, then

$$\frac{t^n}{1-t} \to \infty \quad \text{as} \quad t \to 1$$

and

$$\sup_{t\in[0,1)} \frac{t^n}{1-t} \geq 1.$$

Therefore,

$$\sup_{t\in[0,1)} \frac{t^n}{1-t} \not\to 0 \quad \text{as} \quad n \to \infty.$$

Hence, the series (4.7) is not uniformly convergent to F on $\mathbb{T}$.

Example 4.28 Let $\mathbb{T} = \mathbb{Z}$. Consider the series

$$\sum_{l=1}^{\infty} \frac{(-1)^{l-1}}{t^2 + l}.$$

By the Leibniz criterion, we have that this series is pointwise convergent on $\mathbb{T}$. Moreover,

$$\begin{aligned} \sup_{t\in\mathbb{T}} |F_n(t) - F(t)| &\leq \sup_{t\in\mathbb{T}} \frac{1}{t^2 + n + 1} \\ &\leq \frac{1}{n+1} \to 0 \quad \text{as} \quad n \to \infty. \end{aligned}$$

Therefore, the considered series is uniformly convergent on $\mathbb{T}$.

Theorem 4.29 (Cauchy Criterion for Uniform Convergence of a Function Series) *The function series $\sum_{n=1}^{\infty} f_n$ converges uniformly on D if and only if for any given $\varepsilon > 0$, there exists $N \in \mathbb{N}$ such that*

$$|f_{n+1}(t) + \cdots + f_m(t)| < \varepsilon \tag{4.8}$$

for all $m, n \in \mathbb{N}$ satisfying $m > n$ and every point $t \in D$.

Proof 1. Necessity. Suppose that the function series $\sum_{n=1}^{\infty} f_n$ converges uniformly on D and its sum function is F. Then, for any given $\varepsilon > 0$, there exists $N \in \mathbb{N}$ such that $n > N$ implies

$$\left|F(t) - \sum_{k=1}^{n} f_k(t)\right| < \frac{\varepsilon}{2} \quad \text{for all} \quad t \in D.$$

Hence, for all $m > n > N$ and all $t \in D$, we have

$$\begin{aligned}
|f_{n+1}(t) + \cdots + f_m(t)| &= \left|\sum_{k=1}^{m} f_k(t) - \sum_{k=1}^{n} f_k(t)\right| \\
&= \left|\sum_{k=1}^{m} f_k(t) - F(t) - \sum_{k=1}^{n} f_k(t) + F(t)\right| \\
&\leq \left|\sum_{k=1}^{m} f_k(t) - F(t)\right| + \left|F(t) - \sum_{k=1}^{n} f_k(t)\right| \\
&< \frac{\varepsilon}{2} + \frac{\varepsilon}{2} \\
&= \varepsilon.
\end{aligned}$$

2. Sufficiency. Suppose that for any given $\varepsilon > 0$, there exists $N \in \mathbb{N}$ such that (4.8) holds for all $m > n > N$ and all $t \in D$. Fix $t_0 \in D$. Then the numerical series $\sum_{n=1}^{\infty} f_n(t_0)$ satisfies the Cauchy criterion for convergence of a numerical series. Thus, the numerical series $\sum_{n=1}^{\infty} f_n(t_0)$ is convergent. Because $t_0 \in D$ was arbitrarily chosen, we conclude that the series $\sum_{n=1}^{\infty} f_n$ is pointwise convergent on D. Let F be its sum function. Choose n for

$$\left|\sum_{k=1}^{m} f_k(t) - \sum_{k=1}^{n} f_k(t)\right| < \frac{\varepsilon}{2} \quad \text{for all} \quad t \in \mathbb{T}.$$

If $m \to \infty$, then we get

$$\left|F(t) - \sum_{k=1}^{n} f_k(t)\right| \leq \frac{\varepsilon}{2} < \varepsilon \quad \text{for all} \quad t \in D.$$

Therefore, $\sum_{k=1}^{\infty} f_k$ converges uniformly to F on D.

This completes the proof. □

Corollary 4.30 (Necessary Condition for Uniform Convergence of a Function Series) *A necessary condition for the series $\sum_{n=1}^{\infty} f_n$ to converge uniformly on D is that $f_n \to 0$ uniformly on D as $n \to \infty$.*

Example 4.31 Let $\mathbb{T} = \left\{\frac{1}{\sqrt[3]{n}} : n \in \mathbb{N}\right\} \cup \{0\}$. Consider the series

$$\sum_{k=1}^{\infty} \frac{t^3}{(1+t^3)^k}.$$

Thus,

$$\sum_{k=n+1}^{3n} \frac{t^3}{(1+t^3)^k} = \frac{t^3}{(1+t^3)^{n+1}} + \frac{t^3}{(1+t^3)^{n+2}} + \cdots + \frac{t^3}{(1+t^3)^{3n}}$$
$$> \frac{3nt^3}{(1+t^3)^{3n}}.$$

Let $\varepsilon = \frac{3}{e^3}$. For any $N \in \mathbb{N}$, we choose $m = 3n$, $n > N$, and $t_n = \frac{1}{\sqrt[3]{n}} \in \mathbb{T}$, so that

$$\sum_{k=n+1}^{3n} \frac{t^3}{(1+t^3)^k} > \frac{3n\frac{1}{n}}{\left(1+\frac{1}{n}\right)^{3n}}$$
$$= \frac{3}{\left(1+\frac{1}{n}\right)^{3n}}$$
$$> \frac{3}{e^3} = \varepsilon.$$

Then, using Theorem 4.29, we conclude that the considered series is nonuniformly convergent on $\mathbb{T}$.

Theorem 4.32 (Weierstraß M -Test for Uniform Convergence of a Function Series) *Suppose that every term f_n of the function series $\sum_{n=1}^{\infty} f_n$ satisfies*

$$|f_n(t)| \leq a_n \text{ for all } n \in \mathbb{N} \text{ and all } t \in D,$$

and the numerical series $\sum_{n=1}^{\infty} a_n$ is convergent. Then the function series $\sum_{n=1}^{\infty} f_n$ is uniformly convergent on D.

Proof Since the numerical series $\sum_{k=1}^{\infty} a_k$ is convergent, for any $\varepsilon > 0$, there exists $N \in \mathbb{N}$ such that $m > n > N$ imply

$$a_{n+1} + a_{n+2} + \cdots + a_m < \varepsilon.$$

Hence,

$$\left|\sum_{k=n+1}^{m} f_k(t)\right| \leq \sum_{k=n+1}^{m} |f_k(t)|$$
$$\leq \sum_{k=n+1}^{m} a_k$$
$$< \varepsilon.$$

Then, using Theorem 4.29, we conclude that $\sum_{n=1}^{\infty} f_n$ is uniformly convergent on D. □

Example 4.33 Let $\mathbb{T} = \mathbb{P}_{1,2} = \bigcup_{k=0}^{\infty}[3k, 3k+1]$. Consider the series

$$\sum_{n=1}^{\infty} \frac{t^2}{1+n^5t^4}.$$

We have $f_n(t) = \frac{t^2}{1+n^5t^4}$ and

$$\begin{aligned} f_n(t) &\le \frac{t^2}{n^{\frac{5}{2}}t^2} \\ &= \frac{1}{n^{\frac{5}{2}}}. \end{aligned}$$

Since $\sum_{n=1}^{\infty} \frac{1}{n^{\frac{5}{2}}}$ is convergent, by Theorem 4.32, we conclude that the considered series is uniformly convergent on $\mathbb{T}$.

Example 4.34 Let $\mathbb{T} = [-3, 0] \cup \left\{\frac{1}{n} : n \in \mathbb{N}\right\}$, where $[-3, 0]$ is the real-valued interval. Consider the series

$$\sum_{n=1}^{\infty} \frac{(t+1)\sin^2(nt)}{n\sqrt{n+1}}.$$

Here, $f_n(t) = \frac{(t+1)\sin^2(nt)}{n\sqrt{n+1}}$. We have

$$\begin{aligned} |f_n(t)| &= \left|\frac{(t+1)\sin^2(nt)}{n\sqrt{n+1}}\right| \\ &= \frac{|t+1|\sin^2(nt)}{n\sqrt{n+1}} \\ &\le \frac{|t|+1}{n\sqrt{n+1}} \\ &\le \frac{4}{n\sqrt{n+1}} \quad \text{for all} \quad t \in \mathbb{T}. \end{aligned}$$

Since $\sum_{n=1}^{\infty} \frac{4}{n\sqrt{n+1}}$ is convergent, by Theorem 4.32, we conclude that the considered series is uniformly convergent on $\mathbb{T}$.

Example 4.35 Let $\mathbb{T} = \mathbb{Z}$. Consider the series

$$\sum_{n=1}^{\infty} \frac{\cos(nt)}{n^4+1}.$$

Here, $f_n(t) = \frac{\cos(nt)}{n^4+1}$. We have

$$\begin{aligned}|f_n(t)| &= \left|\frac{\cos(nt)}{n^4+1}\right| \\ &= \frac{|\cos(nt)|}{n^4+1} \\ &\le \frac{1}{n^4+1}.\end{aligned}$$

Since $\sum_{n=1}^{\infty} \frac{1}{n^4+1}$ is convergent, by Theorem 4.32, we conclude that the considered series is uniformly convergent on $\mathbb{T}$.

Exercise 4.36 Using Theorem 4.32, prove that the following series are uniformly convergent.

1. $\sum_{n=1}^{\infty} \frac{1}{n^2+nt+t^2}$, $\mathbb{T} = \mathbb{Z}$,
2. $\sum_{n=1}^{\infty} \frac{1}{3^n\sqrt{1+(2n+1)t}}$, $\mathbb{T} = \mathbb{N}$,
3. $\sum_{n=1}^{\infty} \frac{1}{n^2} \sin \frac{t}{n}$, $\mathbb{T} = 3^{\mathbb{N}_0}$,
4. $\sum_{n=1}^{\infty} \frac{e^{-n^2t^2}}{1+n^2}$, $\mathbb{T} = \mathbb{Z}$,
5. $\sum_{n=1}^{\infty} \frac{\sqrt{1-t^{2n}}}{2^n}$, $\mathbb{T} = \left\{\frac{1}{n} : n \in \mathbb{N}\right\} \cup \{0\}$,
6. $\sum_{n=1}^{\infty} \frac{t^n}{n\sqrt{n}}$, $\mathbb{T} = \left\{\frac{1}{n} : n \in \mathbb{N}\right\} \cup \{0\}$.

Theorem 4.37 (Abel Test) *Let the function sequence $\{f_n(t)\}_{n\in\mathbb{N}}$ be monotone for each fixed $t \in D$ with respect to n and suppose that $\{f_n(t)\}_{n\in\mathbb{N}}$ is uniformly bounded on D, i.e.,*

$$|f_n(t)| \le M \text{ for all } t \in D \text{ and } n \in \mathbb{N}.$$

If the function series $\sum_{n=1}^{\infty} g_n$ is uniformly convergent on D, then the function series $\sum_{n=1}^{\infty} f_n g_n$ is uniformly convergent on $\mathbb{T}$.

Proof Since $\sum_{n=1}^{\infty} g_n(t)$ is uniformly convergent on D, for any $\varepsilon > 0$, there exists $N \in \mathbb{N}$ such that

$$\left|\sum_{k=n+1}^{m} g_k(t)\right| < \varepsilon$$

for all $m > n > N$ and all $t \in D$. By applying the Abel lemma, we obtain

$$\begin{aligned}\left|\sum_{k=n+1}^{m} f_k(t)g_k(t)\right| &\le \varepsilon\left(|f_{n+1}(t)| + 2|f_m(t)|\right) \\ &\le 3M\varepsilon\end{aligned}$$

for all $m > n > N$ and all $t \in D$. Hence, by Theorem 4.29, it follows that $\sum_{k=1}^{\infty} f_k g_k$ is uniformly convergent on D. □

Example 4.38 Consider the series $\sum_{n=1}^{\infty} \frac{e^{-nt}}{n!}$ on $\mathbb{T} = \mathbb{N}$. We have that the sequence $\left\{e^{-nt}\right\}_{n\in\mathbb{N}}$ is monotone for each fixed $t \in \mathbb{T}$ with respect to n, and it is uniformly bounded by 1 on $\mathbb{T}$. Note that $\sum_{n=1}^{\infty} \frac{1}{n!}$ is a convergent numerical series. Hence, using Theorem 4.37, we conclude that the considered series is uniformly convergent on $\mathbb{T}$.

Example 4.39 Let $\mathbb{T} = \left\{\frac{1}{n} : n \in \mathbb{N}\right\} \cup \{0\}$. Consider the series

$$\sum_{n=1}^{\infty} \frac{(-1)^{n-1} t^n}{n(1+t^n)} \quad \text{on} \quad \mathbb{T}.$$

Note that the series $\sum_{n=1}^{\infty} \frac{(-1)^{n-1}}{n}$ is a convergent numerical series. Also, the sequence $\left\{\frac{t^n}{1+t^n}\right\}_{n\in\mathbb{N}}$ is monotone on $\mathbb{T}$ with respect to n and uniformly bounded by 1 on $\mathbb{T}$. Hence, by Theorem 4.37, it follows that the considered series is uniformly convergent on $\mathbb{T}$.

Exercise 4.40 Using the Abel test, prove that the series

$$\sum_{n=1}^{\infty} \frac{(-1)^n \sqrt{n}}{\sqrt{n+t}\,\log\log(1+2\sqrt{n})}$$

is uniformly convergent on $\mathbb{T} = \mathbb{N}$.

Theorem 4.41 (Dirichlet Test) *Suppose that the function sequence $\{f_n(t)\}_{n\in\mathbb{N}}$ is monotone for each $t \in D$ with respect to n and $\{f_n\}_{n\in\mathbb{N}}$ is uniformly convergent to 0 on D. In addition, assume that the partial sum sequence of $\sum_{n=1}^{\infty} g_n(t)$ is uniformly bounded on D, i.e.,*

$$\left|\sum_{k=1}^{n} g_k(t)\right| \leq M \text{ for all } t \in D \text{ and } n \in \mathbb{N}.$$

Then $\sum_{n=1}^{\infty} f_n g_n$ is uniformly convergent on D.

Proof Since $\{f_n\}_{n\in\mathbb{N}}$ is uniformly convergent to 0 on D, for any given $\varepsilon > 0$, there exists $N \in \mathbb{N}$ such that $n > N$ implies

$$|f_n(t)| < \varepsilon \quad \text{for all} \quad t \in D.$$

Moreover, for all $m > n > N$, we have

$$\begin{aligned}\left|\sum_{k=n+1}^{m} g_k(t)\right| &= \left|\sum_{k=1}^{m} g_k(t) - \sum_{k=1}^{n} g_k(t)\right| \\ &\le \left|\sum_{k=1}^{m} g_k(t)\right| + \left|\sum_{k=1}^{n} g_k(t)\right| \\ &\le 2M.\end{aligned}$$

Using the Abel lemma, we get

$$\begin{aligned}\left|\sum_{k=n+1}^{m} f_k(t)g_k(t)\right| &\le 2M\left(|f_{n+1}(t)| + 2|f_m(t)|\right) \\ &< 6M\varepsilon.\end{aligned}$$

Hence, using Theorem 4.29, it follows that $\sum_{n=1}^{\infty} f_n g_n$ is uniformly convergent on D. □

Example 4.42 Let $\mathbb{T} = \mathbb{N}_0 \cup \left\{1 - \frac{1}{n} : n \in \mathbb{N}\right\}$. Consider the series

$$\sum_{n=1}^{\infty} \frac{(-1)^n}{n + t^2}.$$

We set

$$f_n(t) = \frac{1}{n + t^2}, \quad g_n(t) = (-1)^n, \quad t \in D, \quad n \in \mathbb{N}.$$

Then the sequence $\{f_n(t)\}_{n\in\mathbb{N}}$ is monotone for each $t \in D$ with respect to n, and it is uniformly convergent to 0 on D. Moreover,

$$\left|\sum_{k=1}^{n} g_k(t)\right| \le 1 \quad \text{for all} \quad t \in D, \quad n \in \mathbb{N}.$$

Hence, utilizing the Dirichlet test, we conclude that the considered series is uniformly convergent on D.

Example 4.43 Let $\mathbb{T} = \left\{\frac{1}{n} : n \in \mathbb{N} \setminus \{1\}\right\} \cup \{0\}$. Consider the series

$$\sum_{n=1}^{\infty} \frac{(-1)^{n-1} t^n}{1 + t + \cdots + t^{2n-1}}.$$

We set

$$f_n(t) = \frac{t^n}{1 + t + \cdots + t^{2n-1}}, \quad g_n(t) = (-1)^{n-1}, \quad t \in D, \quad n \in \mathbb{N}.$$

Then

$$\begin{aligned} f_n(t) &< \frac{t^n}{1 + t + \cdots + t^{n-1}} \\ &< \frac{t^n}{nt^n} \\ &= \frac{1}{n} \to 0 \quad \text{as} \quad n \to \infty, \end{aligned}$$

i.e., the sequence $\{f_n\}_{n \in \mathbb{N}}$ is uniformly convergent to 0 on $\mathbb{T}$. Moreover,

$$\frac{t^{n+1}}{1 + t + \cdots + t^{2n-1} + t^{2n+1}} < \frac{t^n}{1 + t + t^2 + \cdots + t^{2n-1}},$$

i.e., the sequence $\{f_n(t)\}_{n \in \mathbb{N}}$ is monotone for each $t \in \mathbb{T}$ with respect to n. Note that

$$\left| \sum_{k=1}^{n} g_k(t) \right| \leq 1 \quad \text{for} \quad t \in \mathbb{T}, \quad n \in \mathbb{N}.$$

Hence, using the Dirichlet test, it follows that the considered series is uniformly convergent on $\mathbb{N}$.

Exercise 4.44 Let $\mathbb{T} = \left\{ \frac{1}{n} : n \in \mathbb{N} \right\} \cup \{0\}$. Using the Dirichlet Test, prove that

$$\sum_{n=1}^{\infty} \frac{(-1)^{n-1} t^{2n}}{2n - 1}$$

is uniformly convergent on $\mathbb{T}$.

By Theorem 4.21, the following result is clear.

Theorem 4.45 *If every term f_n of the function series $\sum_{n=1}^{\infty} f_n$ is rd-continuous and $\sum_{n=1}^{\infty} f_n$ is uniformly convergent to f on D, then f is rd-continuous and*

$$\int_a^b f(t)\Delta t = \sum_{n=1}^{\infty} \int_a^b f_n(t)\Delta t \quad \textit{for every} \quad [a, b] \subset D.$$

By Theorem 4.22, we obtain the following result.

Theorem 4.46 *Suppose that the function series $\sum_{n=1}^{\infty} f_n$ satisfies*

1. f_n is differentiable and its derivative function f_n^{Δ} is rd-continuous on $[a, b]$,

2. $\sum_{n=1}^{\infty} f_n$ *converges pointwise to* f *on* $[a, b]$,
3. $\sum_{n=1}^{\infty} f_n^{\Delta}$ *is uniformly convergent to* g *on* $[a, b]$.

Then $f = \sum_{n=1}^{\infty} f_n$ *is differentiable on* $[a, b]$ *and* $f^{\Delta} = g$ *for all* $t \in [a, b]$.

By Theorem 4.23, we get the following result.

Theorem 4.47 (Dini Theorem) *Assume that the function series* $\sum_{n=1}^{\infty} f_n$ *is pointwise convergent to its sum function* f *on* $[a, b]$. *If the conditions*

1. f_n, $n \in \mathbb{N}$, *are rd-continuous on* $[a, b]$,
2. f *is rd-continuous on* $[a, b]$,
3. *for any given* $t \in [a, b]$, *the function series* $\sum_{n=1}^{\infty} f_n(t)$ *is either a positive term series or a negative term series*

hold, then $\sum_{n=1}^{\infty} f_n$ *is uniformly convergent to* f *on* $[a, b]$.

4.3 Advanced Practical Problems

Problem 4.48 Let

$$f_n(t) = t^2\left(t + 1 + e^{-n^2t^2}\right), \quad f(t) = t^3 + t^2.$$

Prove that $f_n \to f$ pointwise on $\mathbb{T}$.

Problem 4.49 Let $\mathbb{T} = 2^{\mathbb{N}_0}$. Check if the following sequences are uniformly convergent on $\mathbb{T}$.

1. $f_n(t) = ne^{-nt^2}$,
2. $f_n(t) = nte^{-nt^2}$,
3. $f_n(t) = \frac{1}{t^2+nt+1}$.

Solution 1. Uniformly convergent to 0 on $\mathbb{T}$,
2. uniformly convergent to 0 on $\mathbb{T}$,
3. uniformly convergent to 0 on $\mathbb{T}$.

Problem 4.50 Let $\mathbb{T} = \{0\} \cup \left\{\frac{1}{n} : n \in \mathbb{N}\right\} \cup 2^{\mathbb{N}_0}$. Using Theorem 4.11, investigate the sequence

$$\left\{f_n(t) = \frac{nt^2}{n+t}\right\}_{n \in \mathbb{N}}$$

for uniform convergence on $D_1 = [1, \infty)$ and $D_2 = [0, 2]$.

Solution The sequence is not uniformly convergent on D_1, and it is uniformly convergent on D_2.

Problem 4.51 Consider $f_n(t) = 2 + \frac{n^2t^2+nt+2}{3+n^3t^3+n^2t^2+nt}$ on $D = \left\{\frac{1}{n} : n \in \mathbb{N}\right\} \cup \{0\}$. Using Theorem 4.16, prove that this sequence is not uniformly convergent to 2 on D.

Problem 4.52 Using Theorem 4.32, prove that the following series are uniformly convergent.

1. $\sum_{n=1}^{\infty} \frac{t}{(1+nt)(1+(n+1)t)}$, $\mathbb{T} = 2^{\mathbb{N}_0}$,
2. $\sum_{n=1}^{\infty} \frac{(\pi-t)\cos^2(nt)}{\sqrt[5]{n^7+1}}$, $\mathbb{T} = \left\{\frac{1}{n} : n \in \mathbb{N}\right\} \cup \{0\}$,
3. $\sum_{n=1}^{\infty} \arctan \frac{t}{t^2+n^3}$, $\mathbb{T} = \mathbb{Z}$,
4. $\sum_{n=1}^{\infty} \frac{t^3\sin^2(nt)}{2+n^3t^6}$, $\mathbb{T} = \mathbb{N}$,
5. $\sum_{n=1}^{\infty} \frac{1}{2^n} \sin \frac{n^3t}{n^2+1}$, $\mathbb{T} = 2^{\mathbb{N}_0}$,
6. $\sum_{n=1}^{\infty} \frac{\arctan(2n^2t)}{\sqrt[3]{n^7+n+t}}$, $\mathbb{T} = \mathbb{N}$.

Problem 4.53 Using the Abel test, prove that the series

$$\sum_{n=1}^{\infty} \frac{1}{2^n n^t}$$

is uniformly convergent on $\mathbb{T} = \mathbb{N}$.

Problem 4.54 Let $\mathbb{T} = \left\{\frac{1}{n} : n \in \mathbb{N}\right\} \cup \{0\}$. Using the Dirichlet test, prove that

$$\sum_{n=1}^{\infty} \frac{(-1)^n}{n^2t^2+n}$$

is uniformly convergent on $\mathbb{T}$.

4.4 Notes and References

In this chapter, the concept of function series and sequences is extended to time scales. Necessary and sufficient conditions and several criteria for uniform convergence of function series and sequences are presented. Several analytical properties of function series and function sequences on general time scales are given. All results in this chapter are taken from Pang and Wang [36].

Chapter 5
Parameter-Dependent Integrals

5.1 Normal Parameter-Dependent Integrals

Let $\mathbb{T}_1$ and $\mathbb{T}_2$ be time scales.

Definition 5.1 With $\mathrm{CC_{rd}}$, we denote the set of functions f defined on $\mathbb{T}_1 \times \mathbb{T}_2$ with the following properties.

1. f is rd-continuous with respect to the first variable.
2. f is rd-continuous with respect to the second variable.
3. If $(t_1, t_2) \in \mathbb{T}_1 \times \mathbb{T}_2$ with t_1 is right-dense or maximal and t_2 is right-dense or maximal, then f is continuous at (t_1, t_2).
4. If t^1 and t^2 are both left-dense points in $\mathbb{T}_1$ and $\mathbb{T}_2$, respectively, then the limit of $f(t_1, t_2)$ exists as (t_1, t_2) approaches (t^1, t^2) along any path in the region

$$R_{LL}(t^1, t^2) = \{(t_1, t_2) : \ t_1 \in [a_1, t^1] \cap \mathbb{T}_1, \ t_2 \in [a_2, t^2] \cap \mathbb{T}_2\}.$$

Exercise 5.2 Let f and g be real-valued functions on $\mathbb{T}_1 \times \mathbb{T}_2$ such that fg is defined on $\mathbb{T}_1 \times \mathbb{T}_2$.

1. Assume f is continuous. Prove that $f \in \mathrm{CC_{rd}}$.
2. Assume f is continuous and $g \in \mathrm{CC_{rd}}$. Prove that $fg \in \mathrm{CC_{rd}}$.

Now, we suppose that $[a_1, b_1] \subset \mathbb{T}_1$ and $[a_2, b_2] \subset \mathbb{T}_2$, and the function $f(t_1, t_2)$ is defined on $[a_1, b_1] \times [a_2, b_2]$. In this section, we investigate normal parameter-dependent integrals

$$I(t_2) = \int_{a_1}^{b_1} f(t_1, t_2) \Delta_1 t_1, \quad t_2 \in [a_2, b_2].$$

Theorem 5.3 *If $f : [a_1, b_1] \times [a_2, b_2] \to \mathbb{R}$ and $f \in \mathrm{CC_{rd}}$, then $I(t_2)$ is rd-continuous on $[a_2, b_2]$.*

M. Bohner and S.G. Georgiev, *Multivariable Dynamic Calculus on Time Scales*,
DOI 10.1007/978-3-319-47620-9_5

Proof Since $f \in \mathcal{CC}_{\mathrm{rd}}$, we have that $f(t_1, t_2)$ is rd-continuous in t_2. Therefore, f is continuous at right-dense points $t_2 \in [a_2, b_2]$. Let $t_2^0 \in [a_2, b_2]$ be a left-dense and right-scattered point. Then, for every $\varepsilon > 0$, there exists $\delta = \delta(\varepsilon) > 0$ such that

$$|f(t_1, t_2^1) - f(t_1, t_2^2)| < \frac{\varepsilon}{b_1 - a_1} \quad \text{for all} \quad t_1^1, t_2^2 \in (t_2^0 - \delta, t_2^0).$$

Hence,

$$\begin{aligned} |I(t_2^1) - I(t_2^2)| &= \left| \int_{a_1}^{b_1} f(t_1, t_2^1) \Delta_1 t_1 - \int_{a_1}^{b_1} f(t_1, t_2^2) \Delta_1 t_1 \right| \\ &= \left| \int_{a_1}^{b_1} \left(f(t_1, t_2^1) - f(t_1, t_2^2) \right) \Delta_1 t_1 \right| \\ &\leq \int_{a_1}^{b_1} \left| f(t_1, t_2^1) - f(t_1, t_2^2) \right| \Delta_1 t_1 \\ &< \frac{\varepsilon}{b_1 - a_1} (b_1 - a_1) \\ &= \varepsilon. \end{aligned}$$

This proves that $\lim\limits_{t_2 \to t_2^0 -} I(t_2)$ exists and is finite, which completes the proof. □

Theorem 5.4 *If $f \in \mathcal{CC}_{\mathrm{rd}}$ and $f_{t_2}^{\Delta_2}(t_1, t_2)$ is continuous, then*

$$I(t_2) = \int_{a_1}^{b_1} f(t_1, t_2) \Delta_1 t_1$$

is differentiable on $[a_2, b_2]$ and satisfies

$$I^{\Delta_2}(t_2) = \int_{a_1}^{b_1} f_{t_2}^{\Delta_2}(t_1, t_2) \Delta_1 t_1, \quad t_2 \in [a_2, b_2].$$

Proof Since $f_{t_2}^{\Delta_2}(t_1, t_2)$ is continuous, it is iterated Riemann Δ-integrable in either order. Hence, by Theorem 7.58 (see later in the book in Chapter 7), we obtain

$$\begin{aligned} \int_{a_2}^{t_2} \int_{a_1}^{b_1} f_{t_2}^{\Delta_2}(t_1, \tau_2) \Delta_1 t_1 \Delta_2 \tau_2 &= \int_{a_1}^{b_1} \int_{a_2}^{t_2} f_{t_2}^{\Delta_2}(t_1, \tau_2) \Delta_2 \tau_2 \Delta_1 t_1 \\ &= \int_{a_1}^{b_1} \left(f(t_1, t_2) - f(t_1, a_2) \right) \Delta_1 t_1 \\ &= I(t_2) - I(a_2). \end{aligned}$$

Note that $\int_{a_1}^{b_1} f_{t_2}^{\Delta_2}(t_1, t_2)\Delta_1 t_1$ is rd-continuous. Therefore, the left-hand side of the above formula is differentiable, which implies that the right-hand side is also differentiable, and this leads to the desired result. □

Example 5.5 Let $\mathbb{T}_1 = \mathbb{T}_2 = \mathbb{Z}$, $[a_1, b_1] = [a_2, b_2] = [0, 2]$. We consider

$$I(t_2) = \int_0^2 (t_1 + t_2)\Delta_1 t_1.$$

We have

$$\begin{aligned} I(t_2) &= \int_0^2 \left(\frac{1}{2}(t_1 + \sigma_1(t_1)) - \frac{1}{2} + t_2\right)\Delta_1 t_1 \\ &= \frac{1}{2}t_1^2\Big|_{t_1=0}^{t_1=2} + \left(-\frac{1}{2} + t_2\right)t_1\Big|_{t_1=0}^{t_1=2} \\ &= 2 + 2\left(-\frac{1}{2} + t_2\right) \\ &= 2 - 1 + 2t_2 \\ &= 1 + 2t_2. \end{aligned}$$

Hence,

$$I^{\Delta_2}(t_2) = 2.$$

On the other hand, using Theorem 5.4, we have

$$I^{\Delta_2}(t_2) = \int_0^2 \Delta_1 t_1 = 2.$$

Example 5.6 Let $\mathbb{T}_1 = \mathbb{Z}$, $\mathbb{T}_2 = 2^{\mathbb{N}_0}$, $[a_1, b_1] = [-1, 1]$, $[a_2, b_2] = [1, 4]$. We consider

$$I(t_2) = \int_{-1}^1 (t_1^2 - 2t_1 t_2 + t_2^2)\Delta_1 t_1.$$

We have

$$\begin{aligned} I(t_2) &= \int_{-1}^1 \left(\frac{t_1^2 + t_1\sigma_1(t_1) + (\sigma_1(t_1))^2}{3} - \frac{(1 + 2t_2)(t_1 + \sigma_1(t_1))}{2} + \frac{1}{6} + t_2 + t_2^2\right)\Delta_1 t_1 \\ &= \frac{1}{3}t_1^3\Big|_{t_1=-1}^{t_1=1} - \frac{1}{2}(1 + 2t_2)t_1^2\Big|_{t_1=-1}^{t_1=1} + \left(\frac{1}{6} + t_2 + t_2^2\right)t_1\Big|_{t_1=-1}^{t_1=1} \end{aligned}$$

$$= \frac{2}{3} + \frac{1}{3} + 2t_2 + 2t_2^2$$

$$= 1 + 2t_2 + 2t_2^2.$$

Hence,

$$I^{\Delta_2}(t_2) = 2 + 2(\sigma_2(t_2) + t_2)$$

$$= 2 + 2(2t_2 + t_2)$$

$$= 2 + 6t_2.$$

On the other hand, using Theorem 5.4, we get

$$I^{\Delta_2}(t_2) = \int_{-1}^{1} (-2t_1 + \sigma_2(t_2) + t_2)\Delta_1 t_1$$

$$= \int_{-1}^{1} (-2t_1 + 3t_2)\Delta_1 t_1$$

$$= \int_{-1}^{1} (-(t_1 + \sigma_1(t_1)) + 1 + 3t_2)\,\Delta_1 t_1$$

$$= -t_1^2\Big|_{t_1=-1}^{t_1=1} + (1 + 3t_2)t_1\Big|_{t_1=-1}^{t_1=1}$$

$$= 2 + 6t_2.$$

Example 5.7 Let $\mathbb{T}_1 = \mathbb{T}_2 = 2^{\mathbb{N}_0}$, $[a_1, b_1] = [a_2, b_2] = [1, 4]$. We consider the integral

$$I(t_2) = \int_1^4 \left(t_1 e_f(t_1, 1) + t_1^2 t_2\right) \Delta_1 t_1 \quad \text{with} \quad f(t_1) = t_1.$$

We have

$$I(t_2) = \int_1^4 \left(e_f^{\Delta_1}(t_1, 1) + \frac{t_1^2 + t_1\sigma_1(t_1) + (\sigma_1(t_1))^2}{7} t_2\right) \Delta_1 t_1$$

$$= e_f(t_1, 1)\Big|_{t_1=1}^{t_1=4} + \frac{1}{7} t_1^3 t_2\Big|_{t_1=1}^{t_1=4}$$

$$= e_f(4, 1) + 9t_2.$$

Hence,

$$I^{\Delta_2}(t_2) = 9.$$

On the other hand, using Theorem 5.4, we get

$$\begin{aligned} I^{\Delta_2}(t_2) &= \int_1^4 t_1^2 \Delta_1 t_1 \\ &= \frac{1}{7}\int_1^4 (t_1^2 + t_1\sigma_1(t_1) + (\sigma_1(t_1))^2)\Delta_1 t_1 \\ &= \frac{1}{7} t_1^3 \Big|_{t_1=1}^{t_1=4} \\ &= 9. \end{aligned}$$

Exercise 5.8 Find $I^{\Delta_2}(t_2)$, where

1. $I(t_2) = \int_0^2 \left(t_1^2 t_2 + t_2^2 + t_1^4 + t_1^5\right) \Delta_1 t_1$, $\mathbb{T}_1 = \mathbb{T}_2 = \mathbb{Z}$,
2. $I(t_2) = \int_0^2 \left(t_1^3 + \sin_f^2(t_1, 1) + t_1^4 - 2t_1 t_2^2\right) \Delta_1 t_1$, $\mathbb{T}_1 = \mathbb{Z}$, $\mathbb{T}_2 = 2^{\mathbb{N}_0}$, $f(t_1) = t_1$,
3. $I(t_2) = \int_1^2 \left(t_1^4 - 2t_1 t_2^2 + t_1^3 - t_2\right) \Delta_1 t_1$, $\mathbb{T}_1 = \mathbb{T}_2 = 2^{\mathbb{N}_0}$.

Solution 1. $3 + 4t_2$,
2. $-6t_2$,
3. $-6t_2 - 1$.

Theorem 5.9 *Let $\mathbb{T}$ be a time scale. If $f : \mathbb{T} \times \mathbb{T} \to \mathbb{R}$ is such that both $f(t, s)$ and $f_t^{\Delta}(t, s)$ are continuous on $\mathbb{T} \times \mathbb{T}$, then*

$$g(t) = \int_a^t f(t, s)\Delta s$$

satisfies

$$g^{\Delta}(t) = \int_a^t f^{\Delta}(t, s)\Delta s + f(\sigma(t), t),$$

where f^{Δ} denotes the derivative of f with respect to the first variable.

Proof 1. If t is right-scattered, then

$$\begin{aligned} g^{\Delta}(t) &= \frac{g(\sigma(t)) - g(t)}{\sigma(t) - t} \\ &= \frac{1}{\sigma(t) - t}\left(\int_a^{\sigma(t)} f(\sigma(t), s)\Delta s - \int_a^t f(t, s)\Delta t\right) \end{aligned}$$

$$= \frac{1}{\sigma(t)-t}\left(\int_a^t f(\sigma(t),s)\Delta s + \int_t^{\sigma(t)} f(\sigma(t),s)\Delta s - \int_a^t f(t,s)\Delta s\right)$$

$$= \frac{1}{\sigma(t)-t}\int_a^t \left(f(\sigma(t),s) - f(t,s)\right)\Delta s + \frac{1}{\sigma(t)-t}\int_t^{\sigma(t)} f(\sigma(t),s)\Delta s$$

$$= \int_a^t f^{\Delta}(t,s)\Delta s + f(\sigma(t),t).$$

2. If t is right-dense, then

$$g^{\Delta}(t) = \lim_{r\to t}\frac{1}{t-r}\left(\int_a^t f(t,s)\Delta s - \int_a^r f(r,s)\Delta s\right)$$

$$= \lim_{r\to t}\frac{1}{t-r}\left(\int_a^t f(t,s)\Delta s - \int_a^t f(r,s)\Delta s + \int_a^t f(r,s)\Delta s - \int_a^r f(r,s)\Delta s\right)$$

$$= \lim_{r\to t}\frac{1}{t-r}\left(\int_a^t (f(t,s) - f(r,s))\Delta s + \int_r^t f(r,s)\Delta s\right)$$

$$= \lim_{r\to t}\int_a^t \frac{f(t,s)-f(r,s)}{t-r}\Delta s + \lim_{r\to t}\frac{1}{t-r}\int_r^t f(r,s)\Delta s. \tag{5.1}$$

Since f is continuous, it is uniformly continuous on any compact subset of $\mathbb{T}\times\mathbb{T}$. Therefore, for every $\varepsilon > 0$, there exists a neighbourhood U of t such that

$$|f(r,s) - f(t,t)| < \varepsilon \quad \text{for all} \quad r, s \in U.$$

Thus,

$$\begin{aligned}\left|\frac{1}{t-r}\int_r^t f(r,s)\Delta s - f(t,t)\right| &= \left|\frac{1}{t-r}\int_r^t (f(r,s) - f(t,t))\Delta s\right|\\ &\le \frac{1}{|t-r|}\left|\int_r^t |f(r,s) - f(t,t)|\Delta s\right|\\ &< \frac{\varepsilon}{|t-r|}\left|\int_r^t \Delta s\right|\\ &= \varepsilon \quad \text{for all} \quad r, t \in U.\end{aligned}$$

Therefore,

$$\lim_{r\to t}\int_a^t \frac{f(r,s)}{t-r}\Delta s = f(t,t). \tag{5.2}$$

For fixed s, we apply the mean value theorem to the function $f(t, s)$ with respect to the variable t. Thus, we can write

$$f^{\Delta}(\tau, s) \le \frac{f(t, s) - f(r, s)}{t - r} \le f^{\Delta}(\xi, s), \tag{5.3}$$

where τ and ξ are between r and t. Note that $\xi, \tau \to t$ as $r \to t$. Since f^{Δ} is continuous, using (5.3), we get

$$\lim_{r \to t} \frac{f(t, s) - f(r, s)}{t - r} = f^{\Delta}(t, s).$$

From here and from (5.1) and (5.2), we get the desired result. □

Example 5.10 Let $\mathbb{T} = \mathbb{Z}$. Consider

$$I(t) = \int_0^t (t^2 + st + s^2)\Delta s.$$

We will compute $I^{\Delta}(t)$. We have

$$\begin{aligned}
I(t) &= \int_0^t \left(t^2 - \frac{t}{2} + \frac{1}{6} + \frac{t-1}{2}(s + \sigma(s)) + \frac{s^2 + s\sigma(s) + (\sigma(s))^2}{3}\right) \Delta s \\
&= \left(t^2 - \frac{t}{2} + \frac{1}{6}\right) s\Big|_{s=0}^{s=t} + \left(\frac{t}{2} - \frac{1}{2}\right) s^2\Big|_{s=0}^{s=t} + \frac{1}{3}s^3\Big|_{s=0}^{s=t} \\
&= \left(t^2 - \frac{t}{2} + \frac{1}{6}\right) t + \left(\frac{t}{2} - \frac{1}{2}\right) t^2 + \frac{1}{3}t^3 \\
&= \left(t^2 - \frac{t}{2} + \frac{1}{6}\right) t + \left(\frac{t}{2} - \frac{1}{2}\right) t^2 + \frac{1}{3}t^3 \\
&= t^3 - \frac{t^2}{2} + \frac{t}{6} + \frac{t^3}{2} - \frac{t^2}{2} + \frac{t^3}{3} \\
&= \frac{11}{6}t^3 - t^2 + \frac{t}{6}.
\end{aligned}$$

Hence,

$$\begin{aligned}
I^{\Delta}(t) &= \frac{11}{6}\left((\sigma(t))^2 + t\sigma(t) + t^2\right) - (\sigma(t) + t) + \frac{1}{6} \\
&= \frac{11}{6}\left((t+1)^2 + t(t+1) + t^2\right) - (t + 1 + t) + \frac{1}{6}
\end{aligned}$$

$$= \frac{11}{6}(t^2 + 2t + 1 + t^2 + t + t^2) - (2t + 1) + \frac{1}{6}$$

$$= \frac{11}{6}(3t^2 + 3t + 1) - 2t - 1 + \frac{1}{6}$$

$$= \frac{11}{2}t^2 + \frac{11}{2}t + \frac{11}{6} - 2t - \frac{5}{6}$$

$$= \frac{11}{2}t^2 + \frac{7}{2}t + 1.$$

Now, we will use Theorem 5.9 to compute $I^{\Delta}(t)$. Here, $f(s, t) = t^2 + st + s^2$ so that

$$\begin{aligned}
I^{\Delta}(t) &= \int_0^t (\sigma(t) + t + s)\Delta s + (\sigma(t))^2 + t\sigma(t) + t^2 \\
&= \int_0^t (2t + 1 + s)\Delta s + (t + 1)^2 + t(t + 1) + t^2 \\
&= \int_0^t (2t + 1 + s)\Delta s + t^2 + 2t + 1 + t^2 + t + t^2 \\
&= \int_0^t \left(2t + 1 + \frac{1}{2}(s + \sigma(s)) - \frac{1}{2}\right) \Delta s + 3t^2 + 3t + 1 \\
&= \left(2t + \frac{1}{2}\right) s\Big|_{s=0}^{s=t} + \frac{1}{2}s^2\Big|_{s=0}^{s=t} + 3t^2 + 3t + 1 \\
&= \left(2t + \frac{1}{2}\right) t + \frac{1}{2}t^2 + 3t^2 + 3t + 1 \\
&= 2t^2 + \frac{1}{2}t + \frac{1}{2}t^2 + 3t^2 + 3t + 1 \\
&= \frac{11}{2}t^2 + \frac{7}{2}t + 1.
\end{aligned}$$

Example 5.11 Let $\mathbb{T} = 2^{\mathbb{N}_0}$. Consider

$$I(t) = \int_1^t \left(s e_g(s, 1) + st + t^2\right) \Delta s, \quad \text{where} \quad g(t) = t.$$

We will find $I^{\Delta}(t)$. We have

$$\begin{aligned}
I(t) &= \int_1^t \left(e_g^{\Delta}(s,1) + \frac{1}{3}t(s+\sigma(s)) + t^2 \right) \Delta s \\
&= e_g(s,1)\Big|_{s=1}^{s=t} + \frac{1}{3}ts^2\Big|_{s=1}^{s=t} + t^2 s\Big|_{s=1}^{s=t} \\
&= e_g(t,1) - 1 + \frac{1}{3}t^3 - \frac{1}{3}t + t^3 - t^2 \\
&= e_g(t,1) - 1 + \frac{4}{3}t^3 - t^2 - \frac{1}{3}t.
\end{aligned}$$

Hence,

$$\begin{aligned}
I^{\Delta}(t) &= te_g(t,1) + \frac{4}{3}((\sigma(t))^2 + t\sigma(t) + t^2) - (t + \sigma(t)) - \frac{1}{3} \\
&= te_g(t,1) + \frac{4}{3}(4t^2 + 2t^2 + t^2) - 3t - \frac{1}{3} \\
&= te_g(t,1) + \frac{28}{3}t^2 - 3t - \frac{1}{3}.
\end{aligned}$$

Now, we will compute $I^{\Delta}(t)$ using Theorem 5.9. Here, $f(s,t) = se_g(s,1) + st + t^2$ so that

$$\begin{aligned}
I^{\Delta}(t) &= \int_1^t (s + \sigma(t) + t)\,\Delta s + te_g(t,1) + t\sigma(t) + (\sigma(t))^2 \\
&= \int_1^t (s + 3t)\Delta s + te_g(t,1) + 6t^2 \\
&= \int_1^t \left(\frac{1}{3}(s + \sigma(s)) + 3t \right) \Delta s + te_g(t,1) + 6t^2 \\
&= \frac{1}{3}s^2\Big|_{s=1}^{s=t} + 3ts\Big|_{s=1}^{s=t} + te_g(t,1) + 6t^2 \\
&= \frac{1}{3}t^2 - \frac{1}{3} + 3t^2 - 3t + te_g(t,1) + 6t^2 \\
&= te_g(t,1) + \frac{28}{3}t^2 - 3t - \frac{1}{3}.
\end{aligned}$$

Example 5.12 Let $\mathbb{T} = 3^{\mathbb{N}_0}$. Consider the integral

$$I(t) = \int_1^t (st - s^2)\Delta s.$$

We will compute $I^{\Delta}(t)$. We have

$$\begin{aligned} I(t) &= \int_1^t \left(t\frac{s+\sigma(s)}{4} - \frac{s^2+s\sigma(s)+(\sigma(s))^2}{13} \right) \Delta s \\ &= \frac{t}{4}s^2\Big|_{s=1}^{s=t} - \frac{1}{13}s^3\Big|_{s=1}^{s=t} \\ &= \frac{t^3}{4} - \frac{t}{4} - \frac{t^3}{13} + \frac{1}{13} \\ &= \frac{9}{52}t^3 - \frac{1}{4}t + \frac{1}{13}. \end{aligned}$$

Hence,

$$\begin{aligned} I^{\Delta}(t) &= \frac{9}{52}((\sigma(t))^2 + t\sigma(t) + t^2) - \frac{1}{4} \\ &= \frac{9}{52}(9t^2 + 3t^2 + t^2) - \frac{1}{4} \\ &= \frac{9}{4}t^2 - \frac{1}{4}. \end{aligned}$$

Now, we will use Theorem 5.9 to compute $I^{\Delta}(t)$. Here, $f(s,t) = st - s^2$ so that

$$\begin{aligned} I^{\Delta}(t) &= \int_1^t s\Delta s + t\sigma(t) - t^2 \\ &= \int_1^t s\Delta s + 3t^2 - t^2 \\ &= \frac{1}{4}\int_1^t (s+\sigma(s))\Delta s + 2t^2 \\ &= \frac{1}{4}s^2\Big|_{s=1}^{s=t} + 2t^2 \\ &= \frac{1}{4}t^2 - \frac{1}{4} + 2t^2 \\ &= \frac{9}{4}t^2 - \frac{1}{4}. \end{aligned}$$

Exercise 5.13 Compute $I^{\Delta}(t)$, where

1. $I(t) = \int_0^t (st - s^2 - t^2)\Delta s$, $\mathbb{T} = \mathbb{Z}$,

2. $I(t) = \int_1^t (s + st - s^3)\Delta s, \mathbb{T} = 2^{\mathbb{N}_0}$,
3. $I(t) = \int_1^t (s^2 t - s^4)\Delta s, \mathbb{T} = 3^{\mathbb{N}_0}$.

Solution 1. $-\frac{3}{2}t^2 - \frac{5}{2}t - 1$,
2. $-t^3 + \frac{7}{3}t^2 + t - \frac{1}{3}$,
3. $-t^4 + \frac{40}{13}t^3 - \frac{1}{13}$.

Exercise 5.14 Let $\mathbb{T}$ be a time scale and $a \in \mathbb{T}$. Suppose that $f(t, s)$ is defined and continuous on $\mathbb{T} \times \mathbb{T}$ and $f_t^\Delta(t, s)$ is continuous on $\mathbb{T} \times \mathbb{T}$. Compute $I^{\Delta\Delta}(t)$, where

$$I(t) = \int_a^t \int_a^{t_1} f(t_1, s)\Delta s \Delta t_1.$$

Solution $\int_a^t f_t^\Delta(t, s)\Delta s + f(\sigma(t), t)$.

5.2 Improper Parameter-Dependent Integrals of the First Kind

Let $\mathbb{T}_1$ and $\mathbb{T}_2$ be time scales and assume that $\mathbb{T}_1$ is unbounded above. Suppose that the real-valued function $f(t_1, t_2)$ is defined on $[a, \infty) \times \mathbb{T}_2$ and is Δ_1-integrable with respect to t_1 from a to any point $A \in \mathbb{T}_1$, $A \geq a$.

Definition 5.15 The *improper parameter-dependent integral of the first kind* is written as

$$I(t_2) = \int_a^\infty f(t_1, t_2)\Delta_1 t_1, \tag{5.4}$$

which is the limit of $\int_a^A f(t_1, t_2)\Delta_1 t_1$ as $A \to \infty$. If the limit exists (finite), then the improper parameter-dependent integral of the first kind is called *convergent*. Otherwise, it is called *divergent*.

Definition 5.16 For $t_2^0 \in [a_2, b_2]$, if the improper parameter-dependent integral of the first kind $\int_a^\infty f(t_1, t_2^0)\Delta_1 t_1$ is convergent, then we say that $\int_a^\infty f(t_1, t_2^0)\Delta_1 t_1$ is convergent at t_2^0, and t_2^0 is called a *point of convergence* of the integral (5.4). With C, we will denote the set of all points of convergence of the integral (5.4).

Example 5.17 Let $\mathbb{T}_1 = \mathbb{T}_2 = \mathbb{Z}$. Consider the integral

$$I(t_2) = \int_1^\infty \frac{2t_1 + 1 + 2t_2^2}{(t_1 + t_2^2)^2 (t_1 + 1 + t_2^2)^2}\Delta_1 t_1, \quad t_2 \in \mathbb{T}_2.$$

Here, $\sigma_1(t_1) = t_1 + 1$. Note that $f(t_1, t_2) = \frac{1}{(t_1 + t_2^2)^2}$ satisfies

$$f_{t_1}^{\Delta_1}(t_1, t_2) = -\frac{(\sigma_1(t_1) + t_2^2) + (t_1 + t_2^2)}{(t_1 + t_2^2)^2(\sigma_1(t_1) + t_2^2)^2}$$

$$= -\frac{2t_1 + 1 + 2t_2^2}{(t_1 + t_2^2)^2(t_1 + 1 + t_2^2)^2}.$$

Therefore,

$$I(t_2) = -\lim_{A\to\infty} \int_1^A f_{t_1}^{\Delta}(t_1, t_2)\Delta_1 t_1$$

$$= -\lim_{A\to\infty} (f(A, t_2) - f(1, t_2))$$

$$= \frac{1}{(1 + t_2^2)^2} < \infty \quad \text{for any} \quad t_2 \in \mathbb{T}_2.$$

Here, $C = \mathbb{Z}$.

Example 5.18 Let $\mathbb{T}_1 = \mathbb{T}_2 = \mathbb{Z}$. Consider the integral

$$I(t_2) = \int_1^\infty \frac{1}{(t_1 + t_2 + 1)(t_1 + t_2 + 2)} \Delta_1 t_1, \quad t_2 \in \mathbb{T}_2.$$

Here, $\sigma_1(t_1) = t_1 + 1$. Note that $f(t_1, t_2) = \frac{1}{t_1+t_2+4}$ satisfies

$$f_{t_1}^{\Delta_1}(t_1, t_2) = -\frac{1}{(t_1 + t_2 + 1)(\sigma_1(t_1) + t_2 + 1)}$$

$$= -\frac{1}{(t_1 + t_2 + 1)(t_1 + t_2 + 2)}.$$

Therefore,

$$I(t_2) = \lim_{A\to\infty} \int_1^A \frac{1}{(t_1 + t_2 + 1)(t_1 + t_2 + 2)} \Delta_1 t_1$$

$$= -\lim_{A\to\infty} \int_1^A f_{t_1}^{\Delta}(t_1, t_2)\Delta_1 t_1$$

$$= -\lim_{A\to\infty} (f(A, t_2) - f(1, t_2))$$

$$= \frac{1}{2 + t_2} < \infty \quad \text{for} \quad t_2 \neq -2.$$

Hence, $I(t_2)$ is convergent for $t_2 \neq -2$ and divergent for $t_2 = -2$. Here, $C = \mathbb{Z} \setminus \{-2\}$.

Example 5.19 Let $\mathbb{T}_1 = 2^{\mathbb{N}_0}$ and $\mathbb{T}_2 = 3^{\mathbb{N}_0}$. Consider the integral

$$I(t_2) = \int_1^\infty \frac{\Delta_1 t_1}{\sqrt{t_1 + t_2 + 1} + \sqrt{2t_1 + t_2 + 1}}, \quad t_2 \in \mathbb{T}_2.$$

Here,

$$\sigma_1(t_1) = 2t_1, \quad t_1 \in \mathbb{T}_1, \quad \sigma_2(t_2) = 3t_2, \quad t_2 \in \mathbb{T}_2.$$

With $f(t_1, t_2) = \sqrt{t_1 + t_2 + 1}$, we have

$$\begin{aligned}
f_{t_1}^{\Delta_1}(t_1, t_2) &= \frac{\sqrt{\sigma_1(t_1) + t_2 + 1} - \sqrt{t_1 + t_2 + 1}}{\sigma_1(t_1) - t_1} \\
&= \frac{\sqrt{2t_1 + t_2 + 1} - \sqrt{t_1 + t_2 + 1}}{t_1} \\
&= \frac{2t_1 + t_2 + 1 - (t_1 + t_2 + 1)}{t_1\left(\sqrt{2t_1 + t_2 + 1} + \sqrt{t_1 + t_2 + 1}\right)} \\
&= \frac{1}{\sqrt{2t_1 + t_2 + 1} + \sqrt{t_1 + t_2 + 1}}.
\end{aligned}$$

Hence,

$$\begin{aligned}
I(t_2) &= \lim_{A \to \infty} \int_1^A \frac{\Delta_1 t_1}{\sqrt{2t_1 + t_2 + 1} + \sqrt{t_1 + t_2 + 1}} \\
&= \lim_{A \to \infty} \int_1^A f_{t_1}^{\Delta_1}(t_1, t_2) \Delta_1 t_1 \\
&= \lim_{A \to \infty} (f(A, t_2) - f(1, t_2)) \\
&= \infty.
\end{aligned}$$

Therefore, $I(t_2)$ is divergent for any $t_2 \in \mathbb{T}_2$.

Exercise 5.20 Let $\mathbb{T}_1 = \mathbb{Z}$ and $\mathbb{T}_2 = 2^{\mathbb{N}_0}$. Investigate the integral

$$I(t_2) = \int_1^\infty \sqrt{t_1^2 + t_2^2 + 1} \Delta_1 t_1, \quad t_2 \in \mathbb{T}_2$$

for convergence and divergence.

Solution Divergent for any $t_2 \in \mathbb{T}_2$.

Definition 5.21 Suppose that $I(t_2)$ exists for any $t_2 \in (a_2, b_2)$. Assume that for every $\varepsilon > 0$, there exists $A_0 = A_0(\varepsilon)$ such that $A > A_0$ implies

$$\left| \int_A^{\infty} f(t_1, t_2)\Delta_1 t_1 \right| < \varepsilon \quad \text{for any} \quad t_2 \in [a_2, b_2].$$

In this case, we say $I(t_2)$ is *uniformly convergent* in t_2 on $[a_2, b_2]$.

Theorem 5.22 (Cauchy's Criterion) *The improper integral* $\int_a^{\infty} f(t_1, t_2)\Delta_1 t_1$ *converges uniformly in* $t_2 \in [a_2, b_2]$ *if and only if for any given* $\varepsilon > 0$*, there exists* $A_0 = A_0(\varepsilon)$ *such that*

$$\left| \int_{A_1}^{A_2} f(t_1, t_2)\Delta_1 t_1 \right| < \varepsilon, \quad t_2 \in [a_2, b_2] \tag{5.5}$$

for any $A_1, A_2 > A_0$.

Proof 1. Sufficiency. Suppose that for any $\varepsilon > 0$, there exists $A_0 = A_0(\varepsilon)$ such that (5.5) holds. Fix $t_2 \in [a_2, b_2]$. Thus, we obtain the improper integral without parameter $\int_a^{\infty} f(t_1, t_2)\Delta_1 t_1$. By the Cauchy criterion, Theorem 3.72, we obtain that the integral $\int_a^{\infty} f(t_1, t_2)\Delta_1 t_1$ is convergent. We take $A_2 \to \infty$ in (5.5) and obtain

$$\left| \int_{A_1}^{\infty} f(t_1, t_2)\Delta_1 t_1 \right| = \left| \int_a^{\infty} f(t_1, t_2)\Delta_1 t_1 - \int_a^{A_1} f(t_1, t_2)\Delta_1 t_1 \right| < \varepsilon$$

for $t_2 \in [a_2, b_2]$, which means that $\int_a^{\infty} f(t_1, t_2)\Delta_1 t_1$ converges uniformly in $t_2 \in [a_2, b_2]$.

2. Necessity. Suppose that $\int_a^{\infty} f(t_1, t_2)\Delta_1 t_1$ converges uniformly in $t_2 \in [a_2, b_2]$. Then, for any $\varepsilon > 0$, there exists $A_0 = A_0(\varepsilon)$ such that for any $A_1, A_2 > A_0$, we have

$$\left| \int_{A_1}^{\infty} f(t_1, t_2)\Delta_1 t_1 \right| < \frac{\varepsilon}{2} \quad \text{and} \quad \left| \int_{A_2}^{\infty} f(t_1, t_2)\Delta_1 t_1 \right| < \frac{\varepsilon}{2}$$

for any $t_2 \in [a_2, b_2]$. Hence,

$$\begin{aligned}
\left| \int_{A_1}^{A_2} f(t_1, t_2)\Delta_1 t_1 \right| &= \left| \int_{A_1}^{\infty} f(t_1, t_2)\Delta_1 t_1 - \int_{A_2}^{\infty} f(t_1, t_2)\Delta_1 t_1 \right| \\
&\leq \left| \int_{A_1}^{\infty} f(t_1, t_2)\Delta_1 t_1 \right| + \left| \int_{A_2}^{\infty} f(t_1, t_2)\Delta_1 t_1 \right| \\
&< \frac{\varepsilon}{2} + \frac{\varepsilon}{2} \\
&= \varepsilon
\end{aligned}$$

for any $t_2 \in [a_2, b_2]$, which completes the proof. □

Example 5.23 Let $\mathbb{T}_1 = \mathbb{T}_2 = \mathbb{Z}$. Consider the integral

$$I(t_2) = \int_1^\infty \frac{1}{(t_1 + t_2)(t_1 + t_2 + 1)} \Delta_1 t_1, \quad t_2 \in [1, 4].$$

Let $\varepsilon > 0$ be arbitrarily chosen. We take $A_0 > \frac{\varepsilon}{2}$. Then, for every $A_1, A_2 > A_0$, we have, with $f(t_1, t_2) = \frac{1}{t_1 + t_2}$,

$$\begin{aligned}
\left| \int_{A_1}^{A_2} \frac{\Delta_1 t_1}{(t_1 + t_2)(t_1 + t_2 + 1)} \right| &= \left| - \int_{A_1}^{A_2} f_{t_1}^{\Delta_1}(t_1, t_2) \Delta_1 t_1 \right| \\
&= |-(f(A_2, t_2) - f(A_1, t_2))| \\
&= \left| \frac{1}{A_1 + t_2} - \frac{1}{A_2 + t_2} \right| \\
&\leq \frac{1}{A_2 + t_2} + \frac{1}{A_1 + t_2} \\
&\leq \frac{1}{A_1} + \frac{1}{A_2} \\
&< \frac{1}{A_0} + \frac{1}{A_0} \\
&= \frac{2}{A_0} \\
&< \varepsilon
\end{aligned}$$

for any $t_2 \in [1, 4]$. Hence, using the Cauchy criterion, Theorem 5.22, it follows that $I(t_2)$ is uniformly convergent in $t_2 \in [1, 4]$.

Example 5.24 Let $\mathbb{T}_1 = 2^{\mathbb{N}_0}$ and $\mathbb{T}_2 = \mathbb{Z}$. Consider the integral

$$I(t_2) = \int_1^\infty \frac{1}{t_1^3 t_2} \Delta_1 t_1, \quad t_2 \in [1, 10].$$

Let $\varepsilon > 0$ be arbitrarily chosen. We take $A_0 > \sqrt{\frac{8}{3\varepsilon}}$. Note that, with $f(t_1) = \frac{1}{t_1^2}$, we have

$$f^{\Delta_1}(t_1) = -\frac{t_1 + \sigma_1(t_1)}{t_1^2 (\sigma_1(t_1))^2} = -\frac{3}{4t_1^3},$$

whereupon

$$\frac{1}{t_1^3} = -\frac{4}{3} f^{\Delta_1}(t_1).$$

Hence, for any $A_1, A_2 > A_0$, we have

$$\begin{aligned}
\left| \int_{A_1}^{A_2} \frac{1}{t_2 t_1^3} \Delta_1 t_1 \right| &= \left| -\frac{4}{3t_2} \int_{A_1}^{A_2} f^{\Delta_1}(t_1) \Delta_1 t_1 \right| \\
&= \left| -\frac{4}{3t_2} (f(A_2) - f(A_1)) \right| \\
&= \left| \frac{4}{3t_2 A_2^2} - \frac{4}{3t_2 A_1^2} \right| \\
&\le \frac{4}{3t_2 A_2^2} + \frac{4}{3t_2 A_1^2} \\
&\le \frac{4}{3A_0^2} + \frac{4}{3A_0^2} \\
&= \frac{8}{3A_0^2} \\
&< \varepsilon
\end{aligned}$$

for any $t_2 \in [1, 10]$. Hence, using the Cauchy criterion, Theorem 5.22, we get that $I(t_2)$ is uniformly convergent in $t_2 \in [1, 10]$.

Example 5.25 Let $\mathbb{T}_1 = \mathbb{T}_2 = 2^{\mathbb{N}_0}$. Consider the integral

$$I(t_2) = \int_2^\infty \frac{\cos(2t_1) - 2\cos(t_1)}{2t_1^2 t_2^3} \Delta_1 t_1, \quad t_2 \in [2, 8].$$

Let $\varepsilon > 0$ be arbitrarily chosen. We take $A_0 > \frac{1}{4\varepsilon}$. Note that, for $f(t_1) = \frac{\cos t_1}{t_1}$, we have

$$\begin{aligned}
f^{\Delta_1}(t_1) &= \frac{t_1 \frac{\cos(2t_1) - \cos(t_1)}{t_1} - \cos(t_1)}{2t_1^2} \\
&= \frac{\cos(2t_1) - 2\cos(t_1)}{2t_1^2}.
\end{aligned}$$

Then, for any $A_1, A_2 > A_0$, we have

$$
\begin{aligned}
\left|\int_{A_1}^{A_2} \frac{\cos(2t_1) - 2\cos(t_1)}{2t_1^2 t_2^3} \Delta_1 t_1\right| &= \left|\frac{1}{t_2^3}\int_{A_1}^{A_2} f^{\Delta_1}(t_1)\Delta_1 t_1\right| \\
&= \left|\frac{1}{t_2^3}(f(A_2) - f(A_1))\right| \\
&= \left|\frac{1}{t_2^3}\frac{\cos(A_2)}{A_2} - \frac{1}{t_2^3}\frac{\cos(A_1)}{A_1}\right| \\
&\le \frac{1}{t_2^3}\frac{|\cos(A_2)|}{A_2} + \frac{1}{t_2^3}\frac{|\cos(A_1)|}{A_1} \\
&\le \frac{1}{8A_0} + \frac{1}{8A_0} \\
&= \frac{1}{4A_0} \\
&< \varepsilon
\end{aligned}
$$

for any $t_2 \in [2, 8]$. Hence, by the Cauchy criterion, Theorem 5.22, we get that $I(t_2)$ is uniformly convergent in $t_2 \in [2, 8]$.

Exercise 5.26 Let $\mathbb{T}_1 = 2^{\mathbb{N}_0}$ and $\mathbb{T}_2 = 3^{\mathbb{N}_0}$. Using the Cauchy criterion, Theorem 5.22, prove that

$$
I(t_2) = \int_1^{\infty} \frac{7 + t_2}{t_1^4 t_2^3} \Delta_1 t_1, \quad t_2 \in [3, 27]
$$

is uniformly convergent in $t_2 \in [3, 27]$.

Theorem 5.27 (Weierstraß Test) *If there exists a function F such that*

1. $|f(t_1, t_2)| \le F(t_1)$ *for any* $t_1 \in [a, \infty)$, $t_2 \in [a_2, b_2]$,
2. $\int_a^{\infty} F(t_1)\Delta_1 t_1$ *is convergent,*

then $\int_a^{\infty} f(t_1, t_2)\Delta_1 t_1$ *converges uniformly in* $t_2 \in [a_2, b_2]$.

Proof Since $\int_a^{\infty} F(t_1)\Delta_1 t_1$ is convergent, by the Cauchy criterion, Theorem 3.72, it follows that for any $\varepsilon > 0$, there exists $A_0 = A_0(\varepsilon)$ so that $A_1, A_2 > A_0$ implies

$$
\left|\int_{A_1}^{A_2} F(t_1)\Delta_1 t_1\right| < \varepsilon.
$$

Hence, for any $\varepsilon > 0$, there exists $A_0 = A_0(\varepsilon)$ such that $A_1, A_2 > A_0$ implies

$$\left|\int_{A_1}^{A_2} f(t_1,t_2)\Delta_1 t_1\right| \le \left|\int_{A_1}^{A_2} |f(t_1,t_2)|\Delta_1 t_1\right|$$

$$\le \left|\int_{A_1}^{A_2} F(t_1)\Delta_1 t_1\right|$$

$$< \varepsilon.$$

From here, by the Cauchy criterion, Theorem 5.27, it follows that $\int_a^\infty f(t_1,t_2)\Delta_1 t_1$ is uniformly convergent in $t_2 \in [a_2, b_2]$. □

Example 5.28 Let $\mathbb{T}_1 = \mathbb{T}_2 = \mathbb{Z}$. Consider

$$I(t_2) = \int_1^\infty \frac{1}{(t_1+t_2^2+3)(t_1+t_2^7+t_2^4+1)}\Delta_1 t_1, \quad t_2 \in [1,4].$$

Here,

$$f(t_1,t_2) = \frac{1}{(t_1+t_2^2+3)(t_1+t_2^7+t_2^4+3)}.$$

We have

$$0 \le f(t_1,t_2) \le \frac{1}{(t_1+1)(t_1+2)}$$

for any $t_1 \in [1,\infty)$ and any $t_2 \in [1,4]$. Let

$$F(t_1) = \frac{1}{(t_1+1)(t_1+2)}.$$

Note that, for $g(t_1) = \frac{1}{t_1+4}$, we have

$$g^{\Delta_1}(t_1) = -\frac{1}{(t_1+1)(\sigma_1(t_1)+1)} = -F(t_1)$$

and

$$\int_1^\infty F(t_1)\Delta_1 t_1 = \lim_{A\to\infty}\int_1^A F(t_1)\Delta_1 t_1$$

$$= -\lim_{A\to\infty}\int_1^A g^{\Delta_1}(t_1)\Delta_1 t_1$$

$$= -\lim_{A\to\infty}(g(A)-g(1))$$

$$= \frac{1}{2},$$

i.e., the integral $\int_1^\infty F(t_1)\Delta_1 t_1$ is convergent. Hence, utilizing the Weierstraß test, Theorem 5.27, we conclude that $I(t_2)$ is uniformly convergent in $t_2 \in [1, 4]$.

Example 5.29 Let $\mathbb{T}_1 = 2^{\mathbb{N}_0}$ and $\mathbb{T}_2 = \mathbb{Z}$. Consider the integral

$$I(t_2) = \int_1^\infty \frac{t_2}{(t_2^2+1)\sqrt{1+t_1t_2}\sqrt{1+2t_1t_2^2}\left(\sqrt{1+t_1}+\sqrt{1+2t_1}\right)}\Delta_1 t_1, \quad t_2 \in [1, 10].$$

Here,

$$f(t_1, t_2) = \frac{t_2}{(t_2^2+1)\sqrt{1+t_1t_2}\sqrt{1+2t_1t_2^2}\left(\sqrt{1+t_1}+\sqrt{1+2t_1}\right)}.$$

We have

$$f(t_1, t_2) \le \frac{1}{\sqrt{1+t_1}\sqrt{1+2t_1}\left(\sqrt{1+t_1}+\sqrt{1+2t_1}\right)}.$$

Let

$$F(t_1) = \frac{1}{\sqrt{1+t_1}\sqrt{1+2t_1}\left(\sqrt{1+t_1}+\sqrt{1+2t_1}\right)}.$$

Note that, for $g(t_1) = \frac{1}{\sqrt{1+t_1}}$, we have

$$\begin{aligned}
g^{\Delta_1}(t_1) &= \frac{\frac{1}{\sqrt{1+\sigma_1(t_1)}} - \frac{1}{\sqrt{1+t_1}}}{\sigma_1(t_1) - t_1} \\
&= \frac{\frac{1}{\sqrt{1+2t_1}} - \frac{1}{\sqrt{1+t_1}}}{2t_1 - t_1} \\
&= \frac{\sqrt{1+t_1}-\sqrt{1+2t_1}}{t_1\sqrt{1+t_1}\sqrt{1+2t_1}} \\
&= \frac{\left(\sqrt{1+t_1}-\sqrt{1+2t_1}\right)\left(\sqrt{1+t_1}+\sqrt{1+2t_1}\right)}{t_1\sqrt{1+t_1}\sqrt{1+2t_1}\left(\sqrt{1+t_1}+\sqrt{1+2t_1}\right)} \\
&= \frac{1+t_1-1-2t_1}{t_1\sqrt{1+t_1}\sqrt{1+2t_1}\left(\sqrt{1+t_1}+\sqrt{1+2t_1}\right)} \\
&= -\frac{1}{\sqrt{1+t_1}\sqrt{1+2t_1}\left(\sqrt{1+t_1}+\sqrt{1+2t_1}\right)} \\
&= -F(t_1).
\end{aligned}$$

Therefore,

$$\begin{aligned}\int_1^\infty F(t_1)\Delta_1 t_1 &= \lim_{A\to\infty}\int_1^A F(t_1)\Delta_1 t_1\\ &= -\lim_{A\to\infty}\int_1^A g^{\Delta_1}(t_1)\Delta_1 t_1\\ &= -\lim_{A\to\infty}(g(A)-g(1))\\ &= \frac{1}{\sqrt{2}},\end{aligned}$$

i.e., $\int_1^\infty F(t_1)\Delta_1 t_1$ is convergent. Hence, by the Weierstraß test, Theorem 5.27, it follows that $I(t_2)$ is uniformly convergent in $t_2 \in [1, 10]$.

Example 5.30 Let $\mathbb{T}_1 = \mathbb{Z}$ and $\mathbb{T}_2 = 2^{\mathbb{N}_0}$. Consider the integral

$$I(t_2) = \int_1^\infty \frac{t_1^2+t_1-1}{\left(1+(t_1+t_2)^2\right)\left(2+2t_1t_2+t_1^2t_2^2\right)}\Delta_1 t_1, \quad t_1 \in [1,4].$$

Here,

$$f(t_1,t_2) = \frac{t_1^2+t_1-1}{\left(1+(t_1+t_2)^2\right)\left(2+2t_1t_2+t_1^2t_2^2\right)}.$$

We have

$$f(t_1,t_2) \le \frac{t_1^2+t_1-1}{(1+t_1^2)(2+2t_1+t_1^2)}.$$

Let

$$F(t_1) = \frac{t_1^2+t_1-1}{(1+t_1^2)(2+2t_1+t_1^2)}.$$

Note that, with $g(t_1) = \frac{t_1}{1+t_1^2}$, we have

$$\begin{aligned}g^{\Delta_1}(t_1) &= \frac{1+t_1^2-t_1(\sigma_1(t_1)+t_1)}{1+t_1^2\left(1+(\sigma_1(t_1))^2\right)}\\ &= \frac{1+t_1^2-t_1(t_1+1+t_1)}{(1+t_1^2)(1+1+2t_1+t_1^2)}\\ &= \frac{1+t_1^2-2t_1^2-t_1}{(1+t_1^2)(2+2t_1+t_1^2)}\end{aligned}$$

$$= \frac{1 - t_1 - t_1^2}{(1 + t_1^2)(2 + 2t_1 + t_1^2)}$$

$$= -F(t_1).$$

Thus,

$$\begin{aligned}
\int_1^\infty F(t_1)\Delta_1 t_1 &= \lim_{A \to \infty} \int_1^A F(t_1)\Delta_1 t_1 \\
&= -\lim_{A \to \infty} \int_1^A g^{\Delta_1}(t_1)\Delta_1 t_1 \\
&= -\lim_{A \to \infty} (g(A) - g(1)) \\
&= \frac{1}{2},
\end{aligned}$$

i.e., $\int_1^\infty F(t_1)\Delta_1 t_1$ is convergent. Hence, using the Weierstraß test, Theorem 5.27, it follows that $I(t_2)$ is uniformly convergent in $t_2 \in [1, 4]$.

Exercise 5.31 Let $\mathbb{T}_1 = 2^{\mathbb{N}_0}$ and $\mathbb{T}_2 = 3^{\mathbb{N}_0}$. Using the Weierstraß test, Theorem 5.27, prove that

$$\int_1^\infty \frac{t_1^2 + t_2^2}{(1 + t_1^2 t_2^2)(2 + t_1^4 + t_2^4)} \Delta_1 t_1, \quad t_2 \in [1, 27]$$

is uniformly convergent in $t_2 \in [1, 27]$.

Theorem 5.32 (Abel Test) *If f and g satisfy the conditions*

1. *$\int_a^\infty f(t_1, t_2)\Delta_1 t_1$ converges uniformly in $t_2 \in [a_2, b_2]$,*
2. *for every fixed $t_2 \in [a_2, b_2]$, $g(t_1, t_2)$ is monotone with respect to t_1,*
3. *$g(t_1, t_2)$ is uniformly bounded, i.e., there exists $L > 0$ such that*

$$|g(t_1, t_2)| \le L, \quad a \le t_1 < \infty, \quad a_2 \le t_2 \le b_2,$$

then $\int_a^\infty f(t_1, t_2) g(t_1, t_2)\Delta_1 t_1$ converges uniformly in $t_2 \in [a_2, b_2]$.

Proof Since $\int_a^\infty f(t_1, t_2)\Delta_1 t_1$ is uniformly convergent in $t_2 \in [a_2, b_2]$, for any $\varepsilon > 0$, there exists $A_0 = A_0(\varepsilon)$ such that $A_1, A_2 > A_0$ implies

$$\left| \int_{A_1}^{A_2} f(t_1, t_2)\Delta_1 t_1 \right| < \frac{\varepsilon}{3L}. \tag{5.6}$$

Define

$$m_f = \inf_{A_1 \le t \le A_2, a_2 \le t_2 \le b_2} \int_{A_1}^{t} f(t_1, t_2)\Delta_1 t_1, \quad M_f = \sup_{A_1 \le t \le A_2, a_2 \le t_2 \le b_2} \int_{A_1}^{t} f(t_1, t_2)\Delta_1 t_1.$$

Hence, using (5.6), we get

$$|m_f| \le \frac{\varepsilon}{3L}, \quad |M_f| \le \frac{\varepsilon}{3L}.$$

From the second mean value theorem, Theorem 3.71, there exists $\Lambda \in [m_f, M_f]$ such that

$$\int_{A_1}^{A_2} f(t_1, t_2)g(t_1, t_2)\Delta_1 t_1 = (g(A_1, t_2) - g(A_2, t_2))\,\Lambda + g(A_2, t_2)\int_{A_1}^{A_2} f(t_1, t_2)\Delta_1 t_1.$$

Therefore,

$$\begin{aligned}\left|\int_{A_1}^{A_2} f(t_1, t_2)g(t_1, t_2)\Delta_1 t_1\right| &= \left|(g(A_1, t_2) - g(A_2, t_2))\,\Lambda + g(A_2, t_2)\int_{A_1}^{A_2} f(t_1, t_2)\Delta_1 t_1\right| \\ &\le (|g(A_1, t_2)| + |g(A_2, t_2)|)\,|\Lambda| + |g(A_2, t_2)|\left|\int_{A_1}^{A_2} f(t_1, t_2)\Delta_1 t_1\right| \\ &< 2L\frac{\varepsilon}{3L} + L\frac{\varepsilon}{3L} \\ &= \varepsilon.\end{aligned}$$

From here and from the Cauchy criterion, Theorem 5.22, it follows that

$$\int_{a}^{\infty} f(t_1, t_2)g(t_1, t_2)\Delta_1 t_1$$

converges uniformly in $t_2 \in [a_2, b_2]$. □

Example 5.33 Let $\mathbb{T}_1 = \mathbb{T}_2 = \mathbb{Z}$. Consider the integral

$$I(t_2) = \int_{1}^{\infty} \frac{t_1 t_2}{\sqrt{t_1 + 2t_2}\sqrt{t_1 + 1 + 2t_2}\left(\sqrt{t_1 + 1 + 2t_2} + \sqrt{t_1 + 2t_2}\right)(t_1 + 1)(t_2 + 1)}\Delta_1 t_1,$$

where $t_2 \in [1, 10]$. Let

$$f(t_1, t_2) = \frac{1}{\sqrt{t_1 + 2t_2}\sqrt{t_1 + 1 + 2t_2}\left(\sqrt{t_1 + 1 + 2t_2} + \sqrt{t_1 + 2t_2}\right)},$$

$$g(t_1, t_2) = \frac{t_1 t_2}{(t_1 + 1)(t_2 + 1)}.$$

Note that

$$f(t_1, t_2) \leq \frac{1}{\sqrt{t_1 + 3}\sqrt{t_1 + 2}\left(\sqrt{t_1 + 3} + \sqrt{t_1 + 2}\right)}$$

and, with $h(t_1) = \frac{1}{\sqrt{t_1+2}}$, we have

$$\begin{aligned} h^{\Delta_1}(t_1) &= -\frac{\sqrt{\sigma_1(t_1) + 2} - \sqrt{t_1 + 2}}{\sqrt{t_1 + 2}\sqrt{\sigma_1(t_1) + 2}} \\ &= -\frac{\sqrt{t_1 + 3} - \sqrt{t_1 + 2}}{\sqrt{t_1 + 2}\sqrt{t_1 + 3}} \\ &= -\frac{\left(\sqrt{t_1 + 3} - \sqrt{t_1 + 2}\right)\left(\sqrt{t_1 + 3} + \sqrt{t_1 + 2}\right)}{\sqrt{t_1 + 2}\sqrt{t_1 + 3}\left(\sqrt{t_1 + 3} + \sqrt{t_1 + 2}\right)} \\ &= -\frac{t_1 + 3 - t_1 - 2}{\sqrt{t_1 + 2}\sqrt{t_1 + 3}\left(\sqrt{t_1 + 2} + \sqrt{t_1 + 3}\right)} \\ &= -\frac{1}{\sqrt{t_1 + 3}\sqrt{t_1 + 2}\left(\sqrt{t_1 + 3} + \sqrt{t_1 + 2}\right)}. \end{aligned}$$

Hence,

$$\begin{aligned} \int_1^\infty \frac{1}{\sqrt{t_1 + 2}\sqrt{t_1 + 3}\left(\sqrt{t_1 + 2} + \sqrt{t_1 + 3}\right)} \Delta_1 t_1 &= -\lim_{A \to \infty} \int_1^A h^{\Delta_1}(t_1) \Delta_1 t_1 \\ &= -\lim_{A \to \infty} (h(A) - h(1)) \\ &= \frac{1}{\sqrt{3}}. \end{aligned}$$

From here, using the Weierstraß test, Theorem 5.27, it follows that $\int_1^\infty f(t_1, t_2) \Delta_1 t_1$ is uniformly convergent in $t_2 \in [1, 10]$. Moreover,

$$\begin{aligned} g_{t_1}^{\Delta_1}(t_1, t_2) &= \frac{t_2}{t_2 + 1} \frac{t_1 + 1 - t_1}{(t_1 + 1)(\sigma_1(t_1) + 1)} \\ &= \frac{t_2}{t_2 + 1} \frac{t_1 + 1 - t_1}{(t_1 + 1)(t_1 + 2)} \end{aligned}$$

$$= \frac{t_2}{t_2+1}\frac{1}{(t_1+1)(t_1+2)}$$

$$> 0 \quad \text{for} \quad t_1 \in [1,\infty), \quad t_2 \in [1,10].$$

Therefore, $g(t_1, t_2)$ is monotone with respect to $t_1 \in [1,\infty)$ for every $t_2 \in [1, 10]$. Note that

$$|g(t_1,t_2) = \left|\frac{t_1 t_2}{(t_1+1)(t_2+1)}\right|$$

$$\leq 1 \quad \text{for any} \quad t_1 \in [1,\infty), \quad t_2 \in [1,10],$$

i.e., $g(t_1, t_2)$ is uniformly bounded on $[1,\infty) \times [1, 10]$. Hence, using the Abel test, Theorem 5.32, it follows that $I(t_2)$ is uniformly convergent in $t_2 \in [1, 10]$.

Example 5.34 Let $\mathbb{T}_1 = \mathbb{T}_2 = 2^{\mathbb{N}_0}$. Consider the integral

$$I(t_2) = \int_1^\infty \frac{\left(2\sin(t_1)\cos(3t_1)(t_1+1)^2 - t_1(3t_1+2)\sin(2t_1)\right)\left(e^{-t_1}t_2+1\right)}{t_1(t_1+1)^2(2t_1+1)^2(t_2^4+t_2^3+t_2)}\Delta_1 t_1$$

for $t_2 \in [1, 4]$. Let

$$f(t_1,t_2) = \frac{2\sin(t_1)\cos(3t_1)(t_1+1)^2 - t_1(3t_1+2)\sin(2t_1)}{t_1(t_1+1)^2(2t_1+1)^2},$$

$$g(t_1,t_2) = \frac{e^{-t_1}t_2+1}{t_2^4+t_2^3+t_2}.$$

Note that, with $h(t_1) = \frac{\sin(2t_1)}{(t_1+1)^2}$, we have

$$h^{\Delta_1}(t_1) = \frac{\frac{\sin(2\sigma_1(t_1))-\sin(2t_1)}{\sigma_1(t_1)-t_1}(t_1+1)^2 - \sin(2t_1)\frac{(\sigma_1(t_1)+1)^2-(t_1+1)^2}{t_1}}{(t_1+1)^2(\sigma_1(t_1)+1)^2}$$

$$= \frac{\frac{\sin(4t_1)-\sin(2t_1)}{t_1}(t_1+1)^2 - \sin(2t_1)\frac{(2t_1+1)^2-(t_1+1)^2}{t_1}}{(t_1+1)^2(2t_1+1)^2}$$

$$= \frac{2\sin(t_1)\cos(3t_1)(t_1+1)^2 - t_1(3t_1+2)\sin(2t_1)}{t_1(t_1+1)^2(2t_1+1)^2}$$

$$= f(t_1,t_2).$$

Thus,

$$\begin{aligned}\int_1^\infty f(t_1,t_2)\Delta_1 t_1 &= \lim_{A\to\infty}\int_1^A f(t_1,t_2)\Delta_1 t_1\\ &= \lim_{A\to\infty}\int_1^A h^{\Delta_1}(t_1)\Delta_1 t_1\\ &= \lim_{A\to\infty}(h(A)-h(1))\\ &= -\frac{\sin 2}{4}.\end{aligned}$$

Therefore, $\int_1^\infty f(t_1,t_2)\Delta_1 t_1$ is uniformly convergent in $t_2 \in [1,4]$. Moreover,

$$|g(t_1,t_2)| = \left|\frac{e^{-t_1}t_2+1}{t_2^4+t_2^3+t_2}\right| \le e^{-t_1}+1 \le 2,$$

i.e., $g(t_1,t_2)$ is uniformly bounded in $[1,\infty)\times[1,4]$. Next,

$$\begin{aligned}g_{t_1}^{\Delta_1}(t_1,t_2) &= \frac{t_2}{t_2^4+t_2^3+t_2}\frac{e^{-\sigma_1(t_1)}-e^{-t_1}}{\sigma_1(t_1)-t_1}\\ &= \frac{t_2}{t_2^4+t_2^3}\frac{e^{-2t_1}-e^{-t_1}}{t_1}\\ &\le 0 \quad \text{for} \quad t_1\in[1,\infty), \quad t_2\in[1,4].\end{aligned}$$

From here, $g(t_1,t_2)$ is monotone with respect to t_1. Hence, using the Abel test, Theorem 5.32, it follows that $I(t_2)$ is uniformly convergent in $t_2 \in [1,4]$.

Example 5.35 Let $\mathbb{T}_1 = \mathbb{Z}$ and $\mathbb{T}_2 = 2^{\mathbb{N}_0}$. Consider the integral

$$I(t_2) = \int_1^\infty \frac{t_2}{t_1(t_1+1)^{10}(t_2^4+t_2^2+1)}\Delta_1 t_1, \quad t_2\in[1,27].$$

Let

$$f(t_1,t_2) = \frac{1}{(t_1+1)^{10}}, \quad g(t_1,t_2) = \frac{t_2}{t_1(t_2^4+t_2^2+1)}.$$

We have that $\int_1^\infty f(t_1,t_2)\Delta_1 t_1$ is uniformly convergent in $t_2\in[1,27]$. Moreover,

$$|g(t_1,t_2)| = \left|\frac{t_2}{t_1(t_2^4+t_2^2+1)}\right| \le 1$$

for any $t_1\in[1,\infty)$, $t_2\in[1,27]$, and

$$g_{t_1}^{\Delta_1}(t_1, t_2) = -\frac{t_2}{t_1(t_1+1)(t_2^4+t_2^2+1)} \leq 0,$$

i.e., $g(t_1, t_2)$ is monotone with respect t_1. Hence, using the Abel test, Theorem 5.32, it follows that $I(t_2)$ is uniformly convergent in $t_2 \in [1, 27]$.

Exercise 5.36 Let $\mathbb{T}_1 = \mathbb{T}_2 = 2^{\mathbb{N}_0}$. Prove that the integral

$$\int_1^\infty \frac{t_1^2 + t_2^2 + t_1 t_2}{(t_1+1)^2(t_1+2)^2(t_2^4+t_2^2+t_2)} \Delta_1 t_1$$

is uniformly convergent in $t_2 \in [1, 8]$.

Theorem 5.37 (Dirichlet Test) *If the function*

$$F(A, t_2) = \int_a^A f(t_1, t_2)\Delta_1 t_1$$

is uniformly bounded on $[a, \infty) \times [a_2, b_2]$*,* $g(t_1, t_2)$ *is monotone with respect to* t_1*, and* $g(t_1, t_2)$ *converges uniformly to* 0 *in* $t_2 \in [a_2, b_2]$ *as* $t_1 \to \infty$*, then the integral*

$$\int_a^\infty f(t_1, t_2) g(t_1, t_2)\Delta_1 t_1$$

is uniformly convergent in $t_2 \in [a_2, b_2]$.

Proof Let $L > 0$ be a constant such that

$$\left| \int_a^A f(t_1, t_2)\Delta_1 t_1 \right| \leq L \quad \text{for any} \quad A \in [a, \infty) \quad \text{and any} \quad t_2 \in [a_2, b_2].$$

We take $\varepsilon > 0$ arbitrarily. Since $g(t_1, t_2)$ converges uniformly to 0 in $t_2 \in [a_2, b_2]$ as $t_1 \to \infty$, there exists $A_0 = A_0(\varepsilon)$ such that for $t_1 > A_0$, we have

$$|g(t_1, t_2)| < \frac{\varepsilon}{6L} \quad \text{for any} \quad t_2 \in [a_2, b_2].$$

Let $A_1, A_2 > A_0$. We set

$$\begin{aligned} m_F &= \inf_{\substack{A_1 \leq t_1 \leq A_2 \\ a_2 \leq t_2 \leq b_2}} \int_{A_1}^{A_2} f(s, t_2)\Delta_1 s \\ &= \inf_{\substack{A_1 \leq t_1 \leq A_2 \\ a_2 \leq t_2 \leq b_2}} \left(\int_{A_1}^{t_1} f(s, t_2)\Delta_1 s + \int_a^{A_1} f(s, t_2)\Delta_1 s - \int_a^{A_1} f(s, t_2)\Delta_1 s \right) \end{aligned}$$

$$= \inf_{\substack{A_1 \le t_1 \le A_2 \\ a_2 \le t_2 \le b_2}} \left(\int_a^{t_1} f(s, t_2) \Delta_1 s - \int_a^{A_1} f(s, t_2) \Delta_1 s \right)$$

and

$$\begin{aligned} M_F &= \sup_{\substack{A_1 \le t_1 \le A_2 \\ a_2 \le t_2 \le b_2}} \int_{A_1}^{A_2} f(s, t_2) \Delta_1 s \\ &= \sup_{\substack{A_1 \le t_1 \le A_2 \\ a_2 \le t_2 \le b_2}} \left(\int_a^{t_1} f(s, t_2) \Delta_1 s - \int_a^{A_1} f(s, t_2) \Delta_1 s \right). \end{aligned}$$

Note that

$$\begin{aligned} |m_F| &= \left| \inf_{\substack{A_1 \le t_1 \le A_2 \\ a_2 \le t_2 \le b_2}} \left(\int_a^{t_1} f(s, t_2) \Delta_1 s - \int_a^{A_1} f(s, t_2) \Delta_1 s \right) \right| \\ &\le \inf_{\substack{A_1 \le t_1 \le A_2 \\ a_2 \le t_2 \le b_2}} \left(\left| \int_a^{t_1} f(s, t_2) \Delta_1 s \right| + \left| \int_a^{A_1} f(s, t_2) \Delta_1 s \right| \right) \\ &\le 2L. \end{aligned}$$

Similarly, we obtain

$$|M_F| \le 2L.$$

From the second mean value theorem, Theorem 3.71, we have that there exists $\Lambda \in [m_F, M_F]$ such that

$$\int_{A_1}^{A_2} f(t_1, t_2) g(t_1, t_2) \Delta_1 t_1 = (g(A_1, t_2) - g(A_2, t_2)) \Lambda + g(A_2, t_2) \int_{A_1}^{A_2} f(t_1, t_2) \Delta_1 t_1.$$

Hence,

$$\left| \int_{A_1}^{A_2} f(t_1, t_2) g(t_1, t_2) \Delta_1 t_1 \right|$$

$$
\begin{aligned}
&= \left| (g(A_1,t_2) - g(A_2,t_2))\,\Lambda + g(A_2,t_2)\int_{A_1}^{A_2} f(t_1,t_2)\Delta_1 t_1 \right| \\
&\le (|g(A_1,t_2)| + |g(A_2,t_2)|)\,|\Lambda| + |g(A_1,A_2)| \left| \int_{A_1}^{A_2} f(t_1,t_2)\Delta_1 t_1 \right| \\
&\le 2\frac{\varepsilon}{6L}\Lambda + \frac{\varepsilon}{6L}\left| \int_a^{A_2} f(t_1,t_2)\Delta_1 t_1 - \int_a^{A_1} f(t_1,t_2)\Delta_1 t_1 \right| \\
&\le 4\frac{\varepsilon}{6L}L + \frac{\varepsilon}{6L}\left(\left| \int_a^{A} f(t_1,t_2)\Delta_1 t_1 \right| + \left| \int_a^{A_1} f(t_1,t_2)\Delta_1 t_1 \right| \right) \\
&\le \frac{2}{3}\varepsilon + \frac{2\varepsilon L}{6L} \\
&= \varepsilon.
\end{aligned}
$$

Then, using the Cauchy criterion, Theorem 5.22, it follows that

$$
\int_a^{\infty} f(t_1,t_2)g(t_1,t_2)\Delta_1 t_1
$$

is uniformly convergent in $t_2 \in [a_2, b_2]$. □

Example 5.38 Let $\mathbb{T}_1 = \mathbb{T}_2 = \mathbb{Z}$. Consider the integral

$$
I(t_2) = \int_1^{\infty} \frac{t_1 t_2}{(t_1+2t_2)(t_1+3t_2)(t_1^2+1)(t_2^2+1)}\Delta_1 t_1, \quad t_2 \in [1, 10].
$$

Let

$$
f(t_1,t_2) = \frac{1}{(t_1+2t_2)(t_1+3t_2)}, \quad g(t_1,t_2) = \frac{t_1 t_2}{(t_1^2+1)(t_2^2+1)}.
$$

For $A > 1$, we have

$$
\begin{aligned}
\left| \int_1^A f(t_1,t_2)\Delta_1 t_1 \right| &= \left| \int_1^A \frac{1}{(t_1+2t_2)(t_1+3t_2)}\Delta_1 t_1 \right| \\
&= \int_1^A \frac{1}{(t_1+2t_2)(t_1+3t_2)}\Delta_1 t_1 \\
&\le \int_1^A \frac{1}{(t_1+1)(t_1+2)}\Delta_1 t_1 \\
&= -\frac{1}{t_1+1}\Big|_{t_1=1}^{t_1=A}
\end{aligned}
$$

$$= \frac{1}{2} - \frac{1}{A+1}$$
$$\leq \frac{1}{2}.$$

Therefore, $F(A, t_2) = \int_1^A f(t_1, t_2)\Delta_1 t_1$ is uniformly bounded on $[1, \infty) \times [1, 10]$. Now, we consider $g(t_1, t_2)$. We have

$$\begin{aligned} g_{t_1}^{\Delta_1}(t_1, t_2) &= \frac{t_2}{t_2^2+1} \frac{t_1^2+1-t_1(\sigma_1(t_1)+t_1)}{(t_1^2+1)(\sigma_1^2(t_1)+1)} \\ &= \frac{t_2}{t_2^2+1} \frac{t_1^2+1-t_1(2t_1+1)}{(t_1^2+1)((t_1+1)^2+1)} \\ &= \frac{t_2}{t_2^2+1} \frac{t_1^2+1-t_1(2t_1+1)}{(t_1^2+1)(t_1^2+2t_1+2)} \\ &= \frac{t_2}{t_2^2+1} \frac{1-t_1-t_1^2}{(t_1^2+1)(t_1^2+2t_1+2)} \\ &\leq 0 \quad \text{on} \quad [1, \infty) \times [1, 10]. \end{aligned}$$

Therefore, $g(t_1, t_2)$ is monotone with respect to t_1 on $[1, \infty) \times [1, 10]$. Moreover,

$$\begin{aligned} |g(t_1, t_2)| &= \left| \frac{t_1 t_2}{(t_1^2+1)(t_2^2+1)} \right| \\ &\leq \frac{t_1}{t_1^2+1} \\ &\to 0 \quad \text{as} \quad t_1 \to \infty, \end{aligned}$$

i.e., $g(t_1, t_2)$ converges uniformly to 0 in t_2 as $t_1 \to \infty$. Hence, using the Dirichlet test, Theorem 5.37, it follows that $I(t_2)$ is uniformly convergent in $t_2 \in [1, 10]$.

Example 5.39 Let $\mathbb{T}_1 = 2^{\mathbb{N}_0}$ and $\mathbb{T}_2 = \mathbb{Z}$. Consider the integral

$$I(t_2) = \int_1^\infty \frac{e^{-t_1}(t_2+1)(t_2^4+1)}{(t_1+t_2+1)^{10}(t_2^{10}+3)} \Delta_1 t_1, \quad t_2 \in [1, 10].$$

We set

$$f(t_1, t_2) = \frac{1}{(t_1+t_2+1)^{10}}, \quad g(t_1, t_2) = \frac{e^{-t_1}(t_2+1)(t_2^4+1)}{t_2^{10}+3}.$$

For $A > 1$, we have

$$
\begin{aligned}
F(A, t_2) &= \int_1^A f(t_1, t_2)\Delta_1 t_1 \\
&= \int_1^A \frac{1}{(t_1 + t_2 + 1)^{10}} \Delta_1 t_1 \\
&\le \int_1^A \frac{1}{t_1^{10}} \Delta_1 t_1 \\
&= -\frac{2^9}{2^8 + 2^7 + \cdots + 2 + 1} \frac{1}{t_1^9}\Big|_{t_1=1}^{t_1=A} \\
&\le \frac{2^9}{2^8 + 2^7 + \cdots + 2 + 1} \quad \text{for any} \quad A \in [1, \infty), \quad t_2 \in [1, 10],
\end{aligned}
$$

i.e., $F(A, t_2)$ is uniformly bounded on $[1, \infty) \times [1, 10]$. Now, we consider $g(t_1, t_2)$. We have

$$
\begin{aligned}
g_{t_1}^{\Delta_1}(t_1, t_2) &= \frac{(t_2 + 1)(t_2^4 + 1)}{t_2^{10} + 3} \frac{e^{-\sigma_1(t_1)} - e^{-t_1}}{t_1} \\
&= \frac{(t_2 + 1)(t_2^4 + 1)}{t_2^{10} + 3} \frac{e^{-2t_1} - e^{-t_1}}{t_1} \\
&\le 0 \quad \text{on} \quad [1, \infty) \times [1, 10],
\end{aligned}
$$

i.e., $g(t_1, t_2)$ is monotone with respect to t_1 on $[1, \infty) \times [1, 10]$. Moreover,

$$
\begin{aligned}
g(t_1, t_2) &= e^{-t_1} \frac{(t_2 + 1)(t_2^4 + 1)}{t_2^{10} + 3} \\
&\le e^{-t_1} \\
&\to 0 \quad \text{as} \quad t_1 \to \infty \quad \text{for any} \quad t_2 \in [1, 10].
\end{aligned}
$$

Hence, utilizing the Dirichlet test, Theorem 5.37, we conclude that $I(t_2)$ converges uniformly in $t_2 \in [1, 10]$.

Example 5.40 Let $\mathbb{T}_1 = \mathbb{T}_2 = 2^{\mathbb{N}_0}$. Consider the integral

$$
I(t_2) = \int_1^\infty \frac{t_2}{t_1^2(t_1 t_2 + 1)(2t_1 t_2 + 1)(t_2^4 + 7)} \Delta_1 t_1, \quad t_2 \in [1, 4].
$$

We set

$$f(t_1,t_2)=\frac{1}{(t_1t_2+1)(2t_1t_2+1)},\quad g(t_1,t_2)=\frac{t_2}{t_1^2(t_2^4+7)}.$$

For $A>1$, we have

$$\begin{aligned}
F(A,t_2)&=\int_1^A \frac{1}{(t_1t_2+1)(2t_1t_2+1)}\Delta_1 t_1\\
&\le \int_1^A \frac{1}{(t_1+1)(2t_1+1)}\Delta_1 t_1\\
&=-\frac{1}{t_1+1}\Big|_{t_1=1}^{t_1=A}\\
&=\frac{1}{2}-\frac{1}{A+1}\\
&\le \frac{1}{2}\quad \text{for any}\quad t_2\in[1,4],
\end{aligned}$$

i.e., $F(A,t_2)$ is uniformly bounded on $[1,\infty)\times[1,4]$. Now, we consider $g(t_1,t_2)$. We have

$$\begin{aligned}
g_{t_1}^{\Delta_1}(t_1,t_2)&=-\frac{t_2}{t_2^4+7}\frac{\sigma_1(t_1)+t_1}{t_1^2(\sigma_1(t_1))^2(t_1)}\\
&=-\frac{t_2}{t_2^4+7}\frac{3t_1}{2t_1^4}\\
&=-\frac{3t_2}{2t_1^3(t_2^4+7)}\\
&\le 0\quad \text{for}\quad t_1\in[1,\infty),\quad t_2\in[1,4],
\end{aligned}$$

i.e., $g(t_1,t_2)$ is monotone with respect to t_1 on $[1,\infty)\times[1,4]$. Also,

$$\begin{aligned}
g(t_1,t_2)&=\frac{t_2}{t_1^2(t_2^4+7)}\\
&\le \frac{1}{t_1^2}\\
&\to 0\quad \text{as}\quad t_1\to\infty.
\end{aligned}$$

Hence, using the Dirichlet test, Theorem 5.37, we conclude that $I(t_2)$ converges uniformly in $t_2\in[1,4]$.

Exercise 5.41 Let $\mathbb{T}_1 = \mathbb{T}_2 = \mathbb{Z}$. Using the Dirichlet test, Theorem 5.37, prove that

$$\int_1^\infty \frac{e^{-t_1} t_2}{(t_1 t_2 + 1)^3 (t_2^4 + t_2^2 + 1)} \Delta_1 t_1, \quad t_2 \in [1, 4],$$

is uniformly convergent in $t_2 \in [1, 4]$.

Theorem 5.42 (Dini Test) *Suppose $f \in \mathcal{CC}_{rd}$ is defined on $[a, \infty) \times [a_2, b_2]$ and is of one sign. If*

$$I(t_2) = \int_a^\infty f(t_1, t_2) \Delta_1 t_1$$

is rd-continuous on $[a_2, b_2]$, then $I(t_2)$ is uniformly convergent in $t_2 \in [a_2, b_2]$.

Proof We suppose the contrary. Without loss of generality, we assume that

$$f(t_1, t_2) \geq 0.$$

Thus, there exists $\varepsilon_0 > 0$ such that for any $N > a$, there exist $n > N$ and $t_{2n} \in [a_2, b_2]$ such that

$$\int_n^\infty f(t_1, t_{2n}) \Delta_1 t_1 \geq \varepsilon_0.$$

In this way, we get the sequence $\{t_{2n}\}$, $t_{2n} \in [a_2, b_2]$, $n > a$. Since the sequence $\{t_{2n}\}$ is a bounded sequence, there exists a subsequence $\{t_{2n_k}\}$ converging to $t_{20} \in [a_2, b_2]$. We note that $\int_a^\infty f(t_1, t_{20}) \Delta_1 t_1$ is convergent. Hence, there exists $A > a$ such that

$$\int_A^\infty f(t_1, t_{20}) \Delta_1 t_1 < \frac{\varepsilon_0}{2}.$$

1. Assume t_{20} is left-dense and right-scattered. Since $t_{2n_k} \to t_{20}$ as $k \to \infty$, we have that $t_{2n_k} \leq t_{20}$. Thus, there exists K so that $k > K$ implies $|t_{2n_k} - t_{20}| < \varepsilon_0$. From Theorem 5.3, we have that $\int_a^A f(t_1, t_2) \Delta_1 t_1$ is rd-continuous on $[a_2, b_2]$. Hence, using that $I(t_2)$ is rd-continuous on $[a_2, b_2]$, there exists K_1 so that $k > K_1$ implies

$$\Bigg| \int_a^\infty f(t_1, t_{2n_k}) \Delta_1 t_1 - \int_a^\infty f(t_1, t_{20}) \Delta_1 t_1 \\ - \left(\int_a^A f(t_1, t_{2n_k}) \Delta_1 t_1 - \int_a^A f(t_1, t_{20}) \Delta_1 t_1 \right) \Bigg| < \frac{\varepsilon_0}{2},$$

   whereupon

$$\left| \int_a^\infty f(t_1, t_{2n_k}) \Delta_1 t_1 - \int_a^A f(t_1, t_{2n_k}) \Delta_1 t_1 \right|$$

$$\begin{aligned}
&< \frac{\varepsilon_0}{2} + \left| \int_a^\infty f(t_1, t_{20}) \Delta_1 t_1 - \int_a^A f(t_1, t_{20}) \Delta_1 t_1 \right| \\
&= \frac{\varepsilon_0}{2} + \left| \int_A^\infty f(t_1, t_{20}) \Delta_1 t_1 \right| \\
&\leq \frac{\varepsilon_0}{2} + \frac{\varepsilon_0}{2} \\
&= \varepsilon_0.
\end{aligned}$$

2. Assume that t_{20} is not left-dense and right-scattered. Thus, t_{20} is a point of continuity of $\int_a^A f(t_1, t_2) \Delta_1 t_1$ and $\int_a^\infty f(t_1, t_2) \Delta_1 t_1$. Since $t_{2n_k} \to t_{20}$ as $k \to \infty$, there exists K_2 such that $k > K_2$ implies

$$\left| \int_a^\infty f(t_1, t_{2n_k}) \Delta_1 t_1 - \int_a^A f(t_1, t_{2n_k}) \Delta_1 t_1 \right| < \varepsilon_0.$$

Hence, for $n_k > A$ and $k > \max\{K_1, K_2\}$, we have

$$\begin{aligned}
\int_{n_k}^\infty f(t_1, t_{2n_k}) \Delta_1 t_1 &\leq \int_A^\infty f(t_1, t_{2n_k}) \Delta_1 t_1 \\
&= \left| \int_a^\infty f(t_1, t_{2n_k}) \Delta_1 t_1 - \int_a^A f(t_1, t_{2n_k}) \Delta_1 t_1 \right| \\
&< \varepsilon_0,
\end{aligned}$$

which is a contradiction. □

For any strictly increasing sequence $\{a_n\}_{n \in \mathbb{N}_0}$ such that $a_0 = a$ and $a_n \to \infty$ as $n \to \infty$, we define

$$I_n(t_2) = \int_{a_{n-1}}^{a_n} f(t_1, t_2) \Delta_1 t_1.$$

Thus,

$$I(t_2) = \int_a^\infty f(t_1, t_2) \Delta_1 t_1$$

$$= \sum_{n=1}^{\infty} \int_{a_{n-1}}^{a_n} f(t_1, t_2)\Delta_1 t_1$$

$$= \sum_{n=1}^{\infty} I_n(t_2).$$

Theorem 5.43 *If $\int_a^\infty f(t_1, t_2)\Delta_1 t_1$ is uniformly convergent in $t_2 \in [a_2, b_2]$, then the function series $\sum\limits_{n=1}^{\infty} I_n(t_2)$ converges uniformly on $[a_2, b_2]$.*

Proof Let $\varepsilon > 0$ be arbitrarily chosen. Thus, there exists $A_0 = A_0(\varepsilon)$ so that $A > A_0$ implies

$$\left| \int_A^\infty f(t_1, t_2)\Delta_1 t_1 \right| < \varepsilon \quad \text{for any} \quad t_2 \in [a_2, b_2].$$

Hence, there exists $N \in \mathbb{N}$ such that $a_n > A_0$ and $a_m > A_0$ for $m, n > N$. Then, for $m, n > N$, we have

$$\begin{aligned}
|I_{n+1}(t_2) + \cdots + I_m(t_2)| &= \left| \int_{a_n}^{a_{n+1}} f(t_1, t_2)\Delta_1 t_1 + \cdots + \int_{a_{m-1}}^{a_m} f(t_1, t_2)\Delta_1 t_1 \right| \\
&= \left| \int_{a_n}^{a_m} f(t_1, t_2)\Delta_1 t_1 \right| \\
&= \left| \int_{a_n}^{\infty} f(t_1, t_2)\Delta_1 t_1 - \int_{a_m}^{\infty} f(t_1, t_2)\Delta_1 t_1 \right| \\
&\le \left| \int_{a_n}^{\infty} f(t_1, t_2)\Delta_1 t_1 \right| + \left| \int_{a_m}^{\infty} f(t_1, t_2)\Delta_1 t_1 \right| \\
&< \varepsilon + \varepsilon \\
&= 2\varepsilon.
\end{aligned}$$

Applying the Cauchy criterion for uniform convergence of function series, we conclude that $\sum_{n=1}^{\infty} I_n(t_2)$ is uniformly convergent on $[a_2, b_2]$. □

Theorem 5.44 *Assume that $f \in \mathrm{CC}_{\mathrm{rd}}$ is defined on $[a, \infty) \times [a_2, b_2]$. If*

$$I(t_2) = \int_a^\infty f(t_1, t_2)\Delta_1 t_1$$

is uniformly convergent in $t_2 \in [a_2, b_2]$, then $I(t_2)$ is rd-continuous in $t_2 \in [a_2, b_2]$.

Proof We have that

$$I(t_2) = \sum_{n=1}^{\infty} I_n(t_2).$$

Since $I(t_2)$ is uniformly convergent in $t_2 \in [a_2, b_2]$, the function series $\sum\limits_{n=1}^{\infty} I_n(t_2)$ is uniformly convergent in $t_2 \in [a_2, b_2]$. Note that $I_n(t_2)$ is rd-continuous on $[a_2, b_2]$ for any $n \in \mathbb{N}$. Hence, $I(t_2)$ is rd-continuous on $[a_2, b_2]$. □

Theorem 5.45 *Assume that* $f : [a, \infty) \times [a_2, b_2] \to \mathbb{R}$ *is continuous. If*

$$\int_a^{\infty} f(t_1, t_2)\Delta_1 t_1$$

is uniformly convergent in $t_2 \in [a_2, b_2]$*, then*

$$\int_{a_2}^{b_2} \Delta_2 t_2 \int_a^{\infty} f(t_1, t_2)\Delta_1 t_1 = \int_a^{\infty} \Delta_1 t_1 \int_{a_2}^{b_2} f(t_1, t_2)\Delta_2 t_2.$$

Proof Note that for any given $A > a$, using Theorem 7.58, we have

$$\int_{a_2}^{b_2} \Delta_2 t_2 \int_a^{A} f(t_1, t_2)\Delta_1 t_1 = \int_a^{A} \Delta_1 t_1 \int_{a_2}^{b_2} f(t_1, t_2)\Delta_2 t_2. \tag{5.7}$$

Also,

$$\left| \int_{a_2}^{b_2} \Delta_2 t_2 \int_a^{\infty} f(t_1, t_2)\Delta_1 t_1 - \int_{a_2}^{b_2} \Delta_2 t_2 \int_a^{A} f(t_1, t_2)\Delta_1 t_1 \right|$$

$$= \left| \int_{a_2}^{b_2} \Delta_2 t_2 \int_A^{\infty} f(t_1, t_2)\Delta_1 t_1 \right| \tag{5.8}$$

$$\leq \int_{a_2}^{b_2} \Delta_2 t_2 \left| \int_A^{\infty} f(t_1, t_2)\Delta_1 t_1 \right|.$$

Let $\varepsilon > 0$ be arbitrarily chosen. Since $\int_a^{\infty} f(t_1, t_2)\Delta_1 t_1$ is uniformly convergent, there exists $A_0 = A_0(\varepsilon)$ such that $A > A_0$ implies

$$\left| \int_A^{\infty} f(t_1, t_2)\Delta_1 t_1 \right| < \frac{\varepsilon}{b_2 - a_2}.$$

Hence, by (5.8), we get

$$\left| \int_{a_2}^{b_2} \Delta_2 t_2 \int_a^\infty f(t_1, t_2) \Delta_1 t_1 - \int_{a_2}^{b_2} \Delta_2 t_2 \int_a^A f(t_1, t_2) \Delta_1 t_1 \right|$$

$$\leq \int_{a_2}^{b_2} \Delta_2 t_2 \left| \int_A^\infty f(t_1, t_2) \Delta_1 t_1 \right|$$

$$< \frac{\varepsilon}{b_2 - a_2} \int_{a_2}^{b_2} \Delta_2 t_2$$

$$= \frac{\varepsilon}{b_2 - a_2} (b_2 - a_2)$$

$$= \varepsilon.$$

Consequently,

$$\lim_{A \to \infty} \int_{a_2}^{b_2} \Delta_2 t_2 \int_a^A f(t_1, t_2) \Delta_1 t_1 = \int_{a_2}^{b_2} \Delta_2 t_2 \int_a^\infty f(t_1, t_2) \Delta_1 t_1. \tag{5.9}$$

Hence, using (5.7), we find

$$\lim_{A \to \infty} \int_{a_2}^{b_2} \Delta_2 t_2 \int_a^A f(t_1, t_2) \Delta_1 t_1 = \lim_{A \to \infty} \int_a^A \Delta_1 t_1 \int_{a_2}^{b_2} f(t_1, t_2) \Delta_2 t_2.$$

Since

$$\lim_{A \to \infty} \int_{a_2}^{b_2} \Delta_2 t_2 \int_a^A f(t_1, t_2) \Delta_1 t_1$$

exists, we conclude that

$$\lim_{A \to \infty} \int_a^A \Delta_1 t_1 \int_{a_2}^{b_2} f(t_1, t_2) \Delta_2 t_2$$

exists. By the definition of the improper parameter-dependent integral, we have

$$\lim_{A \to \infty} \int_a^A \Delta_1 t_1 \int_{a_2}^{b_2} f(t_1, t_2) \Delta_2 t_2 = \int_a^\infty \Delta_1 t_1 \int_{a_2}^{b_2} f(t_1, t_2) \Delta_2 t_2. \tag{5.10}$$

By (5.7), (5.9), and (5.10), we get the assertion. □

Example 5.46 Let $\mathbb{T}_1 = \mathbb{T}_2 = \mathbb{Z}$. Consider

$$I(t_2) = \int_1^\infty \frac{1}{(t_1 + 1)(t_1 + 2)(t_2 + 1)(t_2 + 2)} \Delta_1 t_1, \quad t_2 \in [1, 4].$$

We have

$$\begin{aligned} I(t_2) &\le \int_1^\infty \frac{1}{(t_1+1)(t_1+2)} \Delta_1 t_1 \\ &= -\lim_{A\to\infty} \frac{1}{t_1+1}\Big|_{t_1=1}^{t_1=A} \\ &= \frac{1}{2} \quad \text{for any} \quad t_2 \in [1,4]. \end{aligned}$$

Therefore, $I(t_2)$ is uniformly convergent in $t_2 \in [1,4]$. Hence, employing Theorem 5.45, we get

$$\begin{aligned} &\int_1^4 \Delta_2 t_2 \int_1^\infty \frac{1}{(t_1+1)(t_1+2)(t_2+1)(t_2+2)} \Delta_1 t_1 \\ &\qquad = \int_1^\infty \Delta_1 t_1 \int_1^4 \frac{1}{(t_1+1)(t_1+2)(t_2+1)(t_2+2)} \Delta_2 t_2. \end{aligned}$$

Indeed,

$$\begin{aligned} &\int_1^\infty \frac{\Delta_1 t_1}{(t_1+1)(t_1+2)(t_2+1)(t_2+2)} \\ &= \frac{1}{(t_2+1)(t_2+2)} \int_1^\infty \frac{\Delta_1 t_1}{(t_1+1)(t_1+2)} \\ &= \frac{1}{(t_2+1)(t_2+2)} \lim_{A\to\infty} \int_1^A \frac{1}{(t_1+1)(t_1+2)} \Delta_1 t_1 \\ &= -\frac{1}{(t_2+1)(t_2+2)} \lim_{A\to\infty} \frac{1}{t_1+1}\Big|_{t_1=1}^{t_1=A} \\ &= \frac{1}{2(t_2+1)(t_2+2)}, \end{aligned}$$

$$\begin{aligned} &\int_1^4 \Delta_2 t_2 \int_1^\infty \frac{\Delta_1 t_1}{(t_1+1)(t_1+2)(t_2+1)(t_2+2)} \\ &= \frac{1}{2} \int_1^4 \frac{1}{(t_2+1)(t_2+2)} \Delta_2 t_2 \\ &= -\frac{1}{2} \frac{1}{t_2+1}\Big|_{t_2=1}^{t_2=4} \end{aligned}$$

$$
\begin{aligned}
&= -\frac{1}{2}\left(\frac{1}{5}-\frac{1}{2}\right)\\
&= \frac{3}{20},\\
&\int_1^4 \frac{\Delta_2 t_2}{(t_1+1)(t_1+2)(t_2+1)(t_2+2)}\\
&= \frac{1}{(t_1+1)(t_1+2)}\int_1^4 \frac{\Delta_2 t_2}{(t_2+1)(t_2+2)}\\
&= -\frac{1}{(t_1+1)(t_1+2)}\frac{1}{t_2+1}\Big|_{t_2=1}^{t_2=4}\\
&= -\frac{1}{(t_1+1)(t_1+2)}\left(\frac{1}{5}-\frac{1}{2}\right)\\
&= \frac{3}{10}\frac{1}{(t_1+1)(t_1+2)},\\
&\int_1^\infty \Delta_1 t_1 \int_1^4 \frac{\Delta_2 t_2}{(t_1+1)(t_1+2)(t_2+1)(t_2+2)}\\
&= \frac{3}{10}\lim_{A\to\infty}\int_1^A \frac{1}{(t_1+1)(t_1+2)}\Delta_1 t_1\\
&= -\frac{3}{10}\lim_{A\to\infty}\frac{1}{t_1+1}\Big|_{t_1=1}^{t_1=A}\\
&= \frac{3}{20}.
\end{aligned}
$$

Theorem 5.47 *Suppose that* $f \in \mathrm{CC}_{\mathrm{rd}}$ *and*

$$f_{t_2}^{\Delta_2}(t_1,t_2) : [a,\infty)\times[a_2,b_2] \to \mathbb{R}$$

is continuous. In addition, we assume that

$$\int_a^\infty f(t_1,t_2)\Delta_1 t_1$$

is convergent in $t_2 \in [a_2,b_2]$ *and*

$$\int_a^\infty f_{t_2}^{\Delta_2}(t_1,t_2)\Delta_1 t_1$$

is uniformly convergent in $t_2 \in [a_2, b_2]$. *Then*

$$I(t_2) = \int_a^{\infty} f(t_1, t_2)\Delta_1 t_1$$

is differentiable on $[a_2, b_2]$ *and satisfies*

$$I^{\Delta_2}(t_2) = \int_a^{\infty} f_{t_2}^{\Delta_2}(t_1, t_2)\Delta_1 t_1, \quad t_2 \in [a_2, b_2].$$

Proof Let

$$J(t_2) = \int_a^{\infty} f_{t_2}^{\Delta_2}(t_1, t_2)\Delta_1 t_1, \quad t_2 \in [a_2, b_2].$$

Since $J(t_2)$ is uniformly convergent in $t_2 \in [a_2, b_2]$, using Theorem 5.44, we conclude that it is rd-continuous in $t_2 \in [a_2, b_2]$. Therefore, $J(t_2)$ is integrable on $[a_2, b_2]$. Hence, using Theorem 5.45, we have

$$\begin{aligned}
\int_{a_2}^{t_2} J(s)\Delta_2 s &= \int_{a_2}^{t_2}\int_a^{\infty} f_{t_2}^{\Delta_2}(t_1, s)\Delta_1 t_1 \Delta_2 s \\
&= \int_a^{\infty}\int_{a_2}^{t_2} f_{t_2}^{\Delta_2}(t_1, s)\Delta_2 s \Delta_1 t_1 \\
&= \int_a^{\infty} f(t_1, s)\Big|_{s=a_2}^{s=t_2} \Delta_1 t_1 \\
&= \int_a^{\infty} f(t_1, t_2)\Delta_1 t_1 - \int_a^{\infty} f(t_1, a_2)\Delta_1 t_1 \\
&= I(t_2) - I(a_2).
\end{aligned}$$

From here, using that $\int_{a_2}^{t_2} J(s)\Delta_2 s$ is differentiable, we get

$$I^{\Delta_2}(t_2) = J(t_2),$$

which completes the proof. □

Example 5.48 Let $\mathbb{T}_1 = \mathbb{T}_2 = 2^{\mathbb{N}_0}$ and

$$f(t_1, t_2) = \frac{1}{2t_1^2 t_2}, \quad t_1 \in [1, \infty), \quad t_2 \in [1, 4].$$

We have that $f \in \mathcal{C}\mathcal{C}_{rd}$,

$$f_{t_2}^{\Delta_2}(t_1, t_2) = -\frac{1}{4t_1^2 t_2^2},$$

and $f_{t_2}^{\Delta_2}(t_1, t_2) : [1, \infty) \times [1, 4] \to \mathbb{R}$ is continuous. Moreover,

$$\begin{aligned} I(t_2) &= \int_1^\infty f(t_1, t_2)\Delta_1 t_1 \\ &= \lim_{A\to\infty} \int_1^A \frac{1}{2t_1^2 t_2}\Delta_1 t_1 \\ &= \frac{1}{t_2} \lim_{A\to\infty} \int_1^A \frac{1}{2t_1^2}\Delta_1 t_1 \\ &= -\frac{1}{t_2} \lim_{A\to\infty} \frac{1}{t_1}\Big|_{t_1=1}^{t_1=A} \\ &= \frac{1}{t_2}, \end{aligned}$$

i.e., $I(t_2)$ is convergent in $t_2 \in [1, 4]$. We have

$$\begin{aligned} \left| \int_1^\infty f_{t_2}^{\Delta_2}(t_1, t_2)\Delta_1 t_1 \right| &= \left| -\frac{1}{4} \int_1^\infty \frac{1}{t_1^2 t_2^2}\Delta_1 t_1 \right| \\ &\le \frac{1}{4} \int_1^\infty \frac{1}{t_1^2}\Delta_1 t_1 \\ &= -\frac{1}{4} \lim_{A\to\infty} \frac{1}{t_1}\Big|_{t_1=1}^{t_1=A} \\ &= \frac{1}{4}. \end{aligned}$$

Therefore, $\int_1^\infty f_{t_2}^{\Delta_2}(t_1, t_2)\Delta_1 t_1$ is uniformly convergent in $t_2 \in [1, 4]$. Hence, using Theorem 5.47, we obtain

$$I^{\Delta_2}(t_2) = \int_1^\infty f_{t_2}^{\Delta_2}(t_1, t_2)\Delta_1 t_1.$$

Indeed,

$$\begin{aligned} \int_1^\infty f_{t_2}^{\Delta_2}(t_1, t_2)\Delta_1 t_1 &= -\frac{1}{4} \lim_{A\to\infty} \int_1^A \frac{1}{t_1^2 t_2^2}\Delta_1 t_1 \\ &= \frac{1}{2t_2^2} \lim_{A\to\infty} \frac{1}{t_1}\Big|_{t_1=1}^{t_1=A} \end{aligned}$$

$$= -\frac{1}{2t_2^2},$$

$$I^{\Delta_2}(t_2) = -\frac{1}{2t_2^2}.$$

5.3 Advanced Practical Problems

Problem 5.49 Find $I^{\Delta_2}(t_2)$, where

1. $I(t_2) = \int_{-1}^{1} \left(t_1^4 - 2t_1^5 + t_1^2 t_2^2 + t_1 t_2^3\right) \Delta_1 t_1$, $\mathbb{T}_1 = \mathbb{T}_2 = \mathbb{Z}$,
2. $I(t_2) = \int_1^2 \left(t_1^3 + t_1 t_2 + t_1^2 t_2^2\right) \Delta_1 t_1$, $\mathbb{T}_1 = 2^{\mathbb{N}_0} \cup \{0\}$, $\mathbb{T}_2 = \mathbb{Z}$,
3. $I(t_2) = \int_1^2 \left(t_1^2 \sin_f(t_1, 1) + t_1 t_2 - t_1^2 t_2^2\right) \Delta_1 t_1$, $\mathbb{T}_1 = \mathbb{T}_2 = 2^{\mathbb{N}_0} \cup \{0\}$, $f(t_1) = t_1^2$.

Solution 1. $-t_2 - 3t_2^2$,
2. $2t_2 + 2$,
3. $1 - 3t_2$.

Problem 5.50 Compute $I^{\Delta}(t)$, where

1. $I(t) = \int_0^t (s^3 - s^2 t + s)\Delta s$, $\mathbb{T} = \mathbb{Z}$,
2. $I(t) = \int_1^t (s^2 e_f(s, 1) - st)\Delta s$, $\mathbb{T} = 2^{\mathbb{N}_0}$, $f(t) = t^2$,
3. $I(t) = \int_1^t (s^3 e_g(s, 1) - s^2 t)\Delta s$, $\mathbb{T} = 3^{\mathbb{N}_0}$, $g(t) = t^3$.

Solution 1. $-\frac{t^3}{3} - \frac{t^2}{2} + \frac{5t}{6}$,
2. $-\frac{7}{3}t^2 + \frac{1}{3} + t^2 e_f(t, 1)$,
3. $-\frac{40}{13}t^3 + \frac{1}{13} + t^3 e_g(t, 1)$.

Problem 5.51 Let $\mathbb{T}_1 = 2^{\mathbb{N}_0}$, $\mathbb{T}_2 = 3^{\mathbb{N}_0}$. Investigate the integral

$$I(t_2) = \int_4^{\infty} \frac{1}{(t_1 - t_2 + 3)(2t_1 - t_2 + 3)} \Delta_1 t_1, \quad t_2 \in \mathbb{T}_2$$

for convergence and divergence.

Solution Convergent for $t_2 \neq 9$, $t_2 \in \mathbb{T}_2$, divergent for $t_2 = 9$.

Problem 5.52 Let $\mathbb{T}_1 = \mathbb{T}_2 = 2^{\mathbb{N}_0}$. Using the Cauchy criterion, Theorem 5.22, prove that

$$I(t_2) = \int_2^{\infty} \frac{\sin t_2}{t_1^3 t_2^3} \Delta_1 t_1, \quad t_2 \in [1, 8],$$

is uniformly convergent in $t_2 \in [1, 8]$.

Problem 5.53 Let $\mathbb{T}_1 = \mathbb{N}_0$, $\mathbb{T}_2 = 2^{\mathbb{N}_0}$. Using the Weierstraß test, Theorem 5.27, prove that

$$\int_1^\infty \frac{t_1 + t_2}{(3 + t_1^7 t_2^7)(1 + t_1^2 + t_2^3)} \Delta_1 t_1, \quad t_2 \in [1, 8],$$

is uniformly convergent in $t_2 \in [1, 8]$.

Problem 5.54 Let $\mathbb{T}_1 = \mathbb{Z}$ and $\mathbb{T}_2 = 3^{\mathbb{N}_0}$. Using the Abel test, Theorem 5.32, prove that the integral

$$\int_1^\infty \frac{t_1 t_2^2}{(2t_1 + 4)^2 (2t_1 + 6)^2 (t_2 + 1)^{10}} \Delta_1 t_1$$

is uniformly convergent in $t_2 \in [1, 81]$.

Problem 5.55 Let $\mathbb{T}_1 = \mathbb{T}_2 = 2^{\mathbb{N}_0}$. Using the Dirichlet test, Theorem 5.37, prove that

$$\int_1^\infty \frac{t_2}{(t_1 t_2 + 1)^2 (2t_1 t_2 + 1)^2 t_1^2 (t_2^7 + 1)} \Delta_1 t_1, \quad t_2 \in [1, 8],$$

is uniformly convergent in $t_2 \in [1, 8]$.

5.4 Notes and References

In this chapter, the theory of parameter-dependent integrals is extended to general time scales. By combining multiple integration theory on time scales and usual parameter-dependent integration theory in classical mathematical analysis, both normal parameter-dependent integrals on time scales and improper parameter-dependent integrals on time scales are investigated. Some criteria for uniform convergence are given. Analytical properties such as continuity, differentiability, and integrability are established. All results in this chapter are taken from Zhang and Wang [40].

Chapter 6
Partial Differentiation on Time Scales

6.1 Basic Definitions

Let $n \in \mathbb{N}$ be fixed. For each $i \in \{1, 2, \dots, n\}$, we denote by $\mathbb{T}_i$ a time scale.

Definition 6.1 The set

$$\Lambda^n = \mathbb{T}_1 \times \mathbb{T}_2 \times \cdots \times \mathbb{T}_n = \{t = (t_1, t_2, \dots, t_n) : \ t_i \in \mathbb{T}_i, \ i = 1, 2, \dots, n\}$$

is called an *n-dimensional time scale*.

Example 6.2 $(\mathbb{R}, \mathbb{Z}, \mathbb{N})$ is a 3-dimensional time scale.

Example 6.3 $(\mathbb{Z}, \mathbb{N}_0^2, 2^{\mathbb{N}}, \mathbb{N})$ is a 4-dimensional time scale.

Example 6.4 $(3^{\mathbb{N}}, 4^{\mathbb{N}}, \mathbb{Q})$ is not a 3-dimensional time scale because $\mathbb{Q}$ is not a time scale.

Remark 6.5 We equip Λ^n with the metric

$$d(t,s) = \left(\sum_{i=1}^{n} |t_i - s_i|^2 \right)^{\frac{1}{2}} \quad \text{for} \quad t, s \in \Lambda^n.$$

The set Λ^n with this metric is a complete metric space. Therefore, we have for Λ^n the fundamental concepts such as open balls, neighbourhoods of points, open sets, compact sets, and so on. Also, we have for functions $f : \Lambda^n \to \mathbb{R}$ the concepts of limits, continuity, and properties of continuous functions on general metric spaces.

Definition 6.6 Let σ_i, $i \in \{1, 2, \dots, n\}$, be the forward jump operator in $\mathbb{T}_i$. The operator $\sigma : \Lambda^n \to \mathbb{R}^n$ defined by

$$\sigma(t) = (\sigma_1(t), \sigma_2(t), \dots, \sigma_n(t))$$

is said to be the *forward jump operator* in Λ^n.

M. Bohner and S.G. Georgiev, *Multivariable Dynamic Calculus on Time Scales*,
DOI 10.1007/978-3-319-47620-9_6

Example 6.7 Let $n = 4$ and $\Lambda^4 = \mathbb{R} \times \mathbb{Z} \times \mathbb{Z} \times \mathbb{N}_0^2$. Then

$$\mathbb{T}_1 = \mathbb{R}, \quad \mathbb{T}_2 = \mathbb{Z}, \quad \mathbb{T}_3 = \mathbb{Z}, \quad \mathbb{T}_4 = \mathbb{N}_0^2.$$

We have that

$$\sigma_1(t_1) = t_1, \quad t_1 \in \mathbb{R}, \quad \sigma_2(t_2) = t_2 + 1, \quad t_2 \in \mathbb{Z}$$

and

$$\sigma_3(t_3) = t_3 + 1, \quad t_3 \in \mathbb{Z}, \quad \sigma_4(t_4) = \left(1 + \sqrt{t_4}\right)^2, \quad t_4 \in \mathbb{N}_0^2.$$

Hence,

$$\sigma(t) = \sigma(t_1, t_2, t_3, t_4) = \left(t_1, t_2 + 1, t_3 + 1, \left(1 + \sqrt{t_4}\right)^2\right),$$

$$(t_1, t_2, t_3, t_4) \in \mathbb{T}_1 \times \mathbb{T}_2 \times \mathbb{T}_3 \times \mathbb{T}_4.$$

Example 6.8 Let $n = 3$ and $\Lambda^3 = \mathbb{R} \times \mathbb{N}_0^2 \times 3\mathbb{Z}$. Then

$$\sigma_1(t_1) = t_1, \quad t_1 \in \mathbb{T}_1 = \mathbb{R}, \quad \sigma_2(t_2) = \left(1 + \sqrt{t_2}\right)^2, \quad t_2 \in \mathbb{T}_2 = \mathbb{N}_0^2,$$

and

$$\sigma_3(t_3) = t_3 + 3, \quad t_3 \in \mathbb{T}_3 = 3\mathbb{Z}.$$

Hence,

$$\sigma(t) = \sigma(t_1, t_2, t_3) = \left(t_1, \left(1 + \sqrt{t_2}\right)^2, t_3 + 3\right), \quad (t_1, t_2, t_3) \in \mathbb{T}_1 \times \mathbb{T}_2 \times \mathbb{T}_3.$$

Example 6.9 Let $n = 2$ and $\Lambda^2 = 2\mathbb{Z} \times 2^{\mathbb{N}}$. Here,

$$\mathbb{T}_1 = 2\mathbb{Z}, \quad \mathbb{T}_2 = 2^{\mathbb{N}}.$$

Then

$$\sigma_1(t_1) = t_1 + 2, \quad t_1 \in \mathbb{T}_1, \quad \sigma_2(t_2) = 2t_2, \quad t_2 \in \mathbb{T}_2.$$

Hence,

$$\sigma(t) = \sigma(t_1, t_2) = (t_1 + 2, 2t_2), \quad (t_1, t_2) \in \mathbb{T}_1 \times \mathbb{T}_2.$$

Exercise 6.10 Let $n = 3$ and $\Lambda^3 = \mathbb{Z} \times \mathbb{N} \times 4^{\mathbb{N}}$. Find $\sigma(t)$.

Solution $(t_1 + 1, t_2 + 1, 4t_3)$, $(t_1, t_2, t_3) \in \mathbb{Z} \times \mathbb{N} \times 4^{\mathbb{N}}$.

Definition 6.11 Let ρ_i, $i \in \{1, 2, \ldots, n\}$, be the backward jump operator in $\mathbb{T}_i$. The operator $\rho : \Lambda^n \to \mathbb{R}^n$ defined by

$$\rho(t) = (\rho_1(t_1), \rho_2(t_2), \ldots, \rho_n(t_n)), \quad t = (t_1, t_2, \ldots, t_n) \in \Lambda^n,$$

is said to be the *backward jump operator* in Λ^n.

Example 6.12 Let $n = 4$ and $\Lambda^4 = \mathbb{R} \times \mathbb{Z} \times 3^{\mathbb{N}} \times \mathbb{N}_0^3$. Here,

$$\mathbb{T}_1 = \mathbb{R}, \quad \mathbb{T}_2 = \mathbb{Z}, \quad \mathbb{T}_3 = 3^{\mathbb{N}}, \quad \mathbb{T}_4 = \mathbb{N}_0^3.$$

Then

$$\rho_1(t_1) = t_1, \quad t_1 \in \mathbb{T}_1, \quad \rho_2(t_2) = t_2 - 1, \quad t_2 \in \mathbb{T}_2,$$

$$\rho_3(t_3) = \frac{t_3}{3}, \quad t_3 \in \mathbb{T}_3 \setminus \{3\}, \quad \rho(3) = 3,$$

$$\rho_4(t_4) = \left(\sqrt[3]{t_4} - 1\right)^3, \quad t_4 \in \mathbb{T}_4 \setminus \{0\}, \quad \rho_4(0) = 0.$$

Hence,

$$\rho(t) = (\rho_1(t_1), \rho_2(t_2), \rho_3(t_3), \rho_4(t_4))$$

$$= \begin{cases} \left(t_1, t_2 - 1, \frac{t_3}{3}, \left(\sqrt[3]{t_4} - 1\right)^3\right) \\ \quad \text{if } t_1 \in \mathbb{T}_1, \quad t_2 \in \mathbb{T}_2, \quad t_3 \in \mathbb{T}_3 \setminus \{3\}, \quad t_4 \in \mathbb{T}_4 \setminus \{0\}, \\ (t_1, t_2 - 1, 3, (\sqrt[3]{t_4} - 1)^3) \\ \quad \text{if } t_1 \in \mathbb{T}_1, \quad t_2 \in \mathbb{T}_2, \quad t_3 = 3, \quad t_4 \in \mathbb{T}_4 \setminus \{0\}, \\ \left(t_1, t_2 - 1, \frac{t_3}{3}, 0\right) \\ \quad \text{if } t_1 \in \mathbb{T}_1, \quad t_2 \in \mathbb{T}_2, \quad t_3 \in \mathbb{T}_3 \setminus \{3\}, \quad t_4 = 0, \\ (t_1, t_2 - 1, 3, 0) \\ \quad \text{if } t_1 \in \mathbb{T}_1, \quad t_2 \in \mathbb{T}_2, \quad t_3 = 3, \quad t_4 = 0. \end{cases}$$

Example 6.13 Let $n = 2$ and $\Lambda^2 = (4\mathbb{Z}, \mathbb{R})$. Here,

$$\mathbb{T}_1 = 4\mathbb{Z}, \quad \mathbb{T}_2 = \mathbb{R}.$$

Then

$$\rho_1(t_1) = t_1 - 4, \quad t_1 \in \mathbb{T}_1, \quad \rho_2(t_2) = t_2, \quad t_2 \in \mathbb{T}_2.$$

Hence,

$$\rho(t) = (\rho_1(t_1), \rho_2(t_2)) = (t_1 - 4, t_2), \quad t = (t_1, t_2) \in \mathbb{T}_1 \times \mathbb{T}_2.$$

Example 6.14 Let $n = 3$ and $\Lambda^3 = \mathbb{Z} \times \mathbb{Z} \times 7^{\mathbb{N}}$. Here,

$$\mathbb{T}_1 = \mathbb{Z}, \quad \mathbb{T}_2 = \mathbb{Z}, \quad \mathbb{T}_3 = 7^{\mathbb{N}}.$$

Then

$$\rho_1(t_1) = t_1 - 1, \quad t_1 \in \mathbb{T}_1, \quad \rho_2(t_2) = t_2 - 1, \quad t_2 \in \mathbb{T}_2,$$

$$\rho_3(t_3) = \frac{t_3}{7}, \quad t_3 \in \mathbb{T}_3 \setminus \{7\}, \quad \rho_3(7) = 7.$$

Hence,

$$\rho(t) = (\rho_1(t_1), \rho_2(t_2), \rho_3(t_3))$$

$$= \begin{cases} \left(t_1 - 1, t_2 - 1, \frac{t_3}{7}\right) & \text{if } \ t_1 \in \mathbb{T}_1, \quad t_2 \in \mathbb{T}_2, \quad t_3 \in \mathbb{T}_3 \setminus \{7\}, \\ (t_1 - 1, t_2 - 1, 7) & \text{if } \ t_2 \in \mathbb{T}_1, \quad t_2 \in \mathbb{T}_2, \quad t_3 = 7. \end{cases}$$

Exercise 6.15 Let $n = 2$ and $\Lambda^2 = \mathbb{Z} \times \mathbb{Z}$. Find $\rho(t)$.

Solution $(t_1 - 1, t_2 - 1)$, $t = (t_1, t_2) \in \mathbb{Z} \times \mathbb{Z}$.

Definition 6.16 For $x = (x_1, x_2, \ldots, x_n) \in \mathbb{R}^n$ and $y = (y_1, y_2, \ldots, y_n) \in \mathbb{R}^n$, we write

$$x \geq y$$

whenever

$$x_i \geq y_i \quad \text{for all} \quad i = 1, 2, \ldots, n.$$

In a similar way, we understand $x > y$ and $x < y$ and $x \leq y$.

Definition 6.17 Let $t \in \Lambda^n$, $t = (t_1, t_2, \ldots, t_n)$.

1. If $\sigma(t) > t$, then t is called *strictly right-scattered.*
2. If $\sigma(t) \geq t$ and there are $j, l \in \{1, 2, \ldots, n\}$ such that $\sigma_j(t_j) > t_j$ and $\sigma_l(t_l) = t_l$, then t is called *right-scattered.*
3. If $t < \sup \Lambda^n$ and $\sigma(t) = t$, then t is called *right-dense.*

4. If $\rho(t) < t$, then t is called *strictly left-scattered.*
5. If $\rho(t) \le t$ and there are $l, j \in \{1, 2, \ldots, n\}$ such that $\rho_j(t_j) < t_j$ and $\rho_l(t_l) = t_l$, then t is called *left-scattered.*
6. If $t > \inf \Lambda^n$ and $t = \rho(t)$, then t is called *left-dense*.
7. If t is strictly right-scattered and strictly left-scattered, then t is said to be *strictly isolated.*
8. If t is right-dense and left-dense, then t is said to be *dense*.
9. If t is right-scattered and left-scattered, then t is said to be *isolated.*

Example 6.18 Let $\Lambda^3 = \mathbb{Z} \times \mathbb{Z} \times \mathbb{Z}$. Then, for $t = (t_1, t_2, t_3) \in \Lambda^3$, we have

$$\sigma(t) = (\sigma_1(t_1), \sigma_2(t_2), \sigma_3(t_3)) = (t_1 + 1, t_2 + 1, t_3 + 1) > (t_1, t_2, t_3),$$

i.e., all points of Λ^3 are strictly right-scattered. Also,

$$\rho(t) = (\rho_1(t_1), \rho_2(t_2), \rho_3(t_3)) = (t_1 - 1, t_2 - 1, t_3 - 1) < (t_1, t_2, t_3),$$

i.e., all points of Λ^3 are strictly left-scattered. Consequently, all points of Λ^3 are strictly isolated.

Example 6.19 Let $\Lambda^4 = 2\mathbb{Z} \times \mathbb{R} \times 2^{\mathbb{N}} \times \left(\frac{1}{3}\right)^{\mathbb{N}}$. Then

$$\sigma(t) = (\sigma_1(t_1), \sigma_2(t_2), \sigma_3(t_3), \sigma_4(t_4))$$

$$= (t_1 + 2, t_2, 2t_3, 3t_4) \ge (t_1, t_2, t_3, t_4), \quad t_4 \ne \frac{1}{3}.$$

Therefore, all points $(t_1, t_2, t_3, t_4) \in \Lambda^4$, $t_4 \ne \frac{1}{3}$, are right-scattered. We note that

$$\left(\sigma_1(t_1), \sigma_2(t_2), \sigma_3(t_3), \sigma_4\left(\frac{1}{3}\right)\right) = \left(t_1 + 1, t_2, 2t_3, \frac{1}{3}\right) \ge \left(t_1, t_2, t_3, \frac{1}{3}\right).$$

From here, all points $\left(t_1, t_2, t_3, \frac{1}{3}\right)$ are right-scattered. Also,

$$\rho(t) = (\rho_1(t_1), \rho_2(t_2), \rho_3(t_3), \rho_4(t_4))$$

$$= \left(t_1 - 2, t_2, \frac{t_3}{2}, \frac{t_4}{3}\right) \le (t_1, t_2, t_3, t_4) \text{ if } t_3 \ne 2,$$

i.e., all points $(t_1, t_2, t_3, t_4) \in \Lambda^4$, $t_3 \ne 2$, are left-scattered. We note that

$$\begin{aligned}\rho(t) &= (\rho_1(t_1), \rho_2(t_2), \rho_3(2), \rho_4(t_4)) \\ &= \left(t_1 - 2, t_2, 2, \frac{t_4}{3}\right) \\ &\le (t_1, t_2, 2, t_4).\end{aligned}$$

Therefore, all points $(t_1, t_2, 2, t_4) \in \Lambda^4$ are left-scattered. Moreover, the points

$$(t_1, t_2, t_3, t_4) \in \Lambda^4$$

are isolated.

Example 6.20 Let $\Lambda^3 = \mathbb{N} \times \mathbb{H} \times \mathbb{R}$. Then

$$\begin{aligned}\sigma(t) &= (\sigma_1(t_1), \sigma_2(H_n), \sigma_3(t_3)) \\ &= (t_1 + 1, H_{n+1}, t_3) \\ &\ge (t_1, H_n, t_3) \quad \text{for} \quad (t_1, H_n, t_3) \in \Lambda^3,\end{aligned}$$

i.e., any point $(t_1, H_n, t_3) \in \Lambda^3$ is right-scattered. Also,

$$\begin{aligned}\rho(t) &= (\rho_1(t_1), \rho_2(H_n), \rho_3(t_3)) \\ &= (t_1 - 1, H_{n-1}, t_3) \\ &\le (t_1, H_n, t_3) \quad \text{for} \quad (t_1, H_n, t_3) \in \Lambda^3, \quad t_1 \ne 1, \quad n \ne 0.\end{aligned}$$

Therefore, all points $(t_1, H_n, t_3) \in \Lambda^3$, $t_1 \ne 1$, $n \ne 0$, are left-scattered. We note that

$$\begin{aligned}(\rho_1(1), \rho_2(H_n), \rho_3(t_3)) &= (1, H_{n-1}, t_3) \\ &\le (1, H_n, t_3) \in \Lambda^3, \quad n \in \mathbb{N},\end{aligned}$$

$$(\rho_1(t_1), \rho_2(H_0), \rho_3(t_3)) = (t_1 - 1, H_0, t_3) \le (t_1, H_0, t_3), \quad t_1 \ne 1,$$

$$(\rho_1(1), \rho_2(H_0), \rho_3(t_3)) = (1, H_0, t_3) \le (1, H_0, t_3) \in \Lambda^3.$$

Therefore, the points

$$(t_1, H_0, t_3), \quad t_1 \neq 1; \quad (1, H_n, t_3), \quad n \in \mathbb{N}; \quad (1, H_0, t_3)$$

are left-scattered. Consequently, all points $(t_1, H_n, t_3) \in \Lambda^3$ are isolated.

Exercise 6.21 Classify each point $t \in \Lambda^2 = \mathbb{N}_0^3 \times \mathbb{R}$ as strictly right-scattered, right-scattered, right-dense, strictly left-scattered, left-scattered, left-dense, strictly isolated, dense, and isolated, respectively.

Solution All points are isolated.

Definition 6.22 The *graininess* function $\mu : \Lambda^n \to [0, \infty)^n$ is defined by

$$\mu(t) = (\mu_1(t_1), \mu_2(t_2), \ldots, \mu_n(t_n)), \quad t = (t_1, t_2, \ldots, t_n) \in \Lambda^n.$$

Example 6.23 Let $\Lambda^3 = 3\mathbb{Z} \times \mathbb{R} \times \mathbb{N}_0^4$. Then

$$\mathbb{T}_1 = 3\mathbb{Z}, \quad \mathbb{T}_2 = \mathbb{R}, \quad \mathbb{T}_3 = \mathbb{N}_0^4,$$

$$\sigma_1(t_1) = t_1 + 3, \quad t_1 \in \mathbb{T}_1, \quad \sigma_2(t_2) = t_2, \quad t_2 \in \mathbb{T}_2,$$

$$\sigma_3(t_3) = \left(1 + \sqrt[4]{t_3}\right)^4, \quad t_3 \in \mathbb{T}_3.$$

Hence,

$$\mu_1(t_1) = \sigma_1(t_1) - t_1 = t_1 + 3 - t_1 = 3, \quad t_1 \in \mathbb{T}_1,$$

$$\mu_2(t_2) = \sigma_2(t_2) - t_2 = t_2 - t_2 = 0, \quad t_2 \in \mathbb{T}_2,$$

$$\begin{aligned}\mu_3(t_3) &= \sigma_3(t_3) - t_3 \\ &= \left(1 + \sqrt[4]{t_3}\right)^4 - t_3 \\ &= t_3 + 4\sqrt[4]{t_3^3} + 6\sqrt[4]{t_3^2} + 4\sqrt[4]{t_3} + 1 - t_3 \\ &= 6\sqrt{t_3} + 4\sqrt[4]{t_3} + 4\sqrt[4]{t_3^3} + 1, \quad t_3 \in \mathbb{T}_3,\end{aligned}$$

$$\begin{aligned}\mu(t) &= (\mu_1(t_1), \mu_2(t_2), \mu_3(t_3)) \\ &= \left(3, 0, 6\sqrt{t_3} + 4\sqrt[4]{t_3} + 4\sqrt[4]{t_3^3} + 1\right), \quad t = (t_1, t_2, t_3) \in \Lambda^3.\end{aligned}$$

Example 6.24 Let $\Lambda^3 = \mathbb{Z} \times 2^{\mathbb{N}} \times 3^{\mathbb{N}}$. Then

$$\mathbb{T}_1 = \mathbb{Z}, \quad \mathbb{T}_2 = 2^{\mathbb{N}}, \quad \mathbb{T}_3 = 3^{\mathbb{N}},$$

$$\sigma_1(t_1) = t_1 + 1, \quad t_1 \in \mathbb{T}_1, \quad \sigma_2(t_2) = 2t_2, \quad t_2 \in \mathbb{T}_2,$$

$$\sigma_3(t_3) = 3t_3, \quad t_3 \in \mathbb{T}_3.$$

Hence,

$$\mu_1(t_1) = \sigma_1(t_1) - t_1 = t_1 + 1 - t_1 = 1, \quad t_1 \in \mathbb{T}_1,$$

$$\mu_2(t_2) = \sigma_2(t_2) - t_2 = 2t_2 - t_2 = t_2, \quad t_2 \in \mathbb{T}_2,$$

$$\mu_3(t_3) = \sigma_3(t_3) - t_3 = 3t_3 - t_3 = 2t_3, \quad t_3 \in \mathbb{T}_3,$$

$$\begin{aligned} \mu(t) &= (\mu_1(t_1), \mu_2(t_2), \mu_3(t_3)) \\ &= (1, t_2, 2t_3), \quad t = (t_1, t_2, t_3) \in \Lambda^3. \end{aligned}$$

Example 6.25 Let $\Lambda^4 = 2\mathbb{Z} \times \mathbb{Z} \times \mathbb{N}_0^2 \times 4^{\mathbb{N}}$. Then

$$\mathbb{T}_1 = 2\mathbb{Z}, \quad \mathbb{T}_2 = \mathbb{Z}, \quad \mathbb{T}_3 = \mathbb{N}_0^2, \quad \mathbb{T}_4 = 4^{\mathbb{N}},$$

$$\sigma_1(t_1) = t_1 + 2, \quad t_1 \in \mathbb{T}_1, \quad \sigma_2(t_2) = t_2 + 1, \quad t_2 \in \mathbb{T}_2,$$

$$\sigma_3(t_3) = \left(1 + \sqrt{t_3}\right)^2, \quad t_3 \in \mathbb{T}_3, \quad \sigma_4(t_4) = 4t_4, \quad t_4 \in 4^{\mathbb{N}}.$$

Hence,

$$\mu_1(t_1) = \sigma_1(t_1) - t_1 = t_1 + 2 - t_1 = 2, \quad t_1 \in \mathbb{T}_1,$$

$$\mu_2(t_2) = \sigma_2(t_2) - t_2 = t_2 + 1 - t_2 = 1, \quad t_2 \in \mathbb{T}_2,$$

$$\mu_3(t_3) = \sigma_3(t_3) - t_3 = \left(1 + \sqrt{t_3}\right)^2 - t_3 = 1 + 2\sqrt{t_3}, \quad t_3 \in \mathbb{T}_3,$$

$$\mu_4(t_4) = \sigma_4(t_4) - t_4 = 4t_4 - t_4 = 3t_4, \quad t_4 \in \mathbb{T}_4,$$

$$\mu(t) = (\mu_1(t_1), \mu_2(t_2), \mu_3(t_3), \mu_4(t_4))$$

$$= (2, 1, 1 + 2\sqrt{t_3}, 3t_4), \quad t = (t_1, t_2, t_3, t_4) \in \Lambda^4.$$

Exercise 6.26 Let $\Lambda^3 = 2\mathbb{Z} \times 3\mathbb{Z} \times 4\mathbb{Z}$. Find $\mu(t)$.

Solution $(2, 3, 4)$.

Definition 6.27 Let $f : \Lambda \to \mathbb{R}$. We introduce the following notations.

$$f^{\sigma}(t) = f\left(\sigma_1(t_1), \sigma_2(t_2), \ldots, \sigma_n(t_n)\right),$$

$$f_i^{\sigma_i}(t) = f\left(t_1, \ldots, t_{i-1}, \sigma_i(t_i), t_{i+1}, \ldots, t_n\right),$$

$$f_{i_1 i_2 \ldots i_l}^{\sigma_{i_1}\sigma_{i_2}\ldots\sigma_{i_l}}(t) = f\left(\ldots, \sigma_{i_1}(t_{i_1}), \ldots, \sigma_{i_2}(t_{i_2}), \ldots, \sigma_{i_l}(t_{i_l}), \ldots\right),$$

where $1 \le i_1 < i_2 < \ldots < i_l \le n$, $i_m \in \mathbb{N}$, $m \in \{1, 2, \ldots, l\}$, $l \in \mathbb{N}$.

Example 6.28 Let $\Lambda^2 = 2\mathbb{Z} \times 2^{\mathbb{N}}$. Here,

$$\mathbb{T}_1 = 2\mathbb{Z}, \quad \mathbb{T}_2 = 2^{\mathbb{N}},$$

$$\sigma_1(t_1) = t_1 + 2, \quad t_1 \in \mathbb{T}_1, \quad \sigma_2(t_2) = 2t_2, \quad t_2 \in \mathbb{T}_2.$$

Let

$$f(t_1, t_2) = t_1^2 + t_2.$$

Hence,

$$f^{\sigma}(t) = f(\sigma_1(t_1), \sigma_2(t_2))$$

$$= (\sigma_1(t_1))^2 + \sigma_2(t_2)$$

$$= (t_1 + 2)^2 + 2t_2$$

$$= t_1^2 + 4t_1 + 2t_2 + 4,$$

$$f_1^{\sigma_1}(t) = f(\sigma_1(t_1), t_2)$$

$$= (\sigma_1(t_1))^2 + t_2$$

$$= (t_1 + 2)^2 + t_2$$

$$= t_1^2 + 4t_1 + t_2 + 4,$$

$$f_2^{\sigma_2}(t) = f(t_1, \sigma_2(t_2))$$

$$= t_1^2 + \sigma_2(t_2)$$

$$= t_1^2 + 2t_2, \quad t \in \Lambda^2.$$

Example 6.29 Let $\Lambda^3 = 3\mathbb{Z} \times \mathbb{N} \times 4^{\mathbb{N}}$ and $f : \Lambda^3 \to \mathbb{R}$ be defined by

$$f(t) = t_1 t_2 + t_2 t_3, \quad t = (t_1, t_2, t_3) \in \Lambda^3.$$

We will find

$$f^\sigma(t), \quad f_1^{\sigma_1}(t), \quad f_2^{\sigma_2}(t), \quad f_3^{\sigma_3}(t), \quad f_{12}^{\sigma_1\sigma_2}(t), \quad f_{13}^{\sigma_1\sigma_3}(t), \quad f_{23}^{\sigma_2\sigma_3}(t)$$

and

$$g(t) = f^\sigma(t) + 2 f_{12}^{\sigma_1\sigma_2}(t) - f_{23}^{\sigma_2\sigma_3}(t), \quad t \in \Lambda^3.$$

We have

$$\mathbb{T}_1 = 3\mathbb{Z}, \quad \mathbb{T}_2 = \mathbb{N}, \quad \mathbb{T}_3 = 4^{\mathbb{N}}$$

and

$$\sigma_1(t_1) = t_1 + 3, \quad t_1 \in \mathbb{T}_1, \quad \sigma_2(t_2) = t_2 + 1, \quad t_2 \in \mathbb{T}_2, \quad \sigma_3(t_3) = 4t_3, \quad t_3 \in \mathbb{T}_3.$$

Hence,

$$f^\sigma(t) = f(\sigma_1(t_1), \sigma_2(t_2), \sigma_3(t_3))$$

$$= \sigma_1(t_1)\sigma_2(t_2) + \sigma_2(t_2)\sigma_3(t_3)$$

$$= (t_1 + 3)(t_2 + 1) + (t_2 + 1)4t_3$$

$$= t_1 t_2 + t_1 + 3t_2 + 3 + 4t_2 t_3 + 4t_3$$

$$= t_1 t_2 + 4t_2 t_3 + t_1 + 3t_2 + 4t_3 + 3,$$

$$
\begin{aligned}
f_1^{\sigma_1}(t) &= f(\sigma_1(t_1), t_2, t_3) \\
&= \sigma_1(t_1)t_2 + t_2t_3 \\
&= (t_1 + 3)t_2 + t_2t_3 \\
&= t_1t_2 + t_2t_3 + 3t_2, \\
f_2^{\sigma_2}(t) &= f(t_1, \sigma_2(t_2)) \\
&= t_1\sigma_2(t_2) + \sigma_2(t_2)t_3 \\
&= t_1(t_2 + 1) + (t_2 + 1)t_3 \\
&= t_1t_2 + t_2t_3 + t_1 + t_3, \\
f_3^{\sigma_3}(t) &= f(t_1, t_2, \sigma_3(t_3)) \\
&= t_1t_2 + t_2\sigma_3(t_3) \\
&= t_1t_2 + 4t_2t_3, \\
f_{12}^{\sigma_1\sigma_2}(t) &= f(\sigma_1(t_1), \sigma_2(t_2), t_3) \\
&= \sigma_1(t_1)\sigma_2(t_2) + \sigma_2(t_2)t_3 \\
&= (t_1 + 3)(t_2 + 1) + (t_2 + 1)t_3 \\
&= t_1t_2 + t_2t_3 + t_1 + 3t_2 + t_3 + 3, \\
f_{13}^{\sigma_1\sigma_3}(t) &= f(\sigma_1(t_1), t_2, \sigma_3(t_3)) \\
&= \sigma_1(t_1)t_2 + t_2\sigma_3(t_3) \\
&= (t_1 + 3)t_2 + 4t_2t_3
\end{aligned}
$$

$$= t_1t_2 + 4t_2t_3 + 3t_2,$$

$$f_{23}^{\sigma_2\sigma_3}(t) = f(t_1, \sigma_2(t_2), \sigma_3(t_3))$$

$$= t_1\sigma_2(t_2) + \sigma_2(t_2)\sigma_3(t_3)$$

$$= t_1(t_2 + 1) + (t_2 + 1)4t_3$$

$$= t_1t_2 + 4t_2t_3 + t_1 + 4t_3,$$

$$g(t) = f^{\sigma}(t) + 2f_{12}^{\sigma_1\sigma_2}(t) - f_{23}^{\sigma_2\sigma_3}(t)$$

$$= t_1t_2 + 4t_2t_3 + t_1 + 3t_2 + 4t_3 + 3$$

$$+2t_1t_2 + 2t_2t_3 + 2t_1 + 6t_2 + 2t_3 + 6 - t_1t_2 - 4t_2t_3 - t_1 - 4t_3$$

$$= 2t_1t_2 + 2t_2t_3 + 2t_1 + 9t_2 + 2t_3 + 9, \quad t \in \Lambda^3.$$

Example 6.30 Let $\Lambda^3 = \mathbb{N} \times \mathbb{N} \times \mathbb{N}_0^2$ and $f : \Lambda^3 \to \mathbb{R}$ be defined by

$$f(t) = t_1t_2 + t_3, \quad t = (t_1, t_2, t_3) \in \Lambda^3.$$

We will find

$$f^{\sigma}(t), \quad f_1^{\sigma_1}(t), \quad f_2^{\sigma_2}(t), \quad f_3^{\sigma_3}(t), \quad f_{12}^{\sigma_1\sigma_2}(t), \quad f_{13}^{\sigma_1\sigma_3}(t), \quad f_{23}^{\sigma_2\sigma_3}(t)$$

and

$$g(t) = f^{\sigma}(t) + f_2^{\sigma_2}(t) - 3f_{23}^{\sigma_2\sigma_3}(t), \quad t \in \Lambda^3.$$

We have

$$\sigma_1(t_1) = t_1 + 1, \quad t_1 \in \mathbb{T}_1, \quad \sigma_2(t_2) = t_2 + 1, \quad t_2 \in \mathbb{T}_2,$$

$$\sigma_3(t_3) = \left(1 + \sqrt{t_3}\right)^2, \quad t_3 \in \mathbb{T}_3.$$

Hence,

$$\begin{aligned}
f^{\sigma}(t) &= f(\sigma_1(t_1), \sigma_2(t_2), \sigma_3(t_3)) \\
&= \sigma_1(t_1)\sigma_2(t_2) + \sigma_3(t_3) \\
&= (t_1+1)(t_2+1) + \left(1+\sqrt{t_3}\right)^2 \\
&= t_1t_2 + t_1 + t_2 + 1 + t_3 + 2\sqrt{t_3} + 1 \\
&= t_1t_2 + t_1 + t_2 + t_3 + 2\sqrt{t_3} + 2, \\
f_1^{\sigma_1}(t) &= f(\sigma_1(t_1), t_2, t_3) \\
&= \sigma_1(t_1)t_2 + t_3 \\
&= (t_1+1)t_2 + t_3 \\
&= t_1t_2 + t_2 + t_3, \\
f_2^{\sigma_2}(t) &= f(t_1, \sigma_2(t_2), t_3) \\
&= t_1\sigma_2(t_2) + t_3 \\
&= t_1(t_2+1) + t_3 \\
&= t_1t_2 + t_1 + t_3, \\
f_3^{\sigma_3}(t) &= f(t_1, t_2, \sigma_3(t_3)) \\
&= t_1t_2 + \sigma_3(t_3) \\
&= t_1t_2 + \left(1+\sqrt{t_3}\right)^2 \\
&= t_1t_2 + t_3 + 2\sqrt{t_3} + 1,
\end{aligned}$$

$$\begin{aligned}
f_{12}^{\sigma_1\sigma_2}(t) &= f(\sigma_1(t_1), \sigma_2(t_2), t_3)\\
&= \sigma_1(t_1)\sigma_2(t_2) + t_3\\
&= (t_1+1)(t_2+1) + t_3\\
&= t_1t_2 + t_1 + t_2 + t_3 + 1,\\
f_{13}^{\sigma_1\sigma_3}(t) &= f(\sigma_1(t_1), t_2, \sigma_3(t_3))\\
&= \sigma_1(t_1)t_2 + \sigma_3(t_3)\\
&= (t_1+1)t_2 + \left(1+\sqrt{t_3}\right)^2\\
&= t_1t_2 + t_2 + t_3 + 2\sqrt{t_3} + 1,\\
f_{23}^{\sigma_2\sigma_3}(t) &= f(t_1, \sigma_2(t_2), \sigma_3(t_3))\\
&= t_1\sigma_2(t_2) + \sigma_3(t_3)\\
&= t_1(t_2+1) + \left(1+\sqrt{t_3}\right)^2\\
&= t_1t_2 + t_1 + t_3 + 2\sqrt{t_3} + 1,\\
g(t) &= f^{\sigma}(t) + f_2^{\sigma_2}(t) - 3f_{23}^{\sigma_2\sigma_3}(t)\\
&= t_1t_2 + t_1 + t_2 + t_3\\
&\quad +2\sqrt{t_3} + 2 + t_1t_2 + t_1 + t_3 - 3t_1t_2 - 3t_1 - 3t_3 - 6\sqrt{t_3} - 3\\
&= -t_1t_2 - t_1 + t_2 - t_3 - 4\sqrt{t_3} - 1.
\end{aligned}$$

Exercise 6.31 Let $\Lambda^3 = \mathbb{N} \times \mathbb{N} \times 3\mathbb{Z}$ and $f : \Lambda^3 \to \mathbb{R}$ be defined as

$$f(t) = t_1t_2t_3, \quad t = (t_1, t_2, t_3) \in \Lambda^3.$$

Find

1. $f^{\sigma}(t)$,
2. $f_1^{\sigma_1}(t)$,
3. $f_2^{\sigma_2}(t)$,
4. $f_3^{\sigma_3}(t)$,
5. $f_{12}^{\sigma_1\sigma_2}(t)$,
6. $f_{13}^{\sigma_1\sigma_3}(t)$,
7. $f_{23}^{\sigma_2\sigma_3}(t)$,
8. $g(t) = f^{\sigma}(t) + f_1^{\sigma_1}(t)$.

Solution 1. $t_1t_2t_3 + 3t_1t_2 + t_1t_3 + t_2t_3 + 3t_1 + 3t_2 + t_3 + 3$,
2. $t_1t_2t_3 + t_2t_3$,
3. $t_1t_2t_3 + t_1t_3$,
4. $t_1t_2t_3 + 3t_1t_2$,
5. $t_1t_2t_3 + t_1t_3 + t_2t_3 + t_3$,
6. $t_1t_2t_3 + 3t_1t_2 + t_2t_3 + 3t_2$,
7. $t_1t_2t_3 + 3t_1t_2 + t_1t_3 + 3t_1$,
8. $2t_1t_2t_3 + 3t_1t_2 + t_1t_3 + 2t_2t_3 + 3t_1 + 3t_2 + t_3 + 3$.

Definition 6.32 Let $f : \Lambda^n \to \mathbb{R}$. We introduce the following notations.

$$f^{\rho}(t) = f\left(\rho_1(t_1), \rho_2(t_2), \ldots, \rho_n(t_n)\right),$$

$$f_i^{\rho_i}(t) = f\left(t_1, \ldots, t_{i-1}, \rho_i(t_i), t_{i+1}, \ldots, t_n\right),$$

$$f_{i_1i_2\ldots i_l}^{\rho_{i_1}\rho_{i_2}\ldots\rho_{i_l}}(t) = f\left(\ldots, \rho_{i_1}(t_{i_1}), \ldots, \rho_{i_2}(t_{i_2}), \ldots, \rho_{i_l}(t_{i_l}), \ldots\right),$$

where $1 \le i_1 < i_2 < \ldots < i_l \le n$, $i_m \in \mathbb{N}$, $m \in \{1, 2, \ldots, n\}$.

Example 6.33 Let $\Lambda^3 = \mathbb{Z} \times \mathbb{Z} \times \mathbb{R}$ and consider the function $f : \Lambda^3 \to \mathbb{R}$ defined by

$$f(t) = t_1 - t_2 - 2t_3 + t_3^2, \quad t = (t_1, t_2, t_3) \in \Lambda^3.$$

Then

$$\mathbb{T}_1 = \mathbb{Z}, \quad \mathbb{T}_2 = \mathbb{Z}, \quad \mathbb{T}_3 = \mathbb{R},$$

$$\rho_1(t_1) = t_1 - 1, \quad t_1 \in \mathbb{T}_1, \quad \rho_2(t_2) = t_2 - 1, \quad t_2 \in \mathbb{T}_2,$$

$$\rho_3(t_3) = t_3, \quad t_3 \in \mathbb{T}_3.$$

We will find

$$f^{\rho}(t), \quad f_1^{\rho_1}(t), \quad f_2^{\rho_2}(t), \quad f_3^{\rho_3}(t), \quad f_{12}^{\rho_1\rho_2}(t), \quad f_{13}^{\rho_1\rho_3}(t), \quad f_{23}^{\rho_2\rho_3}(t)$$

and

$$g(t) = f^{\rho}(t) - f_1^{\rho_1}(t).$$

We have

$$\begin{aligned}
f^{\rho}(t) &= f(\rho_1(t), \rho_2(t), \rho_3(t)) \\
&= \rho_1(t_1) - \rho_2(t_2) - 2\rho_3(t_3) + (\rho_3(t_3))^2 \\
&= t_1 - 1 - (t_2 - 1) - 2t_3 + t_3^2 \\
&= t_1 - t_2 - 2t_3 + t_3^2, \\
f_1^{\rho_1}(t) &= f(\rho_1(t_1), t_2, t_3) \\
&= \rho_1(t_1) - t_2 - 2t_3 + t_3^2 \\
&= t_1 - t_2 - 2t_3 + t_3^2 - 1, \\
f_2^{\rho_2}(t) &= f(t_1, \rho_2(t_2), t_3) \\
&= t_1 - \rho_2(t_2) - 2t_3 + t_3^2 \\
&= t_1 - t_2 - 2t_3 + t_3^2 + 1, \\
f_3^{\rho_3}(t) &= f(t_1, t_2, \rho_3(t_3)) \\
&= t_1 - t_2 - 2\rho_3(t_3) + (\rho_3(t_3))^2 \\
&= t_1 - t_2 - 2t_3 + t_3^2, \\
f_{12}^{\rho_1\rho_2}(t) &= f(\rho_1(t_1), \rho_2(t_2), t_3) \\
&= \rho_1(t_1) - \rho_2(t_2) - 2t_3 + t_3^2
\end{aligned}$$

$$= t_1 - 1 - (t_2 - 1) - 2t_3 + t_3^2$$

$$= t_1 - t_2 - 2t_3 + t_3^2,$$

$$f_{13}^{\rho_1\rho_3}(t) = f(\rho_1(t_1), t_2, \rho_3(t_3))$$

$$= \rho_1(t_1) - t_2 - 2\rho_3(t_3) + (\rho_3(t_3))^2$$

$$= t_1 - 1 - t_2 - 2t_3 + t_3^2,$$

$$f_{23}^{\rho_2\rho_3}(t) = f(t_1, \rho_2(t_2), \rho_3(t_3))$$

$$= t_1 - \rho_2(t_2) - 2\rho_3(t_3) + (\rho_3(t_3))^2$$

$$= t_1 - t_2 - 2t_3 + t_3^2 + 1,$$

$$g(t) = f^{\rho}(t) - f_1^{\rho_1}(t)$$

$$= t_1 - t_2 - 2t_3 + t_3^2 - (t_1 - t_2 - 2t_3 + t_3^2 - 1)$$

$$= 1.$$

Example 6.34 Let $\Lambda^3 = 2^{\mathbb{N}} \times \mathbb{N} \times \mathbb{Z}$ and $f : \Lambda^3 \to \mathbb{R}$ be defined by

$$f(t) = t_1 + t_2 + t_3, \quad t = (t_1, t_2, t_3) \in \Lambda^3.$$

We will find

$$f^{\rho}(t), \quad f_1^{\rho_1}(t), \quad f_2^{\rho_2}(t), \quad f_3^{\rho_3}(t), \quad f_{12}^{\rho_1\rho_2}(t), \quad f_{13}^{\rho_1\rho_3}(t), \quad f_{23}^{\rho_2\rho_3}(t)$$

and

$$g(t) = f(t) - f_1^{\rho_1}(t) - f_{23}^{\rho_2\rho_3}(t), \quad t \in \Lambda^3.$$

We have

$$\mathbb{T}_1 = 2^{\mathbb{N}}, \quad \mathbb{T}_2 = \mathbb{N}, \quad \mathbb{T}_3 = \mathbb{Z}.$$

1. Let $t = (t_1, t_2, t_3) \in \Lambda^3$, $t_1 \neq 2$, $t_2 \neq 1$. Then

$$\rho_1(t_1) = \frac{t_1}{2}, \quad t_1 \in \mathbb{T}_1, \quad \rho_2(t_2) = t_2 - 1, \quad t_2 \in \mathbb{T}_2, \quad \rho_3(t_3) = t_3 - 1, \quad t_3 \in \mathbb{T}_3.$$

Hence,

$$\begin{aligned}
f^{\rho}(t) &= f(\rho_1(t_1), \rho_2(t_2), \rho_3(t_3)) \\
&= \rho_1(t_1) + \rho_2(t_2) + \rho_3(t_3) \\
&= \frac{t_1}{2} + t_2 - 1 + t_3 - 1 \\
&= \frac{t_1}{2} + t_2 + t_3 - 2, \\
f_1^{\rho_1}(t) &= f(\rho_1(t_1), t_2, t_3) \\
&= \rho_1(t_1) + t_2 + t_3 \\
&= \frac{t_1}{2} + t_2 + t_3, \\
f_2^{\rho_2}(t) &= f(t_1, \rho_2(t_2), t_3) \\
&= t_1 + \rho_2(t_2) + t_3 \\
&= t_1 + t_2 + t_3 - 1, \\
f_3^{\rho_3}(t) &= f(t_1, t_2, \rho_3(t_3)) \\
&= t_1 + t_2 + \rho_3(t_3) \\
&= t_1 + t_2 + t_3 - 1, \\
f_{12}^{\rho_1\rho_2}(t) &= f(\rho_1(t_1), \rho_2(t_2), t_3) \\
&= \rho_1(t_1) + \rho_2(t_2) + t_3
\end{aligned}$$

$$= \frac{t_1}{2} + t_2 + t_3 - 1,$$

$$f_{13}^{\rho_1\rho_3}(t) = f(\rho_1(t_1), t_2, \rho_3(t_3))$$

$$= \rho_1(t_1) + t_2 + \rho_3(t_3)$$

$$= \frac{t_1}{2} + t_2 + t_3 - 1,$$

$$f_{23}^{\rho_2\rho_3}(t) = f(t_1, \rho_2(t_2), \rho_3(t_3))$$

$$= t_1 + \rho_2(t_2) + \rho_3(t_3)$$

$$= t_1 + t_2 + t_3 - 2,$$

$$g(t) = f(t) - f_1^{\rho_1}(t) - f_{23}^{\rho_2\rho_3}(t)$$

$$= t_1 + t_2 + t_3 - \left(\frac{t_1}{2} + t_2 + t_3\right) - (t_1 + t_2 + t_3 - 2)$$

$$= -\frac{t_1}{2} - t_2 - t_3 + 2.$$

2. Let $t = (t_1, t_2, t_3) \in \Lambda^3$, $t_1 = 2$, $t_2 \neq 1$. Then

$$\rho_1(t_1) = 2, \quad \rho_2(t_2) = t_2 - 1, \quad t_2 \in \mathbb{T}_2, \quad \rho_3(t_3) = t_3 - 1, \quad t_3 \in \mathbb{T}_3.$$

Hence,

$$f^\rho(t) = f(\rho_1(t_1), \rho_2(t_2), \rho_3(t_3))$$

$$= \rho_1(t_1) + \rho_2(t_2) + \rho_3(t_3)$$

$$= 2 + t_2 - 1 + t_3 - 1$$

$$= t_2 + t_3,$$

$$f_1^{\rho_1}(t) = f(\rho_1(t_1), t_2, t_3)$$

$$
\begin{aligned}
&= \rho_1(t_1) + t_2 + t_3 \\
&= t_2 + t_3 + 2, \\
f_2^{\rho_2}(t) &= f(t_1, \rho_2(t_2), t_3) \\
&= t_1 + \rho_2(t_2) + t_3 \\
&= 2 + t_2 - 1 + t_3 \\
&= t_2 + t_3 + 1, \\
f_3^{\rho_3}(t) &= f(t_1, t_2, \rho_3(t_3)) \\
&= t_1 + t_2 + \rho_3(t_3) \\
&= 2 + t_2 + t_3 - 1 \\
&= t_2 + t_3 + 1, \\
f_{12}^{\rho_1\rho_2}(t) &= f(\rho_1(t_1), \rho_2(t_2), t_3) \\
&= \rho_1(t_1) + \rho_2(t_2) + t_3 \\
&= 2 + t_2 + t_3 - 1 \\
&= t_2 + t_3 + 1, \\
f_{13}^{\rho_1\rho_3}(t) &= f(\rho_1(t_1), t_2, \rho_3(t_3)) \\
&= \rho_1(t_1) + t_2 + \rho_3(t_3) \\
&= 2 + t_2 - 1 + t_3 \\
&= t_2 + t_3 + 1,
\end{aligned}
$$

$$
\begin{aligned}
f_{23}^{\rho_2\rho_3} &= f(t_1, \rho_2(t_2), \rho_3(t_3)) \\
&= t_1 + \rho_2(t_2) + \rho_3(t_3) \\
&= 2 + t_2 - 1 + t_3 - 1 \\
&= t_2 + t_3, \\
g(t) &= f(t) - f_1^{\rho_1}(t) - f_{23}^{\rho_2\rho_3}(t) \\
&= 2 + t_2 + t_3 - (t_2 + t_3 + 2) - (t_2 + t_3) \\
&= -t_2 - t_3.
\end{aligned}
$$

3. Let $t = (t_1, t_2, t_3) \in \Lambda^3$, $t_1 \neq 2$, $t_2 = 1$. Then

$$
\rho_1(t_1) = \frac{t_1}{2}, \quad t_1 \in \mathbb{T}_1, \quad \rho_2(t_2) = 1, \quad \rho_3(t_3) = t_3 - 1, \quad t_3 \in \mathbb{T}_3.
$$

Hence,

$$
\begin{aligned}
f^{\rho}(t) &= f(\rho_1(t_1), \rho_2(t_2), \rho_3(t_3)) \\
&= \rho_1(t_1) + \rho_2(t_2) + \rho_3(t_3) \\
&= \frac{t_1}{2} + 1 + t_3 - 1 \\
&= \frac{t_1}{2} + t_3, \\
f_1^{\rho_1}(t) &= f(\rho_1(t_1), t_2, t_3) \\
&= \rho_1(t_1) + t_2 + t_3 \\
&= \frac{t_1}{2} + t_3 + 1, \\
f_2^{\rho_2}(t) &= f(t_1, \rho_2(t_2), t_2) \\
&= t_1 + \rho_2(t_2) + t_3
\end{aligned}
$$

$$
\begin{aligned}
&= t_1 + t_3 + 1,\\
f_3^{\rho_3}(t) &= f(t_1, t_2, \rho_3(t_3))\\
&= t_1 + t_2 + \rho_3(t_3)\\
&= t_1 + 1 + t_3 - 1\\
&= t_1 + t_3,\\
f_{12}^{\rho_1\rho_2}(t) &= f(\rho_1(t_1), \rho_2(t_2), t_3)\\
&= \rho_1(t_1) + \rho_2(t_2) + t_3\\
&= \frac{t_1}{2} + t_3 + 1,\\
f_{13}^{\rho_1\rho_3}(t) &= f(\rho_1(t_1), t_2, \rho_3(t_3))\\
&= \rho_1(t_1) + t_2 + \rho_3(t_3)\\
&= \frac{t_1}{2} + 1 + t_3 - 1\\
&= \frac{t_1}{2} + t_3,\\
f_{23}^{\rho_2\rho_3}(t) &= f(t_1, \rho_2(t_2), \rho_3(t_3))\\
&= t_1 + \rho_2(t_2) + \rho_3(t_3)\\
&= t_1 + 1 + t_3 - 1\\
&= t_1 + t_3,
\end{aligned}
$$

$$
g(t) = f(t) - f_1^{\rho_1}(t) - f_{23}^{\rho_2\rho_3}(t)
$$

$$= t_1 + t_2 + t_3 - \left(\frac{t_1}{2} + t_3 + 1\right) - (t_1 + t_3)$$

$$= -\frac{t_1}{2} - t_3.$$

4. Let $t = (t_1, t_2, t_3) \in \Lambda^3$, $t_1 = 2$, $t_2 = 1$. Then

$$\rho_1(t_1) = 2, \quad \rho_2(t_2) = 1, \quad \rho_3(t_3) = t_3 - 1, \quad t_3 \in \mathbb{T}_3.$$

Hence,

$$\begin{aligned}
f^{\rho}(t) &= f(\rho_1(t_1), \rho_2(t_2), \rho_3(t_3)) \\
&= \rho_1(t_1) + \rho_2(t_2) + \rho_3(t_3) \\
&= 2 + 1 + t_3 - 1 \\
&= t_3 + 2, \\
f_1^{\rho_1}(t) &= f(\rho_1(t_1), t_2, t_3) \\
&= \rho_1(t_1) + t_2 + t_3 \\
&= 2 + 1 + t_3 \\
&= t_3 + 3, \\
f_2^{\rho_2}(t) &= f(t_1, \rho_2(t_2), t_3) \\
&= t_1 + \rho_2(t_2) + t_3 \\
&= 2 + 1 + t_3 \\
&= t_3 + 3, \\
f_3^{\rho_3}(t) &= f(t_1, t_2, \rho_3(t_3)) \\
&= t_1 + t_2 + \rho_3(t_3)
\end{aligned}$$

$$
\begin{aligned}
&= 2 + 1 + t_3 - 1 \\
&= t_3 + 2, \\
f_{12}^{\rho_1\rho_2}(t) &= f(\rho_1(t_1), \rho_2(t_2), t_3) \\
&= \rho_1(t_1) + \rho_2(t_2) + t_3 \\
&= 2 + 1 + t_3 \\
&= t_3 + 3, \\
f_{13}^{\rho_1\rho_3}(t) &= f(\rho_1(t_1), t_2, \rho_3(t_3)) \\
&= \rho_1(t_1) + t_2 + \rho_3(t_3) \\
&= 2 + 1 + t_3 - 1 \\
&= t_3 + 2, \\
f_{23}^{\rho_2\rho_3}(t) &= f(t_1, \rho_2(t_2), \rho_3(t_3)) \\
&= t_1 + \rho_2(t_2) + \rho_3(t_3) \\
&= 2 + 1 + t_3 - 1 \\
&= t_3 + 2, \\
g(t) &= f(t) - f_1^{\rho_1}(t) - f_{23}^{\rho_2\rho_3}(t) \\
&= t_1 + t_2 + t_3 - (t_3 + 3) - (t_3 + 2) \\
&= 2 + 1 + t_3 - t_3 - 3 - t_3 - 2 \\
&= -t_3 - 2.
\end{aligned}
$$

Example 6.35 Let $\Lambda^3 = \mathbb{N}_0 \times 2\mathbb{N} \times \mathbb{N}$ and $f : \Lambda^3 \to \mathbb{R}$ be defined by

$$f(t) = t_1 + t_2 + t_3^2, \quad t = (t_1, t_2, t_3) \in \Lambda^3.$$

We will find

$$f^{\rho}(t), \quad f_1^{\rho_1}(t), \quad f_2^{\rho_2}(t), \quad f_3^{\rho_3}(t), \quad f_{12}^{\rho_1\rho_2}(t), \quad f_{13}^{\rho_1\rho_3}(t), \quad f_{23}^{\rho_2\rho_3}(t)$$

and

$$g(t) = f_1^{\rho_1}(t) + f_2^{\rho_2}(t) + f_3^{\rho_3}(t), \quad t \in \Lambda^3.$$

Here,

$$\mathbb{T}_1 = \mathbb{N}_0, \quad \mathbb{T}_2 = 2\mathbb{N}, \quad \mathbb{T}_3 = \mathbb{N}.$$

1. Let $(t_1, t_2, t_3) \in \Lambda^3$, $t_1 \neq 0$, $t_2 \neq 2$, $t_3 \neq 1$. Then

$$\rho_1(t_1) = t_1 - 1, \quad t_1 \in \mathbb{T}_1, \quad \rho_2(t_2) = t_2 - 2, \quad t_2 \in \mathbb{T}_2,$$

$$\rho_3(t_3) = t_3 - 1, \quad t_3 \in \mathbb{T}_3.$$

Hence,

$$\begin{aligned}
f^{\rho}(t) &= f(\rho_1(t_1), \rho_2(t_2), \rho_3(t_3)) \\
&= \rho_1(t_1) + \rho_2(t_2) + (\rho_3(t_3))^2 \\
&= t_1 - 1 + t_2 - 2 + (t_3 - 1)^2 \\
&= t_1 + t_2 - 3 + t_3^2 - 2t_3 + 1 \\
&= t_1 + t_2 - 2t_3 + t_3^2 - 2, \\
f_1^{\rho_1}(t) &= f(\rho_1(t_1), t_2, t_3) \\
&= \rho_1(t_1) + t_2 + t_3^2 \\
&= t_1 - 1 + t_2 + t_3^2, \\
f_2^{\rho_2}(t) &= f(t_1, \rho_2(t_2), t_3)
\end{aligned}$$

$$
\begin{aligned}
&= t_1 + \rho_2(t_2) + t_3^2 \\
&= t_1 + t_2 + t_3^2 - 2, \\
f_3^{\rho_3}(t) &= f(t_1, t_2, \rho_3(t_3)) \\
&= t_1 + t_2 + (\rho_3(t_3))^2 \\
&= t_1 + t_2 + (t_3 - 1)^2 \\
&= t_1 + t_2 - 2t_3 + t_3^2 + 1, \\
f_{12}^{\rho_1\rho_2}(t) &= f(\rho_1(t_1), \rho_2(t_2), t_3) \\
&= \rho_1(t_1) + \rho_2(t_2) + t_3^2 \\
&= t_1 - 1 + t_2 - 2 + t_3^2 \\
&= t_1 + t_2 + t_3^2 - 3, \\
f_{13}^{\rho_1\rho_3}(t) &= f(\rho_1(t_1), t_2, \rho_3(t_3)) \\
&= \rho_1(t_1) + t_2 + (\rho_3(t_3))^2 \\
&= t_1 - 1 + t_2 + (t_3 - 1)^2 \\
&= t_1 + t_2 - 2t_3 + t_3^2, \\
f_{23}^{\rho_2\rho_3}(t) &= f(t_1, \rho_2(t_2), \rho_3(t_3)) \\
&= t_1 + \rho_2(t_2) + (\rho_3(t_3))^2 \\
&= t_1 + t_2 - 2 + (t_3 - 1)^2 \\
&= t_1 + t_2 - 2t_3 + t_3^2 - 1,
\end{aligned}
$$

$$
\begin{aligned}
g(t) &= f_1^{\rho_1}(t) + f_2^{\rho_2}(t) + f_3^{\rho_3}(t) \\
&= t_1 + t_2 + t_3^2 - 1 + t_1 + t_2 + t_3^2 - 2 + t_1 + t_2 - 2t_3 + t_3^2 + 1 \\
&= 3t_1 + 3t_2 - 2t_3 + 3t_3^2 - 2.
\end{aligned}
$$

2. Let $t = (t_1, t_2, t_3) \in \Lambda^3$, $t_1 = 0$, $t_2 \neq 2$, $t_3 \neq 1$. Then

$$
\rho_1(t_1) = 0, \quad \rho_2(t_2) = t_2 - 2, \quad t_2 \in \mathbb{T}_2, \quad \rho_3(t_3) = t_3 - 1, \quad t_3 \in \mathbb{T}_3
$$

and

$$
\begin{aligned}
f^{\rho}(t) &= f(\rho_1(t_1), \rho_2(t_2), \rho_3(t_3)) \\
&= \rho_1(t_1) + \rho_2(t_2) + (\rho_3(t_3))^2 \\
&= 0 + t_2 - 2 + (t_3 - 1)^2 \\
&= t_2 - 2 + t_3^2 - 2t_3 + 1 \\
&= t_2 - 2t_3 + t_3^2 - 1, \\
f_1^{\rho_1}(t) &= f(\rho_1(t_1), t_2, t_3) \\
&= \rho_1(t_1) + t_2 + t_3^2 \\
&= t_2 + t_3^2, \\
f_2^{\rho_2}(t) &= f(t_1, \rho_2(t_2), t_3) \\
&= t_1 + \rho_2(t_2) + t_3^2 \\
&= t_2 + t_3^2 - 2, \\
f_3^{\rho_3}(t) &= f(t_1, t_2, \rho_3(t_3)) \\
&= t_1 + t_2 + (\rho_3(t_3))^2
\end{aligned}
$$

$$
\begin{aligned}
&= t_2 + (t_3 - 1)^2 \\
&= t_2 - 2t_3 + t_3^2 + 1, \\
f_{12}^{\rho_1\rho_2}(t) &= f(\rho_1(t_1), \rho_2(t_2), t_3) \\
&= \rho_1(t_1) + \rho_2(t_2) + t_3^2 \\
&= t_2 + t_3^2 - 2, \\
f_{13}^{\rho_1\rho_3}(t) &= f(\rho_1(t_1), t_2, \rho_3(t_3)) \\
&= \rho_1(t_1) + t_2 + (\rho_3(t_3))^2 \\
&= t_2 + (t_3 - 1)^2 \\
&= t_2 - 2t_3 + t_3^2 + 1, \\
f_{23}^{\rho_2\rho_3}(t) &= f(t_1, \rho_2(t_2), \rho_3(t_3)) \\
&= \rho_2(t_2) + (\rho_3(t_3))^2 \\
&= t_2 - 2 + (t_3 - 1)^2 \\
&= t_2 + t_3^2 - 2t_3 - 1, \\
g(t) &= f_1^{\rho_1}(t) + f_2^{\rho_2}(t) + f_3^{\rho_3}(t) \\
&= t_2 + t_3^2 + t_2 + t_3^2 - 2 + t_2 - 2t_3 + t_3^2 + 1 \\
&= 3t_2 - 2t_3 + 3t_3^2 - 1.
\end{aligned}
$$

3. Let $t = (t_1, t_2, t_3) \in \Lambda^3$, $t_1 \neq 0$, $t_2 = 2$, $t_3 \neq 1$. Then

$$
\rho_1(t_1) = t_1 - 1, \quad t_1 \in \mathbb{T}_1, \quad \rho_2(t_2) = 2, \quad t_2 \in \mathbb{T}_2, \quad \rho_3(t_3) = t_3 - 1, \quad t_3 \in \mathbb{T}_3
$$

and

$$\begin{aligned}
f^{\rho}(t) &= f(\rho_1(t_1), \rho_2(t_2), \rho_3(t_3)) \\
&= \rho_1(t_1) + \rho_2(t_2) + (\rho_3(t_3))^2 \\
&= t_1 - 1 + 2 + (t_3 - 1)^2 \\
&= t_1 + 1 + t_3^2 - 2t_3 + 1 \\
&= t_1 - 2t_3 + t_3^2 + 2, \\
f_1^{\rho_1}(t) &= f(\rho_1(t_1), t_2, t_3) \\
&= \rho_1(t_1) + t_2 + t_3^2 \\
&= t_1 - 1 + 2 + t_3^2 \\
&= t_1 + t_3^2 + 1, \\
f_2^{\rho_2}(t) &= f(t_1, \rho_2(t_2), t_3) \\
&= t_1 + \rho_2(t_2) + t_3^2 \\
&= t_1 + t_3^2 + 2, \\
f_3^{\rho_3}(t) &= f(t_1, t_2, \rho_3(t_3)) \\
&= t_1 + t_2 + (\rho_3(t_3))^2 \\
&= t_1 + 2 + (t_3 - 1)^2 \\
&= t_1 + 2 + t_3^2 - 2t_3 + 1 \\
&= t_1 - 2t_3 + t_3^2 + 3,
\end{aligned}$$

$$\begin{aligned}
f_{12}^{\rho_1\rho_2}(t) &= f(\rho_1(t_1), \rho_2(t_2), t_3)\\
&= \rho_1(t_1) + \rho_2(t_2) + t_3^2\\
&= t_1 - 1 + 2 + t_3^2\\
&= t_1 + t_3^2 + 1,\\
f_{13}^{\rho_1\rho_3}(t) &= f(\rho_1(t_1), t_2, \rho_3(t_3))\\
&= \rho_1(t_1) + t_2 + (\rho_3(t_3))^2\\
&= t_1 - 1 + 2 + (t_3 - 1)^2\\
&= t_1 + 1 + t_3^2 - 2t_3 + 1\\
&= t_1 - 2t_3 + t_3^2 + 2,\\
f_{23}^{\rho_2\rho_3}(t) &= f(t_1, \rho_2(t_2), \rho_3(t_3))\\
&= t_1 + \rho_2(t_2) + (\rho_3(t_3))^2\\
&= t_1 + 2 + (t_3 - 1)^2\\
&= t_1 + 2 + t_3^2 - 2t_3 + 1\\
&= t_1 - 2t_3 + t_3^2 + 3,\\
g(t) &= f_1^{\rho_1}(t) + f_2^{\rho_2}(t) + f_3^{\rho_3}(t)\\
&= t_1 + t_3^2 + 1 + t_1 + t_3^2 + 2 + t_1 - 2t_3 + t_3^2 + 3\\
&= 3t_1 - 2t_3 + 3t_3^2 + 6.
\end{aligned}$$

4. Let $t = (t_1, t_2, t_3) \in \Lambda^3$, $t_1 \neq 0$, $t_2 \neq 2$, $t_3 = 1$. Then

$$\rho_1(t_1) = t_1 - 1, \quad t_1 \in \mathbb{T}_1, \quad \rho_2(t_2) = t_2 - 2, \quad t_2 \in \mathbb{T}_2, \quad \rho_3(t_3) = 1.$$

Hence,

$$\begin{aligned}
f^{\rho}(t) &= f(\rho_1(t_1), \rho_2(t_2), \rho_3(t_3)) \\
&= \rho_1(t_1) + \rho_2(t_2) + (\rho_3(t_3))^2 \\
&= t_1 - 1 + t_2 - 2 + 1 \\
&= t_1 + t_2 - 2, \\
f_1^{\rho_1}(t) &= f(\rho_1(t_1), t_2, t_3) \\
&= \rho_1(t_1) + t_2 + t_3^2 \\
&= t_1 - 1 + t_2 + 1 \\
&= t_1 + t_2, \\
f_2^{\rho_2}(t) &= f(t_1, \rho_2(t_2), t_3) \\
&= t_1 + \rho_2(t_2) + t_3^2 \\
&= t_1 + t_2 - 2 + 1 \\
&= t_1 + t_2 - 1, \\
f_3^{\rho_3}(t) &= f(t_1, t_2, \rho_3(t_3)) \\
&= t_1 + t_2 + (\rho_3(t_3))^2 \\
&= t_1 + t_2 + 1, \\
f_{12}^{\rho_1\rho_2}(t) &= f(\rho_1(t_1), \rho_2(t_2), t_3) \\
&= \rho_1(t_1) + \rho_2(t_2) + t_3^2
\end{aligned}$$

$$= t_1 - 1 + t_2 - 2 + 1$$

$$= t_1 + t_2 - 2,$$

$$f_{13}^{\rho_1\rho_3}(t) = f(\rho_1(t_1), t_2, \rho_3(t_3))$$

$$= \rho_1(t_1) + t_2 + (\rho_3(t_3))^2$$

$$= t_1 - 1 + t_2 + 1$$

$$= t_1 + t_2,$$

$$f_{23}^{\rho_2\rho_3}(t) = f(t_1, \rho_2(t_2), \rho_3(t_3))$$

$$= t_1 + \rho_2(t_2) + (\rho_3(t_3))^2$$

$$= t_1 + t_2 - 2 + 1$$

$$= t_1 + t_2 - 1,$$

$$g(t) = f_1^{\rho_1}(t) + f_2^{\rho_2}(t) + f_3^{\rho_3}(t)$$

$$= t_1 + t_2 + t_1 + t_2 - 1 + t_1 + t_2 + 1$$

$$= 3t_1 + 3t_2.$$

5. Let $t = (t_1, t_2, t_3) \in \Lambda^3$, $t_1 = 0$, $t_2 = 2$, $t_3 \neq 1$. Then

$$\rho_1(t_1) = 0, \quad \rho_2(t_2) = 2, \quad \rho_3(t_3) = t_3 - 1, \quad t_3 \in \mathbb{T}_3.$$

Hence,

$$f^\rho(t) = f(\rho_1(t_1), \rho_2(t_2), \rho_3(t_3))$$

$$= \rho_1(t_1) + \rho_2(t_2) + (\rho_3(t_3))^2$$

$$= 0 + 2 + (t_3 - 1)^2$$

$$
\begin{aligned}
&= -2t_3 + t_3^2 + 3, \\
f_1^{\rho_1}(t) &= f(\rho_1(t_1), t_2, t_3) \\
&= \rho_1(t_1) + t_2 + t_3^2 \\
&= 2 + t_3^2, \\
f_2^{\rho_2}(t) &= f(t_1, \rho_2(t_2), t_3) \\
&= t_1 + \rho_2(t_2) + t_3^2 \\
&= 2 + t_3^2, \\
f_3^{\rho_3}(t) &= f(t_1, t_2, \rho_3(t_3)) \\
&= t_1 + t_2 + (\rho_3(t_3))^2 \\
&= 2 + (t_3 - 1)^2 \\
&= -2t_3 + t_3^2 + 3, \\
f_{12}^{\rho_1\rho_2}(t) &= f(\rho_1(t_1), \rho_2(t_2), t_3) \\
&= \rho_1(t_1) + \rho_2(t_2) + t_3^2 \\
&= 2 + t_3^2, \\
f_{13}^{\rho_1\rho_3}(t) &= f(\rho_1(t_1), t_2, \rho_3(t_3)) \\
&= \rho_1(t_1) + t_2 + (\rho_3(t_3))^2 \\
&= 2 + (t_3 - 1)^2 \\
&= -2t_3 + t_3^2 + 3,
\end{aligned}
$$

$$\begin{aligned} f_{23}^{\rho_2\rho_3}(t) &= f(t_1, \rho_2(t_2), \rho_3(t_3)) \\ &= t_1 + \rho_2(t_2) + (\rho_3(t_3))^2 \\ &= 2 + (t_3 - 1)^2 \\ &= -2t_3 + t_3^2 + 3, \\ g(t) &= f_1^{\rho_1}(t) + f_2^{\rho_2}(t) + f_3^{\rho_3}(t) \\ &= 2 + t_3^2 + 2 + t_3^2 + t_3^2 - 2t_3 + 3 \\ &= 3t_3^2 - 2t_3 + 7. \end{aligned}$$

6. Let $t = (t_1, t_2, t_3) \in \Lambda^3$, $t_1 = 0$, $t_2 \neq 2$, $t_3 = 1$. Then

$$\rho_1(t_1) = 0, \quad \rho_2(t_2) = t_2 - 2, \quad t_2 \in \mathbb{T}_2, \quad \rho_3(t_3) = 1.$$

Hence,

$$\begin{aligned} f^{\rho}(t) &= f(\rho_1(t_1), \rho_2(t_2), \rho_3(t_3)) \\ &= \rho_1(t_1) + \rho_2(t_2) + (\rho_3(t_3))^2 \\ &= t_2 - 2 + 1 \\ &= t_2 - 1, \\ f_1^{\rho_1}(t) &= f(\rho_1(t_1), t_2, t_3) \\ &= \rho_1(t_1) + t_2 + t_3^2 \\ &= t_2 + 1, \\ f_2^{\rho_2}(t) &= f(t_1, \rho_2(t_2), t_3) \end{aligned}$$

$$= t_1 + \rho_2(t_2) + t_3^2$$

$$= t_2 - 2 + 1$$

$$= t_2 - 1,$$

$$f_3^{\rho_3}(t) = f(t_1, t_2, \rho_3(t_3))$$

$$= t_1 + t_2 + (\rho_3(t_3))^2$$

$$= t_2 + 1,$$

$$f_{12}^{\rho_1\rho_2}(t) = f(\rho_1(t_1), \rho_2(t_2), t_3)$$

$$= \rho_1(t_1) + \rho_2(t_2) + t_3^2$$

$$= t_2 - 1,$$

$$f_{13}^{\rho_1\rho_3}(t) = f(\rho_1(t_1), t_2, \rho_3(t_3))$$

$$= \rho_1(t_1) + t_2 + (\rho_3(t_3))^2$$

$$= t_2 + 1,$$

$$f_{23}^{\rho_2\rho_3}(t) = f(t_1, \rho_2(t_2), \rho_3(t_3))$$

$$= t_1 + \rho_2(t_2) + (\rho_3(t_3))^2$$

$$= t_2 - 2 + 1$$

$$= t_2 - 1,$$

$$g(t) = f_1^{\rho_1}(t) + f_2^{\rho_2}(t) + f_3^{\rho_3}(t)$$

$$= t_2 + 1 + t_2 - 1 + t_2 + 1$$

$$= 3t_2 + 1.$$

7. Let $t = (t_1, t_2, t_3) \in \Lambda^3$, $t_1 \neq 0$, $t_2 = 2$, $t_3 = 1$. Then

$$\rho_1(t_1) = t_1 - 1, \quad t_1 \in \mathbb{T}_1, \quad \rho_2(t_2) = 2, \quad \rho_3(t_3) = 1.$$

Hence,

$$\begin{aligned}
f^{\rho}(t) &= f(\rho_1(t_1), \rho_2(t_2), \rho_3(t_3)) \\
&= \rho_1(t_1) + \rho_2(t_2) + (\rho_3(t_3))^2 \\
&= t_1 - 1 + 2 + 1 \\
&= t_1 + 2, \\
f_1^{\rho_1}(t) &= f(\rho_1(t_1), t_2, t_3) \\
&= \rho_1(t_1) + t_2 + t_3^2 \\
&= t_1 - 1 + 2 + 1 \\
&= t_1 + 2, \\
f_2^{\rho_2}(t) &= f(t_1, \rho_2(t_2), t_3) \\
&= t_1 + \rho_2(t_2) + t_3^2 \\
&= t_1 + 2 + 1 \\
&= t_1 + 3, \\
f_3^{\rho_3}(t) &= f(t_1, t_2, \rho_3(t_3)) \\
&= t_1 + t_2 + (\rho_3(t_3))^2 \\
&= t_1 + 3,
\end{aligned}$$

$$\begin{aligned} f_{12}^{\rho_1\rho_2}(t) &= f(\rho_1(t_1), \rho_2(t_2), t_3) \\ &= \rho_1(t_1) + \rho_2(t_2) + t_3^2 \\ &= t_1 - 1 + 2 + 1 \\ &= t_1 + 2, \\ f_{13}^{\rho_1\rho_3}(t) &= f(\rho_1(t_1), t_2, \rho_3(t_3)) \\ &= \rho_1(t_1) + t_2 + (\rho_3(t_3))^2 \\ &= t_1 - 1 + 2 + 1 \\ &= t_1 + 2, \\ f_{23}^{\rho_2\rho_3}(t) &= f(t_1, \rho_2(t_2), \rho_3(t_3)) \\ &= t_1 + \rho_2(t_2) + (\rho_3(t_3))^2 \\ &= t_1 + 2 + 1 \\ &= t_1 + 3, \\ g(t) &= f_1^{\rho_1}(t) + f_2^{\rho_2}(t) + f_3^{\rho_3}(t) \\ &= t_1 + 2 + t_1 + 3 + t_1 + 3 \\ &= 3t_1 + 8. \end{aligned}$$

8. Let $t = (0, 2, 1)$. Then

$$\rho_1(t_1) = 0, \quad \rho_2(t_2) = 2, \quad \rho_3(t_3) = 1$$

and

$$\begin{aligned}
f^{\rho}(t) &= f_1^{\rho_1}(t) = f_2^{\rho_2}(t) = f_3^{\rho_3}(t_3) \\
&= f_{12}^{\rho_1\rho_2}(t) = f_{13}^{\rho_1\rho_3}(t) = f_{23}^{\rho_2\rho_3}(t) \\
&= f(t) \\
&= 0 + 2 + 1 \\
&= 3, \\
g(t) &= f_1^{\rho_1}(t) + f_2^{\rho_2}(t) + f_3^{\rho_3}(t) \\
&= 9.
\end{aligned}$$

Exercise 6.36 Let $\Lambda^3 = \mathbb{Z} \times \mathbb{N}_0 \times \mathbb{R}$ and $f : \Lambda^3 \to \mathbb{R}$ be defined by

$$f(t) = t_1^2 + t_2^2 + t_3^2, \quad t = (t_1, t_2, t_3) \in \Lambda^3.$$

Find

1. $f^{\rho}(t)$,
2. $f_1^{\rho_1}(t)$,
3. $f_2^{\rho_2}(t)$,
4. $f_3^{\rho_3}(t)$,
5. $f_{12}^{\rho_1\rho_2}(t)$,
6. $f_{13}^{\rho_1\rho_3}(t)$,
7. $f_{23}^{\rho_2\rho_3}(t)$,
8. $g(t) = f_1^{\rho_1}(t) + f_2^{\rho_2}(t) + f_3^{\rho_3}(t), t \in \Lambda^3$.

Solution First, consider $t = (t_1, t_2, t_3) \in \Lambda^3$, $t_2 \neq 0$.

1. $t_1^2 + t_2^2 + t_3^2 - 2t_1 - 2t_2 + 2$,
2. $t_1^2 + t_2^2 + t_3^2 - 2t_1 + 1$,
3. $t_1^2 + t_2^2 + t_3^2 - 2t_2 + 1$,
4. $t_1^2 + t_2^2 + t_3^2$,
5. $t_1^2 + t_2^2 + t_3^2 - 2t_1 - 2t_2 + 2$,
6. $t_1^2 + t_2^2 + t_3^2 - 2t_1 + 1$,
7. $t_1^2 + t_2^2 + t_3^2 - 2t_2 + 1$,
8. $3(t_1^2 + t_2^2 + t_3^2) - 2t_1 - 2t_2 + 2$.

Next, consider $t = (t_1, 0, t_3) \in \Lambda^3$.

1. $t_1^2 + t_3^2 - 2t_1 + 1$,
2. $t_1^2 + t_3^2 - 2t_1 + 1$,

3. $t_1^2 + t_3^2$,
4. $t_1^2 + t_3^2$,
5. $t_1^2 + t_3^2 - 2t_1 + 1$,
6. $t_1^2 + t_3^2 - 2t_1 + 1$,
7. $t_1^2 + t_3^2$,
8. $3(t_1^2 + t_3^2) - 2t_1 + 1$.

Example 6.37 Let $\Lambda^2 = \mathbb{N}_0 \times \mathbb{N}$ and $f : \Lambda^2 \to \mathbb{R}$ be defined by

$$f(t) = t_1 + t_2 + t_1^2 + t_2^2, \quad t = (t_1, t_2) \in \Lambda^2.$$

We will find

$$g(t) = f_1^{\rho_1}(t) + f_2^{\sigma_2}(t), \quad t \in \Lambda^2.$$

1. Let $t = (t_1, t_2) \in \Lambda^2$, $t_1 \neq 0$. Then

$$\rho_1(t_1) = t_1 - 1, \quad t_1 \in \mathbb{N}_0, \quad \sigma_2(t_2) = t_2 + 1, \quad t_2 \in \mathbb{N}.$$

Hence,

$$\begin{aligned} g(t) &= f_1^{\rho_1}(t) + f_2^{\sigma_2}(t) \\ &= f(\rho_1(t_1), t_2) + f(t_1, \sigma_2(t_2)) \\ &= \rho_1(t_1) + t_2 + (\rho_1(t_1))^2 + t_2^2 + t_1 + \sigma_2(t_2) + t_1^2 + (\sigma_2(t_2))^2 \\ &= t_1 - 1 + t_2 + (t_1 - 1)^2 + t_2^2 + t_1 + t_2 + 1 + t_1^2 + (t_2 + 1)^2 \\ &= t_1 - 1 + t_2 + t_1^2 - 2t_1 + 1 + t_2^2 + t_1 + t_2 + 1 + t_1^2 + t_2^2 + 2t_2 + 1 \\ &= 2t_1^2 + 2t_2^2 + 4t_2 + 2. \end{aligned}$$

2. Let $t = (0, t_2) \in \Lambda^2$. Then $\rho_1(t_1) = 0$ and

$$\begin{aligned} g(t) &= \rho_1(t_1) + t_2 + (\rho_1(t_1))^2 + t_2^2 + t_1 + \sigma_2(t_2) + t_1^2 + (\sigma_2(t_2))^2 \\ &= t_2 + t_2^2 + t_2 + 1 + (t_2 + 1)^2 \\ &= t_2^2 + 2t_2 + 1 + t_2^2 + 2t_2 + 1 \\ &= 2t_2^2 + 4t_2 + 2. \end{aligned}$$

6.2 Partial Derivatives and Differentiability

Definition 6.38 We set

$$\Lambda^{\kappa n} = \mathbb{T}_1^{\kappa} \times \mathbb{T}_2^{\kappa} \times \ldots \times \mathbb{T}_n^{\kappa},$$

$$\Lambda_i^{\kappa_i n} = \mathbb{T}_1 \times \ldots \times \mathbb{T}_{i-1} \times \mathbb{T}_i^{\kappa} \times \mathbb{T}_{i+1} \times \ldots \times \mathbb{T}_n, \quad i = 1, 2, \ldots, n,$$

$$\Lambda_{i_1 i_2 \ldots i_l}^{\kappa_{i_1} \kappa_{i_2} \ldots \kappa_{i_l} n} = \ldots \times \mathbb{T}_{i_1}^{\kappa} \times \ldots \times \mathbb{T}_{i_2}^{\kappa} \times \ldots \times \mathbb{T}_{i_l}^{\kappa} \times \ldots,$$

where $1 \le i_1 < i_2 < \ldots < i_l \le n$, $i_m \in \mathbb{N}$, $m = 1, 2, \ldots, l$.

Remark 6.39 If $(i_1, i_2, \ldots, i_l) = (1, 2, \ldots, n)$, then

$$\Lambda_{i_1 i_2 \ldots i_l}^{\kappa_1 \kappa_2 \ldots \kappa_l n} = \Lambda^{\kappa n}.$$

Definition 6.40 Assume that $f : \Lambda^n \to \mathbb{R}$ is a function and let $t \in \Lambda_i^{\kappa_i n}$. We define

$$\frac{\partial f(t_1, t_2, \ldots, t_n)}{\Delta_i t_i} = \frac{\partial f(t)}{\Delta_i t_i} = \frac{\partial f}{\Delta_i t_i}(t) = f_{t_i}^{\Delta_i}(t)$$

to be the number, provided it exists, with the property that for any $\varepsilon_i > 0$, there exists a neighbourhood

$$U_i = (t_i - \delta_i, t_i + \delta_i) \cap \mathbb{T}_i,$$

for some $\delta_i > 0$, such that

$$\Big| f(t_1, \ldots, t_{i-1}, \sigma_i(t_i), t_{i+1}, \ldots, t_n) - f(t_1, \ldots, t_{i-1}, s_i, t_{i+1}, \ldots, t_n)$$
$$- f_{t_i}^{\Delta_i}(t)(\sigma_i(t_i) - s_i)\Big| \le \varepsilon_i |\sigma_i(t_i) - s_i| \quad \text{for all} \quad s_i \in U_i. \tag{6.1}$$

We call $f_{t_i}^{\Delta_i}(t)$ the *partial delta derivative* (or *partial Hilger derivative*) of f with respect to t_i at t. We say that f is *partial delta differentiable* (or *partial Hilger differentiable*) with respect to t_i in $\Lambda_i^{\kappa_i n}$ if $f_{t_i}^{\Delta_i}(t)$ exists for all $t \in \Lambda_i^{\kappa_i n}$. The function $f_{t_i}^{\Delta_i} : \Lambda_i^{\kappa_i n} \to \mathbb{R}$ is said to be the *partial delta derivative* (or *partial Hilger derivative*) with respect to t_i of f in $\Lambda_i^{\kappa_i n}$.

Theorem 6.41 *The partial delta derivative is well defined.*

Proof Let $t \in \Lambda^{\kappa_i n}$ for some $i \in \{1, 2, \ldots, n\}$. We assume that the partial derivative $f_{t_i}^{\Delta_i}(t)$ exists and

$$f_1(t) = f_{t_i}^{\Delta_i}(t), \quad f_2(t) = f_{t_i}^{\Delta_i}(t).$$

Let $\varepsilon_i > 0$ be arbitrarily chosen. Then there exists $\delta_i > 0$ such that for every

$$s_i \in (t_i - \delta_i, t_i + \delta_i) \cap \mathbb{T}_i,$$

we have

$$\Big| f(t_1, t_2, \ldots, t_{i-1}, \sigma_i(t_i), t_{i+1}, \ldots, t_n) - f(t_1, t_2, \ldots, t_{i_1}, s_i, t_{i+1}, \ldots, t_n) \\ - f_1(t)(\sigma_i(t_i) - s_i)\Big| \leq \frac{\varepsilon_i}{2} |\sigma_i(t_i) - s_i| \tag{6.2}$$

and

$$\Big| f(t_1, t_2, \ldots, t_{i-1}, \sigma_i(t_i), t_{i+1}, \ldots, t_n) - f(t_1, t_2, \ldots, t_{i-1}, s_i, t_{i+1}, \ldots, t_n) \\ - f_2(t)(\sigma_i(t_i) - s_i)\Big| \leq \frac{\varepsilon_i}{2} |\sigma_i(t_i) - s_i|. \tag{6.3}$$

From (6.2) and (6.3), for

$$s_i \in (t_i - \delta_i, t_i + \delta_i) \cap \mathbb{T}_i, \quad s_i \neq \sigma_i(t_i),$$

we obtain

$$\begin{aligned} |f_1(t) - f_2(t)| &= \Big| f_1(t) - \frac{f(t_1, t_2, \ldots, t_{i-1}, \sigma_i(t_i), t_{i+1}, \ldots, t_n)}{\sigma_i(t_i) - s_i} \\ &\quad + \frac{f(t_1, t_2, \ldots, t_{i-1}, s_i, t_{i+1}, \ldots, t_n)}{\sigma_i(t_i) - s_i} + \frac{f(t_1, t_2, \ldots, t_{i-1}, \sigma_i(t_i), t_{i+1}, \ldots, t_n)}{\sigma_i(t_i) - s_i} \\ &\quad - \frac{f(t_1, t_2, \ldots, t_{i-1}, s_i, t_{i+1}, \ldots, t_n)}{\sigma_i(t_i) - s_i} - f_2(t)\Big| \\ &\leq \left| f_1(t) - \frac{f(t_1, t_2, \ldots, t_{i-1}, \sigma_i(t_i), t_{i+1}, \ldots, t_n) - f(t_1, t_2, \ldots, t_{i-1}, s_i, t_{i+1}, \ldots, t_n)}{\sigma_i(t_i) - s_i} \right| \\ &\quad + \left| f_2(t) - \frac{f(t_1, t_2, \ldots, t_{i-1}, \sigma_i(t_i), t_{i+1}, \ldots, t_n) - f(t_1, t_2, \ldots, t_{i-1}, s_i, t_{i+1}, \ldots, t_n)}{\sigma_i(t_i) - s_i} \right| \\ &\leq \frac{\varepsilon_i}{2} + \frac{\varepsilon_i}{2} = \varepsilon_i. \end{aligned}$$

Because $\varepsilon_i > 0$ was arbitrarily chosen, we conclude that $f_1(t) = f_2(t)$. □

Example 6.42 Let $f(t) = t_1t_2t_3$, $t = (t_1, t_2, t_3) \in \Lambda^3$. We will prove that

$$f_{t_1}^{\Delta_1}(t) = t_2t_3.$$

Indeed, for every $\varepsilon_1 > 0$, there exists $\delta_1 > 0$ such that for every $s_1 \in (t_1 - \delta_1, t_1 + \delta_1)$, $s_1 \in \mathbb{T}_1$, we have

$$\begin{aligned}
&|f(\sigma_1(t_1), t_2, t_3) - f(s_1, t_2, t_3) - t_2t_3(\sigma_1(t_1) - s_1)| \\
&= |\sigma_1(t_1)t_2t_3 - s_1t_2t_3 - t_2t_3(\sigma_1(t_1) - s_1)| \\
&= 0 \\
&\le \varepsilon_1|\sigma_1(t_1) - s_1|.
\end{aligned}$$

Example 6.43 Let $f(t) = t_2^2 + t_1t_3$, $t = (t_1, t_2, t_3) \in \Lambda^3$. We will prove that

$$f_{t_2}^{\Delta_2}(t) = \sigma_2(t_2) + t_2.$$

Indeed, for every $\varepsilon_2 > 0$, there exists

$$\delta_2 > 0, \quad \delta_2 \le \varepsilon_2,$$

such that for every

$$s_2 \in (t - \delta_2, t + \delta_2), \quad s_2 \in \mathbb{T}_2,$$

we have

$$|s_2 - t_2| \le \delta_2 \le \varepsilon_2$$

and

$$\begin{aligned}
&|f(t_1, \sigma_2(t_2), t_3) - f(t_1, s_2, t_3) - f_{t_2}^{\Delta_2}(t)(\sigma_2(t_2) - s_2)| \\
&= |(\sigma_2(t_2))^2 + t_1t_3 - s_2^2 - t_1t_3 - (\sigma_2(t_2) + t_2)(\sigma_2(t_2) - s_2)| \\
&= |(\sigma_2(t_2))^2 - s_2^2 - (\sigma_2(t_2) + t_2)(\sigma_2(t_2) - s_2)| \\
&= |(\sigma_2(t_2) - s_2)(\sigma_2(t_2) + s_2) - (\sigma_2(t_2) + t_2)(\sigma_2(t_2) - s_2)| \\
&= |(\sigma_2(t_2) - s_2)(\sigma_2(t_2) + s_2 - \sigma_2(t_2) - t_2)|
\end{aligned}$$

$$= |t_2 - s_2||\sigma_2(t_2) - s_2| \le \varepsilon_2|\sigma_2(t_2) - s_2|.$$

Example 6.44 Let $f(t) = t_1t_2\sqrt{t_3}$, $t = (t_1, t_2, t_3) \in \Lambda^3$. We will prove that

$$f_{t_3}^{\Delta_3}(t) = \frac{t_1t_2}{\sqrt{\sigma_3(t_3)} + \sqrt{t_3}}.$$

Indeed, for every $\varepsilon_3 > 0$, there exists $\delta_3 > 0$, $\delta_3 \le \varepsilon_3 \frac{\sigma_3(t_3)\sqrt{t_3}}{1+|t_1t_2|}$, such that for every $s_3 \in (t_3 - \delta_3, t_3 + \delta_3)$, $s_3 \in \mathbb{T}_3$, we have

$$|t_3 - s_3| \le \delta_3$$

and

$$\begin{aligned}
&\left|f(t_1, t_2, \sigma_3(t_3)) - f(t_1, t_2, s_3) - f_{t_3}^{\Delta_3}(t)(\sigma_3(t_3) - s_3)\right| \\
&= \left|t_1t_2\sqrt{\sigma_3(t_3)} - t_1t_2\sqrt{s_3} - \frac{t_1t_2}{\sqrt{\sigma_3(t_3)} + \sqrt{t_3}}(\sigma_3(t_3) - s_3)\right| \\
&= \left|t_1t_2\frac{(\sqrt{\sigma_3(t_3)} - \sqrt{s_3})(\sqrt{\sigma_3(t_3)} + \sqrt{s_3})}{\sqrt{\sigma_3(t_3)} + \sqrt{s_3}} - \frac{t_1t_2}{\sqrt{\sigma_3(t_3)} + \sqrt{t_3}}(\sigma_3(t_3) - s_3)\right| \\
&= |\sigma_3(t_3) - s_3||t_1t_2|\left|\frac{1}{\sqrt{\sigma_3(t_3)} + \sqrt{s_3}} - \frac{1}{\sqrt{\sigma_3(t_3)} + \sqrt{t_3}}\right| \\
&= |\sigma_3(t_3) - s_3||t_1t_2|\frac{|\sqrt{t_3} - \sqrt{s_3}|}{(\sqrt{\sigma_3(t_3)} + \sqrt{t_3})(\sqrt{\sigma_3(t_3)} + \sqrt{s_3})} \\
&= |\sigma_3(t_3) - s_3||t_1t_2|\frac{|t_3 - s_3|}{(\sqrt{\sigma_3(t_3)} + \sqrt{t_3})(\sqrt{\sigma_3(t_3)} + \sqrt{s_3})(\sqrt{t_3} + \sqrt{s_3})} \\
&\le |\sigma_3(t_3) - s_3||t_1t_2|\frac{|t_3 - s_3|}{\sigma_3(t_3)\sqrt{t_3}} \\
&\le \delta_3|\sigma_3(t_3) - s_3||t_1t_2|\frac{1}{\sigma_3(t_3)\sqrt{t_3}} \\
&\le \delta_3|\sigma_3(t_3) - s_3|(1 + |t_1t_2|)\frac{1}{\sigma_3(t_3)\sqrt{t_3}} \\
&\le \varepsilon_3|\sigma_3(t_3) - s_3|.
\end{aligned}$$

Exercise 6.45 Let $f(t) = t_1^2t_2 + t_1t_3$, $t = (t_1, t_2, t_3) \in \Lambda^3$. Prove that

$$f_{t_1}^{\Delta_1}(t) = (\sigma_1(t_1) + t_1)t_2 + t_3.$$

Remark 6.46 For $t \in \Lambda^n$ and $s_i \in \mathbb{T}_i$, we write

$$t_{s_i} = (t_1, \ldots, t_{i-1}, s_i, t_{i+1}, \ldots, t_n), \quad i \in \{1, 2, \ldots, n\}.$$

Then we can rewrite (6.1) in the form

$$|f_i^{\sigma_i}(t) - f(t_{s_i}) - f_{t_i}^{\Delta_i}(t)(\sigma_i(t_i) - s_i)| \le \varepsilon_i |\sigma_i(t_i) - s_i|. \tag{6.4}$$

Theorem 6.47 *Let $f : \Lambda^n \to \mathbb{R}$ be a function and $t \in \Lambda_i^{\kappa_i n}$. If f is delta differentiable with respect to t_i at t, then*

$$\lim_{s_i \to t_i} f(t_{s_i}) = f(t). \tag{6.5}$$

Proof Because f is delta differentiable with respect to t_i at t, we have that for every $\varepsilon > 0$, there exists $\delta > 0$, $\delta < \{1, \varepsilon^*\}$, such that for every $s_i \in (t_i - \delta, t_i + \delta)$, $s_i \in \mathbb{T}_i$, we have

$$|f_i^{\sigma_i}(t) - f(t_{s_i}) - f_{t_i}^{\Delta_i}(t)(\sigma_i(t_i) - s_i)| \le \varepsilon^* |\sigma_i(t_i) - s_i|$$

and

$$|f_i^{\sigma_i}(t) - f(t) - f_{t_i}^{\Delta_i}(t)(\sigma_i(t_i) - t_i)| \le \varepsilon^* |\sigma_i(t_i) - t_i|,$$

for

$$\varepsilon^* = \frac{\varepsilon}{1 + 2\mu_i(t_i) + |f_{t_i}^{\Delta_i}(t)|}.$$

Hence, for every $s_i \in (t_i - \delta, t_i + \delta)$, $s_i \in \mathbb{T}_i$, we have

$$\begin{aligned}
|f(t) - f(t_{s_i})| &= \Big| f_i^{\sigma_i}(t) - f(t_{s_i}) - f_{t_i}^{\Delta_i}(t)(\sigma_i(t_i) - s_i) \\
&\quad -(f_i^{\sigma_i}(t) - f(t) - f_{t_i}^{\Delta_i}(t)(\sigma_i(t_i) - t_i)) + f_{t_i}^{\Delta_i}(t)(t_i - s_i) \Big| \\
&\le |f_i^{\sigma_i}(t) - f(t_{s_i}) - f_{t_i}^{\Delta_i}(t)(\sigma_i(t_i) - s_i)| \\
&\quad + |f_i^{\sigma_i}(t) - f(t) - f_{t_i}^{\Delta_i}(t)(\sigma_i(t_i) - t_i)| + |f_{t_i}^{\Delta_i}(t)||t_i - s_i| \\
&\le \varepsilon^* |\sigma_i(t_i) - s_i| + \varepsilon^* |\sigma_i(t_i) - t_i| + |f_{t_i}^{\Delta_i}(t)||t_i - s_i| \\
&\le \varepsilon^* |\sigma_i(t_i) - s_i| + \varepsilon^* \mu_i(t_i) + \varepsilon^* |f_{t_i}^{\Delta_i}(t)| \\
&= \varepsilon^* \left(\mu_i(t_i) + |\sigma_i(t_i) - s_i| + |f_{t_i}^{\Delta_i}(t)| \right)
\end{aligned}$$

$$= \varepsilon^* \left(\mu_i(t_i) + |\sigma_i(t_i) - t_i + t_i - s_i| + |f_{t_i}^{\Delta_i}(t)| \right)$$

$$\leq \varepsilon^* \left(\mu_i(t_i) + \mu_i(t_i) + |t_i - s_i| + |f_{t_i}^{\Delta_i}(t)| \right)$$

$$\leq \varepsilon^* \left(1 + 2\mu_i(t_i) + |f_{t_i}^{\Delta_i}(t)| \right)$$

$$= \varepsilon.$$

Because $\varepsilon > 0$ was arbitrarily chosen, we obtain (6.5). □

Theorem 6.48 *Let $f : \Lambda^n \to \mathbb{R}$, $t \in \Lambda_i^{\kappa_i n}$, and*

$$\lim_{s_i \to t_i} f(t_{s_i}) = f(t). \tag{6.6}$$

If $t_i < \sigma_i(t_i)$, then f is delta differentiable with respect to t_i at t and

$$f_{t_i}^{\Delta_i}(t) = \frac{f_i^{\sigma_i}(t) - f(t)}{\mu_i(t_i)}. \tag{6.7}$$

Proof When $s_i \to t_i$, using (6.1) and (6.6), we get (6.7). □

Example 6.49 Let $\Lambda^2 = \mathbb{N} \times 2^{\mathbb{N}}$ and define $f : \Lambda^2 \to \mathbb{R}$ by

$$f(t) = t_1^2 t_2 - 2t_1, \quad t = (t_1, t_2) \in \Lambda^2.$$

Then

$$\sigma_1(t_1) = t_1 + 1, \quad t_1 \in \mathbb{T}_1, \quad \sigma_2(t_2) = 2t_2, \quad t_2 \in 2^{\mathbb{N}}.$$

Hence,

$$t_1 < \sigma_1(t_1), \quad t_1 \in \mathbb{N}, \quad t_2 < \sigma_2(t_2), \quad t_2 \in 2^{\mathbb{N}}.$$

Therefore,

$$\begin{aligned} f_{t_1}^{\Delta_1}(t) &= \frac{f_1^{\sigma_1}(t) - f(t)}{\sigma_1(t_1) - t_1} \\ &= \frac{(\sigma_1(t_1))^2 t_2 - 2\sigma_1(t_1) - t_1^2 t_2 + 2t_1}{t_1 + 1 - t_1} \\ &= (t_1 + 1)^2 t_2 - 2(t_1 + 1) - t_1^2 t_2 + 2t_1 \\ &= t_1^2 t_2 + 2t_1 t_2 + t_2 - 2t_1 - 2 - t_1^2 t_2 + 2t_1 \end{aligned}$$

$$= 2t_1t_2 + t_2 - 2, \quad (t_1, t_2) \in \Lambda_1^{\kappa_1 2},$$

$$\begin{aligned} f_{t_2}^{\Delta_2}(t) &= \frac{f_2^{\sigma_2}(t) - f(t)}{\sigma_2(t_2) - t_2} \\ &= \frac{t_1^2\sigma_2(t_2) - 2t_1 - (t_1^2 t_2 - 2t_1)}{2t_2 - t_2} \\ &= \frac{2t_1^2 t_2 - 2t_1 - t_1^2 t_2 + 2t_1}{t_2} \\ &= t_1^2, \quad (t_1, t_2) \in \Lambda_2^{\kappa_2 2}. \end{aligned}$$

Example 6.50 Let $\Lambda^2 = 3^{\mathbb{N}} \times \mathbb{N}_0^2$ and define $f : \Lambda^2 \to \mathbb{R}$ by

$$f(t) = t_1^2 + 2t_1t_2 + t_2, \quad t = (t_1, t_2) \in \Lambda^2.$$

We will find $f_{t_1}^{\Delta_1}$ and $f_{t_2}^{\Delta_2}$. Here,

$$\sigma_1(t_1) = 3t_1, \quad t_1 \in 3^{\mathbb{N}}, \quad \sigma_2(t_2) = \left(1 + \sqrt{t_2}\right)^2, \quad t_2 \in \mathbb{N}_0^2.$$

Then

$$\sigma_1(t_1) > t_1 \quad \text{for all} \quad t_1 \in 3^{\mathbb{N}}, \quad \sigma_2(t_2) > t_2 \quad \text{for all} \quad t_2 \in \mathbb{N}_0^2.$$

Hence,

$$\begin{aligned} f_{t_1}^{\Delta_1}(t) &= \frac{f(\sigma_1(t_1), t_2) - f(t_1, t_2)}{\sigma_1(t_1) - t_1} \\ &= \frac{(\sigma_1(t_1))^2 + 2\sigma_1(t_1)t_2 + t_2 - t_1^2 - 2t_1t_2 - t_2}{\sigma_1(t_1) - t_1} \\ &= \frac{(\sigma_1(t_1) - t_1)(\sigma_1(t_1) + t_1) + 2(\sigma_1(t_1) - t_1)t_2}{\sigma_1(t_1) - t_1} \\ &= \sigma_1(t_1) + t_1 + 2t_2 \\ &= 3t_1 + t_1 + 2t_2 \\ &= 4t_1 + 2t_2, \quad (t_1, t_2) \in \Lambda_1^{\kappa_1 2}, \end{aligned}$$

$$f_{t_2}^{\Delta_2}(t) = \frac{f(t_1, \sigma_2(t_2)) - f(t_1, t_2)}{\sigma_2(t_2) - t_2}$$

$$= \frac{t_1^2 + 2t_1\sigma_2(t_2) + \sigma_2(t_2) - t_1^2 - 2t_1t_2 - t_2}{\sigma_2(t_2) - t_2}$$

$$= \frac{2t_1(\sigma_2(t_2) - t_2) + \sigma_2(t_2) - t_2}{\sigma_2(t_2) - t_2}$$

$$= 2t_1 + 1, \quad (t_1, t_2) \in \Lambda^2, \quad (t_1, t_2) \in \Lambda_2^{\kappa_2 2}.$$

Example 6.51 Let $\Lambda^2 = \mathbb{N} \times 4^{\mathbb{N}}$ and define $f : \Lambda^2 \to \mathbb{R}$ by

$$f(t) = t_1^3 + t_2^3 + t_1^2t_2^2 + t_2^4, \quad (t_1, t_2) \in \Lambda^2.$$

We will find $f_{t_1}^{\Delta_1}$ and $f_{t_2}^{\Delta_2}$. We have

$$\sigma_1(t_1) = t_1 + 1, \quad t_1 \in \mathbb{N}, \quad \sigma_2(t_2) = 4t_2, \quad t_2 \in 4^{\mathbb{N}}.$$

Therefore,

$$\sigma_1(t_1) > t_1 \quad \text{for all} \quad t_1 \in \mathbb{N}, \quad \sigma_2(t_2) > t_2 \quad \text{for all} \quad t_2 \in 4^{\mathbb{N}}.$$

Hence,

$$f_{t_1}^{\Delta_1}(t) = \frac{f(\sigma_1(t_1), t_2) - f(t_1, t_2)}{\sigma_1(t_1) - t_1}$$

$$= \frac{(\sigma_1(t_1))^3 + t_2^3 + (\sigma_1(t_1))^2t_2^2 + t_2^4 - t_1^3 - t_2^3 - t_1^2t_2^2 - t_2^4}{\sigma_1(t_1) - t_1}$$

$$= \frac{(\sigma_1(t_1) - t_1)((\sigma_1(t_1))^2 + t_1\sigma_1(t_1) + t_1^2) + t_2^2(\sigma_1(t_1) - t_1)(\sigma_1(t_1) + t_1)}{\sigma_1(t_1) - t_1}$$

$$= (\sigma_1(t_1))^2 + \sigma_1(t_1)t_1 + t_1^2 + t_2^2(\sigma_1(t_1) + t_1)$$

$$= (t_1 + 1)^2 + t_1(t_1 + 1) + t_1^2 + t_2^2(2t_1 + 1)$$

$$= t_1^2 + 2t_1 + 1 + t_1^2 + t_1 + t_1^2 + 2t_1t_2^2 + t_2^2$$

$$= 3t_1^2 + 3t_1 + 2t_1t_2^2 + t_2^2 + 1, \quad (t_1, t_2) \in \Lambda_1^{\kappa_1 2},$$

$$f_{t_2}^{\Delta_2}(t) = \frac{f(t_1, \sigma_2(t_2)) - f(t_1, t_2)}{\sigma_2(t_2) - t_2}$$

$$= \frac{t_1^3 + (\sigma_2(t_2))^3 + t_1^2(\sigma_2(t_2))^2 + (\sigma_2(t_2))^4 - t_1^3 - t_2^3 - t_1^2t_2^2 - t_2^4}{\sigma_2(t_2) - t_2}$$

$$= \frac{(\sigma_2(t_2)-t_2)((\sigma_2(t_2))^2+t_2\sigma_2(t_2)+t_2^2)+t_1^2(\sigma_2(t_2)-t_2)(\sigma_2(t_2)+t_2)}{\sigma_2(t_2)-t_2}$$

$$+\frac{(\sigma_2(t_2)-t_2)((\sigma_2(t_2))^3+t_2(\sigma_2(t_2))^2+t_2^2\sigma_2(t_2)+t_2^3)}{\sigma_2(t_2)-t_2}$$

$$= (\sigma_2(t_2))^2+t_2\sigma_2(t_2)+t_2^2+t_1^2(\sigma_2(t_2)+t_2)+(\sigma_2(t_2))^3$$

$$+t_2(\sigma_2(t_2))^2+t_2^2\sigma_2(t_2)+t_2^3$$

$$= 16t_2^2+4t_2^2+t_2^2+5t_1^2t_2+64t_2^3+16t_2^3+4t_2^3+t_2^3$$

$$= 5t_1^2t_2+21t_2^2+85t_2^3, \quad (t_1,t_2)\in \Lambda_2^{\kappa_2 2}.$$

Exercise 6.52 Let $\Lambda^2 = \mathbb{N}\times\mathbb{N}$ and define $f:\Lambda^2\to\mathbb{R}$ by

$$f(t) = t_1^3+t_1t_2+t_1^2-2t_1.$$

Find $f_{t_1}^{\Delta_1}(t)$, $t\in\Lambda_1^{\kappa_1 2}$, and $f_{t_2}^{\Delta_2}(t)$, $t\in\Lambda_2^{\kappa_2 2}$.

Solution $3t_1^2+5t_1+t_2$, t_1.

Theorem 6.53 *Let $t\in\Lambda_i^{\kappa_i n}$ and $t_i=\sigma_i(t_i)$. Then f is partial delta differentiable with respect to t_i at t if and only if the limit*

$$\lim_{s_i\to t_i}\frac{f(t)-f(t_{s_i})}{t_i-s_i}$$

exists as a finite number. In this case,

$$f_{t_i}^{\Delta_i}(t) = \lim_{s_i\to t_i}\frac{f(t)-f(t_{s_i})}{t_i-s_i}. \tag{6.8}$$

Proof 1. Let f be partial delta differentiable with respect to t_i at t. Then, for every $\varepsilon>0$, there exists $\delta>0$ such that for every $s_i\in(t-\delta_i,t+\delta_i)$, $s_i\in\mathbb{T}_i$, we have (6.4). Hence, using that $\sigma_i(t_i)=t_i$, we get

$$\left|\frac{f(t)-f(t_{s_i})}{t_i-s_i}-f_{t_i}^{\Delta_i}(t)\right|\le\varepsilon.$$

Because $\varepsilon>0$ was arbitrarily chosen, we conclude (6.8).

2. Suppose (6.8) holds. Then, for $\varepsilon>0$, there exists $\delta>0$ such that for every $s_i\in(t_i-\delta,t_i+\delta)$, $s_i\in\mathbb{T}_i$, we have

$$|f(t) - f(t_{s_i}) - f_{t_i}^{\Delta_i}(t)(t_i - s_i)| \le \varepsilon |t_i - s_i|.$$

Hence, using $t_i = \sigma_i(t_i)$, we get (6.4).
The proof is complete. □

Example 6.54 Let

$$\Lambda^2 = \left(\left(\frac{1}{2}\right)^{\mathbb{N}} \cup \left[\frac{1}{2}, 1\right]\right) \times \mathbb{N},$$

where $\left[\frac{1}{2}, 1\right]$ is the real number interval, and define $f : \Lambda^2 \to \mathbb{R}$ by

$$f(t) = t_1^2 t_2, \quad t = (t_1, t_2) \in \Lambda^2.$$

We will find $f_{t_1}^{\Delta_1}\left(\frac{1}{2}, t_2\right)$, $t_2 \in \mathbb{N}$. We note that

$$\mathbb{T}_1 = \left(\frac{1}{2}\right)^{\mathbb{N}}, \quad \mathbb{T}_2 = \mathbb{N}.$$

Hence, $\sigma_1\left(\frac{1}{2}\right) = \frac{1}{2}$. Thus,

$$\begin{aligned}
\lim_{s_1 \to \frac{1}{2}} \frac{f\left(\frac{1}{2}, t_2\right) - f(s_1, t_2)}{\frac{1}{2} - s_1} &= \lim_{s_1 \to \frac{1}{2}} \frac{\frac{1}{4} t_2 - s_1^2 t_2}{\frac{1}{2} - s_1} \\
&= \lim_{s_1 \to \frac{1}{2}} \frac{\left(\frac{1}{2} - s_1\right)\left(\frac{1}{2} + s_1\right) t_2}{\frac{1}{2} - s_1} \\
&= \lim_{s_1 \to \frac{1}{2}} \left(\frac{1}{2} + s_1\right) t_2 \\
&= t_2.
\end{aligned}$$

Consequently,

$$f_{t_1}^{\Delta_1}\left(\frac{1}{2}, t_2\right) = t_2.$$

Example 6.55 Let

$$\Lambda^2 = \mathbb{Z} \times ((-2\mathbb{N}) \cup [-2, -1]),$$

where $[-2, -1]$ is the real number interval, and define $f : \Lambda^2 \to \mathbb{R}$ by

$$f(t) = t_1 t_2 + t_2^2, \quad t = (t_1, t_2) \in \Lambda^2.$$

We will find $f_{t_2}^{\Delta_2}(t_1, -2)$, $t_1 \in \mathbb{Z}$. We note that $\mathbb{T}_1 = \mathbb{Z}$, $\mathbb{T}_2 = -2\mathbb{N}$. Hence, $\sigma_2(-2) = -2$ and

$$\begin{aligned}\lim_{t_2 \to -2} \frac{f(t_1, -2) - f(t_1, t_2)}{-2 - t_2} &= \lim_{t_2 \to -2} \frac{-2t_1 + 4 - t_1 t_2 - t_2^2}{-2 - t_2} \\ &= \lim_{t_2 \to -2} \frac{(2 + t_2)t_1 + (t_2 - 2)(t_2 + 2)}{2 + t_2} \\ &= \lim_{t_2 \to -2} (t_1 + t_2 - 2) \\ &= t_1 - 4.\end{aligned}$$

Consequently,

$$f_{t_2}^{\Delta_2}(t_1, -2) = t_1 - 4.$$

Example 6.56 Let

$$\Lambda^2 = \mathbb{Z} \times \left(\left(\frac{1}{3}\right)^{\mathbb{N}_0} \cup [1, 2] \right),$$

where $[1, 2]$ is the real number interval, and define $f : \Lambda^2 \to \mathbb{R}$ by

$$f(t) = t_1 + t_2 + t_2^2, \quad t = (t_1, t_2) \in \Lambda^2.$$

We will find $f_{t_2}^{\Delta_2}(t_1, 1)$, $t_1 \in \mathbb{Z}$. We note that

$$\mathbb{T}_1 = \mathbb{Z}, \quad \mathbb{T}_2 = \left(\frac{1}{3}\right)^{\mathbb{N}_0}.$$

Hence, $\sigma_2(1) = 1$ and

$$\begin{aligned}\lim_{s_2 \to 1} \frac{f(t_1, 1) - f(t_1, s_2)}{1 - s_2} &= \lim_{s_2 \to 1} \frac{t_1 + 2 - t_1 - s_2 - s_2^2}{1 - s_2} \\ &= \lim_{s_2 \to 1} \frac{(1 - s_2) + (1 - s_2)(1 + s_2)}{1 - s_2} \\ &= \lim_{s_2 \to 1} (2 + s_2) = 3.\end{aligned}$$

Consequently, $f_{t_2}^{\Delta_2}(t_1, 1) = 3$.

Exercise 6.57 Let $\Lambda^2 = \mathbb{N} \times 4\mathbb{Z}$ and define $f : \Lambda^2 \to \mathbb{R}$ by

$$f(t) = t_1^3 + 3t_1^2 + t_1 t_2, \quad t = (t_1, t_2) \in \Lambda^2.$$

Find $f_{t_1}^{\Delta_1}(1, t_2)$.

Solution $16 + t_2$.

Theorem 6.58 *Let $t \in \Lambda_i^{\kappa_i n}$. Suppose $f : \Lambda^n \to \mathbb{R}$ is a function that is partial delta differentiable with respect to t_i at t. If $\alpha \in \mathbb{R}$, then αf is partial delta differentiable with respect to t_i at t and*

$$(\alpha f)_{t_i}^{\Delta_i}(t) = \alpha f_{t_i}^{\Delta_i}(t).$$

Proof Without loss of generality, we may assume $\alpha \neq 0$. Since f is partial delta differentiable with respect to t_i at t, for every $\varepsilon > 0$, there exists $\delta > 0$ such that for every $s_i \in (t_i - \delta, t_i + \delta)$, $s_i \in \mathbb{T}_i$, we have

$$|f_i^{\sigma_i}(t) - f(t_{s_i}) - f_{t_i}^{\Delta_i}(t)(\sigma_i(t_i) - s_i)| \leq \frac{\varepsilon}{|\alpha|}|\sigma_i(t_i) - s_i|.$$

Hence,

$$\begin{aligned}
&|(\alpha f)_i^{\sigma_i}(t) - (\alpha f)(t_{s_i}) - \alpha f_{t_i}^{\Delta_i}(t)(\sigma_i(t_i) - s_i)| \\
&= |\alpha||f_i^{\sigma_i}(t) - f(t_{s_i}) - f_{t_i}^{\Delta_i}(t)(\sigma_i(t_i) - s_i)| \\
&\leq |\alpha|\frac{\varepsilon}{|\alpha|}|\sigma_i(t_i) - s_i| \\
&= \varepsilon|\sigma_i(t_i) - s_i|,
\end{aligned}$$

which completes the proof. □

Theorem 6.59 *Let $t \in \Lambda_i^{\kappa_i n}$. Assume $f, g : \Lambda^n \to \mathbb{R}$ are partial delta differentiable with respect to t_i at t. Then $f + g$ is partial delta differentiable with respect to t_i at t and*

$$(f + g)_{t_i}^{\Delta_i}(t) = f_{t_i}^{\Delta_i}(t) + g_{t_i}^{\Delta_i}(t).$$

Proof Since f and g are partial delta differentiable with respect to t_i at t, for any $\varepsilon > 0$, there exists $\delta > 0$ such that for any $s_i \in (t_i - \delta, t_i + \delta)$, $s_i \in \mathbb{T}_i$, we have

$$|f_i^{\sigma_i}(t) - f(t_{s_i}) - f_{t_i}^{\Delta_i}(t)(\sigma_i(t_i) - s_i)| \leq \frac{\varepsilon}{2}|\sigma_i(t_i) - s_i|$$

and

$$|g_i^{\sigma_i}(t) - g(t_{s_i}) - g_{t_i}^{\Delta_i}(t)(\sigma_i(t_i) - s_i)| \leq \frac{\varepsilon}{2}|\sigma_i(t_i) - s_i|,$$

whereupon

$$
\begin{aligned}
&|(f+g)_i^{\sigma_i}(t) - (f+g)(t_{s_i}) - (f_{t_i}^{\Delta_i}(t) + g_{t_i}^{\Delta_i}(t))(\sigma_i(t_i) - s_i)| \\
&= |f_i^{\sigma_i}(t) + g_i^{\sigma_i}(t) - f(t_{s_i}) - g(t_{s_i}) - f_{t_i}^{\Delta_i}(t)(\sigma_i(t_i) - s_i) - g_{t_i}^{\Delta_i}(t)(\sigma_i(t_i) - s_i)| \\
&\leq |f_i^{\sigma_i}(t) - f(t_{s_i}) - f_{t_i}^{\Delta_i}(t)(\sigma_i(t_i) - s_i)| + |g_i^{\sigma_i}(t) - g(t_{s_i}) - g_{t_i}^{\Delta_i}(\sigma_i(t_i) - s_i)| \\
&\leq \frac{\varepsilon}{2}|\sigma_i(t_i) - s_i| + \frac{\varepsilon}{2}|\sigma_i(t_i) - s_i| \\
&= \varepsilon|\sigma_i(t_i) - s_i|,
\end{aligned}
$$

which completes the proof. □

Theorem 6.60 *Let $t \in \Lambda_i^{\kappa_i n}$. Assume $f, g : \Lambda^n \to \mathbb{R}$ are partial delta differentiable with respect to t_i at t. Then fg is partial delta differentiable with respect to t_i at t and*

$$
(fg)_{t_i}^{\Delta_i}(t) = f_{t_i}^{\Delta_i}(t)g(t) + f_i^{\sigma_i}(t)g_{t_i}^{\Delta_i}(t) = f(t)g_{t_i}^{\Delta_i}(t) + f_{t_i}^{\Delta_i}(t)g_i^{\sigma_i}(t).
$$

Proof Since f and g are partial delta differentiable at t, for any $\varepsilon > 0$, there exists $\delta > 0$ such that for any $s_i \in (t_i - \delta, t_i + \delta)$, $s_i \in \mathbb{T}_i$, we have

$$
|f_i^{\sigma_i}(t) - f(t_{s_i}) - f_{t_i}^{\Delta_i}(t)(\sigma_i(t_i) - s_i)| \leq \varepsilon^*|\sigma_i(t_i) - s_i|,
$$

$$
|g_i^{\sigma_i}(t) - g(t_{s_i}) - g_{t_i}^{\Delta_i}(t)(\sigma_i(t_i) - s_i)| \leq \varepsilon^*|\sigma_i(t_i) - s_i|,
$$

$$
|f(t_{s_i}) - f(t)| \leq \varepsilon^*,
$$

$$
|g(t_{s_i}) - g(t)| \leq \varepsilon^*,
$$

where

$$
\varepsilon^* < \min\left\{\frac{\varepsilon}{1 + |f_i^{\sigma_i}(t)| + |g(t_{s_i})| + |f_{t_i}^{\Delta_i}(t)|}, \frac{\varepsilon}{1 + |g_i^{\sigma_i}(t)| + |f(t_{s_i})| + |g_{t_i}^{\Delta_i}(t)|}\right\}.
$$

Hence,

$$
\begin{aligned}
&|(fg)_i^{\sigma_i}(t) - (fg)(t_{s_i}) - (f_{t_i}^{\Delta_i}(t)g(t) + f_i^{\sigma_i}(t)g_{t_i}^{\Delta_i}(t))(\sigma_i(t_i) - s_i)| \\
&= |f_i^{\sigma_i}(t)g_i^{\sigma_i}(t) - f(t_{s_i})g(t_{s_i}) - (f_{t_i}^{\Delta_i}(t)g(t) + f_i^{\sigma_i}(t)g_{t_i}^{\Delta_i}(t))(\sigma_i(t_i) - s_i)|
\end{aligned}
$$

$$= |f_i^{\sigma_i}(t)(g_i^{\sigma_i}(t) - g(t_{s_i}) - g_{t_i}^{\Delta_i}(t)(\sigma_i(t_i) - s_i))$$

$$+f_i^{\sigma_i}(t)g(t_{s_i}) - f(t_{s_i})g(t_{s_i}) - f_{t_i}^{\Delta_i}(t)g(t)(\sigma_i(t_i) - s_i)|$$

$$= |f_i^{\sigma_i}(t)(g_i^{\sigma_i}(t) - g(t_{s_i}) - g_{t_i}^{\Delta_i}(t)(\sigma_i(t_i) - s_i))$$

$$+g(t_{s_i})(f_i^{\sigma_i}(t) - f(t_{s_i}) - f_{t_i}^{\Delta_i}(t)(\sigma_i(t_i) - s_i))$$

$$+(g(t_{s_i}) - g(t))f_{t_i}^{\Delta_i}(t)(\sigma_i(t_i) - s_i)|$$

$$\leq \varepsilon^*|\sigma_i(t_i) - s_i||f_i^{\sigma_i}(t)| + \varepsilon^*|g(t_{s_i})||\sigma_i(t_i) - s_i|$$

$$+|g(t_{s_i}) - g(t)||f_{t_i}^{\Delta_i}(t)||\sigma_i(t_i) - s_i|$$

$$\leq \varepsilon^*|\sigma_i(t_i) - s_i||f_i^{\sigma_i}(t)| + \varepsilon^*|g(t_{s_i})||\sigma_i(t_i) - s_i| + \varepsilon^*|f_{t_i}^{\Delta_i}(t)||\sigma_i(t_i) - s_i|$$

$$= \varepsilon^*(|f_i^{\sigma_i}(t)| + |g(t_{s_i})| + |f_{t_i}^{\Delta_i}(t)|)|\sigma_i(t_i) - s_i|$$

$$\leq \varepsilon^*(1 + |f_i^{\sigma_i}(t)| + |g(t_{s_i})| + |f_{t_i}^{\Delta_i}(t)|)|\sigma_i(t_i) - s_i|$$

$$\leq \varepsilon|\sigma_i(t_i) - s_i|$$

and

$$|(fg)_i^{\sigma_i}(t) - (fg)(t_{s_i}) - (f(t)g_{t_i}^{\Delta_i}(t) + f_{t_i}^{\Delta_i}(t)g_i^{\sigma_i}(t))(\sigma_i(t_i) - s_i)|$$

$$= |f_i^{\sigma_i}(t)g_i^{\sigma_i}(t) - f(t_{s_i})g(t_{s_i}) - (f(t)g_{t_1}^{\Delta_i}(t) + f_{t_i}^{\Delta_i}(t)g_i^{\sigma_i}(t))(\sigma_i(t_i) - s_i)|$$

$$= |g_i^{\sigma_i}(t)(f_i^{\sigma_i}(t) - f(t_{s_i}) - f_{t_i}^{\Delta_i}(\sigma_i(t_i) - s_i))$$

$$+g_i^{\sigma_i}(t)f(t_{s_i}) - f(t_{s_i})g(t_{s_i}) - f(t)g_{t_i}^{\Delta_i}(t)(\sigma_i(t_i) - s_i)|$$

$$= |g_i^{\sigma_i}(t)(f_i^{\sigma_i}(t) - f(t_{s_i}) - f_{t_i}^{\Delta_i}(t)(\sigma_i(t_i) - s_i))$$

$$+f(t_{s_i})(g_i^{\sigma_i}(t) - g(t_{s_i}) - g_{t_i}^{\Delta_i}(t)(\sigma_i(t_i) - s_i))$$

$$+f(t_{s_i})g_{t_i}^{\Delta_i}(t)(\sigma_i(t_i)-s_i)-f(t)g_{t_i}^{\Delta_i}(t)(\sigma_i(t_i)-s_i)|$$

$$\leq |g_i^{\sigma_i}(t)||f_i^{\sigma_i}(t)-f(t_{s_i})-f_{t_i}^{\Delta_i}(t)(\sigma_i(t_i)-s_i)|$$

$$+|f(t_{s_i})||g_i^{\sigma_i}(t)-g(t_{s_i})-g_{t_i}^{\Delta_i}(t)(\sigma_i(t_i)-s_i)|$$

$$+|g_{t_i}^{\Delta_i}(t)||f(t_{s_i})-f(t)||\sigma_i(t_i)-s_i|$$

$$\leq \varepsilon^*|g_i^{\sigma_i}(t)||\sigma_i(t_i)-s_i|+\varepsilon^*|f(t_{s_i})||\sigma_i(t_i)-s_i|+\varepsilon^*|g_{t_i}^{\Delta_i}(t)||\sigma_i(t_i)-s_i|$$

$$\leq \varepsilon^*(1+|g_i^{\sigma_i}(t)|+|f(t_{s_i})|+|g_{t_i}^{\Delta_i}(t)|)|\sigma_i(t_i)-s_i|$$

$$\leq \varepsilon|\sigma_i(t_i)-s_i|.$$

This completes the proof. □

Example 6.61 Let $\Lambda^2 = \mathbb{N} \times \mathbb{N}_0^2$ and define $h : \Lambda^2 \to \mathbb{R}$ by

$$h(t) = (t_1^2 + 2t_1)(t_1^3 + t_2), \quad t \in \Lambda^2.$$

Here, $\mathbb{T}_1 = \mathbb{N}$ and $\mathbb{T}_2 = \mathbb{N}_0^2$. Then

$$\sigma_1(t_1) = t_1 + 1, \quad t_1 \in \mathbb{T}_1, \quad \sigma_2(t_2) = (1+\sqrt{t_2})^2, \quad t_2 \in \mathbb{T}_2.$$

We will find $h_{t_1}^{\Delta_1}(t)$, $t \in \Lambda_1^{\kappa_1 2}$. Let

$$f(t) = t_1^2 + 2t_1, \quad g(t) = t_1^3 + t_2, \quad t \in \Lambda^2.$$

Hence, $h(t) = f(t)g(t)$, $t \in \Lambda^2$. Then, for $t \in \Lambda_1^{\kappa_1 2}$, we have

$$h_{t_1}^{\Delta_1}(t) = f_{t_1}^{\Delta_1}(t)g(t) + f_1^{\sigma_1}(t)g_{t_1}^{\Delta_1}(t)$$

$$= (\sigma_1(t_1)+t_1+2)g(t) + ((\sigma_1(t_1))^2 + 2\sigma_1(t_1))((\sigma_1(t_1))^2 + t_1\sigma_1(t_1) + t_1^2)$$

$$= (t_1+1+t_1+2)(t_1^3+t_2) + ((t_1+1)^2+2t_1+2)((t_1+1)^2+t_1(t_1+1)+t_1^2)$$

$$= (2t_1+3)(t_1^3+t_2) + (t_1^2+4t_1+3)(3t_1^2+3t_1+1)$$

$$= 2t_1^4 + 2t_1t_2 + 3t_1^3 + 3t_2 + 3t_1^4 + 3t_1^3 + t_1^2 + 12t_1^3 + 12t_1^2 + 4t_1 + 9t_1^2 + 9t_1 + 3$$

$$= 5t_1^4 + 18t_1^3 + 22t_1^2 + 13t_1 + 2t_1t_2 + 3t_2 + 3.$$

Example 6.62 Let $\Lambda^2 = 2^{\mathbb{N}} \times \mathbb{N}$ and define $h : \Lambda^2 \to \mathbb{R}$ by

$$h(t) = (t_1^3 + t_2)(t_1^2 + t_1t_2 + t_2^2), \quad t \in \Lambda^2.$$

We will find $h_{t_1}^{\Delta_1}(t)$ for $t \in \Lambda_1^{\kappa_1 2}$ and $h_{t_2}^{\Delta_2}(t)$ for $t \in \Lambda_2^{\kappa_2 2}$. Let

$$f(t) = t_1^3 + t_2, \quad g(t) = t_1^2 + t_1t_2 + t_2^2, \quad t = (t_1, t_2) \in \Lambda^2.$$

Here,

$$\mathbb{T}_1 = 2^{\mathbb{N}}, \quad \mathbb{T}_2 = \mathbb{N}.$$

Then

$$\sigma_1(t_1) = 2t_1, \quad t_1 \in \mathbb{T}_1, \quad \sigma_2(t_2) = t_2 + 1, \quad t_2 \in \mathbb{T}_2.$$

Then, for $t \in \Lambda_1^{\kappa_1 2}$, we have

$$h_{t_1}^{\Delta_1}(t) = f_{t_1}^{\Delta_1}(t)g(t) + f_1^{\sigma_1}(t)g_{t_1}^{\Delta_1}(t)$$

$$= ((\sigma_1(t_1))^2 + t_1\sigma_1(t_1) + t_1^2)(t_1^2 + t_1t_2 + t_2^2)$$

$$+((\sigma_1(t_1))^3 + t_2)(\sigma_1(t_1) + t_1 + t_2)$$

$$= (4t_1^2 + 2t_1^2 + t_1^2)(t_1^2 + t_1t_2 + t_2^2)$$

$$+(8t_1^3 + t_2)(2t_1 + t_1 + t_2)$$

$$= 7t_1^2(t_1^2 + t_1t_2 + t_2^2)$$

$$+(8t_1^3 + t_2)(3t_1 + t_2)$$

$$= 7t_1^4 + 7t_1^3t_2 + 7t_1^2t_2^2 + 24t_1^4$$

$$+8t_1^3t_2 + 3t_1t_2 + t_2^2$$

$$= 31t_1^4 + 15t_1^3t_2 + 7t_1^2t_2^2 + 3t_1t_2 + t_2^2.$$

For $t \in \Lambda_2^{\kappa_2 2}$, we have

$$\begin{aligned} h_{t_2}^{\Delta_2}(t) &= f_{t_2}^{\Delta_2}(t)g(t) + f_2^{\sigma_2}(t)g_{t_2}^{\Delta_2}(t) \\ &= g(t) + (t_1^3 + \sigma_2(t_2))(t_1 + \sigma_2(t_2) + t_2) \\ &= t_1^2 + t_1t_2 + t_2^2 + (t_1^3 + t_2 + 1)(t_1 + t_2 + 1 + t_2) \\ &= t_1^2 + t_1t_2 + t_2^2 + (t_1^3 + t_2 + 1)(t_1 + 2t_2 + 1) \\ &= t_1^2 + t_1t_2 + t_2^2 + t_1^4 + 2t_1^3t_2 + t_1^3 + t_1t_2 + 2t_2^2 + t_2 + t_1 + 2t_2 + 1 \\ &= t_1^4 + t_1^3(2t_2 + 1) + t_1^2 + t_1(2t_2 + 1) + 3t_2^2 + 3t_2 + 1. \end{aligned}$$

Example 6.63 Let $\Lambda^3 = \mathbb{N} \times \mathbb{Z} \times \mathbb{R}$ and define $h : \Lambda^3 \to \mathbb{R}$ by

$$h(t) = t_1^2t_2t_3(3t_1t_2 + t_1^2 + t_3^2), \quad t \in \Lambda^3.$$

We will find $h_{t_1}^{\Delta_1}(t)$ for $t \in \Lambda_1^{\kappa_1 3}$, $h_{t_2}^{\Delta_2}(t)$ for $t \in \Lambda_2^{\kappa_2 3}$, and $h_{t_3}^{\Delta_3}(t)$ for $t \in \Lambda_3^{\kappa_3 3}$. Let

$$f(t) = t_1^2t_2t_3, \quad g(t) = 3t_1t_2 + t_1^2 + t_3^2, \quad t \in \Lambda^3.$$

Here,

$$\sigma_1(t_1) = t_1 + 1, \quad t_1 \in \mathbb{T}_1 = \mathbb{N}, \quad \sigma_2(t_2) = t_2 + 1, \quad t_2 \in \mathbb{T}_2 = \mathbb{Z},$$

$$\sigma_3(t_3) = t_3, \quad t_3 \in \mathbb{T}_3 = \mathbb{R}.$$

For $t \in \Lambda_1^{\kappa_1 3}$, we have

$$\begin{aligned} h_{t_1}^{\Delta_1}(t) &= f_{t_1}^{\Delta_1}(t)g(t) + f_1^{\sigma_1}(t)g_{t_1}^{\Delta_1}(t) \\ &= (\sigma_1(t_1) + t_1)t_2t_3(3t_1t_2 + t_1^2 + t_3^2) + (\sigma_1(t_1))^2t_2t_3(3t_2 + \sigma_1(t_1) + t_1) \\ &= (2t_1 + 1)t_2t_3(3t_1t_2 + t_1^2 + t_3^2) + (t_1 + 1)^2t_2t_3(3t_2 + 2t_1 + 1) \\ &= (2t_1 + 1)(3t_1t_2^2t_3 + t_1^2t_2t_3 + t_2t_3^3) + (t_1^2 + 2t_1 + 1)(3t_2^2t_3 + 2t_1t_2t_3 + t_2t_3) \\ &= 6t_1^2t_2^2t_3 + 2t_1^3t_2t_3 + 2t_1t_2t_3^3 + 3t_1t_2^2t_3 + t_1^2t_2t_3 + t_2t_3^3 + 3t_1^2t_2^2t_3 \end{aligned}$$

$$+2t_1^3t_2t_3 + t_1^2t_2t_3 + 6t_1t_2^2t_3 + 4t_1^2t_2t_3 + 2t_1t_2t_3 + 3t_2^2t_3 + 2t_1t_2t_3 + t_2t_3$$

$$= 4t_1^3t_2t_3 + 9t_1^2t_2^2t_3 + 6t_1^2t_2t_3 + 9t_1t_2^2t_3 + 2t_1t_2t_3^3$$

$$+4t_1t_2t_3 + 3t_2^2t_3 + t_2t_3 + t_2t_3^3.$$

For $t \in \Lambda_2^{\kappa_2 3}$, we have

$$h_{t_2}^{\Delta_2}(t) = f_{t_2}^{\Delta_2}(t)(g(t) + f_2^{\sigma_2}(t)g_{t_2}^{\Delta_2}(t)$$

$$= t_1^2t_3(3t_1t_2 + t_1^2 + t_3^2) + 3t_1^2\sigma_2(t_2)t_3t_1$$

$$= t_1^2t_3(3t_1t_2 + t_1^2 + t_3^2) + 3t_1^2(t_2 + 1)t_3t_1$$

$$= 3t_1^3t_2t_3 + t_1^4t_3 + t_1^2t_3^3 + 3t_1^3t_2t_3 + 3t_1^3t_3$$

$$= 6t_1^3t_2t_3 + 3t_1^3t_3 + t_1^2t_3^3 + t_1^4t_3.$$

For $t \in \Lambda_3^{\kappa_3 3}$, we have

$$h_{t_3}^{\Delta_3}(t) = f_{t_3}^{\Delta_3}(t)g(t) + f_3^{\sigma_3}(t)g_{t_3}^{\Delta_3}(t)$$

$$= t_1^2t_2(3t_1t_2 + t_1^2 + t_3^2) + t_1^2t_2\sigma_3(t_3)(\sigma_3(t_3) + t_3)$$

$$= t_1^2t_2(3t_1t_2 + t_1^2 + t_3^2) + 2t_1^2t_2t_3t_3$$

$$= 3t_1^3t_2^2 + t_1^4t_2 + t_1^2t_2t_3^2 + 2t_1^2t_2t_3^2$$

$$= t_1^4t_2 + 3t_1^3t_2^2 + 3t_1^2t_2t_3^2.$$

Exercise 6.64 Let $\Lambda^2 = \mathbb{N}_0^2 \times \mathbb{N}$ and define $h : \Lambda^2 \to \mathbb{R}$ by

$$h(t) = (t_1 + t_2^2)(t_1^2 + 2t_1t_2), \quad t = (t_1, t_2) \in \Lambda^2.$$

Find $h_{t_1}^{\Delta_1}(t)$ for $t \in \Lambda_1^{\kappa_1 2}$ and $h_{t_2}^{\Delta_2}(t)$ for $t \in \Lambda_2^{\kappa_2 2}$.

Solution

$$h_{t_1}^{\Delta_1}(t) = 3t_1^2 + 6t_1^{\frac{3}{2}} + t_1(7 + 4t_2 + 2t_2^2) + 2\sqrt{t_1}(2 + 2t_2 + t_2^2)$$

$$+2t_2^3 + t_2^2 + 2t_2 + 1, \quad t \in \Lambda_1^{\kappa_1 2},$$

$$h_{t_2}^{\Delta_2}(t) = t_1^2(3 + 2t_2) + t_1(6t_2^2 + 6t_2 + 2), \quad t \in \Lambda_2^{\kappa_2 2}.$$

Example 6.65 Let $f : \Lambda^n \to \mathbb{R}$ be partial delta differentiable with respect to t_i at $t \in \Lambda_i^{\kappa_i n}$. Then

$$\begin{aligned}
\left(f^2\right)_{t_i}^{\Delta_i} &= (ff)_{t_i}^{\Delta_i} \\
&= f_{t_i}^{\Delta_i} f + f_i^{\sigma_i} f_{t_i}^{\Delta_i} \\
&= f_{t_i}^{\Delta_i}(f_i^{\sigma_i} + f), \\
(f^3)_{t_i}^{\Delta_i} &= (ff^2)_{t_i}^{\Delta_i} \\
&= f_{t_i}^{\Delta_i} f^2 + f_i^{\sigma_i}(f^2)_{t_i}^{\Delta_i} \\
&= f_{t_i}^{\Delta_i} f^2 + f_i^{\sigma_i} f_{t_i}^{\Delta_i}(f_i^{\sigma_i} + f) \\
&= f_{t_i}^{\Delta_i}(f^2 + ff_i^{\sigma_i} + (f_i^{\sigma_i})^2).
\end{aligned}$$

We assume that

$$(f^n)_{t_i}^{\Delta_i} = f_{t_i}^{\Delta_i} \sum_{k=0}^{n-1} f^k (f_i^{\sigma_i})^{n-1-k}$$

for some $n \in \mathbb{N}$. We now prove that

$$(f^{n+1})_{t_i}^{\Delta_i} = f_{t_i}^{\Delta_i} \sum_{k=0}^{n} f^k (f_i^{\sigma_i})^{n-k}.$$

Indeed,

$$
\begin{aligned}
(f^{n+1})_{t_i}^{\Delta_i} &= (ff^n)_{t_i}^{\Delta_i} \\
&= f_{t_i}^{\Delta_i} f^n + f_i^{\sigma_i} (f^n)_{t_i}^{\Delta_i} \\
&= f_{t_i}^{\Delta_i} f^n + f_{t_i}^{\Delta_i} (f^{n-1} + f^{n-2} f_i^{\sigma_i} + \cdots \\
&\quad + f(f_i^{\sigma_i})^{n-2} + (f_i^{\sigma_i})^{n-1}) f_i^{\sigma_i} \\
&= f_{t_i}^{\Delta_i} \left(f^n + f^{n-1} f_i^{\sigma_i} + f^{n-2} (f_i^{\sigma_i})^2 + \cdots + (f_i^{\sigma_i})^n\right) \\
&= f_{t_i}^{\Delta_i} \sum_{k=0}^{n} f^k (f_i^{\sigma_i})^{n-k}.
\end{aligned}
$$

Theorem 6.66 *Let $f, g : \Lambda^n \to \mathbb{R}$ be partial delta differentiable with respect to t_i at $t \in \Lambda_i^{\kappa_i n}$. Assume $g_i^{\sigma_i}(t) g(t) \neq 0$. Then $\frac{f}{g}$ is partial delta differentiable with respect to t_i at t and*

$$
\left(\frac{f}{g}\right)_{t_i}^{\Delta_i} (t) = \frac{f_{t_i}^{\Delta_i}(t) g(t) - f(t) g_{t_i}^{\Delta_i}(t)}{g_i^{\sigma_i}(t) g(t)}.
$$

Proof Since $f, g : \Lambda^2 \to \mathbb{R}$ are partial delta differentiable with respect to t_i at t, for every $\varepsilon > 0$, there exists $\delta > 0$ such that for every $s_i \in (t_i - \delta, t_i + \delta)$, $s_i \in \mathbb{T}_i$, we have

$$
|f_i^{\sigma_i}(t) - f(t_{s_i}) - f_{t_i}^{\Delta_i}(t)(\sigma_i(t_i) - s_i)| \leq \varepsilon^* |\sigma_i(t_i) - s_i|,
$$

$$
|g_i^{\sigma_i}(t) - g(t_{s_i}) - g_{t_i}^{\Delta_i}(t)(\sigma_i(t_i) - s_i)| \leq \varepsilon^* |\sigma_i(t_i) - s_i|,
$$

$$
|f(t) g(t_{s_i}) - f(t_{s_i}) g(t)| \leq \varepsilon^*,
$$

where

$$
0 < \varepsilon^* < \varepsilon \frac{|g_i^{\sigma_i}(t)||g(t_{s_i})||g(t)|}{1 + |g(t)||g(t_{s_i})| + |f(t_{s_i})||g(t)| + |g_{t_i}^{\Delta_i}(t)|}.
$$

Hence,

$$
\left| \left(\frac{f}{g}\right)_i^{\sigma_i} (t) - \left(\frac{f}{g}\right)(t_{s_i}) - \frac{f_{t_i}^{\Delta_i}(t) g(t) - f(t) g_{t_i}^{\Delta_i}(t)}{g_i^{\sigma_i}(t) g(t)} (\sigma_i(t) - s_i) \right|
$$

$$
\begin{aligned}
&= \left| \frac{f_i^{\sigma_i}(t)}{g_i^{\sigma_i}(t)} - \frac{f(t_{s_i})}{g(t_{s_i})} - \frac{f_{t_i}^{\Delta_i}(t)g(t) - f(t)g_{t_i}^{\Delta_i}(t)}{g_i^{\sigma_i}(t)g(t)}(\sigma_i(t_i) - s_i) \right| \\
&= \frac{1}{|g_i^{\sigma_i}(t)||g(t_{s_i})||g(t)|} \Big| f_i^{\sigma_i}(t)g(t)g(t_{s_i}) - f(t_{s_i})g(t)g_i^{\sigma_i}(t) \\
&\quad - f_{t_i}^{\Delta_i}(t)g(t)g(t_{s_i})(\sigma_i(t_i) - s_i) + f(t)g(t_{s_i})g_{t_i}^{\Delta_i}(t)(\sigma_i(t_i) - s_i) \Big| \\
&= \frac{1}{|g_i^{\sigma_i}(t)||g(t_{s_i})||g(t)|} \Big| f_i^{\sigma_i}(t)g(t)g(t_{s_i}) - f(t_{s_i})g(t)g(t_{s_i}) \\
&\quad + f(t_{s_i})g(t)g(t_{s_i}) - f(t_{s_i})g(t)g_i^{\sigma_i}(t) \\
&\quad - f_{t_i}^{\Delta_i}(t)g(t)g(t_{s_i})(\sigma_i(t_i) - s_i) + f(t)g(t_{s_i})g_{t_i}^{\Delta_i}(t)(\sigma_i(t_i) - s_i) \Big| \\
&= \frac{1}{|g_i^{\sigma_i}(t)||g(t_{s_i})||g(t)|} \Big| g(t)g(t_{s_i})(f_i^{\sigma_i}(t) - f(t_{s_i}) - f_{t_i}^{\Delta_i}(t)(\sigma_i(t_i) - s_i)) \\
&\quad + f(t_{s_i})g(t)g(t_{s_i}) - f(t_{s_i})g(t)g_i^{\sigma_i}(t) + f(t_{s_i})g(t)g_{t_i}^{\Delta_i}(t)(\sigma_i(t_i) - s_i) \\
&\quad - f(t_{s_i})g(t)g_{t_i}^{\Delta_i}(t)(\sigma_i(t_i) - s_i) + f(t)g(t_{s_i})g_{t_i}^{\Delta_i}(t)(\sigma_i(t_i) - s_i) \Big| \\
&= \frac{1}{|g_i^{\sigma_i}(t)||g(t_{s_i})||g(t)|} \Big| g(t)g(t_{s_i})(f_i^{\sigma_i}(t) - f(t_{s_i}) - f_{t_i}^{\Delta_i}(t)(\sigma_i(t_i) - s_i)) \\
&\quad - f(t_{s_i})g(t)(g_i^{\sigma_i}(t) - g(t_{s_i}) - g_{t_i}^{\Delta_i}(t)(\sigma_i(t_i) - s_i)) \\
&\quad + g_{t_i}^{\Delta_i}(t)(f(t)g(t_{s_i}) - f(t_{s_i})g(t))(\sigma_i(t_i) - s_i) \Big| \\
&\le \frac{1}{|g_i^{\sigma_i}(t)||g(t_{s_i})||g(t)|} \Big(|g(t)||g(t_{s_i})||f_i^{\sigma_i}(t) - f(t_{s_i}) - f_{t_i}^{\Delta_i}(t)(\sigma_i(t_i) - s_i)| \\
&\quad + |f(t_{s_i})||g(t)||g_i^{\sigma_i}(t) - g(t_{s_i}) - g_{t_i}^{\Delta_i}(t)(\sigma_i(t_i) - s_i)| \\
&\quad + |g_{t_i}^{\Delta_i}(t)||(f(t)g(t_{s_i}) - f(t_{s_i})g(t)||\sigma_i(t_i) - s_i| \Big) \\
&\le \frac{1}{|g_i^{\sigma_i}(t)||g(t_{s_i})||g(t)|} (\varepsilon^* |g(t)||g(t_{s_i})||\sigma_i(t_i) - s_i| \\
&\quad + \varepsilon^* |f(t_{s_i})||g(t)||\sigma_i(t_i) - s_i| + \varepsilon^* |g_{t_i}^{\Delta_i}(t)||\sigma_i(t_i) - s_i|)
\end{aligned}
$$

$$\leq \frac{1}{|g_i^{\sigma_i}(t)||g(t_{s_i})||g(t)|}\varepsilon^*(|g(t)||g(t_{s_i})||\sigma_i(t_i) - s_i|$$

$$+|f(t_{s_i})||g(t)||\sigma_i(t_i) - s_i| + |g_{t_i}^{\Delta_i}(t)||\sigma_i(t_i) - s_i|)$$

$$\leq \frac{1}{|g_i^{\sigma_i}(t)||g(t_{s_i})||g(t)|}\varepsilon^*(1 + |g(t)||g(t_{s_i})|$$

$$+|f(t_{s_i})||g(t)| + |g_{t_i}^{\Delta_i}(t)|)|\sigma_i(t_i) - s_i|$$

$$\leq \varepsilon|\sigma_i(t_i) - s_i|.$$

This completes the proof. □

Example 6.67 Let $\Lambda^2 = \mathbb{N} \times \mathbb{N}$ and define $h : \Lambda^2 \to \mathbb{R}$ by

$$h(t) = \frac{t_1^2 + 2t_1t_2 + t_2^3}{t_1 + t_2}, \quad t = (t_1, t_2) \in \Lambda^2.$$

We will find $h_{t_1}^{\Delta_1}(t)$ for $t \in \Lambda_1^{\kappa_1 2}$ and $h_{t_2}^{\Delta_2}(t)$ for $t_2 \in \Lambda_2^{\kappa_2 2}$. Let

$$f(t) = t_1^2 + 2t_1t_2 + t_2^3, \quad g(t) = t_1 + t_2, \quad t = (t_1, t_2) \in \Lambda^2.$$

Here,

$$\mathbb{T}_1 = \mathbb{T}_2 = \mathbb{N}$$

and

$$\sigma_1(t_1) = t_1 + 1, \quad t_1 \in \mathbb{T}_1, \quad \sigma_2(t_2) = t_2 + 1, \quad t_2 \in \mathbb{T}_2.$$

We have

$$f_{t_1}^{\Delta_1}(t) = \sigma_1(t_1) + t_1 + 2t_2$$

$$= 2t_1 + 2t_2 + 1, \quad t \in \Lambda_1^{\kappa_1 2},$$

$$f_{t_2}^{\Delta_2}(t) = 2t_1 + (\sigma_2(t_2))^2 + t_2\sigma_2(t_2) + t_2^2$$

$$= 2t_1 + (t_2 + 1)^2 + t_2(t_2 + 1) + t_2^2$$

$$= 2t_1 + t_2^2 + 2t_2 + 1 + t_2^2 + t_2 + t_2^2$$

$$= 2t_1 + 3t_2^2 + 3t_2 + 1, \quad t \in \Lambda_2^{\kappa_2 2},$$

$$g_{t_1}^{\Delta_1}(t) = 1, \quad t \in \Lambda_1^{\kappa_1 2},$$

$$g_{t_2}^{\Delta_2}(t) = 1, \quad t \in \Lambda_2^{\kappa_2 2},$$

$$g_1^{\sigma_1}(t) = g_2^{\sigma_2}(t)$$

$$= t_1 + t_2 + 1, \quad t \in \Lambda^2.$$

Hence, for $t \in \Lambda_1^{\kappa_1 2}$, we have

$$\begin{aligned} h_{t_1}^{\Delta_1}(t) &= \frac{f_{t_1}^{\Delta_1}(t)g(t) - f(t)g_{t_1}^{\Delta_1}(t)}{g_1^{\sigma_1}(t)g(t)} \\ &= \frac{(2t_1 + 2t_2 + 1)(t_1 + t_2) - (t_1^2 + 2t_1t_2 + t_2^3)}{(t_1 + t_2 + 1)(t_1 + t_2)} \\ &= \frac{2t_1^2 + 2t_1t_2 + 2t_1t_2 + 2t_2^2 + t_1 + t_2 - t_1^2 - 2t_1t_2 - t_2^3}{(t_1 + t_2 + 1)(t_1 + t_2)} \\ &= \frac{t_1^2 + 2t_1t_2 + 2t_2^2 - t_2^3 + t_1 + t_2}{(t_1 + t_2 + 1)(t_1 + t_2)}. \end{aligned}$$

For $t \in \Lambda_2^{\kappa_2 2}$, we have

$$\begin{aligned} h_{t_2}^{\Delta_2}(t) &= \frac{f_{t_2}^{\Delta_2}(t)g(t) - f(t)g_{t_2}^{\Delta_2}(t)}{g_2^{\sigma_2}(t)g(t)} \\ &= \frac{(2t_1 + 3t_2^2 + 3t_2 + 1)(t_1 + t_2) - (t_1^2 + 2t_1t_2 + t_2^3)}{(t_1 + t_2 + 1)(t_1 + t_2)} \\ &= \frac{2t_1^2 + 2t_1t_2 + 3t_1t_2^2 + 3t_2^3 + 3t_1t_2 + 3t_2^2 + t_1 + t_2 - t_1^2 - 2t_1t_2 - t_2^3}{(t_1 + t_2 + 1)(t_1 + t_2)} \\ &= \frac{t_1^2 + 3t_1t_2 + 3t_1t_2^2 + 2t_2^3 + 3t_2^2 + t_1 + t_2}{(t_1 + t_2 + 1)(t_1 + t_2)}. \end{aligned}$$

Example 6.68 Let $\Lambda^2 = 2^{\mathbb{N}} \times \mathbb{N}_0$ and define $h : \Lambda^2 \to \mathbb{R}$ by

$$h(t) = \frac{t_1 - t_2}{t_1^2 + t_2}, \quad t \in \Lambda^2.$$

We will find $h_{t_1}^{\Delta_1}(t)$ for $t \in \Lambda_1^{\kappa_1 2}$ and $h_{t_2}^{\Delta_2}(t)$ for $t \in \Lambda_2^{\kappa_2 2}$. Let

$$f(t) = t_1 - t_2, \quad g(t) = t_1^2 + t_2, \quad t \in \Lambda^2.$$

Here,

$$\mathbb{T}_1 = 2^{\mathbb{N}}, \quad \mathbb{T}_2 = \mathbb{N}.$$

$$\sigma_1(t_1) = 2t_1, \quad t_1 \in \mathbb{T}_1, \quad \sigma_2(t_2) = t_2 + 1, \quad t_2 \in \mathbb{T}_2,$$

$$f_{t_1}^{\Delta_1}(t) = 1, \quad g_{t_1}^{\Delta_1}(t) = \sigma_1(t_1) + t_1 = 2t_1 + t_1 = 3t_1, \quad t \in \Lambda_1^{\kappa_1 2},$$

$$f_{t_2}^{\Delta_2}(t) = -1, \quad g_{t_2}^{\Delta_2}(t) = 1, \quad t \in \Lambda_2^{\kappa_2 2},$$

$$g_1^{\sigma_1}(t) = 4t_1^2 + t_2, \quad g_2^{\sigma_2}(t) = t_1^2 + t_2 + 1, \quad t \in \Lambda^2.$$

Hence, for $t \in \Lambda_1^{\kappa_1 2}$, we have

$$\begin{aligned} h_{t_1}^{\Delta_1}(t) &= \frac{f_{t_1}^{\Delta_1}(t)g(t) - f(t)g_{t_1}^{\Delta_1}(t)}{g_1^{\sigma_1}(t)g(t)} \\ &= \frac{(t_1^2 + t_2) - (t_1 - t_2)3t_1}{(4t_1^2 + t_2)(t_1^2 + t_2)} \\ &= \frac{t_1^2 + t_2 - 3t_1^2 + 3t_1 t_2}{(4t_1^2 + t_2)(t_1^2 + t_2)} \\ &= \frac{-2t_1^2 + 3t_1 t_2 + t_2}{(4t_1^2 + t_2)(t_1^2 + t_2)}. \end{aligned}$$

For $t \in \Lambda_2^{\kappa_2 2}$, we get

$$h_{t_2}^{\Delta_2}(t) = \frac{f_{t_2}^{\Delta_2}(t)g(t) - f(t)g_{t_2}^{\Delta_2}(t)}{g_2^{\sigma_2}(t)g(t)}$$

$$= \frac{-(t_1^2 + t_2) - (t_1 - t_2)}{(t_1^2 + t_2 + 1)(t_1^2 + t_2)}$$

$$= -\frac{t_1^2 + t_1}{(t_1^2 + t_2 + 1)(t_1^2 + t_2)}.$$

Example 6.69 Let $\Lambda^2 = \mathbb{Z} \times \mathbb{N}_0$ and define $h : \Lambda^2 \to \mathbb{R}$ by

$$h(t) = \frac{t_1 + 2t_2 + 3}{t_1 - t_2 + 1}, \quad t \in \Lambda^2.$$

We will find

$$h_{t_1}^{\Delta_1}(t), \quad t \in \Lambda_1^{\kappa_1 2}, \quad (t_1 - t_2 + 1)(t_1 - t_2 + 2) \neq 0$$

and

$$h_{t_2}^{\Delta_2}(t), \quad t \in \Lambda_2^{\kappa_2 2}, \quad (t_1 - t_2 + 1)(t_1 - t_2) \neq 0.$$

Here,

$$\mathbb{T}_1 = \mathbb{Z}, \quad \mathbb{T}_2 = \mathbb{N}_0$$

and

$$\sigma_1(t_1) = t_1 + 1, \quad t_1 \in \mathbb{T}_1, \quad \sigma_2(t_2) = t_2 + 1, \quad t_2 \in \mathbb{T}_2.$$

Let

$$f(t) = t_1 + 2t_2 + 3, \quad g(t) = t_1 - t_2 + 1, \quad t \in \Lambda^2.$$

We have

$$f_{t_1}^{\Delta_1}(t) = g_{t_1}^{\Delta_1}(t) = 1, \quad t \in \Lambda_1^{\kappa_1 2},$$

$$f_{t_2}^{\Delta_2} = 2, \quad g_{t_2}^{\Delta_2} = -1, \quad t \in \Lambda_2^{\kappa_2 2},$$

$$g_1^{\sigma_1}(t) = t_1 - t_2 + 2, \quad g_2^{\sigma_2}(t) = t_1 - t_2, \quad t \in \Lambda^2.$$

For $t \in \Lambda_1^{\kappa_1 2}$ with $(t_1 - t_2 + 1)(t_1 - t_2 + 2) \neq 0$, we have

$$h_{t_1}^{\Delta_1}(t) = \frac{f_{t_1}^{\Delta_1}(t)g(t) - f(t)g_{t_1}^{\Delta_1}(t)}{g_1^{\sigma_1}(t)g(t)}$$

$$= \frac{t_1 - t_2 + 1 - (t_1 + 2t_2 + 3)}{(t_1 - t_2 + 2)(t_1 - t_2)}$$

$$= \frac{-3t_2 - 2}{(t_1 - t_2 + 2)(t_1 - t_2)}.$$

For $t \in \Lambda_2^{\kappa_2 2}$ with $(t_1 - t_2 + 1)(t_1 - t_2) \neq 0$, we have

$$\begin{aligned} h_{t_2}^{\Delta_2}(t) &= \frac{f_{t_2}^{\Delta_2}(t)g(t) - f(t)g_{t_2}^{\Delta_2}(t)}{g_2^{\sigma_2}(t)g(t)} \\ &= \frac{2(t_1 - t_2 + 1) + (t_1 + 2t_2 + 3)}{(t_1 - t_2)(t_1 - t_2 + 1)} \\ &= \frac{3t_1 + 5}{(t_1 - t_2)(t_1 - t_2 + 1)}. \end{aligned}$$

Exercise 6.70 Let $\Lambda^2 = \mathbb{N} \times \mathbb{N}_0$ and define $h : \Lambda^2 \to \mathbb{R}$ by

$$h(t) = \frac{t_1 + t_2}{t_1 - t_2 + 1}, \quad t \in \Lambda^2.$$

Find $h_{t_1}^{\Delta_1}(t)$ for $t \in \Lambda_1^{\kappa_1 2}$, $(t_1 - t_2 + 2)(t_1 - t_2 + 1) \neq 0$, and $h_{t_2}^{\Delta_2}(t)$ for $t \in \Lambda_2^{\kappa_2 2}$, $(t_1 - t_2)(t_1 - t_2 + 1) \neq 0$.

Solution

$$h_{t_1}^{\Delta_1}(t) = \frac{1 - 2t_2}{(t_1 - t_2 + 1)(t_1 - t_2 + 2)}, \quad h_{t_2}^{\Delta_2}(t) = \frac{2t_1 + 1}{(t_1 - t_2 + 1)(t_1 - t_2)}.$$

Definition 6.71 For a function $f : \Lambda^n \to \mathbb{R}$, we shall talk about the *second-order partial delta derivative* with respect to t_i and t_j, $i, j \in \{1, 2, \ldots, n\}$, $f_{t_i t_j}^{\Delta_i \Delta_j}$, provided $f_{t_i}^{\Delta_i}$ is partial delta differentiable with respect to t_j on $\Lambda_{ij}^{\kappa_i \kappa_j n} = \left(\Lambda_i^{\kappa_i n}\right)_j^{\kappa_j n}$ with partial delta derivative

$$f_{t_i t_j}^{\Delta_i \Delta_j} = \left(f_{t_i}^{\Delta_i}\right)_{t_j}^{\Delta_j} : \Lambda_{ij}^{\kappa_i \kappa_j n} \to \mathbb{R}.$$

For $i = j$, we will write

$$f_{t_i t_i}^{\Delta_i \Delta_i} = f_{t_i^2}^{\Delta_i^2}.$$

Similarly, we define *higher-order partial delta derivatives*

$$f_{t_i t_j \ldots t_l}^{\Delta_i \Delta_j \ldots \Delta_l} : \Lambda_{ij \ldots l}^{\kappa_i \kappa_j \ldots \kappa_l n} \to \mathbb{R}.$$

For $t \in \Lambda^n$, we define

$$\sigma^2(t) = \sigma(\sigma(t)) = (\sigma_1(\sigma_1(t_1)), \sigma_2(\sigma_2(t_2)), \ldots, \sigma_n(\sigma_n(t_n)))$$

and

$$\rho^2(t) = \rho(\rho(t)) = (\rho_1(\rho_1(t_1)), \rho_2(\rho_2(t_2)), \ldots, \rho_n(\rho_n(t_n))),$$

and $\sigma^m(t)$ and $\rho^m(t)$ for $m \in \mathbb{N}$ are defined accordingly. Finally, we put

$$\rho^0(t) = \sigma^0(t) = t, \quad f_{t_i}^{\Delta_i^0} = f, \quad \Lambda_i^{\kappa_i^0 n} = \Lambda^n.$$

Example 6.72 Let $\Lambda^3 = \mathbb{N} \times \mathbb{N}_0 \times \mathbb{Z}$. Here

$$\mathbb{T}_1 = \mathbb{N}, \quad \mathbb{T}_2 = \mathbb{N}_0, \quad \mathbb{T}_3 = \mathbb{Z}$$

and

$$\sigma_1(t_1) = t_1 + 1, \quad t_1 \in \mathbb{T}_1, \quad \sigma_2(t_2) = t_2 + 1, \quad t_2 \in \mathbb{T}_2,$$

$$\sigma_3(t_3) = t_3 + 1, \quad t_3 \in \mathbb{T}_3.$$

Hence,

$$\begin{aligned}\sigma^2(t) &= (\sigma_1(\sigma_1(t_1)), \sigma_2(\sigma_2(t_2)), \sigma_3(\sigma_3(t_3)))\\ &= (\sigma_1(t_1) + 1, \sigma_2(t_2) + 1, \sigma_3(t_3) + 1)\\ &= (t_1 + 1 + 1, t_2 + 1 + 1, t_3 + 1 + 1)\\ &= (t_1 + 2, t_2 + 2, t_3 + 2).\end{aligned}$$

Example 6.73 Let $\Lambda^2 = 2^{\mathbb{N}} \times \mathbb{N}_0$. Here,

$$\mathbb{T}_1 = 2^{\mathbb{N}}, \quad \mathbb{T}_2 = \mathbb{N}_0.$$

Then

$$\rho_1(t_1) = \frac{t_1}{2}, \quad t_1 \neq 2, \quad t_1 \in \mathbb{T}_1, \quad \rho_1(2) = 2,$$

$$\rho_2(t_2) = t_2 - 1, \quad t_2 \in \mathbb{T}_2, \quad t_2 \neq 0, \quad \rho_2(0) = 0.$$

Therefore,

$$\rho(t) = \begin{cases} \left(\frac{t_1}{2}, t_2 - 1\right) & \text{if } t_1 \geq 4, \quad t_2 \geq 1, \\ (2, t_2 - 1) & \text{if } t_1 = 2, \quad t_2 \geq 1, \\ \left(\frac{t_1}{2}, 0\right) & \text{if } t_1 \geq 4, \quad t_2 = 0, \\ (2, 0) & \text{if } t_1 = 2, \quad t_2 = 0. \end{cases}$$

1. Let $t_1 \geq 4$, $t_2 \geq 1$. Then

$$\begin{aligned} \rho^2(t) &= (\rho_1(\rho_1(t_1)), \rho_2(\rho_2(t_2))) \\ &= \left(\rho_1\left(\frac{t_1}{2}\right), \rho_2(t_2 - 1)\right) \\ &= \begin{cases} \left(\frac{t_1}{4}, t_2 - 2\right) & \text{if } t_1 \geq 8, \quad t_2 \geq 2, \\ \left(\frac{t_1}{4}, 0\right) & \text{if } t_1 \geq 8, \quad t_2 = 1, \\ (2, t_1 - 2) & \text{if } t_1 = 4, \quad t_2 \geq 2, \\ (2, 0) & \text{if } t_1 = 4, \quad t_2 = 1. \end{cases} \end{aligned}$$

2. Let $t_1 = 2$, $t_2 \geq 1$. Then

$$\begin{aligned} \rho^2(t) &= (\rho_1(2), \rho_2(t_2 - 1)) \\ &= \begin{cases} (2, t_2 - 2) & \text{if } t_2 \geq 2, \\ (2, 0) & \text{if } t_2 = 1. \end{cases} \end{aligned}$$

3. Let $t_1 \geq 4$, $t_2 = 0$. Then

$$\begin{aligned} \rho^2(t) &= \left(\rho_1\left(\frac{t_1}{2}\right), \rho_2(0)\right) \\ &= \begin{cases} \left(\frac{t_1}{4}, 0\right) & \text{if } t_1 \geq 8, \\ (2, 0) & \text{if } t_1 = 4. \end{cases} \end{aligned}$$

4. $\rho^2(2,0) = (\rho_1(2), \rho_2(0)) = (2,0)$.

Example 6.74 Let $\Lambda^2 = \mathbb{Z} \times \mathbb{N}$. We will find $(\sigma^2(t)) \cdot (\rho^3(t))$, $t \in \Lambda^2$, where $\cdot$ is the inner product in $\mathbb{R}^2$. Here,

$$\mathbb{T}_1 = \mathbb{Z}, \quad \mathbb{T}_2 = \mathbb{N}$$

and

$$\sigma_1(t_1) = t_1 + 1, \quad t_1 \in \mathbb{T}_1, \quad \sigma_2(t_2) = t_2 + 1, \quad t_2 \in \mathbb{T}_2.$$

Hence,

$$\begin{aligned}\sigma(t) &= (\sigma_1(t_1), \sigma_2(t_2)) \\ &= (t_1 + 1, t_2 + 1), \\ \sigma^2(t) &= (\sigma_1(\sigma_1(t_1)), \sigma_2(\sigma_2(t_2))) \\ &= (\sigma_1(t_1 + 1), \sigma_2(t_2 + 1)) \\ &= (t_1 + 2, t_2 + 2), \quad t \in \Lambda^2.\end{aligned}$$

Also,

$$\begin{aligned}\rho(t) &= (\rho_1(t_1), \rho_2(t_2)) \\ &= \begin{cases} (t_1 - 1, t_2 - 1) & \text{if } \; t_2 \geq 2, \\ (t_1 - 1, 1) & \text{if } \; t_2 = 1, \end{cases} \\ \rho^2(t) &= (\rho_1(\rho_1(t_1)), \rho_2(\rho_2(t_2))) \\ &= \begin{cases} (\rho_1(t_1 - 1), \rho_2(t_2 - 1)) & \text{if } \; t_2 \geq 2, \\ (\rho_1(t_1 - 1), \rho_2(1)) & \text{if } \; t_2 = 1 \end{cases} \\ &= \begin{cases} (t_1 - 2, t_2 - 2) & \text{if } \; t_2 \geq 3, \\ (t_1 - 2, 1) & \text{if } \; t_2 \in \{1, 2\}, \end{cases} \\ \rho^3(t) &= (\rho_1(\rho_1^2(t_1)), \rho_2(\rho_2^2(t_2)))\end{aligned}$$

$$= \begin{cases} (\rho_1(t_1-2), \rho_2(t_2-2) & \text{if } t_2 \geq 3, \\ (\rho_1(t_1-2), \rho_2(1)) & \text{if } t_2 \in \{1,2\} \end{cases}$$

$$= \begin{cases} (t_1-3, t_2-3) & \text{if } t_2 \geq 4, \\ (t_1-3, 1) & \text{if } t_2 \in \{1,2,3\}, \quad t \in \Lambda^2. \end{cases}$$

Hence, for $t \in \Lambda^2$, we have

$$\sigma^2(t) \cdot \rho^3(t) = \begin{cases} (t_1+2, t_2+2) \cdot (t_1-3, t_2-3) & \text{if } t_2 \geq 4, \\ (t_1+2, t_2+2) \cdot (t_1-3, 1) & \text{if } t_2 \in \{1,2,3\} \end{cases}$$

$$= \begin{cases} (t_1+2, t_2+2) \cdot (t_1-3, t_2-3) & \text{if } t_2 \geq 4, \\ (t_1+2, 3) \cdot (t_1-3, 1) & \text{if } t_2 = 1, \\ (t_1+2, 4) \cdot (t_1-3, 1) & \text{if } t_2 = 2, \\ (t_1+2, 5) \cdot (t_1-3, 1) & \text{if } t_2 = 3, \end{cases}$$

$$= \begin{cases} (t_1+2)(t_1-3) + (t_2+2)(t_2-3) & \text{if } t_2 \geq 4, \\ (t_1+2)(t_1-3) + 3 & \text{if } t_2 = 1, \\ (t_1+2)(t_1-3) + 4 & \text{if } t_2 = 2, \\ (t_1+2)(t_1-3) + 5 & \text{if } t_2 = 3 \end{cases}$$

$$= \begin{cases} t_1^2 + t_2^2 - t_1 - t_2 - 12 & \text{if } t_2 \geq 4, \\ t_1^2 - t_1 - 3 & \text{if } t_2 = 1, \\ t_1^2 - t_1 - 2 & \text{if } t_2 = 2, \\ t_1^2 - t_1 - 1 & \text{if } t_2 = 3. \end{cases}$$

Exercise 6.75 Let $\Lambda^2 = 3^{\mathbb{N}} \times \mathbb{Z}$. Find $\sigma^2(t)$ and $\rho^2(t)$, $t \in \Lambda^2$.

Solution

$$\sigma^2(t) = (9t_1, t_2 + 2), \quad \rho^2(t) = \begin{cases} \left(\frac{t_1}{9}, t_2 - 2\right) & \text{if} \quad t_1 \geq 27, \\ \\ (3, t_2 - 2) & \text{if} \quad t_2 \in \{3, 9\}, \quad t \in \Lambda^2. \end{cases}$$

Example 6.76 Let $\Lambda^2 = 2^{\mathbb{N}} \times \mathbb{N}_0$ and define $f : \Lambda^2 \to \mathbb{R}$ by

$$f(t) = t_1^2 t_2 + t_1 t_2^2, \quad t \in \Lambda^2.$$

Here, $\mathbb{T}_1 = 2^{\mathbb{N}}$, $\mathbb{T}_2 = \mathbb{N}_0$, and

$$\sigma_1(t_1) = 2t_1, \quad t_1 \in \mathbb{T}_1, \quad \sigma_2(t_2) = t_2 + 1, \quad t_2 \in \mathbb{T}_2.$$

Thus,

$$\begin{aligned} f_{t_1}^{\Delta_1}(t) &= t_2(\sigma_1(t_1) + t_1) + t_2^2 \\ &= t_2(2t_1 + t_1) + t_2^2 \\ &= 3t_1 t_2 + t_2^2, \quad t \in \Lambda_1^{\kappa_1 2}, \\ f_{t_2}^{\Delta_2}(t) &= t_1^2 + t_1(\sigma_2(t_2) + t_2) \\ &= t_1^2 + t_1(2t_2 + 1) \\ &= t_1^2 + 2t_1 t_2 + t_1, \quad t \in \Lambda_2^{\kappa_2 2}, \\ f_{t_1 t_2}^{\Delta_1 \Delta_2}(t) &= 3t_1 + t_2 + \sigma_2(t_2) \\ &= 3t_1 + 2t_2 + 1, \\ f_{t_2 t_1}^{\Delta_2 \Delta_1}(t) &= \sigma_1(t_1) + t_1 + 2t_2 + 1 \\ &= 3t_1 + 2t_2 + 1, \quad t \in \Lambda_{12}^{\kappa_1 \kappa_2 2}. \end{aligned}$$

Example 6.77 Let $\Lambda^3 = \mathbb{N} \times \mathbb{N}_0 \times \mathbb{Z}$ and define $f : \Lambda^3 \to \mathbb{R}$ by

$$f(t) = t_1 t_2 t_3 + t_1^2 + t_2^2 + t_3^2, \quad t \in \Lambda^3.$$

We will find $f_{t_1 t_2 t_3}^{\Delta_1 \Delta_2 \Delta_3}(t)$ for $t \in \Lambda_{123}^{\kappa_1 \kappa_2 \kappa_3 3}$. Here, $\mathbb{T}_1 = \mathbb{N}$, $\mathbb{T}_2 = \mathbb{N}_0$, $\mathbb{T}_3 = \mathbb{Z}$, and

$$\sigma_1(t_1) = t_1 + 1, \quad t_1 \in \mathbb{T}_1, \quad \sigma_2(t_2) = t_2 + 1, \quad t_2 \in \mathbb{T}_2,$$

$$\sigma_3(t_3) = t_3 + 1, \quad t_3 \in \mathbb{T}_3.$$

Hence,

$$f_{t_1}^{\Delta_1}(t) = t_2 t_3 + \sigma_1(t_1) + t_1$$

$$= t_2 t_3 + 2t_1 + 1, \quad t \in \Lambda_1^{\kappa_1 3},$$

$$f_{t_1 t_2}^{\Delta_1 \Delta_2}(t) = t_3, \quad t \in \Lambda_{12}^{\kappa_1 \kappa_2 3},$$

$$f_{t_1 t_2 t_3}^{\Delta_1 \Delta_2 \Delta_3} = 1, \quad t \in \Lambda_{123}^{\kappa_1 \kappa_2 \kappa_3 3}.$$

Example 6.78 Let $\Lambda^3 = 3^{\mathbb{N}} \times \mathbb{N}_0 \times \mathbb{Z}$ and define $f : \Lambda^3 \to \mathbb{R}$ by

$$f(t) = t_1^2 t_2^2 \sin(t_3), \quad t \in \Lambda^3.$$

Here, $\mathbb{T}_1 = 3^{\mathbb{N}}$, $\mathbb{T}_2 = \mathbb{N}_0$, $\mathbb{T}_3 = \mathbb{Z}$, and

$$\sigma_1(t_1) = 3t_1, \quad t_1 \in \mathbb{T}_1, \quad \sigma_2(t_2) = t_2 + 1, \quad t_2 \in \mathbb{T}_2,$$

$$\sigma_3(t_3) = t_3 + 1, \quad t_3 \in \mathbb{T}_3.$$

Hence,

$$f_{t_1}^{\Delta_1}(t) = (t_1 + \sigma_1(t_1)) t_2^2 \sin(t_3)$$

$$= 4t_1 t_2^2 \sin(t_3), \quad t \in \Lambda_1^{\kappa_1 3},$$

$$f_{t_1 t_2}^{\Delta_1 \Delta_2}(t) = 4t_1 (t_2 + \sigma_2(t_2)) \sin(t_3)$$

$$= 4t_1 (2t_2 + 1) \sin(t_3), \quad t \in \Lambda_{12}^{\kappa_1 \kappa_2 3},$$

$$f_{t_1t_2t_3}^{\Delta_1\Delta_2\Delta_3}(t) = 4t_1(2t_2+1)\frac{\sin(t_3+1)-\sin(t_3)}{\sigma_3(t_3)-t_3}$$
$$= 4t_1(2t_2+1)\left(\sin(t_3+1)-\sin(t_3)\right)$$
$$= 8t_1(2t_2+1)\sin\left(\frac{1}{2}\right)\cos\left(t_3+\frac{1}{2}\right), \quad t \in \Lambda_{123}^{\kappa_1\kappa_2\kappa_3 3}.$$

Exercise 6.79 Let $\Lambda^3 = 4^{\mathbb{N}} \times \mathbb{N}_0 \times \mathbb{R}$ and define $f : \Lambda^3 \to \mathbb{R}$ by

$$f(t) = \cos(t_1) + t_1^2 t_2 t_3, \quad t \in \Lambda^3.$$

Find

$$f_{t_1t_2t_3}^{\Delta_1\Delta_2\Delta_3}(t), \quad t \in \Lambda_{123}^{\kappa_1\kappa_2\kappa_3 3}.$$

Solution $5t_1$.

Theorem 6.80 (Leibniz Formula) *Let $S_{ik}^{(m)}$ be the set consisting of all possible strings of length m, containing exactly k times σ_i and $m-k$ times Δ_i. If $f_{t_i^{m-k}}^{\alpha}$ exists for any $\alpha \in S_{ik}^{(m)}$ and $g_{t_i^k}^{\Delta_i^k}$ exists for any $k \in \{0, 1, \ldots, m\}$, then*

$$(fg)_{t_i^m}^{\Delta_i^m} = \sum_{k=0}^{m}\left(\sum_{\alpha \in S_{ik}^{(m)}} f_{t_i^{m-k}}^{\alpha}\right) g_{t_i^k}^{\Delta_i^k} \tag{6.9}$$

holds for any $m \in \mathbb{N}$.

Proof We will use induction.

1. Since

$$(fg)_{t_i}^{\Delta_i} = f_{t_i}^{\Delta_i} g + f_i^{\sigma_i} g_{t_i}^{\Delta_i},$$

(6.9) holds for $m = 1$.
2. We assume (6.9) holds for some $m \in \mathbb{N}$. We will prove

$$(fg)_{t_i^{m+1}}^{\Delta_i^{m+1}} = \sum_{k=0}^{m+1}\left(\sum_{\alpha \in S_{ik}^{(m+1)}} f_{t_i^{m-k+1}}^{\alpha}\right) g_{t_i^k}^{\Delta_i^k}. \tag{6.10}$$

Indeed,

$$(fg)_{t_i^{m+1}}^{\Delta_i^{m+1}} = \left((fg)_{t_i^m}^{\Delta_i^m}\right)_{t_i}^{\Delta_i}$$

$$= \left(\sum_{k=0}^{m} \left(\sum_{\alpha \in S_{ik}^{(m)}} f_{t_i^{m-k}}^{\alpha} \right) g_{t_i^k}^{\Delta_i^k} \right)_{t_i}^{\Delta_i}$$

$$= \sum_{k=0}^{m} \left(\sum_{\alpha \in S_{ik}^{(m)}} f_{t_i^{m-k}}^{\alpha} \right)_{t_i}^{\Delta_i} g_{t_i^k}^{\Delta_i^k} + \sum_{k=0}^{m} \left(\sum_{\alpha \in S_{ik}^{(m)}} f_{t_i^{m-k}}^{\alpha} \right)^{\sigma_i} g_{t_i^{k+1}}^{\Delta_i^{k+1}}$$

$$= \sum_{k=0}^{m} \left(\sum_{\alpha \in S_{ik}^{(m)}} f_{t_i^{m-k}}^{\alpha} \right)_{t_i}^{\Delta_i} g_{t_i^k}^{\Delta_i^k} + \sum_{k=1}^{m+1} \left(\sum_{\alpha \in S_{ik-1}^{(m)}} f_{t_i^{m-k+1}}^{\alpha\sigma_i} \right) g_{t_i^k}^{\Delta_i^k}$$

$$= \sum_{\alpha \in S_{i0}^{(m)}} f_{t_i^{m+1}}^{\alpha\Delta_i} g + \sum_{k=1}^{m} \left(\sum_{\alpha \in S_{ik}^{(m)}} f_{t_i^{m-k+1}}^{\alpha\Delta_i} \right) g_{t_i^k}^{\Delta_i^k}$$

$$+ \sum_{k=1}^{m} \left(\sum_{\alpha \in S_{ik-1}^{(m)}} f_{t_i^{m-k+1}}^{\alpha\sigma_i} \right) g_{t_i^k}^{\Delta_i^k} + \sum_{\alpha \in S_{im}^{(m)}} f_{t_i^0}^{\alpha\sigma_i} g_{t_i^{m+1}}^{\Delta_i^{m+1}}$$

$$= \sum_{\alpha \in S_{i0}^{(m)}} f_{t_i^{m+1}}^{\alpha\Delta_i} g + \sum_{\alpha \in S_{im}^{(m)}} f_{t_i^0}^{\alpha\sigma_i} g_{t_i^{m+1}}^{\Delta_i^{m+1}}$$

$$+ \sum_{k=1}^{m} \left(\sum_{\alpha \in S_{ik-1}^{(m)}} f_{t_i^{m-k}}^{\alpha\sigma_i} + \sum_{\alpha \in S_{ik}^{(m)}} f_{t_i^{m+1-k}}^{\alpha\Delta_i} \right) g_{t_i^k}^{\Delta_i^k}$$

$$= \left(\sum_{\alpha \in S_{im+1}^{(m+1)}} f^{\alpha} \right) g_{t_i^{m+1}}^{\Delta_i^{m+1}} + \left(\sum_{\alpha \in S_{i0}^{(m+1)}} f_{t_i^{m+1}}^{\alpha} \right) g$$

$$+ \sum_{k=1}^{m} \left(\sum_{\alpha \in S_{ik}^{(m+1)}} f_{t_i^{m+1-k}}^{\alpha} \right) g_{t_i^k}^{\Delta_i^k},$$

i.e., (6.10) holds.

By the principle of mathematical induction, (6.9) holds for any $m \in \mathbb{N}$. □

Remark 6.81 We note that (6.9) holds for $m = 0$ with the convention that

$$\sum_{\alpha \in \emptyset} f_{t_i^k}^{\alpha} = f.$$

Definition 6.82 We put

$$\Lambda_{\kappa}^{n} = \mathbb{T}_{1\kappa} \times \mathbb{T}_{2\kappa} \times \ldots \times \mathbb{T}_{n\kappa},$$

$$\Lambda_{i\kappa_i}^{n} = \mathbb{T}_1 \times \ldots \times \mathbb{T}_{i-1} \times \mathbb{T}_{i\kappa} \times \mathbb{T}_{i+1} \times \ldots \times \mathbb{T}_n, \quad i = 1, 2, \ldots, n,$$

$$\Lambda_{i_1 i_2 \ldots i_l \kappa_{i_1} \kappa_{i_2} \ldots \kappa_{i_l}}^{n} = \ldots \times \mathbb{T}_{i_1\kappa} \times \ldots \times \mathbb{T}_{i_2\kappa} \times \ldots \times \mathbb{T}_{i_l\kappa} \times \ldots,$$

where $1 \le i_1 < i_2 < \ldots < i_l \le n$, $i_m \in \mathbb{N}$, $m = 1, 2, \ldots, l$.

Remark 6.83 If $(i_1, i_2, \ldots, i_l) = (1, 2, \ldots, n)$, then

$$\Lambda_{i_1 i_2 \ldots i_l \kappa_1 \kappa_2 \ldots \kappa_l}^{n} = \Lambda_{\kappa}^{n}.$$

Definition 6.84 Assume that $f : \Lambda^n \to \mathbb{R}$ is a function and let $t \in \Lambda_{i\kappa_i n}$. We define

$$\frac{\partial f(t_1, t_2, \ldots, t_n)}{\nabla_i t_i} = \frac{\partial f(t)}{\nabla_i t_i} = \frac{\partial f}{\nabla_i t_i}(t) = f_{t_i}^{\nabla_i}(t)$$

to be the number, provided it exists, with the property such that for any $\varepsilon_i > 0$, there exists a neighbourhood

$$U_i = (t_i - \delta_i, t_i + \delta_i) \cap \mathbb{T}_i$$

for some $\delta_i > 0$ such that

$$\Big| f(t_1, \ldots, t_{i-1}, \rho_i(t_i), t_{i+1}, \ldots, t_n) - f(t_1, \ldots, t_{i-1}, s_i, t_{i+1}, \ldots, t_n) - f_{t_i}^{\nabla_i}(t)(\rho_i(t_i) - s_i) \Big| \le \varepsilon_i |\rho_i(t_i) - s_i| \tag{6.11}$$

for all $s_i \in U_i$. We call $f_{t_i}^{\nabla_i}(t)$ the *partial nabla derivative* of f with respect to t_i at t. We say that f is *partial nabla differentiable* with respect to t_i in $\Lambda_{i\kappa_i}^{n}$ if $f_{t_i}^{\nabla_i}(t)$ exists for all $t \in \Lambda_{i\kappa_i}^{n}$. The function $f_{t_i}^{\nabla_i} : \Lambda_{i\kappa_i}^{n} \to \mathbb{R}$ is said to be the *partial nabla derivative* with respect to t_i of f in $\Lambda_{i\kappa_i}^{n}$.

Theorem 6.85 *The partial nabla derivative is well defined.*

Proof Let $t \in \Lambda_{i\kappa_i}^{n}$ for some $i \in \{1, 2, \ldots, n\}$. We assume that the partial nabla derivative $f_{t_i}^{\nabla_i}(t)$ exists and

$$g_1(t) = f_{t_i}^{\nabla_i}(t), \quad g_2(t) = f_{t_i}^{\nabla_i}(t).$$

Let $\varepsilon_i > 0$ be arbitrarily chosen. Then there exists $\delta_i > 0$ such that for every

$$s_i \in (t_i - \delta_i, t_i + \delta_i) \cap \mathbb{T}_i,$$

we have

$$\Big| f(t_1, t_2, \dots, t_{i-1}, \rho_i(t_i), t_{i+1}, \dots, t_n) - f(t_1, t_2, \dots, t_{i-1}, s_i, t_{i+1}, \dots, t_n) \\ - g_1(t)(\rho_i(t_i) - s_i)\Big| \le \frac{\varepsilon_i}{2} |\rho_i(t_i) - s_i| \tag{6.12}$$

and

$$\Big| f(t_1, t_2, \dots, t_{i-1}, \rho_i(t_i), t_{i+1}, \dots, t_n) - f(t_1, t_2, \dots, t_{i-1}, s_i, t_{i+1}, \dots, t_n) \\ - g_2(t)(\rho_i(t_i) - s_i)\Big| \le \frac{\varepsilon_i}{2} |\rho_i(t_i) - s_i|. \tag{6.13}$$

From (6.12) and (6.13), we obtain

$$\begin{aligned} |g_1(t) - g_2(t)| &= \Big| g_1(t) - \frac{f(t_1, t_2, \dots, t_{i-1}, \rho_i(t_i), t_{i+1}, \dots, t_n)}{\rho_i(t_i) - s_i} \\ &\quad + \frac{f(t_1, t_2, \dots, t_{i-1}, s_i, t_{i+1}, \dots, t_n)}{\rho_i(t_i) - s_i} + \frac{f(t_1, t_2, \dots, t_{i-1}, \rho_i(t_i), t_{i+1}, \dots, t_n)}{\rho_i(t_i) - s_i} \\ &\quad - \frac{f(t_1, t_2, \dots, t_{i-1}, s_i, t_{i+1}, \dots, t_n)}{\rho_i(t_i) - s_i} - g_2(t)\Big| \\ &\le \left| g_1(t) - \frac{f(t_1, t_2, \dots, t_{i-1}, \rho_i(t_i), t_{i+1}, \dots, t_n) - f(t_1, t_2, \dots, t_{i-1}, s_i, t_{i+1}, \dots, t_n)}{\rho_i(t_i) - s_i} \right| \\ &\quad + \left| g_2(t) - \frac{f(t_1, t_2, \dots, t_{i-1}, \rho_i(t_i), t_{i+1}, \dots, t_n) - f(t_1, t_2, \dots, t_{i-1}, s_i, t_{i+1}, \dots, t_n)}{\rho_i(t_i) - s_i} \right| \\ &\le \frac{\varepsilon_i}{2} + \frac{\varepsilon_i}{2} \\ &= \varepsilon_i. \end{aligned}$$

Because $\varepsilon_i > 0$ was arbitrarily chosen, we conclude that $g_1(t) = g_2(t)$. □

Example 6.86 Let

$$\Lambda^2 = ([0, 1] \cup \mathbb{N}) \times \mathbb{N}_0, \quad f(t) = t_1^2 t_2, \quad t = (t_1, t_2) \in \Lambda^2.$$

Here, $\mathbb{T}_1 = \mathbb{N}$, $\mathbb{T}_2 = \mathbb{N}_0$. We will prove that

$$f_{t_1}^{\nabla_1}(t) = \begin{cases} (2t_1 - 1)t_2 & \text{if } t_1 \in \mathbb{T}_1, \quad t_1 \geq 2, \quad t_2 \in \mathbb{T}_2, \\ \\ 2t_2 & \text{if } t_1 = 1, \quad t_2 \in \mathbb{T}_2. \end{cases}$$

We have $\rho_1(t_1) = t_1 - 1$ for $t_1 \in \mathbb{T}_1$, $t_1 \geq 2$, $\rho_1(1) = 1$. Let $\varepsilon > 0$ be arbitrarily chosen. Then, for every $s_1 \in (t_1 - \varepsilon^*, t_1 + \varepsilon^*)$, $s_1 \in \mathbb{T}_1$, we have, for $\varepsilon^* \leq \frac{\varepsilon}{1+t_2}$,

$$|t_1 - s_1| \leq \varepsilon^*.$$

Hence, for $t = (t_1, t_2) \in \Lambda^2_{1\kappa_1}$, $t_1 \geq 2$, we get

$$\begin{aligned} f(\rho_1(t_1), t_2) &= (\rho_1(t_1))^2 t_2 \\ &= (t_1 - 1)^2 t_2, \end{aligned}$$

$$\begin{aligned} &|f(\rho_1(t_1), t_2) - f(s_1, t_2) - (2t_1 - 1)t_2(\rho_1(t_1) - s_1)| \\ &= |(t_1 - 1)^2 t_2 - s_1^2 t_2 - (2t_1 - 1)t_2(t_1 - 1 - s_1)| \\ &= |(t_1 - 1 - s_1)(t_1 + s_1 - 1)t_2 - (2t_1 - 1)t_2(t_1 - 1 - s_1)| \\ &= |s_1 - t_1||t_2||t_1 - s_1 - 1| \\ &\leq |s_1 - t_1|(1 + t_2)|t_1 - s_1 - 1| \\ &\leq \varepsilon^*(1 + t_2)|t_1 - s_1 - 1| \\ &\leq \varepsilon|t_1 - s_1 - 1|. \end{aligned}$$

For $t \in \Lambda^2_{1\kappa_1}$, $t_1 = 1$, we have

$$f(\rho_1(1), t_2) = (\rho_1(1))^2 t_2 = t_2$$

and

$$\begin{aligned} |f(\rho_1(1), t_2) - f(s_1, t_2) - 2t_2(\rho_1(1) - s_1)| &= |f(1, t_2) - f(s_1, t_2) - 2t_2(1 - s_1)| \\ &= |t_2 - s_1^2 t_2 - 2t_2(1 - s_1)| \end{aligned}$$

$$= |1 - s_1| t_2 |1 + s_1 - 2|$$

$$\leq (1 + t_2)(1 - s_1)^2$$

$$\leq \varepsilon^*(1 + t_2)|1 - s_1|$$

$$\leq \varepsilon |1 - s_1|.$$

Exercise 6.87 Let

$$\Lambda^2 = \left([0, 2] \cup 2^{\mathbb{N}}\right) \times \mathbb{Z},$$

where $[0, 2]$ is the real number interval, and $f(t) = t_1^2 + t_1 t_2 - 2$, $t \in \Lambda^2$. Prove that

$$f_{t_1}^{\nabla_1}(t_1, t_2) = \frac{3}{2} t_1 + t_2, \quad t_1 \geq 4,$$

$$f_{t_1}^{\nabla_1}(2, t_2) = 4 + t_2, \quad t_2 \in \mathbb{T}_2.$$

The proofs of the following theorems repeat the main steps of the proofs of the corresponding theorems for partial delta derivatives. Thus, these proofs are omitted and left to the reader.

Theorem 6.88 *Let $f : \Lambda^n \to \mathbb{R}$ be a function and $t \in \Lambda^n_{i\kappa_i}$. If f is nabla differentiable with respect to t_i at t, then*

$$\lim_{s_i \to t_i} f(t_{s_i}) = f(t).$$

Theorem 6.89 *Let $f : \Lambda^n \to \mathbb{R}$, $t \in \Lambda^n_{i\kappa_i}$, and*

$$\lim_{s_i \to t_i} f(t_{s_i}) = f(t).$$

If $\rho_i(t_i) < t_i$, then f is nabla differentiable with respect to t_i at t and

$$f_{t_i}^{\nabla_i}(t) = \frac{f_i^{\rho_i}(t) - f(t)}{\rho_i(t_i) - t_i}.$$

Theorem 6.90 *Let $t \in \Lambda^n_{i\kappa_i}$ and $t_i = \rho_i(t_i)$. Then f is partial nabla differentiable with respect to t_i at t if and only if the limit*

$$\lim_{s_i \to t_i} \frac{f(t) - f(t_{s_i})}{t_i - s_i}$$

exists as a finite number. In this case,

$$f_{t_i}^{\nabla_i}(t) = \lim_{s_i \to t_i} \frac{f(t) - f(t_{s_i})}{t_i - s_i}.$$

Theorem 6.91 *Let $t \in \Lambda^n_{i\kappa_i}$. Suppose $f : \Lambda^n \to \mathbb{R}$ is a function that is partial nabla differentiable with respect to t_i at t. If $\alpha \in \mathbb{R}$, then αf is partial nabla differentiable with respect to t_i at t and*

$$(\alpha f)_{t_i}^{\nabla_i}(t) = \alpha f_{t_i}^{\nabla_i}(t).$$

Theorem 6.92 *Let $t \in \Lambda^n_{i\kappa_i}$. Suppose $f, g : \Lambda^n \to \mathbb{R}$ are partial nabla differentiable with respect to t_i at t. Then fg is partial nabla differentiable with respect to t_i at t and*

$$(fg)_{t_i}^{\nabla_i}(t) = f_{t_i}^{\nabla_i}(t)g(t) + f_i^{\rho_i}(t)g_{t_i}^{\nabla_i}(t) = f(t)g_{t_i}^{\nabla_i}(t) + f_{t_i}^{\nabla_i}(t)g_i^{\rho_i}(t).$$

Theorem 6.93 *Let $f, g : \Lambda^n \to \mathbb{R}$ be partial nabla differentiable with respect to t_i at $t \in \Lambda^n_{i\kappa_i}$. Assume $g_i^{\rho_i}(t)g(t) \neq 0$. Then $\frac{f}{g}$ is partial nabla differentiable with respect to t_i at t and*

$$\left(\frac{f}{g}\right)_{t_i}^{\nabla_i}(t) = \frac{f_{t_i}^{\nabla_i}(t)g(t) - f(t)g_{t_i}^{\nabla_i}(t)}{g_i^{\rho_i}(t)g(t)}.$$

Theorem 6.94 (Leibniz Formula) *Let $Q_{ik}^{(m)}$ be the set consisting of all possible strings of length m, containing exactly k times ρ_i and $m-k$ times ∇_i. If $f^{\beta}_{t_i^{m-k}}$ exists for any $\beta \in Q_{ik}^{(m)}$ and $g_{t_i^k}^{\nabla_i^k}$ exists for any $k \in \{0, 1, \ldots, m\}$, then*

$$(fg)_{t_i^m}^{\nabla_i^m} = \sum_{k=0}^{m} \left(\sum_{\beta \in Q_{ik}^{(m)}} f^{\beta}_{t_i^{m-k}} \right) g_{t_i^k}^{\nabla_i^k}$$

holds for any $m \in \mathbb{N}$.

We can define higher-order nabla derivatives and also mixed derivatives obtained by combining both delta and nabla differentiations such as, for instance, $f_{t_i t_j}^{\nabla_i \Delta_j}$ or $f_{t_i t_j t_l}^{\nabla_i \nabla_j \Delta_l}$.

Example 6.95 Let $\Lambda^2 = \mathbb{N} \times \mathbb{Z}$, $f(t) = t_1^2 t_2 + t_1 t_2^2 + t_2^2$, $t \in \Lambda^2$. Here, $\mathbb{T}_1 = \mathbb{N}$, $\mathbb{T}_2 = \mathbb{Z}$, and

$$\sigma_1(t_1) = t_1 + 1, \quad t_1 \in \mathbb{T}_1,$$

$$\sigma_2(t_2) = t_2 + 1, \quad t_2 \in \mathbb{T}_2,$$

$$\rho_1(t_1) = \begin{cases} t_1 - 1 & \text{if } t_1 \in \mathbb{T}_1, \quad t_1 \geq 2, \\ 1 & \text{if } t_1 = 1, \end{cases}$$

$$\rho_2(t_2) = t_2 - 1, \quad t_2 \in \mathbb{T}_2.$$

Hence,

$$\begin{aligned} f_{t_1}^{\Delta_1}(t) &= (\sigma_1(t_1) + t_1)t_2 + t_2^2 \\ &= (t_1 + 1 + t_1)t_2 + t_2^2 \\ &= 2t_1t_2 + t_2^2 + t_2, \quad t \in \Lambda_1^{\kappa_1 2}, \\ f_{t_1t_2}^{\Delta_1\nabla_2}(t) &= \left(f_{t_1}^{\Delta_1}\right)_{t_2}^{\nabla_2}(t) \\ &= 2t_1 + \rho_2(t_2) + t_2 + 1 \\ &= 2t_1 + t_2 - 1 + t_2 + 1 \\ &= 2t_1 + 2t_2, \quad t \in \Lambda_{12\kappa_2}^{\kappa_1 2}. \end{aligned}$$

Exercise 6.96 Let $\Lambda^2 = 3^{\mathbb{N}} \times \mathbb{Z}$, $f(t) = t_1^2 t_2^2$, $t \in \Lambda^2$. Find $f_{t_1t_2}^{\Delta_1\nabla_2}(t)$, $t \in \Lambda_{12\kappa_2}^{\kappa_1 2}$.

Solution $4t_1(2t_2 - 1)$.

6.3 Completely Differentiable Functions

Definition 6.97 We say that a function $f : \Lambda^n \to \mathbb{R}$ is *completely delta differentiable* at a point $t^0 \in \Lambda^{\kappa n}$ if there exist numbers a_i and A_{ij}, $i, j \in \{1, 2, \ldots, n\}$ independent of $t \in \Lambda^n$ but, in general, dependent on t^0, such that for all $t \in U_\delta(t_0)$,

$$f(t^0) - f(t) = \sum_{i=1}^{n} a_i(t_i^0 - t_i) + \sum_{i=1}^{n} \alpha_i(t_i^0 - t_i) \tag{6.14}$$

and, for each $i \in \{1, 2, \ldots, n\}$ and all $t \in U_\delta(t^0)$,

$$f_i^{\sigma_i}(t^0) - f(t) = A_{ii}(\sigma_i(t_i^0) - t_i) + \sum_{l=1, l\neq i}^{n} A_{li}(t_l^0 - t_l)$$
$$+ \beta_{ii}(\sigma_i(t_i^0) - t_i) + \sum_{l=1, l\neq i}^{n} \beta_{li}(t_l^0 - t_l), \tag{6.15}$$

where $\delta > 0$ is a sufficiently small real number, $U_\delta(t^0)$ is the δ-neighbourhood of t^0 in Λ^n, $\alpha_i = \alpha_i(t^0, t)$, $\beta_{li} = \beta_{li}(t^0, t)$ are defined in $U_\delta(t^0)$, $l, i \in \{1, 2, \dots, n\}$, such that

$$\lim_{t\to t^0} \alpha_i(t^0, t) = \lim_{t\to t^0} \beta_{ij}(t^0, t) = 0 \quad \text{for all} \quad i, j \in \{1, 2, \dots, n\}.$$

Remark 6.98 If $\Lambda^n = \mathbb{R}^n$, then Definition 6.97 coincides with the classical total differentiability of functions of n real variables.

Example 6.99 Let $\Lambda^2 = 2^{\mathbb{N}} \times \mathbb{N}$, $f(t) = t_1^2 + t_2^2$, $t \in \Lambda^2$. Let $t^0 \in \Lambda^{\kappa 2}$ be arbitrarily chosen. We will prove that f is completely delta differentiable at t^0. Here, $\mathbb{T}_1 = 2^{\mathbb{N}}$, $\mathbb{T}_2 = \mathbb{N}$, and

$$\sigma_1(t_1) = 2t_1, \quad t_1 \in \mathbb{T}_1, \quad \sigma_2(t_2) = t_2 + 1, \quad t_2 \in \mathbb{T}_2.$$

Let $\delta > 0$ be sufficiently small. For every $t \in U_\delta(t^0)$, we have

$$\begin{aligned} f(t^0) - f(t) &= t_1^{02} + t_2^{02} - t_1^2 - t_2^2 \\ &= t_1^{02} + t_2^{02} - t_1^2 - t_2^2 + 2t_1^0(t_1^0 - t_1) - 2t_1^0(t_1^0 - t_1) \\ &\quad + 2t_2^0(t_2^0 - t_2) - 2t_2^0(t_2^0 - t_2) \\ &= 2t_1^0(t_1^0 - t_1) + 2t_2^0(t_2^0 - t_2) + (t_1^{02} - 2t_1^{02} + 2t_1^0 t_1 - t_1^2) \\ &\quad + (t_2^{02} - t_2^2 - 2t_2^{02} + 2t_2^0 t_2) \\ &= 2t_1^0(t_1^0 - t_1) + 2t_2^0(t_2^0 - t_2) - (t_1^0 - t_1)^2 - (t_2^0 - t_2)^2. \end{aligned}$$

Therefore,

$$\alpha_1 = -(t_1^0 - t_1), \quad \alpha_2 = -(t_2^0 - t_2), \quad a_1 = 2t_1^0, \quad a_2 = 2t_2^0.$$

We note that

$$\lim_{t\to t^0} \alpha_1 = \lim_{t\to t^0} \alpha_2 = 0.$$

For $t \in U_\delta(t^0)$, we have

$$f_1^{\sigma_1}(t^0) - f(t) = \sigma_1^2(t_1^0) + t_2^{02} - t_1^2 - t_2^2$$

$$= 4t_1^{02} + t_2^{02} - t_1^2 - t_2^2$$

$$= (2t_1^0 - t_1)(2t_1^0 + t_1) + (t_2^0 - t_2)(t_2^0 + t_2)$$

$$= (2t_1^0 - t_1)(3t_1^0 - t_1^0 + t_1) + (t_2^0 - t_2)(2t_2^0 - t_2^0 + t_2)$$

$$= 3t_1^0(2t_1^0 - t_1) + 2t_2^0(t_2^0 - t_2) - (2t_1^0 - t_1)(t_1^0 - t_1) - (t_2^0 - t_2)^2.$$

Here,

$$A_{11} = 3t_1^0, \quad A_{21} = 2t_2^0, \quad \beta_{11} = -(t_1^0 - t_1), \quad \beta_{21} = -(t_2^0 - t_2).$$

We note that

$$\lim_{t \to t^0} \beta_{11} = \lim_{t \to t^0} \beta_{21} = 0.$$

For all $t \in U_\delta(t^0)$, we have

$$f_2^{\sigma_2}(t^0) - f(t) = t_1^{02} + (t_2^0 + 1)^2 - t_1^2 - t_2^2$$

$$= (t_1^0 - t_1)(t_1^0 + t_1) + (t_2^0 + 1 - t_2)(t_2^0 + 1 + t_2)$$

$$= (t_1^0 - t_1)(2t_1^0 + t_1 - t_1^0) + (t_2^0 + 1 - t_2)(2t_2^0 + 1 + t_2 - t_2^0)$$

$$= 2t_1^0(t_1^0 - t_1) + (2t_2^0 + 1)(t_2^0 + 1 - t_2)$$

$$-(t_1^0 - t_1)^2 - (t_2^0 + 1 - t_2)(t_2^0 - t_2).$$

Thus,

$$A_{12} = 2t_1^0, \quad A_{22} = 2t_2^0 + 1, \quad \beta_{12} = -(t_1^0 - t_1), \quad \beta_{22} = -(t_2^0 - t_2).$$

We note that

$$\lim_{t \to t^0} \beta_{12} = \lim_{t \to t^0} \beta_{22} = 0.$$

Example 6.100 Let $\Lambda^2 = \mathbb{Z} \times 3^{\mathbb{N}}$, $f(t) = t_1 t_2^2$, $t \in \Lambda^2$. Let $t^0 \in \Lambda^2$ be arbitrarily chosen. We will prove that the function f is completely delta differentiable at t_0.

Here, $\mathbb{T}_1 = \mathbb{Z}$, $\mathbb{T}_2 = 3^{\mathbb{N}}$, and

$$\sigma_1(t_1) = t_1 + 1, \quad t_1 \in \mathbb{T}_1, \quad \sigma_2(t_2) = 3t_2, \quad t_2 \in \mathbb{T}_2.$$

Let $\delta > 0$ be arbitrarily chosen. For $t \in U_\delta(t^0)$, we have

$$\begin{aligned}
f(t^0) - f(t) &= t_1^0 t_2^{02} - t_1 t_2^2 \\
&= t_1^0 t_2^{02} - t_1 t_2^{02} + t_1 t_2^{02} - t_1 t_2^2 \\
&= (t_1^0 - t_1) t_2^{02} + t_1 (t_2^0 - t_2)(t_2^0 + t_2) \\
&= (t_1^0 - t_1) t_2^{02} + 2 t_1 t_2^0 (t_2^0 - t_2) - t_1 (t_2^0 - t_2)^2 \\
&= (t_1^0 - t_1) t_2^{02} + 2 t_1^0 t_2^0 (t_2^0 - t_2) - 2 t_1^0 t_2^0 (t_2^0 - t_2) \\
&\quad + 2 t_1 t_2^0 (t_2^0 - t_2) - t_1 (t_2^0 - t_2)^2 \\
&= t_2^{02} (t_1^0 - t_1) + 2 t_1^0 t_2^0 (t_2^0 - t_2) - 2 t_2^0 (t_2^0 - t_2)(t_1^0 - t_1) \\
&\quad - t_1 (t_2^0 - t_2)^2.
\end{aligned}$$

Here,

$$a_1 = t_2^{02}, \quad a_2 = 2 t_1^0 t_2^0, \quad \alpha_1 = -2 t_2^0 (t_2^0 - t_2), \quad \alpha_2 = -t_1 (t_2^0 - t_2).$$

We note that

$$\lim_{t \to t^0} \alpha_1 = \lim_{t \to t^0} \alpha_2 = 0.$$

For $t \in U_\delta(t^0)$, we have

$$\begin{aligned}
f_1^{\sigma_1}(t^0) - f(t) &= (t_1^0 + 1) t_2^{02} - t_1 t_2^2 \\
&= t_2^{02} (t_1^0 + 1 - t_1) + 2 t_1^0 t_2^0 (t_2^0 - t_2) - t_2^{02} (t_1^0 + 1 - t_1) \\
&\quad - 2 t_1^0 t_2^0 (t_2^0 - t_2) + (t_1^0 + 1) t_2^{02} - t_1 t_2^2 \\
&= t_2^{02} (t_1^0 + 1 - t_1) + 2 t_1^0 t_2^0 (t_2^0 - t_2) - t_1^0 t_2^{02}
\end{aligned}$$

$$-t_2^{02} + t_1 t_2^{02} - 2t_1^0 t_2^{02} + 2t_1^0 t_2^0 t_2 + t_1^0 t_2^{02} + t_2^{02} - t_1 t_2^2$$

$$= t_2^{02}(t_1^0 + 1 - t_1) + 2t_1^0 t_2^0 (t_2^0 - t_2) + t_1 (t_2^0 - t_2)(t_2^0 + t_2)$$

$$-2t_1^0 t_2^0 (t_2^0 - t_2)$$

$$= t_2^{02}(t_1^0 + 1 - t_1) + 2t_1^0 t_2^0 (t_2^0 - t_2) + (t_1 t_2^0 + t_1 t_2$$

$$-2t_1^0 t_2^0)(t_2^0 - t_2)$$

$$= t_2^{02}(t_1^0 + 1 - t_1) + 2t_1^0 t_2^0 (t_2^0 - t_2) + (t_1 t_2^0 - t_1^0 t_2^0$$

$$-t_1^0 t_2^0 + t_1 t_2)(t_2^0 - t_2)$$

$$= t_2^{02}(t_1^0 + 1 - t_1) + 2t_1^0 t_2^0 (t_2^0 - t_2)$$

$$-(t_1^0 - t_1) t_2^0 (t_2^0 - t_2) + (t_1 t_2 - t_1^0 t_2 + t_1^0 t_2 - t_1^0 t_2^0)(t_2^0 - t_2)$$

$$= t_2^{02}(t_1^0 + 1 - t_1) + 2t_1^0 t_2^0 (t_2^0 - t_2) - (t_1^0 - t_1) t_2^0 (t_2^0 - t_2)$$

$$+(t_1 - t_1^0) t_2 (t_2^0 - t_2) - t_1^0 (t_2^0 - t_2)^2$$

$$= t_2^{02}(t_1^0 + 1 - t_1) + 2t_1^0 t_2^0 (t_2^0 - t_2) - (t_2 + t_2^0)(t_2^0 - t_2)(t_1^0 - t_1)$$

$$-t_1^0 (t_2^0 - t_2)^2.$$

Here,

$$A_{11} = t_2^{02}, \quad A_{21} = 2t_1^0 t_2^0, \quad \beta_{11} = -(t_2^0 + t_2)(t_2^0 - t_2), \quad \beta_{21} = -t_1^0 (t_2^0 - t_2).$$

We note that

$$\lim_{t \to t^0} \beta_{11} = \lim_{t \to t^0} \beta_{21} = 0.$$

For $t \in U_\delta(t^0)$, we have

$$f_2^{\sigma_2}(t^0) - f(t) = t_1^0 (3t_2^0)^2 - t_1 t_2^2$$

$$= 9t_1^0 t_2^{02} - t_1 t_2^2$$

$$= t_2^{02}(t_1^0 - t_1) + 4t_1^0 t_2^0(3t_2^0 - t_2) - t_2^{02}(t_1^0 - t_1)$$

$$-4t_1^0 t_2^0(3t_2^0 - t_2) + 9t_1^0 t_2^{02} - t_1 t_2^2$$

$$= t_2^{02}(t_1^0 - t_1) + 4t_1^0 t_2^0(3t_2^0 - t_2) - t_2^{02}(t_1^0 - t_1)$$

$$-4t_1^0 t_2^0(3t_2^0 - t_2) + 9t_1^0 t_2^{02} - t_1^0 t_2^2 + t_1^0 t_2^2 - t_1 t_2^2$$

$$= t_2^{02}(t_1^0 - t_1) + 4t_1^0 t_2^0(3t_2^0 - t_2) - t_2^{02}(t_1^0 - t_1)$$

$$-4t_1^0 t_2^0(3t_2^0 - t_2) + t_1^0(3t_2^0 - t_2)(3t_2^0 + t_2) + (t_1^0 - t_1)t_2^2$$

$$= t_2^{02}(t_1^0 - t_1) + 4t_1^0 t_2^0(3t_2^0 - t_2)$$

$$+(t_2^2 - t_2^{02})(t_1^0 - t_1) + (t_1^0 t_2 - t_1^0 t_2^0)(3t_2^0 - t_2).$$

Here,

$$A_{12} = t_2^{02}, \quad A_{22} = 4t_1^0 t_2^0, \quad \beta_{12} = t_2^2 - t_2^{02}, \quad \beta_{22} = t_1^0 t_2 - t_1^0 t_2^0.$$

We note that

$$\lim_{t \to t^0} \beta_{12} = \lim_{t \to t^0} \beta_{22} = 0.$$

Example 6.101 Let $\Lambda^2 = \mathbb{N} \times \mathbb{N}$, $f(t) = t_1 t_2$, $t \in \Lambda^2$. Let $t_0 \in \Lambda^2$ be arbitrarily chosen. Here, $\mathbb{T}_1 = \mathbb{T}_2 = \mathbb{N}$ and

$$\sigma_1(t_1) = t_1 + 1, \quad t_1 \in \mathbb{T}_1, \quad \sigma_2(t_2) = t_2 + 1, \quad t_2 \in \mathbb{T}_2.$$

We will prove that f is completely delta differentiable at t_0. Let $\delta > 0$ be sufficiently small. For $t \in U_\delta(t^0)$, we have

$$f(t^0) - f(t) = t_1^0 t_2^0 - t_1 t_2 = t_2^0(t_1^0 - t_1) + t_1^0(t_2^0 - t_2)$$

$$-t_2^0(t_1^0 - t_1) - t_1^0(t_2^0 - t_2) + t_1^0 t_2^0 - t_1 t_2$$

$$= t_2^0(t_1^0 - t_1) + t_1^0(t_2^0 - t_2) - t_1^0 t_2^0 + t_1 t_2^0 - t_1^0 t_2^0$$

$$+t_1^0 t_2 + t_1^0 t_2^0 - t_1 t_2$$

$$= t_2^0(t_1^0 - t_1) + t_1^0(t_2^0 - t_2) - t_1^0 t_2^0 + t_1 t_2^0 + t_1^0 t_2 - t_1 t_2$$

$$= t_2^0(t_1^0 - t_1) + t_1^0(t_2^0 - t_2) - t_2^0(t_1^0 - t_1) + t_2(t_1^0 - t_1)$$

$$= t_2^0(t_1^0 - t_1) + t_1^0(t_2^0 - t_2) + (t_2 - t_2^0)(t_1^0 - t_1).$$

Here,

$$a_1 = t_2^0, \quad a_2 = t_1^0, \quad \alpha_1 = t_2 - t_2^0, \quad \alpha_2 = 0.$$

We note that

$$\lim_{t \to t^0} \alpha_1 = 0.$$

For $t \in U_\delta(t^0)$, we have

$$f_1^{\sigma_1}(t^0) - f(t) = (t_1^0 + 1)t_2^0 - t_1 t_2$$

$$= t_2^0(t_1^0 + 1 - t_1) + t_1^0(t_2^0 - t_2) - t_2^0(t_1^0 + 1 - t_1)$$

$$-t_1^0(t_2^0 - t_2) + (t_1^0 + 1)t_2^0 - t_1 t_2$$

$$= t_2^0(t_1^0 + 1 - t_1) + t_1^0(t_2^0 - t_2) - t_1^0 t_2^0 - t_2^0 + t_1 t_2^0$$

$$-t_1^0 t_2^0 + t_1^0 t_2 + t_1^0 t_2^0 + t_2^0 - t_1 t_2$$

$$= t_2^0(t_1^0 + 1 - t_1) + t_1^0(t_2^0 - t_2) - t_1^0 t_2^0$$

$$+t_1 t_2^0 + t_1^0 t_2 - t_1 t_2$$

$$= t_2^0(t_1^0 + 1 - t_1) + t_1^0(t_2^0 - t_2) + (t_1 - t_1^0)(t_2^0 - t_2).$$

Here,

$$A_{11} = t_2^0, \quad A_{21} = t_1^0, \quad \beta_{11} = 0, \quad \beta_{21} = t_1 - t_1^0.$$

We note that

$$\lim_{t \to t^0} \beta_{21} = 0.$$

For $t \in U_\delta(t^0)$, we have

$$f_2^{\sigma_2}(t^0) - f(t) = t_1^0(t_2^0 + 1) - t_1 t_2$$

$$= t_2^0(t_1^0 - t_1) + t_1^0(t_2^0 + 1 - t_2) - t_2^0(t_1^0 - t_1)$$

$$-t_1^0(t_2^0 + 1 - t_2) + t_1^0(t_2^0 + 1) - t_1 t_2$$

$$= t_2^0(t_1^0 - t_1) + t_1^0(t_2^0 + 1 - t_2) - t_1^0 t_2^0 + t_1 t_2^0 - t_1^0 t_2^0$$

$$-t_1^0 + t_1^0 t_2 + t_1^0 t_2^0 + t_1^0 - t_1 t_2$$

$$= t_2^0(t_1^0 - t_1) + t_1^0(t_2^0 + 1 - t_2) - t_1^0 t_2^0$$

$$+t_1 t_2^0 + t_1^0 t_2 - t_1 t_2$$

$$= t_2^0(t_1^0 - t_1) + t_1^0(t_2^0 + 1 - t_2) + (t_2 - t_2^0)(t_1^0 - t_1).$$

Here,

$$A_{12} = t_2^0, \quad A_{22} = t_1^0, \quad \beta_{12} = t_2 - t_2^0, \quad \beta_{22} = 0.$$

We note that

$$\lim_{t \to t^0} \beta_{12} = 0.$$

Exercise 6.102 Let $\Lambda^2 = 2^{\mathbb{N}} \times 3^{\mathbb{N}}$, $f(t) = t_1^2 + t_2 - 2$, $t \in \Lambda^2$. Prove that f is completely delta differentiable in Λ^2.

Theorem 6.103 *Every function defined on $\mathbb{Z}^n$ is completely delta differentiable at every point.*

Proof We may use

$$A_{ii} = f_{t_i}^{\Delta_i}(t^0), \quad A_{li} = 0, \quad l \neq i, \quad \beta_{li} = 0, \quad l, i \in \{1, 2, \ldots, n\}. \qquad \square$$

Theorem 6.104 *Let $f : \Lambda^n \to \mathbb{R}$ be completely delta differentiable at $t \in \Lambda^{\kappa n}$. Then $f_{t_i}^{\Delta_i}(t)$ exists and*

$$A_{ii} = f_{t_i}^{\Delta_i}(t) \quad \textit{for all} \quad i \in \{1, 2, \ldots, n\}. \tag{6.16}$$

Proof From (6.15), we get, for $i \in \{1, 2, \ldots, n\}$,

$$\lim_{s_i \to t_i} \frac{f_i^{\sigma_i}(t) - f(t_{s_i})}{\sigma_i(t_i) - s_i} = A_{ii},$$

which implies (6.16). $\square$

Exercise 6.105 Let $f : \Lambda^n \to \mathbb{R}$ be completely delta differentiable at $t \in \Lambda^{\kappa n}$. Prove that f is continuous at t.

Example 6.106 Let $\mathbb{T} = [0, 2] \cup \{4\}$, where $[0, 2]$ is the real number interval. Let $f : \mathbb{T} \to \mathbb{R}$ be defined by $f(t) = t^3$, $t \in [0, 2]$, $f(4) = 2$. Then

$$f^{\Delta}(t) = 3t^2, \quad t \in [0, 2),$$

$$f^{\Delta}(2-) = 12,$$

$$\begin{aligned} f^{\Delta}(2) &= \frac{f(\sigma(2)) - f(2)}{\sigma(2) - 2} \\ &= \frac{f(4) - f(2)}{4 - 2} \\ &= \frac{2 - 8}{2} \\ &= -3. \end{aligned}$$

Therefore, the function f is not completely delta differentiable at the point $t = 2$.

Example 6.107 Let $\mathbb{T} = [2, 3] \cup \{5\}$, where $[2, 3]$ is the real number interval. Let $f : \mathbb{T} \to \mathbb{R}$ be defined by

$$f(t) = t^2 + t + 1, \quad t \in [2, 3] \quad \text{and} \quad f(5) = c,$$

where $c \in \mathbb{R}$. Then

$$f^{\Delta}(t) = 2t + 1, \quad t \in [2, 3),$$

$$f^{\Delta}(3-) = 7,$$

$$\begin{aligned} f^{\Delta}(3) &= \frac{f(\sigma(3)) - f(3)}{\sigma(3) - 3} \\ &= \frac{f(5) - f(3)}{5 - 3} \\ &= \frac{c - (9 + 3 + 1)}{2} \\ &= \frac{c - 13}{2}. \end{aligned}$$

Hence, $f^{\Delta}(3-) = f^{\Delta}(3)$ if

$$\frac{c-13}{2} = 7, \quad \text{i.e.,} \quad c = 27.$$

Consequently, the function f is completely delta differentiable at $t = 3$ if $c = 27$. If $c \neq 27$, then the function f is not completely delta differentiable at $t = 3$.

Example 6.108 Let $\mathbb{T}_1 = [0, 1] \cup \{3\}$, $\mathbb{T}_2 = [0, 1]$, where $[0, 1]$ is the real number interval. Let $f : \mathbb{T}_1 \times \mathbb{T}_2 \to \mathbb{R}$ be defined by $f(t) = t_1^2 + t_2^2$ for $(t_1, t_2) \in [0, 1] \times [0, 1]$ and $f(3, t_2) = a + bt_2$, $t_2 \in [0, 1]$, where a and b are real constants. We have

$$f_{t_1}^{\Delta_1}(t) = 2t_1, \quad (t_1, t_2) \in [0, 1) \times [0, 1],$$

$$f_{t_2}^{\Delta_2}(t) = 2t_2, \quad (t_1, t_2) \in [0, 1) \times [0, 1],$$

$$f_{t_1}^{\Delta_1}(1-, t_2) = 2,$$

$$f_{t_2}^{\Delta_2}(1-, t_2) = 2t_2, \quad t_2 \in [0, 1],$$

$$\begin{aligned}
f_{t_1}^{\Delta_1}(1, t_2) &= \frac{f(\sigma_1(1), t_2) - f(1, t_2)}{\sigma_1(1) - 1} \\
&= \frac{f(3, t_2) - f(1, t_2)}{3 - 1} \\
&= \frac{a + bt_2 - 1 - t_2^2}{2},
\end{aligned}$$

$$f_{t_1}^{\Delta_1}(1, 0) = \frac{a-1}{2},$$

$$\begin{aligned}
f_{t_2}^{\Delta_2}(1, 0) &= f_{t_2}^{\Delta_2}(1-, 0) \\
&= 0.
\end{aligned}$$

Therefore, f is completely delta differentiable at $(1, 0)$ if

$$\frac{a-1}{2} = 2, \quad \text{i.e.,} \quad a = 5.$$

Exercise 6.109 Let $\mathbb{T}_1 = [1, 2] \cup \{3\}$, $\mathbb{T}_2 = [1, 2]$, where $[1, 2]$ is the real number interval. Let $f : \mathbb{T}_1 \times \mathbb{T}_2 \to \mathbb{R}$ be defined by

$$f(t_1, t_2) = 2t_1 + t_1^2 + t_2^2, \quad (t_1, t_2) \in \mathbb{T}_1 \times \mathbb{T}_2, \quad t_1 \neq 3$$

and $f(3, t_2) = a + \sqrt{t_2}$, $t_2 \in [1, 2]$, where $a \in \mathbb{R}$. Find a constant a so that f is completely delta differentiable at the point $(2, 1)$.

Solution $a = 14$.

Definition 6.110 For $i \in \{1, 2, \ldots, n\}$, we say that a function

$$f : \Lambda^n \to \mathbb{R}$$

is σ_i-*completely delta differentiable* at $t^0 \in \Lambda^{\kappa n}$ if it is completely delta differentiable at that point in the sense of the conditions (6.14) and (6.15) and moreover, along with the numbers A_{ii}, there exist numbers B_{li} independent of $t \in \Lambda^n$ (but in general, dependent on t^0), $l \neq i$, $l \in \{1, 2, \ldots, n\}$, such that

$$f^{\sigma}(t^0) - f(t) = A_{ii}(\sigma_i(t_i^0) - t_i) + \sum_{l=1, l\neq i}^{n} B_{li}(\sigma_l(t_l^0) - t_l) + \sum_{l=1}^{n} \gamma_l(\sigma_l(t_l^0) - t_l) \quad (6.17)$$

for all $t \in V^{\sigma_i}(t^0)$, where $V^{\sigma_i}(t^0)$ is a union of some neighbourhoods of t^0 and

$$(t_1^0, \ldots, t_{i-1}^0, \sigma_i(t_i^0), t_{i+1}^0, \ldots, t_n^0),$$

and $\gamma_i = \gamma_i(t^0, t)$, $\gamma_l = \gamma_l(t^0, t_1, \ldots, t_{i-1}, t_{i+1}, \ldots, t_n)$, $l \neq i$, are equal to zero for $t = t^0$ and

$$\lim_{t \to t^0} \gamma_i(t^0, t) = 0, \quad \lim_{t_l \to t_l^0} \gamma_l(t^0, t_1, \ldots, t_{i-1}, t_{i+1}, \ldots, t_n) = 0, \quad l \neq i,$$

where $l \in \{1, 2, \ldots, n\}$.

Remark 6.111 In fact, in (6.17), we have

$$B_{li} = f_{t_l}^{\Delta_l}(\sigma_1(t_1^0), \ldots, \sigma_{l-1}(t_{l-1}^0), t_l^0, \sigma_{l+1}(t_{l+1}^0), \ldots, \sigma_n(t_n^0)), \quad l \neq i,$$

where $l \in \{1, 2, \ldots, n\}$.

Example 6.112 Let $\mathbb{T}_1 = [0, 1] \cup \{2\}$, $\mathbb{T}_2 = [0, 1]$, where $[0, 1]$ is the real interval. Let $f : \mathbb{T}_1 \times \mathbb{T}_2 \to \mathbb{R}$ be defined by

$$f(t_1, t_2) = t_1^2 + t_2^2 \quad \text{for} \quad (t_1, t_2) \in [0, 1] \times [0, 1]$$

and

$$f(2, t_2) = 3 + \sqrt{t_2} \quad \text{for} \quad t_2 \in [0, 1].$$

Thus,

$$
\begin{aligned}
f_{t_1}^{\Delta_1}(t_1,t_2) &= \sigma_1(t_1)+t_1 \\
&= 2t_1, \quad t_1 \in [0,1), \quad t_2 \in [0,1], \\
f_{t_1}^{\Delta_1}(1-,t_2) &= 2, \quad t_2 \in [0,1], \\
f_{t_1}^{\Delta_1}(1-,0) &= 2, \\
f_{t_1}^{\Delta_1}(1,t_2) &= \frac{f(\sigma_1(1),t_2)-f(1,t_2)}{\sigma_1(1)-1} \\
&= \frac{f(2,t_2)-f(1,t_2)}{2-1} \\
&= 3+\sqrt{t_2}-1-t_2^2 \\
&= 2+\sqrt{t_2}-t_2^2, \\
f_{t_1}^{\Delta_1}(1,0) &= 2, \\
f_{t_2}^{\Delta_2}(t_1,t_2) &= \sigma_2(t_2)+t_2 \\
&= 2t_2, \\
f_{t_2}^{\Delta_2}(t_1,0) &= 0.
\end{aligned}
$$

Therefore, f is completely delta differentiable. Also,

$$
\begin{aligned}
f_{t_1}^{\Delta_1}(t_1,\sigma_2(t_2)) &= 2t_1, \quad t_1 \in [0,1), \\
f_{t_1}^{\Delta_1}(1-,\sigma_2(t_2)) &= 2, \\
f_{t_1}^{\Delta_1}(1,\sigma_2(t_2)) &= \frac{f(\sigma_1(1),\sigma_2(t_2))-f(1,\sigma_2(t_2))}{\sigma_1(1)-1} \\
&= \frac{f(2,\sigma_2(t_2))-f(1,\sigma_2(t_2))}{2-1} \\
&= 3+\sqrt{\sigma_2(t_2)}-1-\sigma_2^2(t_2)
\end{aligned}
$$

$$= 2 + \sqrt{t_2} - t_2^2,$$

$$f_{t_1}^{\Delta_1}(1, 0) = 2,$$

$$f_{t_2}^{\Delta_2}(\sigma_1(1), 0) = \lim_{s\to 0} \frac{f(\sigma_1(1), s) - f(\sigma_1(1), 0)}{s - 0}$$

$$= \lim_{s\to 0} \frac{f(2, s) - f(2, 0)}{s}$$

$$= \lim_{s\to 0} \frac{3 + \sqrt{s} - 3}{s}$$

$$= \infty,$$

and thus, while f is σ_1-completely delta differentiable, it is not σ_2-completely delta differentiable. Now, assume that

$$f(2, t_2) = 3 + t_2 \quad \text{for} \quad t_2 \in [0, 1]$$

so that

$$f_{t_1}^{\Delta_1}(1, t_2) = 2 + t_2 - t_2^2,$$

$$f_{t_1}^{\Delta_1}(1, 0) = 2,$$

$$f_{t_2}^{\Delta_2}(t_1, t_2) = 2t_2,$$

$$f_{t_2}^{\Delta_2}(1, 0) = 0,$$

$$f_{t_1}^{\Delta_1}(1, \sigma_2(t_2)) = 2t_1,$$

$$f_{t_2}^{\Delta_2}(1-, 0) = 2,$$

$$f_{t_1}^{\Delta_1}(1, \sigma_2(t_2)) = 2 + t_2 - t_2^2,$$

$$f_{t_1}^{\Delta_1}(1, 0) = 2,$$

$$f_{t_2}^{\Delta_2}(\sigma_1(1), 0) = \lim_{s\to 0} \frac{f(\sigma_1(1), s) - f(\sigma_1(1), 0)}{s}$$

$$= \lim_{s \to 0} \frac{f(2,s) - f(2,0)}{s}$$

$$= \lim_{s \to 0} \frac{3 + s - 3}{s} = 1.$$

Therefore, f is both σ_1-completely delta differentiable and σ_2-completely delta differentiable.

Example 6.113 Let $\mathbb{T}_1 = [1, 2] \cup \{3\}$, $\mathbb{T}_2 = [1, 2]$, where $[1, 2]$ is the real interval. Let $f : \mathbb{T}_1 \times \mathbb{T}_2 \to \mathbb{R}$ be defined by

$$f(t_1, t_2) = t_1 + t_2, \quad (t_1, t_2) \in [1, 2] \times [1, 2]$$

and

$$f(3, t_2) = a + t_2, \quad t_2 \in [1, 2],$$

where a is a real constant. We will find a constant a such that the function f is σ_1-completely delta differentiable at the point (2, 1). We have

$$f_{t_1}^{\Delta_1}(t_1, t_2) = 1, \quad (t_1, t_2) \in [1, 2) \times [1, 2],$$

$$f_{t_2}^{\Delta_2}(t_1, t_2) = 1, \quad (t_1, t_2) \in [1, 2] \times [1, 2],$$

$$f_{t_1}^{\Delta_1}(2, t_2) = \frac{f(3, t_2) - f(2, t_2)}{1}$$

$$= \frac{a + t_2 - 2 - t_2}{1}$$

$$= a - 2,$$

$$f_{t_1}^{\Delta_1}(2, 1) = a - 2,$$

$$f_{t_2}^{\Delta_2}(2, t_2) = 1, \quad t_2 \in [1, 2],$$

$$f_{t_1}^{\Delta_1}(2-, t_2) = 1, \quad t_2 \in [1, 2].$$

Hence, f is σ_1-completely delta differentiable iff $a - 2 = 1$, i.e., $a = 3$.

Example 6.114 Let $\mathbb{T}_1 = [0, 2] \cup \{4\}$, $\mathbb{T}_2 = [0, 2]$, where $[0, 2]$ is the real interval. Let $f : \mathbb{T}_1 \times \mathbb{T}_2 \to \mathbb{R}$ be defined by

$$f(t_1, t_2) = 2t_1 + 3t_2, \quad (t_1, t_2) \in [0, 2] \times [0, 2]$$

and

$$f(4, t_2) = a + \sqrt{t_2}, \quad t_2 \in [0, 2],$$

where a is a real constant. We will find a constant a so that the function f is completely delta differentiable, σ_1-completely delta differentiable, and σ_2-completely delta differentiable at the point (2, 2). We have

$$f_{t_1}^{\Delta_1}(t_1, t_2) = 2, \quad (t_1, t_2) \in [0, 2) \times [0, 2],$$

$$f_{t_1}^{\Delta_1}(2-, t_2) = 2, \quad t_2 \in [0, 2],$$

$$f_{t_2}^{\Delta_1}(t_1, t_2) = 3, \quad (t_1, t_2) \in [0, 2] \times [0, 2],$$

$$\begin{aligned} f_{t_1}^{\Delta_1}(2, t_2) &= \frac{f(\sigma_1(2), t_2) - f(2, t_2)}{\sigma_1(2) - 2} \\ &= \frac{f(4, t_2) - f(2, t_2)}{4 - 2} \\ &= \frac{a + \sqrt{t_2} - (4 + 3t_2)}{2} \\ &= \frac{a - 4}{2} + \frac{1}{2}\sqrt{t_2} - \frac{3}{2}t_2, \end{aligned}$$

$$\begin{aligned} f_{t_1}^{\Delta_1}(2, 2) &= \frac{a - 4}{2} + \frac{\sqrt{2}}{2} - 3 \\ &= \frac{a}{2} + \frac{\sqrt{2}}{2} - 5, \end{aligned}$$

$$f_{t_1}^{\Delta_1}(2-, \sigma_2(2)) = 2,$$

$$\begin{aligned} f_{t_2}^{\Delta_2}(\sigma_1(2), 2) &= \lim_{s \to 2} \frac{f(\sigma_1(2), s) - f(\sigma_1(2), 2)}{s - 2} \\ &= \lim_{s \to 2} \frac{f(4, s) - f(4, 2)}{s - 2} \\ &= \lim_{s \to 2} \frac{a + \sqrt{s} - a - \sqrt{2}}{s - 2} \\ &= \lim_{s \to 2} \frac{1}{\sqrt{s} + \sqrt{2}} \end{aligned}$$

$$= \frac{\sqrt{2}}{4}.$$

Therefore, the function f is completely delta differentiable, σ_1-completely delta differentiable, and σ_2-completely delta differentiable at the point $(2, 2)$ iff

$$\frac{a}{2} + \frac{\sqrt{2}}{2} - 5 = 2, \quad \text{i.e.,} \quad 4 = a + \sqrt{2} - 10, \quad \text{i.e.,} \quad a = 14 - \sqrt{2}.$$

Exercise 6.115 Let $\mathbb{T}_1 = [0, 1] \cup \{5\}$, $\mathbb{T}_2 = [0, 1]$, where $[0, 1]$ is the real interval. Let $f : \mathbb{T}_1 \times \mathbb{T}_2 \to \mathbb{R}$ be defined by

$$f(t) = t_1^2 + 2t_1 + 3t_2, \quad (t_1, t_2) \in [0, 1] \times [0, 1]$$

and

$$f(5, t_2) = a + 3t_2, \quad t_2 \in [0, 1],$$

where a is a real constant. Find a constant a so that the function f is completely delta differentiable, σ_1-completely delta differentiable, and σ_2-completely delta differentiable.

Solution $a = 19$.

Theorem 6.116 (Mean Value Theorem) *Let $f : \Lambda^n \to \mathbb{R}$, $x, y \in \Lambda^n$, $\alpha = \min_{i \in \{1,\ldots,n\}} x_i$, $\beta = \max_{i \in \{1,\ldots,n\}} y_i$. Assume f is continuous on*

$$[\alpha, \beta] \times [\alpha, \beta] \times \cdots \times [\alpha, \beta] \subset \Lambda^n$$

and $f_{x_i}^{\Delta_i}(x)$ exists for all $x \in \Lambda_i^{\kappa_i n}$. Then there exist numbers $\xi_i, \eta_i \in [\alpha, \beta)$, $i \in \{1, 2, \ldots, n\}$, such that

$$\begin{aligned} & f_{x_1}^{\Delta_1}(\xi_1, x_2, \ldots, x_n)(x_1 - y_1) + f_{x_2}^{\Delta_2}(y_1, \xi_2, \ldots, x_n)(x_2 - y_2) \\ & + \cdots + f_{x_n}^{\Delta_n}(y_1, y_2, \ldots, \xi_n)(x_n - y_n) \le f(x) - f(y) \\ & \le f_{x_1}^{\Delta_1}(\eta_1, x_2, \ldots, x_n)(x_1 - y_1) + f_{x_2}^{\Delta_2}(y_1, \eta_2, \ldots, x_n)(x_2 - y_2) \\ & + \cdots + f_{x_n}^{\Delta_n}(y_1, y_2, \ldots, \eta_n)(x_n - y_n). \end{aligned} \tag{6.18}$$

Proof We have

$$
\begin{aligned}
f(x) - f(y) =& f(x_1, x_2, \ldots, x_n) - f(y_1, x_2, \ldots, x_n) + f(y_1, x_2, \ldots, x_n) \\
& - f(y_1, y_2, x_3, \ldots, x_n) + f(y_1, y_2, x_3, \ldots, x_n) \\
& - \cdots + f(y_1, y_2, \ldots, y_{n-1}, x_n) - f(y_1, y_2, \ldots, y_{n-1}, y_n).
\end{aligned} \tag{6.19}
$$

Using the mean value theorem, Theorem 2.41, there exist numbers $\xi_i, \eta_i \in [\alpha, \beta)$, $i \in \{1, 2, \ldots, n\}$, such that

$$
\begin{aligned}
f_{x_1}^{\Delta_1}(\xi_1, x_2, \ldots, x_n)(x_1 - y_1) &\le f(x_1, x_2, \ldots, x_n) - f(y_1, x_2, \ldots, x_n) \\
&\le f_{x_1}^{\Delta_1}(\eta_1, x_2, \ldots, x_n)(x_1 - y_1), \\
f_{x_2}^{\Delta_2}(y_1, \xi_2, \ldots, x_n)(x_2 - y_2) &\le f(y_1, x_2, \ldots, x_n) - f(y_1, y_2, \ldots, x_n) \\
&\le f_{x_2}^{\Delta_2}(y_1, \eta_2, \ldots, x_n)(x_2 - y_2), \\
&\vdots \\
f_{x_n}^{\Delta_n}(y_1, y_2, \ldots, x_n)(x_n - y_n) &\le f(y_1, y_2, \ldots, x_n) - f(y_1, y_2, \ldots, y_n) \\
&\le f_{x_n}^{\Delta_n}(y_1, y_2, \ldots, \eta_n)(x_n - y_n).
\end{aligned}
$$

Thus, using (6.19), we get (6.18). □

Corollary 6.117 *Suppose $f : \Lambda^n \to \mathbb{R}$ is a continuous function and $f_{x_i}^{\Delta_i}(x)$ exists for every $x \in \Lambda_i^{\kappa_i n}$, $i \in \{1, 2, \ldots, n\}$. If these derivatives are identically zero, then f is a constant function on Λ^n.*

6.4 Geometric Sense of Differentiability

Illustrating now the two-variable case, let us consider the "surface" $\mathscr{S}$ represented by a real-valued continuous function $u = f(t, s)$ defined on $\mathbb{T}_1 \times \mathbb{T}_2$, that is, the set of points $\{(t, s, f(t, s)) : (t, s) \in \mathbb{T}_1 \times \mathbb{T}_2\}$ in the xyz-space. Let (t^0, s^0) be a fixed point in $\mathbb{T}_1^\kappa \times \mathbb{T}_2^\kappa$.

Definition 6.118 A plane Ω_0 passing through the point $P_0 = (t^0, s^0, f(t^0, s^0))$ is called the *delta tangent plane* to the surface $\mathscr{S}$ at the point P_0 if

1. Ω_0 passes also through the points

$$P_0^{\sigma_1} = (\sigma_1(t^0), s^0, f(\sigma_1(t^0), s^0)) \quad \text{and} \quad P_0^{\sigma_2} = (t_0, \sigma_2(s^0), f(t^0, \sigma_2(s^0)));$$

2. if P_0 is not an isolated point of the surface $\mathscr{S}$, then

$$\lim_{\substack{P \to P_0 \\ P \neq P_0}} \frac{d(P, \Omega_0)}{d(P, P_0)} = 0, \tag{6.20}$$

where P is the moving point of the surface $\mathscr{S}$, $d(P, \Omega_0)$ is the distance from the point P to the plane Ω_0, and $d(P, P_0)$ is the distance of the point P from the point P_0 (see Figure 6.1).

Theorem 6.119 *If the function f is completely delta differentiable at the point (t^0, s^0), then the surface represented by this function has the uniquely determined*

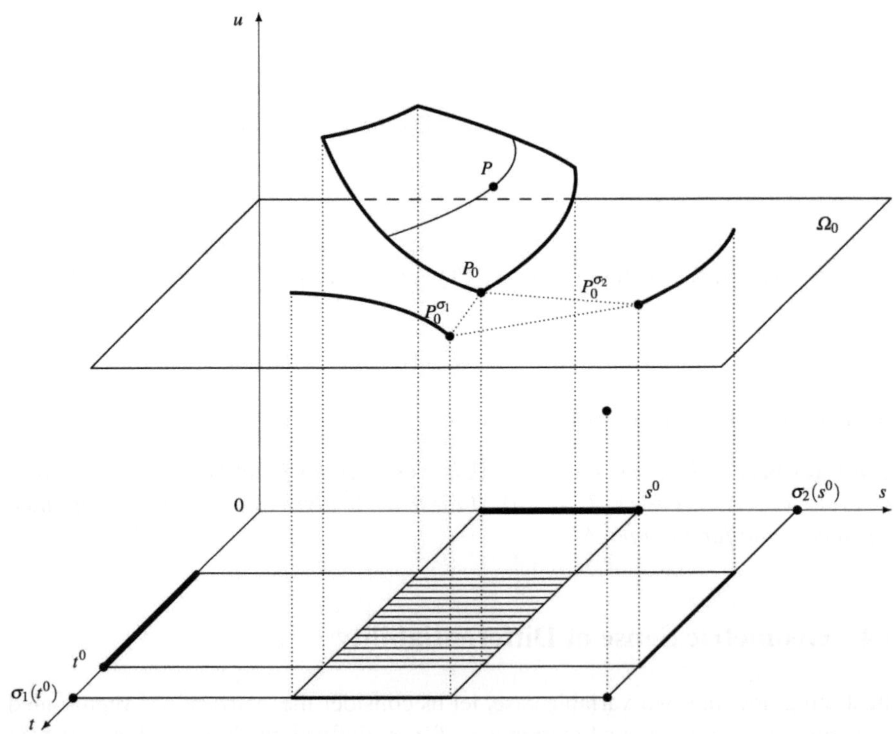

Fig. 6.1 Each of $\mathbb{T}_1$ and $\mathbb{T}_2$ consists of a real number interval and a separate point. Accordingly, the (time scale) surface $\mathscr{S}$ consists of one piece of a usual surface, two arcs of usual curves and one separate point. In order the surface $\mathscr{S}$ to have a delta tangent plane Ω_0 at the point P_0, there must exist the usual "left-sided" tangent plane to $\mathscr{S}$ at P_0 and, moreover, that plane must pass through both points $P_0^{\sigma_1}$ and $P_0^{\sigma_2}$.

delta tangent plane at the point $P_0 = (t^0, s^0, f(t^0, s^0))$ specified by the equation

$$z - f(t^0, s^0) = \frac{\partial f(t^0, s^0)}{\Delta_1 t}(x - t^0) + \frac{\partial f(t^0, s^0)}{\Delta_2 s}(y - s^0), \tag{6.21}$$

where (x, y, z) is the current point of the plane.

Proof Let f be a completely delta differentiable function at a point $(t^0, s^0) \in \mathbb{T}_1^\kappa \times \mathbb{T}_2^\kappa$, $\mathscr{S}$ be the surface represented by this function, and Ω_0 be the plane described by equation (6.21). Let us show that Ω_0 passes also through the point $P_0^{\sigma_1}$. Indeed, if $\sigma_1(t^0) = t^0$, then $P_0^{\sigma_1} = P_0$ and the statement is true. Let now $\sigma_1(t^0) > t^0$. Substituting the coordinates $(\sigma_1(t^0), s^0, f(\sigma_1(t^0), s^0))$ of the point $P_0^{\sigma_1}$ into equation (6.21), we get

$$f(\sigma_1(t^0), s^0) - f(t^0, s^0) = \frac{\partial f(t^0, s^0)}{\Delta_1 t}\left[\sigma_1(t^0) - t^0\right],$$

which is obviously true due to the continuity of f at (t^0, s^0). Similarly, we can see that Ω_0 passes also through the point $P_0^{\sigma_2}$. Now, we check (6.20). Assume that P_0 is not an isolated point of the surface $\mathscr{S}$. The variable point $P \in \mathscr{S}$ has the coordinates $(t, s, f(t, s))$. As is known from analytic geometry, the distance between P and Ω_0 with equation (6.21) is expressed by the formula

$$d(P, \Omega_0) = \frac{1}{N}\left| f(t, s) - f(t^0, s^0) - \frac{\partial f(t^0, s^0)}{\Delta_1 t}(t - t^0) - \frac{\partial f(t^0, s^0)}{\Delta_2 s}(s - s^0)\right|,$$

where

$$N = \sqrt{1 + \left[\frac{\partial f(t^0, s^0)}{\Delta_1 t}\right]^2 + \left[\frac{\partial f(t^0, s^0)}{\Delta_2 s}\right]^2}.$$

Hence, by the differentiability condition (6.14) in which

$$A_1 = \frac{\partial f(t^0, s^0)}{\Delta_1 t} \quad \text{and} \quad A_2 = \frac{\partial f(t^0, s^0)}{\Delta_2 s}$$

due to the other differentiability conditions (6.14) and (6.15),

$$\begin{aligned} d(P, \Omega_0) &= \frac{1}{N}\left|\alpha_1(t - t^0) + \alpha_2(s - s^0)\right| \\ &\le \frac{1}{N}\sqrt{\alpha_1^2 + \alpha_2^2}\sqrt{(t - t^0)^2 + (s - s^0)^2}. \end{aligned}$$

Next,

$$d(P, P_0) = \sqrt{(t - t^0)^2 + (s - s^0)^2 + \left[f(t, s) - f(t^0, s^0)\right]^2}$$

$$\geq \sqrt{(t-t^0)^2+(s-s^0)^2}.$$

Therefore

$$\frac{d(P,\Omega_0)}{d(P,P_0)} \leq \frac{1}{N}\sqrt{\alpha_1^2+\alpha_2^2} \to 0 \quad \text{as} \quad P \to P_0.$$

Thus, we have proved that the plane specified by equation (6.21) is the delta tangent plane to $\mathscr{S}$ at the point P_0.

Now, we must show that there are no other delta tangent planes to $\mathscr{S}$ at the point P_0 distinct from Ω_0. If $\sigma_1(t^0) > t^0$ and $\sigma_2(s^0) > s^0$ at the same time, the points P_0, $P_0^{\sigma_1}$, and $P_0^{\sigma_2}$ are pairwise distinct. In this case the delta tangent plane (provided it exists) is unique as it passes through the three distinct points P_0, $P_0^{\sigma_1}$, and $P_0^{\sigma_2}$. Further, we have to consider the remaining possible cases. Suppose that there is a delta tangent plane Ω to $\mathscr{S}$ at the point P_0 described by an equation

$$a(x-t^0)+b(y-s^0)-c\left[z-f(t^0,s^0)\right]=0 \quad \text{with} \quad a^2+b^2+c^2=1. \tag{6.22}$$

Let $P=(t,s,f(t,s))$ be a variable point on $\mathscr{S}$. Using equation (6.22), we have

$$d(P,\Omega)=\left|a(t-t^0)+b(s-s^0)-c\left[f(t,s)-f(t^0,s^0)\right]\right|.$$

Hence, by the differentiability condition (6.14) in which

$$A_1=\frac{\partial f(t^0,s^0)}{\Delta_1 t} \quad \text{and} \quad A_2=\frac{\partial f(t^0,s^0)}{\Delta_2 s},$$

the latter being a result of the conditions (6.14) and (6.15),

$$d(P,\Omega)=\left|\left[a-c(A_1+\alpha_1)\right](t-t^0)+\left[b-c(A_2+\alpha_2)\right](s-s^0)\right|.$$

Next, by the same differentiability condition,

$$d(P,P_0)=\sqrt{(t-t^0)^2+(s-s^0)^2+\left[f(t,s)-f(t^0,s^0)\right]^2}$$

$$=\sqrt{(t-t^0)^2+(s-s^0)^2+\left[(A_1+\alpha_1)(t-t^0)+(A_2+\alpha_2)(s-s^0)\right]^2}.$$

Hence, we have

$$\frac{d(P,\Omega)}{d(P,P_0)}=\frac{\left|\left[a-c(A_1+\alpha_1)\right](t-t^0)+\left[b-c(A_2+\alpha_2)\right](s-s^0)\right|}{\sqrt{(t-t^0)^2+(s-s^0)^2+\left[(A_1+\alpha_1)(t-t^0)+(A_2+\alpha_2)(s-s^0)\right]^2}}. \tag{6.23}$$

Now, we discuss the remaining possible cases.

(a) Let $\sigma_1(t^0) = t^0$ and $\sigma_2(s^0) = s^0$. By our assumption, the left-hand side of (6.23) tends to zero as $(t, s) \to (t^0, s^0)$, that is, as $P \to P_0$. On putting $s = s^0$ in (6.23), dividing in the right-hand side the numerator and denominator by $|t - t^0|$, and passing then to the limit as $t \to t^0$, we get

$$a - cA_1 = 0, \quad \text{that is,} \quad a - c\frac{\partial f(t^0, s^0)}{\Delta_1 t} = 0. \tag{6.24}$$

Similarly, putting $t = t^0$ in (6.23), cancelling $|s - s^0|$, and passing then to the limit as $s \to s^0$, we obtain

$$b - cA_2 = 0, \quad \text{that is,} \quad b - c\frac{\partial f(t^0, s^0)}{\Delta_2 s} = 0. \tag{6.25}$$

We see that $c \neq 0$ because, if otherwise, we would have $a = b = c = 0$. Hence, the plane Ω is described by equation (6.21).

(b) Let now $\sigma_1(t^0) = t^0$ and $\sigma_2(s^0) > s^0$. In this case, we obtain (6.24) from (6.23) as in case (a). However, now we can, in general, not get (6.25) from (6.23) as in case (a) because the point s^0 may be isolated in $\mathbb{T}_2$, and therefore for all points (t, s) in a sufficiently small neighbourhood of (t^0, s^0), we may have $s = s^0$ (and hence we cannot divide by $|s - s^0|$ in order to pass then to the limit as $s \to s^0$). We proceed as follows. Since by definition of the delta tangent plane, the plane Ω must also pass through the point $P_0^{\sigma_2}$, we also have the equation

$$a(x - t^0) + b\left[y - \sigma_2(s^0)\right] - c\left[z - f(t^0, \sigma_2(s^0))\right] = 0$$

for the same plane Ω. Using this equation, we obtain

$$d(P, \Omega) = \left|a(t - t^0) + b\left[s - \sigma_2(s^0)\right] - c\left[f(t, s) - f(t^0, \sigma_2(s^0))\right]\right|.$$

Hence, by the differentiability condition (6.15), we get

$$d(P, \Omega) = \left|\left[a - c(A_1 + \beta_{21})\right](t - t^0) + \left[b - c(A_2 + \beta_{22})\right]\left[\sigma_2(s^0) - s\right]\right|.$$

Therefore

$$\frac{d(P, \Omega)}{d(P, P_0)} = \frac{\left|\left[a - c(A_1 + \beta_{21})\right](t - t^0) + \left[b - c(A_2 + \beta_{22})\right]\left[\sigma_2(s^0) - s\right]\right|}{\sqrt{(t - t^0)^2 + (s - s^0)^2 + \left[(A_1 + \alpha_1)(t - t^0) + (A_2 + \alpha_2)(s - s^0)\right]^2}}.$$

Setting here $s = s^0$ and taking into account (6.24) proved in the considered case, we obtain

$$\left.\frac{d(P, \Omega)}{d(P, P_0)}\right|_{s=s^0} = \frac{\left|-c\beta_{21}(t - t^0) + \left[b - c(A_2 + \beta_{22})\right]\left[\sigma_2(s^0) - s^0\right]\right|}{|t - t^0|\sqrt{1 + (A_1 + \alpha_1)^2}}.$$

Passing here to the limit as $t \to t^0$, we see that

$$b - cA_2 = 0, \quad \text{that is,} \quad b - c\frac{\partial f(t^0, s^0)}{\Delta_2 s} = 0$$

because, if otherwise, the right-hand side would tend to infinity. Further, the proof is completed as in case (a).

(c) Finally, suppose $\sigma_1(t^0) > t^0$ and $\sigma_2(s^0) = s^0$. In this case the proof is analogous to that in case (b) and uses the equation (Ω also passes through $P_0^{\sigma_1}$)

$$a\left[x - \sigma_1(t^0)\right] + b(y - s^0) - c\left[z - f(\sigma_1(t^0), s^0)\right] = 0$$

and the differentiability condition (6.15). □

Remark 6.120 If $P_0^{\sigma_1} \neq P_0$ and $P_0^{\sigma_2} \neq P_0$ (hence also $P_0^{\sigma_1} \neq P_0^{\sigma_2}$) at the same time and if there is a delta tangent plane at the point P_0 to the surface $\mathscr{S}$, then it coincides with the unique plane through the three points P_0, $P_0^{\sigma_1}$, and $P_0^{\sigma_2}$.

Example 6.121 Let $\mathbb{T}_1 = \mathbb{Z}$ and $\mathbb{T}_2 = \mathbb{N}_0$. We will find the delta tangent plane to the surface

$$\{(t, s, f(t, s)) : (t, s) \in \mathbb{T}_1 \times \mathbb{T}_2\} \quad \text{with} \quad f(t, s) = t^2 + 2st + 2s^2$$

at the point $(-2, 0)$. Here,

$$\sigma_1(t) = t + 1, \quad t \in \mathbb{T}_1, \quad \sigma_2(s) = s + 1, \quad s \in \mathbb{T}_2.$$

Hence,

$$f(-2, 0) = 4,$$

$$f_t^{\Delta_1}(-2, 0) = \sigma_1(t) + t + 2s = 2t + 2s + 1,$$

$$f_s^{\Delta_2}(t, s) = 2t + 2(\sigma_2(s) + s) = 2t + 4s + 2,$$

$$f_t^{\Delta_1}(-2, 0) = -3,$$

$$f_s^{\Delta_2}(-2, 0) = -2.$$

Therefore,

$$z - 4 = -3(x + 2) - 2y, \quad \text{i.e.,} \quad 3x + 2y + z = -2$$

is the delta tangent plane to the considered surface at the point $(-2, 0)$.

Example 6.122 Let $\mathbb{T}_1 = \mathbb{Z}$ and $\mathbb{T}_2 = 2^{\mathbb{N}_0}$. We will find the delta tangent plane to the surface

$$\{(t, s, f(t, s)) : (t, s) \in \mathbb{T}_1 \times \mathbb{T}_2\} \quad \text{with} \quad f(t, s) = t^3 - s^3$$

at the point (0, 2). Here,

$$\sigma_1(t) = t + 1, \quad t \in \mathbb{T}_1, \quad \sigma_2(s) = 2s, \quad s \in \mathbb{T}_2.$$

Thus,

$$\begin{aligned} f(0, 2) &= -8, \\ f_t^{\Delta_1}(t, s) &= (\sigma_1(t))^2 + t\sigma_1(t) + t^2 \\ &= (t+1)^2 + t(t+1) + t^2 \\ &= t^2 + 2t + 1 + t^2 + t + t^2 \\ &= 3t^2 + 3t + 1, \\ f_s^{\Delta_2}(t, s) &= -(\sigma_2(s))^2 - s\sigma_2(s) - s^2 \\ &= -4s^2 - 2s^2 - s^2 \\ &= -7s^2, \\ f_t^{\Delta_1}(0, 2) &= 1, \\ f_s^{\Delta_2}(0, 2) &= -28. \end{aligned}$$

Therefore,

$$z + 8 = x - 28(y - 2), \quad \text{i.e.,} \quad -x + 28y + z = 48$$

is the delta tangent plane to the considered surface at the point (0, 2).

Example 6.123 Let $\mathbb{T}_1 = 2^{\mathbb{N}_0}$ and $\mathbb{T}_2 = 3^{\mathbb{N}_0}$. We will find the delta tangent plane to the surface

$$\{(t, s, f(t, s)) : (t, s) \in \mathbb{T}_1 \times \mathbb{T}_2\} \quad \text{with} \quad f(t, s) = t^2 - s^3 t - s^2$$

at the point (2, 3). Here,

$$\sigma_1(t) = 2t, \quad t \in \mathbb{T}_1, \quad \sigma_2(s) = 3s, \quad s \in \mathbb{T}_2.$$

Thus,

$$\begin{aligned}
f(2,3) &= -59, \\
f_t^{\Delta_1}(t,s) &= \sigma_1(t) + t - s^3 \\
&= 2t + t - s^3 \\
&= 3t - s^3, \\
f_s^{\Delta_2}(t,s) &= -t\left((\sigma_2(s))^2 + s\sigma_2(s) + s^2\right) - (\sigma_2(s) + s) \\
&= -t\left(9s^2 + 3s^2 + s^2\right) - (3s + s) \\
&= -13s^2 t - 4s, \\
f_t^{\Delta_1}(2,3) &= -21, \\
f_s^{\Delta_2}(2,3) &= -246.
\end{aligned}$$

Therefore,

$$z + 59 = -21(x-2) - 246(y-3), \quad \text{i.e.,} \quad 21x + 246y + z = 721$$

is the delta tangent plane to the considered surface at the point (2, 3).

Exercise 6.124 Let $\mathbb{T}_1 = 2^{\mathbb{N}_0}$ and $\mathbb{T}_2 = \mathbb{N}_0$. Find the delta tangent plane to the surface

$$\{(t, s, t^2 - s^2) : (t,s) \in \mathbb{T}_1 \times \mathbb{T}_2\}$$

at the point (2, 0).

Solution $6x - y - z = 8$.

6.5 Sufficient Conditions for Differentiability

Theorem 6.125 *Suppose the function $f : \Lambda^n \to \mathbb{R}$ is continuous and has first-order partial delta derivatives $f_{t_1}^{\Delta_i}(t)$ in some neighbourhood $U_\delta(t^0)$ of the point $t^0 \in \Lambda^{\kappa n}$. If these derivatives are continuous at the point t^0, then f is completely delta differentiable at t^0.*

Proof We have

$$f(t^0) - f(t) = f(t^0) - f(t_1, t_2^0, \dots, t_n^0) + f(t_1, t_2^0, \dots, t_n^0) - \cdots$$

$$+ f(t_1, t_2, \dots, t_{n-1}, t_n^0) - f(t_1, t_2, \dots, t_n).$$

By the one-variable case, we have

$$f(t^0) - f(t_1, t_2^0, \dots, t_n^0) = f_{t_1}^{\Delta_1}(t^0)(t_1^0 - t_1) + \alpha_1(t_1^0 - t_1)$$

for $(t_1, t_2^0, \dots, t_n^0) \in U_\delta(t^0)$, where

$$\alpha_1 = \alpha_1(t^0, t_1), \quad \alpha_1 \to 0 \quad \text{as} \quad (t_1, t_2^0, \dots, t_n^0) \to (t_1^0, t_2^0, \dots, t_n^0).$$

Further, applying the one-variable mean value result, we get

$$\begin{aligned} f_{t_2}^{\Delta_2}(t_1, \xi_1, \dots, t_n^0)(t_2^0 - t_2) \leq & f(t_1, t_2^0, \dots, t_n^0) - f(t_1, t_2, \dots, t_n^0) \\ \leq & f_{t_2}^{\Delta_2}(t_1, \xi_2, \dots, t_n^0)(t_2^0 - t_2), \end{aligned} \tag{6.26}$$

where $\xi_1, \xi_2 \in [\alpha, \beta)$ and $\alpha = \min\{t_2^0, t_2\}$, $\beta = \max\{t_2^0, t_2\}$. Since $\xi_1, \xi_2 \to t_2^0$ as $t_2 \to t_2^0$ and $f_{t_2}^{\Delta_2}(\cdot)$ is continuous at t^0, we get

$$\lim_{t \to t^0} f_{t_2}^{\Delta_2}(t_1, \xi_1, \dots, t_n^0) = \lim_{t \to t^0} f_{t_2}^{\Delta_2}(t_1, \xi_2, \dots, t_n^0) = f_{t_2}^{\Delta_2}(t^0).$$

Thus, using (6.26), we find

$$f(t_1, t_2^0, \dots, t_n^0) - f(t_1, t_2, \dots, t_n^0) = f_{t_2}^{\Delta_2}(t^0)(t_2^0 - t_2) + \alpha_2(t_2^0 - t_2),$$

where $\alpha_2 = \alpha_2(t^0, t_1, t_2)$, $\alpha_2 \to 0$ as $t \to t^0$, and so on,

$$f(t_1, t_2, \dots, t_n^0) - f(t_1, t_2, \dots, t_n) = f_{t_n}^{\Delta_n}(t^0)(t_n^0 - t_n) + \alpha_n(t_n^0 - t_n),$$

where $\alpha_n = \alpha_n(t^0, t_1, t_2, \ldots, t_n)$, $\alpha_n \to 0$ as $t \to t^0$. Consequently,

$$f(t^0) - f(t) = \sum_{i=1}^{n} f_{t_i}^{\Delta_i}(t^0)(t_i^0 - t_i) + \sum_{i=1}^{n} \alpha_i (t_i^0 - t_i). \tag{6.27}$$

Now, we consider the difference

$$\begin{aligned} & f(t_1^0, t_2^0, \ldots, t_{i-1}^0, \sigma_i(t_i^0), t_{i+1}^0, \ldots, t_n^0) - f(t_1, t_2, \ldots, t_{i-1}, t_i, t_{i+1}, \ldots, t_n) \\ &= f(t_1^0, t_2^0, \ldots, t_{i-1}^0, \sigma_i(t_i^0), t_{i+1}^0, \ldots, t_n^0) - f(t_1^0, t_2^0, \ldots, t_{i-1}^0, t_i^0, t_{i+1}^0, \ldots, t_n^0) \\ &\quad + f(t_1^0, t_2^0, \ldots, t_{i-1}^0, t_i^0, t_{i+1}^0, \ldots, t_n^0) - f(t_1, t_2, \ldots, t_{i-1}, t_i, t_{i+1}, \ldots, t_n). \end{aligned} \tag{6.28}$$

Using the definition of the partial delta derivative, we get

$$\begin{aligned} & f(t_1^0, t_2^0, \ldots, t_{i-1}^0, \sigma_i(t_i^0), t_{i+1}^0, \ldots, t_n^0) - f(t_1^0, t_2^0, \ldots, t_{i-1}^0, t_i^0, t_{i+1}^0, \ldots, t_n^0) \\ &\qquad = f_{t_i}^{\Delta_i}(t^0)(\sigma_i(t_i^0) - t_i^0) + \beta_{ii}(\sigma_i(t_i^0) - t_i^0). \end{aligned} \tag{6.29}$$

Substituting (6.27) and (6.29) in (6.28), we obtain

$$\begin{aligned} f(t_1^0, t_2^0, \ldots, t_{i-1}^0, \sigma_i(t_i^0), t_{i+1}^0, \ldots, t_n^0) - f(t) &= f_{t_i}^{\Delta_i}(t^0)(\sigma_i(t_i^0) - t_i) + \beta_{ii}(\sigma_i(t_i^0) - t_i) \\ &\quad + \sum_{i=1}^{n} f_{t_i}^{\Delta_i}(t^0)(t_i^0 - t_i) + \sum_{i=1}^{n} \alpha_i (t_i^0 - t_i). \end{aligned}$$

Therefore, the function f is completely delta differentiable at t^0. □

Theorem 6.126 *Assume $f : \Lambda^n \to \mathbb{R}$ is a continuous function that has partial derivatives $f_{t_i}^{\Delta_i}(t)$, $i \in \{1, 2, \ldots, n\}$, in a union of some neighbourhoods of the points t^0 and*

$$\left(\sigma_1(t_1^0), \ldots, \sigma_i(t_{i-1}^0), t_i^0, \sigma_{i+1}(t_{i+1}^0), \ldots, \sigma_n(t_n^0)\right).$$

If these derivatives are continuous at the point t^0 and, moreover,

$$f_{t_i}^{\Delta_i}(\sigma_1(t_1^0), \ldots, \sigma_{i-1}(t_{i-1}^0), t_i, \sigma_{i+1}(t_{i+1}^0), \ldots, \sigma_n(t_n^0))$$

is continuous at $t_i = t_i^0$, then f is σ_i-completely delta differentiable at t^0.

Proof From Theorem 6.125, it follows that the function f is completely delta differentiable at t^0. Now, we consider the difference

$$f(\sigma_1(t_1^0), \sigma_2(t_2^0), \ldots, \sigma_n(t_n^0)) - f(t^0)$$

$$= f(t_1^0, t_2^0, \ldots, t_{i-1}^0, \sigma_i(t_i^0), t_{i+1}^0, \ldots, t_n^0) - f(t_1^0, t_2^0, \ldots, t_{i-1}^0, t_i^0, t_{i+1}^0, \ldots, t_n^0)$$

$$- f(t_1^0, t_2^0, \ldots, t_{i-1}^0, \sigma_i(t_i^0), t_{i+1}^0, \ldots, t_n^0) + f(\sigma_1(t_1^0), \sigma_2(t_2^0), \ldots, \sigma_n(t_n^0)). \tag{6.30}$$

Using the definition of the partial delta derivative, we have

$$f(t_1^0, t_2^0, \ldots, t_{i-1}^0, \sigma_i(t_i^0), t_{i+1}^0, \ldots, t_n^0) - f(t_1^0, t_2^0, \ldots, t_{i-1}^0, t_i^0, t_{i+1}^0, \ldots, t_n^0)$$

$$= f_{t_i}^{\Delta_i}(t^0)(\sigma_i(t_i^0) - t_i^0). \tag{6.31}$$

Also,

$$f(\sigma_1(t_1^0), \sigma_2(t_2^0), \ldots, \sigma_n(t_n^0)) - f(t_1^0, t_2^0, \ldots, t_{i-1}^0, \sigma_i(t_i^0), t_{i+1}^0, \ldots, t_n^0)$$

$$\begin{aligned}
= \;& f(\sigma_1(t_1^0), \sigma_2(t_2^0), \ldots, \sigma_{i-1}(t_{i-1}^0), \sigma_i(t_i^0), \sigma_{i+1}(t_{i+1}^0), \ldots, \sigma_n(t_n^0)) \\
& - f(\sigma_1(t_1^0), \sigma_2(t_2^0) \ldots, \sigma_{i-1}(t_{i-1}^0), t_i^0, \sigma_{i+1}(t_{i+1}^0), \ldots, \sigma_n(t_n^0)) \\
& + f(\sigma_1(t_1^0), \sigma_2(t_2^0), \ldots, \sigma_{i-1}(t_{i-1}^0), t_i^0, \sigma_{i+1}(t_{i+1}^0), \ldots, \sigma_n(t_n^0)) \\
& - f(t_1^0, t_2^0, \ldots, t_{i-1}^0, \sigma_i(t_i^0), t_{i+1}^0, \ldots, t_n^0) \\
= \;& f_{t_i}^{\Delta_i}(\sigma_1(t_1^0), \sigma_2(t_2^0), \ldots, \sigma_{i-1}(t_{i-1}^0), t_i^0, \sigma_{i+1}(t_{i+1}^0), \ldots, \sigma_n(t_n^0))(\sigma_i(t_i^0) - t_i^0) \\
& + f(\sigma_1(t_1^0), \ldots, \sigma_{i-1}(t_{i-1}^0), t_i^0, \sigma_{i+1}(t_{i+1}^0), \ldots, \sigma_n(t_n^0)) \\
& - f(t_1^0, \sigma_2(t_2^0), \ldots, \sigma_{i-1}(t_{i-1}^0), t_i^0, \sigma_{i+1}(t_{i+1}^0), \ldots, \sigma_n(t_n^0)) \\
& + f(t_1^0, \sigma_2(t_2^0), \ldots, \sigma_{i-1}(t_{i-1}^0), t_i^0, \sigma_{i+1}(t_{i+1}^0), \ldots, \sigma_n(t_n^0)) \\
& - \cdots \\
& + f(t_1^0, t_2^0, \ldots, t_{i-1}^0, t_i^0, t_{i+1}^0, \ldots, t_n^0) \\
& - f(t_1^0, t_2^0, \ldots, t_{i-1}^0, \sigma_i(t_i^0), t_{i+1}^0, \ldots, t_n^0)
\end{aligned}$$

$$= f_{t_1}^{\Delta_1}(\sigma_1(t_1^0), \sigma_2(t_2^0), \ldots, \sigma_{i-1}(t_{i-1}^0), t_i^0, \sigma_{i+1}(t_{i+1}^0), \ldots, \sigma_n(t_n^0))(\sigma_i(t_i^0) - t_i^0)$$

$$+ f_{t_1}^{\Delta_1}(t_1^0, \sigma_2(t_2^0), \ldots, \sigma_{i-1}(t_{i-1}^0), t_i^0, \sigma_{i+1}(t_{i+1}^0), \ldots, \sigma_n(t_n^0))(\sigma_1(t_1^0) - t_1^0)$$

$$+ \cdots$$

$$+ f_{t_n}^{\Delta_n}(t^0)(\sigma_n(t_n^0) - t_n^0) - f_{t_i}^{\Delta_i}(t^0)(\sigma_i(t_i^0) - t_i^0).$$

Hence, using (6.30) and (6.31), we conclude that f is σ_i-completely delta differentiable at t^0. □

6.6 Equality of Mixed Partial Derivatives

Theorem 6.127 *Suppose the function $f : \Lambda^n \to \mathbb{R}$ has mixed partial derivatives*

$$f_{t_i t_j}^{\Delta_i \Delta_j}(t) \quad and \quad f_{t_j t_i}^{\Delta_j \Delta_i}(t)$$

in some neighbourhood of the point $t^0 \in \Lambda_{ij}^{\kappa_i \kappa_j n}$. If these derivatives are continuous at the point t^0, then

$$f_{t_i t_j}^{\Delta_i \Delta_j}(t^0) = f_{t_j t_i}^{\Delta_j \Delta_i}(t^0).$$

Here, $i, j \in \{1, 2, \ldots, n\}$, $i \neq j$.

Proof For convenience, we suppose that $i < j$. Let

$$\Phi(t) = f(t_1, \ldots, t_{i-1}, \sigma_i(t_i^0), t_{i+1}, \ldots, t_{j-1}, \sigma_j(t_j^0), t_{j+1}, \ldots, t_n)$$

$$- f(t_1, \ldots, t_{i-1}, t_i, t_{i+1}, \ldots, t_{j-1}, \sigma_j(t_j^0), t_{j+1}, \ldots, t_n)$$

$$- f(t_1, \ldots, t_{i-1}, \sigma_i(t_i^0), t_{i+1}, \ldots, t_{j-1}, t_j, t_{j+1}, \ldots, t_n) + f(t)$$

and

$$\phi(t_i) = f(t_1, \ldots, t_{i-1}, t_i, t_{i+1}, \ldots, t_{j-1}, \sigma_j(t_j^0), t_{j+1}, \ldots, t_n) - f(t).$$

Then

$$\phi(\sigma_i(t_i^0)) = f(t_1, \ldots, t_{i-1}, \sigma_i(t_i^0), t_{i+1}, \ldots, t_{j-1}, \sigma_j(t_j^0), t_{j+1}, \ldots, t_n)$$

$$- f(t_1, \ldots, t_{i-1}, \sigma_i(t_i^0), t_{i+1}, \ldots, t_n).$$

Therefore,

$$\Phi(t) = \phi(\sigma_i(t_i^0)) - \phi(t_i).$$

Hence, using the mean value theorem, there exist points $\xi_i^1, \xi_i^2 \in [\alpha_i, \beta_i)$, where

$$\alpha_i = \min\{t_i, \sigma_i(t_i^0)\}, \quad \beta_i = \max\{t_i, \sigma_i(t_i^0)\},$$

such that

$$\phi^{\Delta}(\xi_i^1)(\sigma_i(t_i^0) - t_i) \le \phi(\sigma_i(t_i^0)) - \phi(t_i) \le \phi^{\Delta}(\xi_i^2)(\sigma_i(t_i^0) - t_i),$$

i.e.,

$$\begin{aligned}
&\Big(f_{t_i}^{\Delta_i}(t_1, \ldots, t_{i-1}, \xi_i^1, t_{i+1}, \ldots, t_{j-1}, \sigma_j(t_j^0), t_{j+1}, \ldots, t_n)\\
&\quad - f_{t_i}^{\Delta_i}(t_1, \ldots, t_{i-1}, \xi_i^1, t_{i+1}, \ldots, t_n)\Big)(\sigma_i(t_i^0) - t_i)\\
&\le \Phi(t)\\
&\le \Big(f_{t_i}^{\Delta_i}(t_1, \ldots, t_{i-1}, \xi_i^2, t_{i+1}, \ldots, t_{j-1}, \sigma_j(t_j^0), t_{j+1}, \ldots, t_n)\\
&\quad - f_{t_i}^{\Delta_i}(t_1, \ldots, t_{i-1}, \xi_i^2, t_{i+1}, \ldots, t_n)\Big)(\sigma_i(t_i^0) - t_i),
\end{aligned}$$

whereupon

$$\begin{aligned}
&f_{t_i}^{\Delta_i}(t_1, \ldots, t_{i-1}, \xi_i^1, t_{i+1}, \ldots, t_{j-1}, \sigma_j(t_j^0), t_{j+1}, \ldots, t_n)\\
&\quad - f_{t_i}^{\Delta_i}(t_1, \ldots, t_{i-1}, \xi_i^1, t_{i+1}, \ldots, t_n)\\
&\le \frac{\Phi(t)}{\sigma_i(t_i^0) - t_i}\\
&\le f_{t_i}^{\Delta_i}(t_1, \ldots, t_{i-1}, \xi_i^2, t_{i+1}, \ldots, t_{j-1}, \sigma_j(t_j^0), t_{j+1}, \ldots, t_n)\\
&\quad - f_{t_i}^{\Delta_i}(t_1, \ldots, t_{i-1}, \xi_i^2, t_{i+1}, \ldots, t_n)
\end{aligned}$$

if $\sigma_i(t_i^0) > t_i$, and

$$f_{t_i}^{\Delta_i}(t_1, \ldots, t_{i-1}, \xi_i^2, t_{i+1}, \ldots, t_{j-1}, \sigma_j(t_j^0), t_{j+1}, \ldots, t_n)$$

$$-f_{t_i}^{\Delta_i}(t_1, \ldots, t_{i-1}, \xi_i^2, t_{i+1}, \ldots, t_n)$$

$$\le \frac{\Phi(t)}{\sigma_i(t_i^0) - t_i}$$

$$\le f_{t_i}^{\Delta_i}(t_1, \ldots, t_{i-1}, \xi_i^1, t_{i+1}, \ldots, t_{j-1}, \sigma_j(t_j^0), t_{j+1}, \ldots, t_n)$$

$$-f_{t_i}^{\Delta_i}(t_1, \ldots, t_{i-1}, \xi_i^2, t_{i+1}, \ldots, t_n)$$

if $\sigma_i(t_i^0) < t_i$. Without loss of generality, we assume that $\sigma_i(t_i^0) > t_i$. Again, we apply the mean value theorem and find that there exist $\xi_j^1, \xi_j^2 \in [\alpha_j, \beta_j)$, where

$$\alpha_j = \min\{\sigma_j(t_j^0), t_j\}, \quad \beta_j = \max\{\sigma_j(t_j^0), t_j\},$$

such that

$$f_{t_j t_i}^{\Delta_j \Delta_i}(t_1, \ldots, t_{i-1}, \xi_i^1, t_{i+1}, \ldots, t_{j-1}, \xi_j^1, t_{j+1}, \ldots, t_n)(\sigma_j(t_j^0) - t_j) \le \frac{\Phi(t)}{\sigma_i(t_i^0) - t_i}$$

$$\le f_{t_j t_i}^{\Delta_j \Delta_i}(t_1, \ldots, t_{i-1}, \xi_i^2, t_{i+1}, \ldots, t_{j-1}, \xi_j^2, t_{j+1}, \ldots, t_n)(\sigma_j(t_j^0) - t_j),$$

from where

$$f_{t_j t_i}^{\Delta_j \Delta_i}(t_1, \ldots, t_{i-1}, \xi_i^1, t_{i+1}, \ldots, t_{j-1}, \xi_j^1, t_{j+1}, \ldots, t_n)$$

$$\le \frac{\Phi(t)}{(\sigma_j(t_j^0) - t_j)(\sigma_i(t_i^0) - t_i)}$$

$$\le f_{t_j t_i}^{\Delta_j \Delta_i}(t_1, \ldots, t_{i-1}, \xi_i^2, t_{i+1}, \ldots, t_{j-1}, \xi_j^2, t_{j+1}, \ldots, t_n)$$

if $\sigma_j(t_j^0) > t_j$, and

$$f_{t_j t_i}^{\Delta_j \Delta_i}(t_1, \ldots, t_{i-1}, \xi_i^2, t_{i+1}, \ldots, t_{j-1}, \xi_j^2, t_{j+1}, \ldots, t_n)$$

$$\le \frac{\Phi(t)}{(\sigma_j(t_j^0) - t_j)(\sigma_i(t_i^0) - t_i)}$$

$$\le f_{t_j t_i}^{\Delta_j \Delta_i}(t_1, \ldots, t_{i-1}, \xi_i^1, t_{i+1}, \ldots, t_{j-1}, \xi_j^1, t_{j+1}, \ldots, t_n)$$

if $\sigma_j(t_j^0) < t_j$. Without loss of generality, we assume that $\sigma_j(t_j^0) > t_j$. Since $f_{t_j t_i}^{\Delta_j \Delta_i}$ is continuous at t^0, we get

$$\begin{aligned}
\lim_{\substack{t \to t^0 \\ t_i \neq \sigma_i(t_i^0) \\ t_j \neq \sigma_j(t_j^0)}} f_{t_j t_i}^{\Delta_j \Delta_i}(t_1, \ldots, t_{i-1}, \xi_i^1, t_{i+1}, \ldots, t_{j-1}, \xi_j^1, t_{j+1}, \ldots, t_n) = f_{t_j t_i}^{\Delta_j \Delta_i}(t^0) \\
\leq \lim_{\substack{t \to t^0 \\ t_i \neq \sigma_i(t_i^0) \\ t_j \neq \sigma_j(t_j^0)}} \frac{\Phi(t)}{(\sigma_j(t_j^0) - t_j)(\sigma_i(t_i^0) - t_i)} \\
\leq \lim_{\substack{t \to t^0 \\ t_i \neq \sigma_i(t_i^0) \\ t_j \neq \sigma_j(t_j^0)}} f_{t_j t_i}^{\Delta_j \Delta_i}(t_1, \ldots, t_{i-1}, \xi_i^2, t_{i+1}, \ldots, t_{j-1}, \xi_j^2, t_{j+1}, \ldots, t_n) \\
= f_{t_j t_i}^{\Delta_j \Delta_i}(t^0),
\end{aligned}$$

i.e.,

$$f_{t_j t_i}^{\Delta_j \Delta_i}(t^0) = \lim_{\substack{t \to t^0 \\ t_i \neq \sigma_i(t_i^0) \\ t_j \neq \sigma_j(t_j^0)}} \frac{\Phi(t)}{(\sigma_j(t_i^0) - t_j)(\sigma_i(t_i^0) - t_i)}. \tag{6.32}$$

Let

$$\psi(t_j) = f(t_1, \ldots, t_{i-1}, \sigma_i(t_i^0), t_{i+1}, \ldots, t_{j-1}, t_j, t_{j+1}, \ldots, t_n) - f(t).$$

Thus,

$$\begin{aligned}
\psi(\sigma_j(t_j^0)) = f(t_1, \ldots, t_{i-1}, \sigma_i(t_i^0), t_{i+1}, \ldots, t_{j-1}, \sigma_j(t_j^0), t_{j+1}, \ldots, t_n) \\
- f(t_1, \ldots, t_{j-1}, \sigma_j(t_j^0), t_{j+1}, \ldots, t_n).
\end{aligned}$$

Hence,

$$\Phi(t) = \psi(\sigma_j(t_j^0)) - \psi(t_j).$$

From the mean value theorem, it follows that there exist $\eta_j^1, \eta_j^2 \in [\alpha_j, \beta_j)$ such that

$$\psi^{\Delta}(\eta_j^1)(\sigma_j(t_j^0) - t_j) \leq \psi(\sigma_j(t_j^0)) - \psi(t_j) \leq \psi^{\Delta}(\eta_j^2)(\sigma_j(t_j^0) - t_j),$$

i.e.,

$$
\begin{aligned}
&\Big(f_{t_j}^{\Delta_j}(t_1,\ldots,t_{i-1},\sigma_i(t_i^0),t_{i+1},\ldots,t_{j-1},\eta_j^1,t_{j+1},\ldots,t_n)\\
&\quad -f_{t_j}^{\Delta_j}(t_1,\ldots,t_{j-1},\eta_j^1,t_{j+1},\ldots,t_n)\Big)(\sigma_j(t_j^0)-t_j)\\
&\le \Phi(t)\\
&\le \Big(f_{t_j}^{\Delta_j}(t_1,\ldots,t_{i-1},\sigma_i(t_i^0),t_{i+1},\ldots,t_{j-1},\eta_j^2,t_{j+1},\ldots,t_n)\\
&\quad -f_{t_j}^{\Delta_j}(t_1,\ldots,t_{j-1},\eta_j^2,t_{j+1},\ldots,t_n)\Big)(\sigma_j(t_j^0)-t_j),
\end{aligned}
$$

whereupon

$$
\begin{aligned}
&f_{t_j}^{\Delta_j}(t_1,\ldots,t_{i-1},\sigma_i(t_i^0),t_{i+1},\ldots,t_{j-1},\eta_j^1,t_{j+1},\ldots,t_n)\\
&\quad -f_{t_j}^{\Delta_j}(t_1,\ldots,t_{j-1},\eta_j^1,t_{j+1},\ldots,t_n)\\
&\le \frac{\Phi(t)}{(\sigma_j(t_j^0)-t_j)}\\
&\le f_{t_j}^{\Delta_j}(t_1,\ldots,t_{i-1},\sigma_i(t_i^0),t_{i+1},\ldots,t_{j-1},\eta_j^2,t_{j+1},\ldots,t_n)\\
&\quad -f_{t_j}^{\Delta_j}(t_1,\ldots,t_{j-1},\eta_j^2,t_{j+1},\ldots,t_n)
\end{aligned}
$$

if $\sigma_j(t_j^0)>t_j$, and

$$
\begin{aligned}
&f_{t_j}^{\Delta_j}(t_1,\ldots,t_{i-1},\sigma_i(t_i^0),t_{i+1},\ldots,t_{j-1},\eta_j^2,t_{j+1},\ldots,t_n)\\
&\quad -f_{t_j}^{\Delta_j}(t_1,\ldots,t_{j-1},\eta_j^2,t_{j+1},\ldots,t_n)\\
&\le \frac{\Phi(t)}{(\sigma_j(t_j^0)-t_j)}\\
&\le f_{t_j}^{\Delta_j}(t_1,\ldots,t_{i-1},\sigma_i(t_i^0),t_{i+1},\ldots,t_{j-1},\eta_j^1,t_{j+1},\ldots,t_n)\\
&\quad -f_{t_j}^{\Delta_j}(t_1,\ldots,t_{j-1},\eta_j^1,t_{j+1},\ldots,t_n)
\end{aligned}
$$

if $\sigma_j(t_j^0) < t_j$. Without loss of generality, we suppose that $\sigma_j(t_j^0) > t_j$. Again, we apply the mean value theorem and get $\eta_i^1, \eta_i^2 \in [\alpha_i, \beta_i)$ such that

$$\begin{aligned}
&f_{t_it_j}^{\Delta_i\Delta_j}(t_1, \ldots, t_{i-1}, \eta_i^1, t_{i+1}, \ldots, t_{j-1}, \eta_j^1, t_{j+1}, \ldots, t_n)(\sigma_i(t_i^0) - t_i)\\
&\le \frac{\Phi(t)}{(\sigma_j(t_j^0) - t_j)}\\
&\le f_{t_it_j}^{\Delta_i\Delta_j}(t_1, \ldots, t_{i-1}, \eta_i^2, t_{i+1}, \ldots, t_{j-1}, \eta_j^2, t_{j+1}, \ldots, t_n)(\sigma_i(t_i^0) - t_i),
\end{aligned}$$

from where

$$\begin{aligned}
&f_{t_it_j}^{\Delta_i\Delta_j}(t_1, \ldots, t_{i-1}, \eta_i^1, t_{i+1}, \ldots, t_{j-1}, \eta_j^1, t_{j+1}, \ldots, t_n)\\
&\le \frac{\Phi(t)}{(\sigma_i(t_i^0) - t_i)(\sigma_j(t_j^0) - t_j)}\\
&\le f_{t_it_j}^{\Delta_i\Delta_j}(t_1, \ldots, t_{i-1}, \eta_i^2, t_{i+1}, \ldots, t_{j-1}, \eta_j^2, t_{j+1}, \ldots, t_n)
\end{aligned}$$

if $\sigma_i(t_i^0) > t_i$, and

$$\begin{aligned}
&f_{t_it_j}^{\Delta_i\Delta_j}(t_1, \ldots, t_{i-1}, \eta_i^2, t_{i+1}, \ldots, t_{j-1}, \eta_j^2, t_{j+1}, \ldots, t_n)\\
&\le \frac{\Phi(t)}{(\sigma_i(t_i^0) - t_i)(\sigma_j(t_j^0) - t_j)}\\
&\le f_{t_it_j}^{\Delta_i\Delta_j}(t_1, \ldots, t_{i-1}, \eta_i^1, t_{i+1}, \ldots, t_{j-1}, \eta_j^1, t_{j+1}, \ldots, t_n)
\end{aligned}$$

if $\sigma_i(t_i^0) < t_i$. Without loss of generality, we assume that $\sigma_i(t_i^0) > t_i$. Then

$$\begin{aligned}
&\lim_{\substack{t \to t^0\\ t_i \neq \sigma_i(t_i^0)\\ t_j \neq \sigma_j(t_j^0)}} f_{t_it_j}^{\Delta_i\Delta_j}(t_1, \ldots, t_{i-1}, \eta_i^1, t_{i+1}, \ldots, t_{j-1}, \eta_j^1, t_{j+1}, \ldots, t_n) = f_{t_it_j}^{\Delta_i\Delta_j}(t^0)\\
&\le \lim_{\substack{t \to t^0\\ t_i \neq \sigma_i(t_i^0)\\ t_j \neq \sigma_j(t_j^0)}} \frac{\Phi(t)}{(\sigma_i(t_i^0) - t_i)(\sigma_j(t_j^0) - t_j)}
\end{aligned}$$

$$\leq \lim_{\substack{t \to t^0 \\ t_i \neq \sigma_i(t_i^0) \\ t_j \neq \sigma_j(t_j^0)}} f_{t_i t_j}^{\Delta_i \Delta_j}(t_1, \ldots, t_{i-1}, \eta_i^2, t_{i+1}, \ldots, t_{j-1}, \eta_j^2, t_{j+1}, \ldots, t_n) = f_{t_i t_j}^{\Delta_i \Delta_j}(t^0),$$

i.e.,

$$f_{t_i t_j}^{\Delta_i \Delta_j}(t^0) = \lim_{\substack{t \to t^0 \\ t_i \neq \sigma_i(t_i^0) \\ t_j \neq \sigma_j(t_j^0)}} \frac{\Phi(t)}{(\sigma_i(t_i^0) - t_i)(\sigma_j(t_j^0) - t_j)}.$$

From the last equality and from (6.32), we find

$$f_{t_i t_j}^{\Delta_i \Delta_j}(t^0) = f_{t_j t_i}^{\Delta_j \Delta_i}(t^0),$$

completing the proof. □

Example 6.128 Let $\Lambda^2 = \mathbb{N} \times \mathbb{Z}$ and define $f : \Lambda^2 \to \mathbb{R}$ by

$$f(t) = t_1^2 t_2 + t_1 t_2^2, \quad t \in \Lambda^2.$$

Here,

$$\mathbb{T}_1 = \mathbb{N}, \quad \mathbb{T}_2 = \mathbb{Z}$$

and

$$\sigma_1(t_1) = t_1 + 1, \quad t_1 \in \mathbb{T}_1, \quad \sigma_2(t_2) = t_2 + 1, \quad t_2 \in \mathbb{T}_2.$$

Hence,

$$f_{t_1}^{\Delta_1}(t) = (\sigma_1(t_1) + t_1)t_2 + t_2^2$$

$$= (2t_1 + 1)t_2 + t_2^2,$$

$$f_{t_1 t_2}^{\Delta_1 \Delta_2}(t) = 2t_1 + 1 + \sigma_2(t_2) + t_2$$

$$= 2t_1 + 2t_2 + 2,$$

$$f_{t_2}^{\Delta_2}(t) = t_1^2 + (\sigma_2(t_2) + t_2)t_1$$

$$= t_1^2 + (2t_2 + 1)t_1,$$

$$f_{t_2 t_1}^{\Delta_2 \Delta_1}(t) = \sigma_1(t_1) + t_1 + 2t_2 + 1$$

$$= 2t_1 + 2t_2 + 2, \quad t \in \Lambda^2.$$

Consequently,

$$f_{t_1 t_2}^{\Delta_1 \Delta_2}(t) = f_{t_2 t_1}^{\Delta_2 \Delta_1}(t), \quad t \in \Lambda^2.$$

Example 6.129 Let $\Lambda^2 = 2^{\mathbb{N}} \times \mathbb{N}$ and define $f : \Lambda^2 \to \mathbb{R}$ by

$$f(t) = (\log t_2)t_1 + \sin t_1, \quad t \in \Lambda^2.$$

Here,

$$\mathbb{T}_1 = 2^{\mathbb{N}}, \quad \mathbb{T}_2 = \mathbb{N}$$

and

$$\sigma_1(t_1) = 2t_1, \quad t_1 \in \mathbb{T}_1, \quad \sigma_2(t_2) = t_2 + 1, \quad t_2 \in \mathbb{T}_2.$$

Hence,

$$\begin{aligned}
f_{t_1}^{\Delta_1}(t) &= \log t_2 + \frac{\sin(\sigma_1(t_1)) - \sin t_1}{\sigma_1(t_1) - t_1} \\
&= \log t_2 + \frac{\sin(2t_1) - \sin(t_1)}{2t_1 - t_1} \\
&= \log t_2 + 2\frac{\sin \frac{t_1}{2} \cos \frac{3t_1}{2}}{t_1}, \\
f_{t_1 t_2}^{\Delta_1 \Delta_2}(t) &= \frac{\log(\sigma_2(t_2)) - \log(t_2)}{\sigma_2(t_2) - t_2} \\
&= \frac{\log(t_2 + 1) - \log(t_2)}{t_2 + 1 - t_2} \\
&= \log \frac{t_2 + 1}{t_2}, \\
f_{t_2}^{\Delta_2}(t) &= \frac{\log(\sigma_2(t_2)) - \log t_2}{\sigma_2(t_2) - t_2} t_1 \\
&= \frac{\log(t_2 + 1) - \log(t_2)}{t_2 + 1 - t_2} t_1 \\
&= \log \frac{t_2 + 1}{t_2} t_1, \\
f_{t_2 t_1}^{\Delta_2 \Delta_1}(t) &= \log \frac{t_2 + 1}{t_2}, \quad t \in \Lambda^2.
\end{aligned}$$

Consequently,

$$f_{t_1 t_2}^{\Delta_1 \Delta_2}(t) = f_{t_2 t_1}^{\Delta_2 \Delta_1}(t), \quad t \in \Lambda^2.$$

Example 6.130 Let

$$\mathbb{T}_1 = [0, 2] \times \{4\}, \quad \mathbb{T}_2 = [0, 2],$$

where [0, 2] is real number interval. Define $f : \mathbb{T}_1 \times \mathbb{T}_2 \to \mathbb{R}$ by

$$f(t) = t_1 + t_1 t_2, \quad t \in [0, 2] \times [0, 2]$$

and

$$f(4, t_2) = t_2^2, \quad t_2 \in [0, 2].$$

This yields

$$\begin{aligned}
f_{t_1}^{\Delta_1}(2, t_2) &= \frac{f(\sigma_1(2), t_2) - f(2, t_2)}{\sigma_1(2) - 2} \\
&= \frac{f(4, t_2) - f(2, t_2)}{2} \\
&= \frac{t_2^2 - 2 - 2t_2}{2} \\
&= \frac{1}{2}(t_2^2 - 2t_2) - 1, \\
f_{t_1 t_2}^{\Delta_1 \Delta_2}(2, t_2) &= \frac{1}{2}(\sigma_2(t_2) + t_2 - 2) \\
&= \frac{1}{2}(2t_2 - 2) \\
&= t_2 - 1, \\
f_{t_2}^{\Delta_2}(2, t_2) &= 2, \\
f_{t_2}^{\Delta_2}(4, t_2) &= 2t_2, \\
f_{t_2 t_1}^{\Delta_2 \Delta_1}(2, t_2) &= \frac{f_{t_2}^{\Delta_2}(4, t_2) - f_{t_2}^{\Delta_2}(2, t_2)}{2} \\
&= \frac{2t_2 - 2}{2}
\end{aligned}$$

$$= t_2 - 1, \quad t_2 \in [0, 2].$$

Therefore,

$$f_{t_1 t_2}^{\Delta_1 \Delta_2}(2, t_2) = f_{t_2 t_1}^{\Delta_2 \Delta_1}(2, t_2), \quad t_2 \in [0, 2].$$

Exercise 6.131 Let

$$\mathbb{T}_1 = [0, 1] \times \{4\}, \quad \mathbb{T}_2 = [0, 1],$$

where [0, 1] is the real interval. Define $f : \mathbb{T}_1 \times \mathbb{T}_2 \to \mathbb{R}$ by

$$f(t) = t_2 + t_1 t_2 + 4, \quad t \in [0, 1] \times [0, 1]$$

and

$$f(4, t_2) = t_2^3, \quad t_2 \in [0, 1].$$

Check if $f_{t_1 t_2}^{\Delta_1 \Delta_2}(1, t_2) = f_{t_2 t_1}^{\Delta_2 \Delta_1}(1, t_2)$, $t_2 \in [0, 1]$.

6.7 The Chain Rule

Let $\mathbb{T}$ be a time scale with forward jump operator σ and delta operator Δ. Suppose $\phi_i : \mathbb{T} \to \mathbb{R}$ is such that $\phi_i(\mathbb{T}) = \mathbb{T}_i$, where $\mathbb{T}_i$ is a time scale with forward jump operator σ_i and delta operator Δ_i, $i \in \{1, 2, \ldots, n\}$. Let $\xi^0 \in \mathbb{T}^\kappa$, $t_i^0 = \phi_i(\xi^0)$, $i \in \{1, 2, \ldots, n\}$. We assume that

$$\phi_i(\sigma(\xi^0)) = \sigma_i(\phi_i(\xi^0)), \quad i = 1, 2, \ldots, n.$$

Theorem 6.132 *Let $i \in \{1, 2, \ldots, n\}$ be fixed. Assume*

$$f : \mathbb{T}_1 \times \mathbb{T}_2 \times \cdots \times \mathbb{T}_n \to \mathbb{R}$$

is σ_i-completely delta differentiable at the point t^0. If the functions

$$\phi_j, \quad j \in \{1, 2, \ldots, n\}$$

have delta derivatives at the point ξ^0, then the composite function

$$F(\xi) = f(\phi_1(\xi), \phi_2(\xi), \ldots, \phi_n(\xi)), \quad \xi \in \mathbb{T},$$

has a delta derivative at that point, which is expressed by the formula

$$F^\Delta(\xi^0) = f_{t_i}^{\Delta_i}(t^0)\phi_i^\Delta(\xi^0)$$

$$+f_{t_{i-1}}^{\Delta_{i-1}}(\phi_1(\xi^0), \ldots, \phi_{i-1}(\xi^0), \sigma_i(\phi_i(\xi^0)), \phi_{i+1}(\xi^0), \ldots, \phi_n(\xi^0))\phi_{i-1}^{\Delta}(\xi^0)$$

$$+\cdots$$

$$+f_{t_1}^{\Delta_1}(\phi_1(\xi^0), \sigma_2(\phi_2(\xi^0)), \ldots, \sigma_i(\phi_i(\xi^0)), \phi_{i+1}(\xi^0), \ldots, \phi_n(\xi^0))\phi_1^{\Delta}(\xi^0)$$

$$+f_{t_{i+1}}^{\Delta_{i+1}}(\sigma_1(\phi_1(\xi^0)), \ldots, \sigma_i(\phi_i(\xi^0)), \phi_{i+1}(\xi^0), \ldots, \phi_n(\xi^0))\phi_{i+1}^{\Delta}(\xi^0)$$

$$+\cdots$$

$$+f_{t_n}^{\Delta_n}(\sigma_1(\phi_1(\xi^0)), \ldots, \sigma_{n-1}(\phi_{n-1}(\xi^0)), \phi_n(\xi^0))\phi_n^{\Delta}(\xi^0).$$

Proof We have

$$F(\sigma(\xi^0)) - F(\xi) = f(\phi_1(\sigma(\xi^0)), \ldots, \phi_n(\sigma(\xi^0)) - f(\phi_1(\xi), \ldots, \phi_n(\xi))$$

$$= f(\sigma_1(\phi_1(\xi^0)), \sigma_2(\phi_2(\xi^0)), \ldots, \sigma_n(\phi_n(\xi^0)))$$

$$-f(\phi_1(\xi), \phi_2(\xi), \ldots, \phi_n(\xi))$$

$$= f(\phi_1(\xi), \ldots, \phi_{i-1}(\xi), \sigma_i(\phi_i(\xi^0)), \phi_{i+1}(\xi), \ldots, \phi_n(\xi))$$

$$-f(\phi_1(\xi), \ldots, \phi_{i-1}(\xi), \phi_i(\xi), \phi_{i+1}(\xi), \ldots, \phi_n(\xi))$$

$$+f(\phi_1(\xi), \ldots, \sigma_{i-1}(\phi_{i-1}(\xi^0)), \sigma_i(\phi_i(\xi^0)), \phi_{i+1}(\xi), \ldots, \phi_n(\xi))$$

$$-f(\phi_1(\xi), \ldots, \phi_{i-1}(\xi), \sigma_i(\phi_i(\xi^0)), \phi_{i+1}(\xi), \ldots, \phi_n(\xi))$$

$$+\cdots$$

$$+f(\sigma_1(\phi_1(\xi^0)), \ldots, \sigma_{i-1}(\phi_{i-1}(\xi^0)), \sigma_i(\phi_i(\xi^0)), \phi_{i+1}(\xi), \ldots, \phi_n(\xi))$$

$$-f(\phi_1(\xi), \sigma_2(\phi_2(\xi^0)), \ldots, \sigma_{i-1}(\phi_{i-1}(\xi^0)), \sigma_i(\phi_i(\xi^0)), \phi_{i+1}(\xi), \ldots, \phi_n(\xi))$$

$$+f(\sigma_1(\phi_1(\xi^0)), \ldots, \sigma_i(\phi_i(\xi^0)), \sigma_{i+1}(\phi_{i+1}(\xi^0)), \phi_{i+2}(\xi), \ldots, \phi_n(\xi))$$

$$-f(\sigma_1(\phi_1(\xi^0)), \ldots, \sigma_{i-1}(\phi_{i-1}(\xi^0)), \sigma_i(\phi_i(\xi^0)), \phi_{i+1}(\xi), \ldots, \phi_n(\xi))$$

$$
\begin{aligned}
&+\cdots \\
&+f(\sigma_1(\phi_1(\xi^0)),\ldots,\sigma_n(\phi_n(\xi^0))) \\
&-f(\sigma_1(\phi_1(\xi^0)),\ldots,\sigma_{n-1}(\phi_{n-1}(\xi^0)),\phi_n(\xi)) \\
=\ & f_{t_i}^{\Delta_i}(\phi_1(\xi^0),\ldots,\phi_n(\xi^0))(\sigma_i(\phi_i(\xi^0))-\phi_i(\xi)) \\
&+f_{t_{i-1}}^{\Delta_{i-1}}(\phi_1(\xi^0),\ldots,\phi_{i-1}(\xi^0),\sigma_i(\phi_i(\xi^0)),\phi_{i+1}(\xi^0),\ldots,\phi_n(\xi^0)) \\
&\times(\sigma_{i-1}(\phi_{i-1}(\xi^0))-\phi_{i-1}(\xi)) \\
&+\cdots \\
&+f_{t_1}^{\Delta_1}(\phi_1(\xi^0),\sigma_2(\phi_2(\xi^0)),\ldots,\sigma_i(\phi_i(\xi^0)),\phi_{i+1}(\xi^0),\ldots,\phi_n(\xi^0)) \\
&\times(\sigma_1(\phi_1(\xi^0))-\phi_1(\xi)) \\
&+f_{t_{i+1}}^{\Delta_{i+1}}(\sigma_1(\phi_1(\xi^0)),\ldots,\sigma_i(\phi_i(\xi^0)),\phi_{i+1}(\xi^0),\ldots,\phi_n(\xi^0)) \\
&\times(\sigma_{i+1}(\phi_{i+1}(\xi^0))-\phi_{i+1}(\xi)) \\
&+\cdots \\
&+f_{t_n}^{\Delta_n}(\sigma_1(\phi_1(\xi^0)),\ldots,\sigma_{n-1}(\phi_{n-1}(\xi^0)),\phi_n(\xi^0))(\sigma_n(\phi_n(\xi^0))-\phi_n(\xi)) \\
&+\sum_{i=1}^{n}\alpha_i(\sigma_i(\phi_i(\xi^0))-\phi_i(\xi)) \\
=\ & f_{t_i}^{\Delta_i}(\phi_1(\xi^0),\ldots,\phi_n(\xi^0))(\phi_i(\sigma(\xi^0))-\phi_i(\xi)) \\
&+f_{t_{i-1}}^{\Delta_{i-1}}(\phi_1(\xi^0),\ldots,\phi_{i-1}(\xi^0),\sigma_i(\phi_i(\xi^0)),\phi_{i+1}(\xi^0),\ldots,\phi_n(\xi^0)) \\
&\times(\phi_{i-1}(\sigma(\xi^0))-\phi_{i-1}(\xi))
\end{aligned}
$$

$$+\cdots$$

$$+f_{t_1}^{\Delta_1}(\phi_1(\xi^0), \sigma_2(\phi_2(\xi^0)), \ldots, \sigma_i(\phi_i(\xi^0)), \phi_{i+1}(\xi^0), \ldots, \phi_n(\xi^0))$$

$$\times(\phi_1(\sigma(\xi^0)) - \phi_1(\xi))$$

$$+f_{t_{i+1}}^{\Delta_{i+1}}(\sigma_1(\phi_1(\xi^0)), \ldots, \sigma_i(\phi_i(\xi^0)), \phi_{i+1}(\xi^0), \ldots, \phi_n(\xi^0))$$

$$\times(\phi_{i+1}(\sigma(\xi^0)) - \phi_{i+1}(\xi))$$

$$+\cdots$$

$$+f_{t_n}^{\Delta_n}(\sigma_1(\phi_1(\xi^0)), \ldots, \sigma_{n-1}(\phi_{n-1}(\xi^0)), \phi_n(\xi^0))$$

$$\times(\phi_n(\sigma(\xi^0)) - \phi_n(\xi))$$

$$+\sum_{i=1}^{n} \alpha_i(\sigma_i(\phi_i(\xi^0)) - \phi_i(\xi)),$$

where $\alpha_i \to 0$ as $\xi \to \xi^0$, $i \in \{1, 2, \ldots, n\}$. Dividing both sides of this equality by $\sigma(\xi^0) - \xi$ and then passing to the limit as $\xi \to \xi^0$ completes the proof. □

Example 6.133 Let $\mathbb{T} = \mathbb{N}$, $\phi_1 : \mathbb{T} \to \mathbb{R}$, $\phi_1(\xi) = 2^\xi$, $\xi \in \mathbb{T}$, $\phi_2(\xi) = \xi$, $\xi \in \mathbb{T}$. Then

$$\mathbb{T}_1 = \phi_1(\mathbb{T}) = 2^{\mathbb{N}}, \quad \mathbb{T}_2 = \phi_2(\mathbb{T}) = \mathbb{N},$$

and

$$\sigma(\xi) = \xi + 1, \quad \xi \in \mathbb{T}, \quad \sigma_1(\xi) = 2\xi, \quad \xi \in \mathbb{T}_1, \quad \sigma_2(\xi) = \xi + 1, \quad \xi \in \mathbb{T}_2.$$

Let

$$f(\phi_1, \phi_2) = \log(\phi_1(\xi) + \phi_2(\xi)) = \log(2^\xi + \xi)$$

and

$$t_1 = \phi_1(\xi), \quad t_2 = \phi_2(\xi).$$

From here, $f(t) = \log(t_1 + t_2)$. We note that

$$\phi_1(\sigma(\xi)) = \phi_1(\xi + 1) = 2^{\xi+1}, \quad \sigma_1(\phi(\xi)) = \sigma_1(2^\xi) = 2^{\xi+1},$$

i.e., $\phi_1(\sigma(\xi)) = \sigma_1(\phi(\xi))$, and

$$\phi_2(\sigma(\xi)) = \phi_2(\xi + 1) = \xi + 1, \quad \sigma_2(\phi(\xi)) = \sigma_2(\xi) = \xi + 1,$$

i.e., $\phi_2(\sigma(\xi)) = \sigma_2(\phi(\xi))$. We will find $f^{\Delta}(\xi)$.

1. First way. We have

$$f^{\Delta}(\xi) = f_{t_1}^{\Delta_1}(t)\phi_1^{\Delta}(\xi) + f_{t_2}^{\Delta_2}(\sigma_1(t_1), t_2)\phi_2^{\Delta}(\xi),$$

$$\begin{aligned}
f_{t_1}^{\Delta_1}(t) &= \frac{f(\sigma_1(t_1), t_2) - f(t_1, t_2)}{\sigma_1(t_1) - t_1} \\
&= \frac{f(2t_1, t_2) - f(t_1, t_2)}{2t_1 - t_1} \\
&= \frac{\log(2t_1 + t_2) - \log(t_1 + t_2)}{t_1} \\
&= \frac{1}{t_1} \log \frac{2t_1 + t_2}{t_1 + t_2}, \\
\phi_1^{\Delta}(\xi) &= \frac{\phi_1(\sigma(\xi)) - \phi_1(\xi)}{\sigma(\xi) - \xi} \\
&= \frac{2^{\xi+1} - 2^{\xi}}{1} \\
&= 2^{\xi+1} - 2^{\xi}, \\
f_{t_2}^{\Delta_2}(\sigma_1(t_1), t_2) &= \frac{f(\sigma_1(t_1), \sigma_2(t_2)) - f(\sigma_1(t_1), t_2)}{\sigma_2(t_2) - t_2} \\
&= \frac{\log(2t_1 + t_2 + 1) - \log(2t_1 + t_2)}{t_2 + 1 - t_2} \\
&= \log \frac{2t_1 + t_2 + 1}{2t_1 + t_2}, \\
\phi_2^{\Delta}(\xi) &= 1.
\end{aligned}$$

Therefore,

$$\begin{aligned}
f^{\Delta}(\xi) &= \frac{1}{t_1} \log \frac{2t_1 + t_2}{t_1 + t_2}(2^{\xi+1} - 2^{\xi}) + \log \frac{2t_1 + t_2 + 1}{2t_1 + t_2} \\
&= 2^{-\xi} \log \frac{2^{\xi+1} + \xi}{2^{\xi} + \xi}(2^{\xi+1} - 2^{\xi}) + \log \frac{2^{\xi+1} + \xi + 1}{2^{\xi+1} + \xi}
\end{aligned}$$

$$= \log \frac{2^{\xi+1} + \xi}{2^{\xi} + \xi} + \log \frac{2^{\xi+1} + \xi + 1}{2^{\xi+1} + \xi}$$

$$= \log \frac{2^{\xi+1} + \xi + 1}{2^{\xi} + \xi}.$$

2. Second way. Here, $f(\xi) = \log(2^{\xi} + \xi)$. Hence,

$$f^{\Delta}(\xi) = \log(2^{\xi+1} + \xi + 1) - \log(2^{\xi} + \xi)$$

$$= \log \frac{2^{\xi+1} + \xi + 1}{2^{\xi} + 1}.$$

Example 6.134 Let

$$\mathbb{T} = 2^{\mathbb{N}}, \quad \phi_1 : \mathbb{T} \to \mathbb{R}, \quad \phi_1(\xi) = \xi, \quad \xi \in \mathbb{T},$$

$$\phi_2 : \mathbb{T} \to \mathbb{R}, \quad \phi_2(\xi) = \xi + 1, \quad \xi \in \mathbb{T}$$

so that

$$\mathbb{T}_1 = \phi_1(\mathbb{T}) = \mathbb{T}, \quad \mathbb{T}_2 = \phi_2(\mathbb{T}) = 2^{\mathbb{N}} + 1,$$

$$\sigma(\xi) = 2\xi, \quad \xi \in \mathbb{T}, \quad \sigma_1(\xi) = 2\xi, \quad \xi \in \mathbb{T}_1,$$

$$\sigma_2(\xi) = 2\xi - 1, \quad \xi \in \mathbb{T}_2.$$

Let

$$f(\phi_1, \phi_2) = \sin(\phi_1(\xi) + \phi_2(\xi))$$

$$= \sin(\xi + \xi + 1)$$

$$= \sin(2\xi + 1)$$

and

$$t_1 = \phi_1(\xi) = \xi, \quad t_2 = \phi_2(\xi) = \xi + 1.$$

Then $f(\phi_1, \phi_2) = \sin(t_1 + t_2)$. We will find $f^{\Delta}(\xi)$. We note that

$$\phi_1(\sigma(\xi)) = 2\xi,$$

$$\sigma_1(\phi(\xi)) = 2\xi,$$

$$\begin{aligned}\phi_2(\sigma(\xi)) &= \sigma(\xi) + 1\\ &= 2\xi + 1,\end{aligned}$$

$$\begin{aligned}\sigma_2(\phi_2(\xi)) &= 2\phi_2(\xi) - 1\\ &= 2(\xi + 1) - 1\\ &= 2\xi + 1.\end{aligned}$$

1. First way. We have

$$\begin{aligned}f^{\Delta}(\xi) &= f_{t_1}^{\Delta_1}(t)\phi_1^{\Delta}(\xi) + f_{t_2}^{\Delta_2}(\sigma_1(t_1), t_2)\phi_2^{\Delta}(\xi),\\ f_{t_1}^{\Delta_1}(t_1, t_2) &= \frac{\sin(\sigma_1(t_1) + t_2) - \sin(t_1 + t_2)}{\sigma_1(t_1) - t_1}\\ &= \frac{\sin(2t_1 + t_2) - \sin(t_1 + t_2)}{2t_1 - t_1}\\ &= \frac{2\sin\frac{t_1}{2}\cos\frac{3t_1+2t_2}{2}}{t_1}\\ &= \frac{2}{\xi}\sin\frac{\xi}{2}\cos\frac{5\xi + 2}{2},\\ \phi_1^{\Delta}(\xi) &= 1,\\ f_{t_2}^{\Delta_2}(\sigma_1(t_1), t_2) &= \frac{f(\sigma_1(t_1), \sigma_2(t_2)) - f(\sigma_1(t_1), t_2)}{\sigma_2(t_2) - t_2}\\ &= \frac{\sin(2t_1 + 2t_2 - 1) - \sin(2t_1 + t_2)}{2t_2 - t_2 - 1}\\ &= \frac{1}{t_2 - 1}2\sin\frac{t_2 - 1}{2}\cos\frac{4t_1 + 3t_2 - 1}{2}\\ &= \frac{2}{\xi}\sin\frac{\xi}{2}\cos\frac{7\xi + 2}{2},\\ \phi_2^{\Delta}(\xi) &= 1.\end{aligned}$$

Therefore,

$$\begin{aligned} f^{\Delta}(\xi) &= \frac{2}{\xi}\sin\frac{\xi}{2}\cos\frac{5\xi+2}{2} + \frac{2}{\xi}\sin\frac{\xi}{2}\cos\frac{7\xi+2}{2} \\ &= \frac{2}{\xi}\sin\frac{\xi}{2}\left(\cos\frac{5\xi+2}{2} + \cos\frac{7\xi+2}{2}\right) \\ &= \frac{4}{\xi}\sin\frac{\xi}{2}\cos\frac{\xi}{2}\cos(3\xi+1) \\ &= \frac{2}{\xi}\sin\xi\cos(3\xi+1). \end{aligned}$$

2. Second way. We have

$$\begin{aligned} f^{\Delta}(\xi) &= (\sin(2\xi+1))^{\Delta} \\ &= \frac{\sin(4\xi+1) - \sin(2\xi+1)}{\xi} \\ &= \frac{2}{\xi}\sin\xi\cos(3\xi+1). \end{aligned}$$

Example 6.135 Let $\mathbb{T} = 2\mathbb{Z}$ and define $\phi_1 : \mathbb{T} \to \mathbb{R}$ and $\phi_2 : \mathbb{T} \to \mathbb{R}$ by

$$\phi_1(\xi) = 3\xi, \quad \xi \in \mathbb{T}, \quad \phi_2(\xi) = 2\xi, \quad \xi \in \mathbb{T}$$

so that

$$\mathbb{T}_1 = \phi_1(\mathbb{T}) = 6\mathbb{Z}, \quad \mathbb{T}_2 = \phi_2(\mathbb{Z}) = 4\mathbb{Z},$$

$$\sigma(\xi) = \xi + 2, \quad \xi \in \mathbb{T}, \quad \sigma_1(\xi) = \xi + 6, \quad \xi \in \mathbb{T}_1, \quad \sigma_2(\xi) = \xi + 4, \quad \xi \in \mathbb{T}_2.$$

Therefore,

$$\begin{aligned} \phi_1(\sigma(\xi)) &= 3\sigma(\xi) = 3\xi + 6, \\ \sigma_1(\phi_1(\xi)) &= \phi_1(\xi) + 6 = 3\xi + 6, \\ \phi_2(\sigma(\xi)) &= 2\sigma(\xi) = 2\xi + 4, \\ \sigma_2(\phi_2(\xi)) &= \phi_2(\xi) + 4 = 2\xi + 4. \end{aligned}$$

Define

$$f(\phi_1, \phi_2) = \cos(\phi_1 - \phi_2) = \cos\xi,$$

$$t_1 = \phi_1 = 3\xi, \quad t_2 = \phi_2 = 2\xi, \quad f(t_1, t_2) = \cos(t_1 - t_2).$$

We will find $f^{\Delta}(\xi)$.

1. First way. We have

$$f^{\Delta}(\xi) = f_{t_1}^{\Delta_1}(t_1, t_2)\phi_1^{\Delta}(\xi) + f_{t_2}^{\Delta_2}(\sigma_1(t_1), t_2)\phi_2^{\Delta}(\xi),$$

$$\begin{aligned} f_{t_1}^{\Delta_1}(t_1, t_2) &= \frac{f(\sigma_1(t_1), t_2) - f(t_1, t_2)}{\sigma_1(t_1) - t_1} \\ &= \frac{f(t_1 + 6, t_2) - f(t_1, t_2)}{t_1 + 6 - t_1} \\ &= \frac{\cos(t_1 + 6 - t_2) - \cos(t_1 - t_2)}{6} \\ &= -\frac{1}{3}\sin 3 \sin(t_1 - t_2 + 3), \end{aligned}$$

$$\phi_1^{\Delta}(\xi) = 3,$$

$$\begin{aligned} f_{t_2}^{\Delta_2}(\sigma_1(t_1), t_2) &= \frac{f(\sigma_1(t_1), \sigma_2(t_2)) - f(\sigma_1(t_1), t_2)}{\sigma_2(t_2) - t_2} \\ &= \frac{\cos(\sigma_1(t_1) - \sigma_2(t_2)) - \cos(\sigma_1(t_1) - t_2)}{t_2 + 4 - t_2} \\ &= \frac{\cos(t_1 + 6 - t_2 - 4) - \cos(t_1 + 6 - t_2)}{t_2 + 4 - t_2} \\ &= \frac{\cos(t_1 - t_2 + 2) - \cos(t_1 - t_2 + 6)}{4} \\ &= \frac{1}{2}\sin 2 \sin(t_1 - t_2 + 4), \end{aligned}$$

$$\phi_2^{\Delta}(\xi) = 2.$$

Therefore,

$$f^{\Delta}(\xi) = -\sin 3 \sin(t_1 - t_2 + 3) + \sin 2 \sin(t_1 - t_2 + 4)$$

$$
\begin{aligned}
&= -\sin 3 \sin(\xi + 3) + \sin 2 \sin(\xi + 4) \\
&= -\sin 3 \sin(\xi + 3) + \sin 2(\sin(\xi + 3) \cos 1 + \cos(\xi + 3) \sin 1) \\
&= (-\sin 3 + \sin 2 \cos 1) \sin(\xi + 3) + \sin 1 \sin 2 \cos(\xi + 3) \\
&= (-\sin 2 \cos 1 - \sin 1 \cos 2 + \sin 2 \cos 1) \sin(\xi + 3) \\
&\quad + \sin 1 \sin 2 \cos(\xi + 3) \\
&= -\sin 1 \cos 2 \sin(\xi + 3) + \sin 1 \sin 2 \cos(\xi + 3) \\
&= -\sin 1 \sin(\xi + 1).
\end{aligned}
$$

2. Second way. We have

$$
\begin{aligned}
f^{\Delta}(\xi) &= \frac{\cos(\sigma(\xi)) - \cos(\xi)}{\sigma(\xi) - \xi} \\
&= \frac{\cos(\xi + 2) - \cos \xi}{\xi + 2 - \xi} \\
&= -\sin 1 \sin(\xi + 1).
\end{aligned}
$$

Let $\mathbb{T}_{(i)}$, $i \in \{1, 2, \ldots, n\}$ be given time scales and put

$$
\Lambda_{(n)} = \mathbb{T}_{(1)} \times \mathbb{T}_{(2)} \times \cdots \times \mathbb{T}_{(n)},
$$

$\phi_i : \Lambda_{(n)} \to \mathbb{R}$, and

$$
\xi^0 = (\xi_1^0, \xi_2^0, \ldots, \xi_n^0) \in \mathbb{T}_{(1)} \times \mathbb{T}_{(2)} \times \cdots \times \mathbb{T}_{(n)}.
$$

Let

$$
\mathbb{T}_i = \phi_i(\xi_1^0, \ldots, \xi_{i-1}^0, \mathbb{T}_{(i)}, \xi_{i+1}^0, \ldots, \xi_n^0), \quad i \in \{1, \ldots, n\},
$$

$$
t^0 = (\phi_1(\xi^0), \phi_2(\xi^0), \ldots, \phi_n(\xi^0)).
$$

Suppose that $\mathbb{T}_i$, $i = 1, \ldots, n$, are time scales with forward jump operator σ_i and delta operator Δ_i. With $\sigma_{(i)}$ and $\Delta_{(i)}$, we denote the forward jump operator and the delta operator of $\mathbb{T}_{(i)}$, respectively. We assume that

$$\phi_j(\xi_1^0, \ldots, \xi_{i-1}^0, \sigma_{(i)}(\xi_i^0), \ldots, \xi_n^0) = \sigma_j(\phi_j(\xi^0)), \quad i, j = 1, 2, \ldots, n.$$

Theorem 6.136 *Let $i \in \{1, 2, \ldots, n\}$ be fixed. Assume f is σ_i-completely delta differentiable at the point t^0. If the functions ϕ_j, $j \in \{1, 2, \ldots, n\}$, have first-order partial delta derivatives at the point ξ^0, then the composite function*

$$F(\xi) = f(\phi_1(\xi), \phi_2(\xi), \ldots, \phi_n(\xi)), \quad \xi \in \Lambda_{(n)},$$

has first-order partial derivatives at the point ξ^0, which are expressed by the formulas

$$\begin{aligned}
F_{\xi_j}^{\Delta_{(j)}}(\xi^0) &= f_{t_i}^{\Delta_i}(t^0)(\phi_i)_{\xi_j}^{\Delta_{(j)}}(\xi^0) \\
&+ f_{t_{i-1}}^{\Delta_{i-1}}(\phi_1(\xi^0), \ldots, \phi_{i-1}(\xi^0), \sigma_i(\phi_i(\xi^0)), \phi_{i+1}(\xi^0), \ldots, \phi_n(\xi^0))(\phi_{i-1})_{\xi_j}^{\Delta_{(j)}}(\xi^0) \\
&+ \cdots \\
&+ f_{t_1}^{\Delta_1}(\phi_1(\xi^0), \sigma_2(\phi_2(\xi^0)), \ldots, \sigma_i(\phi_i(\xi^0)), \phi_{i+1}(\xi^0), \ldots, \phi_n(\xi^0))(\phi_1)_{\xi_j}^{\Delta_{(j)}}(\xi^0) \\
&+ f_{t_{i+1}}^{\Delta_{i+1}}(\sigma_1(\phi_1(\xi^0)), \ldots, \sigma_i(\phi_i(\xi^0)), \phi_{i+1}(\xi^0), \ldots, \phi_n(\xi^0))(\phi_{i+1})_{\xi_j}^{\Delta_{(j)}}(\xi^0) \\
&+ \cdots \\
&+ f_{t_n}^{\Delta_n}(\sigma_1(\phi_1(\xi^0)), \ldots, \sigma_{n-1}(\phi_{n-1}(\xi^0)), \phi_n(\xi^0))(\phi_n)_{\xi_j}^{\Delta_{(j)}}(\xi^0).
\end{aligned}$$

Proof We have

$$\begin{aligned}
&F(\xi_1^0, \ldots, \xi_{j-1}^0, \sigma_{(j)}(\xi_j^0), \xi_{j+1}^0, \ldots, \xi_n^0) - F(\xi) \\
&= f(\sigma_1(\phi_1(\xi^0)), \sigma_2(\phi_2(\xi^0)), \ldots, \sigma_n(\phi_n(\xi^0))) \\
&\quad - f(\phi_1(\xi), \phi_2(\xi), \ldots, \phi_n(\xi)) \\
&= f(\phi_1(\xi), \ldots, \phi_{i-1}(\xi), \sigma_i(\phi_i(\xi^0)), \phi_{i+1}(\xi), \ldots, \phi_n(\xi)) \\
&\quad - f(\phi_1(\xi), \ldots, \phi_{i-1}(\xi), \phi_i(\xi), \phi_{i+1}(\xi), \ldots, \phi_n(\xi)) \\
&\quad + f(\phi_1(\xi), \ldots, \sigma_{i-1}(\phi_{i-1}(\xi^0)), \sigma_i(\phi_i(\xi^0)), \phi_{i+1}(\xi), \ldots, \phi_n(\xi))
\end{aligned}$$

$$-f(\phi_1(\xi), \ldots, \phi_{i-1}(\xi), \sigma_i(\phi_i(\xi^0)), \phi_{i+1}(\xi), \ldots, \phi_n(\xi))$$

$$+\cdots$$

$$+f(\sigma_1(\phi_1(\xi^0)), \ldots, \sigma_{i-1}(\phi_{i-1}(\xi^0)), \sigma_i(\phi_i(\xi^0)), \phi_{i+1}(\xi), \ldots, \phi_n(\xi))$$

$$-f(\phi_1(\xi), \sigma_2(\phi_2(\xi^0)), \ldots, \sigma_{i-1}(\phi_{i-1}(\xi^0)), \sigma_i(\phi_i(\xi^0)), \phi_{i+1}(\xi), \ldots, \phi_n(\xi))$$

$$+f(\sigma_1(\phi_1(\xi^0)), \ldots, \sigma_i(\phi_i(\xi^0)), \sigma_{i+1}(\phi_{i+1}(\xi^0)), \phi_{i+2}(\xi), \ldots, \phi_n(\xi))$$

$$-f(\sigma_1(\phi_1(\xi^0)), \ldots, \sigma_{i-1}(\phi_{i-1}(\xi^0)), \sigma_i(\phi_i(\xi^0)), \phi_{i+1}(\xi), \ldots, \phi_n(\xi))$$

$$+\cdots$$

$$+f(\sigma_1(\phi_1(\xi^0)), \ldots, \sigma_n(\phi_n(\xi^0)))$$

$$-f(\sigma_1(\phi_1(\xi^0)), \ldots, \sigma_{n-1}(\phi_{n-1}(\xi^0)), \phi_n(\xi))$$

$$= f_{t_i}^{\Delta_i}(\phi_1(\xi^0), \ldots, \phi_n(\xi^0))(\sigma_i(\phi_i(\xi^0)) - \phi_i(\xi))$$

$$+f_{t_{i-1}}^{\Delta_{i-1}}(\phi_1(\xi^0), \ldots, \phi_{i-1}(\xi^0), \sigma_i(\phi_i(\xi^0)), \phi_{i+1}(\xi^0), \ldots, \phi_n(\xi^0))$$

$$\times(\sigma_{i-1}(\phi_{i-1}(\xi^0)) - \phi_{i-1}(\xi))$$

$$+\cdots$$

$$+f_{t_1}^{\Delta_1}(\phi_1(\xi^0), \sigma_2(\phi_2(\xi^0)), \ldots, \sigma_i(\phi_i(\xi^0)), \phi_{i+1}(\xi^0), \ldots, \phi_n(\xi^0))$$

$$\times(\sigma_1(\phi_1(\xi^0)) - \phi_1(\xi))$$

$$+f_{t_{i+1}}^{\Delta_{i+1}}(\sigma_1(\phi_1(\xi^0)), \ldots, \sigma_i(\phi_i(\xi^0)), \phi_{i+1}(\xi^0), \ldots, \phi_n(\xi^0))$$

$$\times(\sigma_{i+1}(\phi_{i+1}(\xi^0)) - \phi_{i+1}(\xi))$$

$$+\cdots$$

$$+f_{t_n}^{\Delta_n}(\sigma_1(\phi_1(\xi^0)), \ldots, \sigma_{n-1}(\phi_{n-1}(\xi^0)), \phi_n(\xi^0))(\sigma_n(\phi_n(\xi^0)) - \phi_n(\xi))$$

$$+\sum_{i=1}^{n} \alpha_i(\sigma_i(\phi_i(\xi^0)) - \phi_i(\xi))$$

$$= f_{t_i}^{\Delta_i}(\phi_1(\xi^0), \ldots, \phi_n(\xi^0))(\phi_i(\xi_1^0, \ldots, \xi_{j-1}^0, \sigma_{(j)}(\xi_j^0), \xi_{j+1}^0, \ldots, \xi_n^0) - \phi_i(\xi))$$

$$+f_{t_{i-1}}^{\Delta_{i-1}}(\phi_1(\xi^0), \ldots, \phi_{i-1}(\xi^0), \sigma_i(\phi_i(\xi^0)), \phi_{i+1}(\xi^0), \ldots, \phi_n(\xi^0))$$

$$\times(\phi_{i-1}(\xi_1^0, \ldots, \xi_{j-1}^0, \sigma_{(j)}(\xi_j^0), \xi_{j+1}^0, \ldots, \xi_n^0) - \phi_{i-1}(\xi))$$

$$+\cdots$$

$$+f_{t_1}^{\Delta_1}(\phi_1(\xi^0), \sigma_2(\phi_2(\xi^0)), \ldots, \sigma_i(\phi_i(\xi^0)), \phi_{i+1}(\xi^0), \ldots, \phi_n(\xi^0))$$

$$\times(\phi_1(\xi_1^0, \ldots, \xi_{j-1}^0, \sigma_{(j)}(\xi_j^0), \xi_{j+1}^0, \ldots, \xi_n^0) - \phi_1(\xi))$$

$$+f_{t_{i+1}}^{\Delta_{i+1}}(\sigma_1(\phi_1(\xi^0)), \ldots, \sigma_i(\phi_i(\xi^0)), \phi_{i+1}(\xi^0), \ldots, \phi_n(\xi^0))$$

$$\times(\phi_{i+1}(\xi_1^0, \ldots, \xi_{j-1}^0, \sigma_{(j)}(\xi_j^0), \xi_{j+1}^0, \ldots, \xi_n^0)) - \phi_{i+1}(\xi))$$

$$+\cdots$$

$$+f_{t_n}^{\Delta_n}(\sigma_1(\phi_1(\xi^0)), \ldots, \sigma_{n-1}(\phi_{n-1}(\xi^0)), \phi_n(\xi^0))$$

$$\times(\phi_n(\sigma(\xi_1^0, \ldots, \xi_{j-1}^0, \sigma_{(j)}(\xi_j^0), \xi_{j+1}^0, \ldots, \xi_n^0)) - \phi_n(\xi))$$

$$+\sum_{i=1}^{n} \alpha_i(\phi_i(\xi_1^0, \ldots, \xi_{j-1}^0, \sigma_{(j)}(\xi_j^0), \xi_{j+1}^0, \ldots, \xi_n^0) - \phi_i(\xi)),$$

where $\alpha_i \to 0$ as $\xi \to \xi^0$, $i \in \{1, 2, \ldots, n\}$. Dividing both sides of this equality by $\sigma_{(j)}(\xi_j^0) - \xi_j$ and then passing to the limit as $\xi \to \xi^0$ completes the proof. □

Example 6.137 Let

$$\mathbb{T}_{(1)} = \mathbb{T}_{(2)} = \mathbb{N},$$

$$\phi_1 : \mathbb{T}_{(1)} \times \mathbb{T}_{(2)} \to \mathbb{R}, \quad \phi_1(t) = t_1^2, \quad t \in \mathbb{T}_{(1)} \times \mathbb{T}_{(2)},$$

$$\phi_2 : \mathbb{T}_{(1)} \times \mathbb{T}_{(2)} \to \mathbb{R}, \quad \phi_2(t) = t_2^3, \quad t \in \mathbb{T}_{(1)} \times \mathbb{T}_{(2)},$$

$$\mathbb{T}_1 = \phi_1(\mathbb{T}_{(1)}, t_2) = \mathbb{N}^2, \quad \mathbb{T}_2 = \phi_2(t_1, \mathbb{T}_{(2)}) = \mathbb{N}^3$$

so that

$$\sigma_{(1)}(t_1) = t_1 + 1, \quad \sigma_{(2)}(t_2) = t_2 + 1,$$

$$\sigma_1(t_1) = (1 + \sqrt{t_1})^2, \quad \sigma_2(t_2) = (1 + \sqrt[3]{t_2})^3,$$

$$\phi_1(\xi_1, \xi_2) = \xi_1^2 = t_1, \quad \phi_2(\xi_1\, \xi_2) = \xi_2^3 = t_2$$

and

$$\begin{aligned}
F(\xi_1, \xi_2) &= \log(\phi_1(\xi_1, \xi_2) + \phi_2(\xi_1, \xi_2)) \\
&= f(\phi_1(\xi_1, \xi_2), \phi_2(\xi_1, \xi_2)) \\
&= \log(\xi_1^2 + \xi_2^3) \\
&= \log(t_1 + t_2)
\end{aligned}$$

$$\begin{aligned}
F_{\xi_1}^{\Delta_{(1)}}(\xi_1, \xi_2) &= f_{t_1}^{\Delta_1}(t_1, t_2)\phi_{1\xi_1}^{\Delta_{(1)}}(\xi_1, \xi_2) + f_{t_2}^{\Delta_2}(\sigma_1(t_1), t_2)\phi_{2\xi_1}^{\Delta_{(1)}}(\xi_1, \xi_2) \\
&= f_{t_1}^{\Delta_1}(t_1, t_2)\phi_{1\xi_1}^{\Delta_{(1)}}(\xi_1, \xi_2), \\
f_{t_1}^{\Delta_1}(t_1, t_2) &= \frac{\log(\sigma_1(t_1) + t_2) - \log(t_1 + t_2)}{\sigma_1(t_1) - t_1} \\
&= \frac{\log((1 + \sqrt{t_1})^2 + t_2) - \log(t_1 + t_2)}{(1 + \sqrt{t_1})^2 - t_1} \\
&= \frac{\log(t_1 + 2\sqrt{t_1} + t_2 + 1) - \log(t_1 + t_2)}{1 + 2\sqrt{t_1}}, \\
\phi_{1\xi_1}^{\Delta_{(1)}}(\xi_1, \xi_2) &= \sigma_{(1)}(\xi_1) + \xi_1
\end{aligned}$$

$$= 2\xi_1 + 1$$

$$= 1 + 2\sqrt{t_1},$$

i.e.,

$$F_{\xi_1}^{\Delta_{(1)}}(\xi_1, \xi_2) = \frac{\log(t_1 + 2\sqrt{t_1} + t_2 + 1) - \log(t_1 + t_2)}{1 + 2\sqrt{t_1}}(1 + 2\sqrt{t_1})$$

$$= \log(t_1 + 2\sqrt{t_1} + t_2 + 1) - \log(t_1 + t_2)$$

$$= \log(\xi_1^2 + 2\xi_1 + \xi_2^3 + 1) - \log(\xi_1^2 + \xi_2^3)$$

$$= \log \frac{\xi_1^2 + 2\xi_1 + \xi_2^3 + 1}{\xi_1^2 + \xi_2^3}.$$

Example 6.138 Let

$$\mathbb{T}_{(1)} = \mathbb{T}_{(2)} = \mathbb{N},$$

$$\sigma_{(1)}(\xi_1) = \xi_1 + 1, \quad \xi_1 \in \mathbb{N}, \sigma_{(2)}(\xi_2) = \xi_2 + 1, \quad \xi_2 \in \mathbb{N},$$

$$F(\xi_1, \xi_2) = \log(2^{\xi_1} + \xi_2^2),$$

$$\phi_1(\xi_1, \xi_2) = 2^{\xi_1} = t_1, \quad \phi_2(\xi_1, \xi_2) = \xi_2^2 = t_2,$$

$$\mathbb{T}_1 = \phi_1(\mathbb{T}_{(1)}, t_2) = 2^{\mathbb{N}}, \quad \mathbb{T}_2 = \phi_2(t_1, \mathbb{T}_{(2)}) = \mathbb{N}^2$$

so that

$$\sigma_1(t_1) = 2t_1, \quad t_1 \in \mathbb{T}_1, \quad \sigma_2(t_2) = (1 + \sqrt{t_2})^2, \quad t_2 \in \mathbb{T}_2$$

and

$$F(\xi_1, \xi_2) = f(\phi_1(\xi_1, \xi_2), \phi_2(\xi_1, \xi_2))$$

$$= \log(\phi_1(\xi_1, \xi_2) + \phi_2(\xi_1, \xi_2))$$

$$= \log(2^{\xi_1} + \xi_2^2)$$

$$= \log(t_1 + t_2).$$

Therefore,

$$\begin{aligned}
F_{\xi_1}^{\Delta_{(1)}}(\xi_1, \xi_2) &= f_{t_1}^{\Delta_1}(t_1, t_2)\phi_{1\xi_1}^{\Delta_{(1)}}(\xi_1, \xi_2) + f_{t_2}^{\Delta_2}(t_1, t_2)\phi_{2\xi_1}^{\Delta_{(1)}}(\xi_1, \xi_2) \\
&= f_{t_1}^{\Delta_1}(t_1, t_2)\phi_{1\xi_1}^{\Delta_{(1)}}(\xi_1, \xi_2), \\
f_{t_1}^{\Delta_1}(t_1, t_2) &= \frac{\log(\sigma_1(t_1) + t_2) - \log(t_1 + t_2)}{\sigma_1(t_1) - t_1} \\
&= \frac{\log(2t_1 + t_2) - \log(t_1 + t_2)}{t_1} \\
&= \frac{1}{t_1} \log \frac{2t_1 + t_2}{t_1 + t_2}, \\
\phi_{1\xi_1}^{\Delta_{(1)}}(\xi_1, \xi_2) &= \frac{\phi_1(\sigma_{(1)}(\xi_1), \xi_2) - \phi_1(\xi_1, \xi_2)}{\sigma_{(1)}(\xi_1) - \xi_1} \\
&= \frac{2^{\xi_1+1} - 2^{\xi_1}}{\xi_1 + 1 - \xi_1} = 2^{\xi_1} = t_1.
\end{aligned}$$

Hence,

$$\begin{aligned}
F_{\xi_1}^{\Delta_{(1)}}(\xi_1, \xi_2) &= \frac{1}{t_1} \log \frac{2t_1 + t_2}{t_1 + t_2} t_1 \\
&= \log \frac{2t_1 + t_2}{t_1 + t_2} \\
&= \log \frac{2^{\xi_1+1} + \xi_2^2}{2^{\xi_1} + \xi_2^2}.
\end{aligned}$$

6.8 The Directional Derivative

Let $\mathbb{T}$ be a time scale with forward jump operator σ and delta operator Δ. Let $x^0 \in \mathbb{T}$, $w = (w_1, w_2, \ldots, w_n) \in \mathbb{R}^n$ be a unit vector, and $t^0 = (t_1^0, t_2^0, \ldots, t_n^0)$ be a fixed point in $\mathbb{R}^n$. We set

$$\mathbb{T}_i = \{t_i = t_i^0 + (\xi - x^0)w_i : \xi \in \mathbb{T}\}, \quad i \in \{1, 2, \ldots, n\}.$$

We note that $\mathbb{T}_i$, $i \in \{1, 2, \dots, n\}$, are time scales and $t_i^0 \in \mathbb{T}_i$, $i \in \{1, 2, \dots, n\}$. Denote the forward jump operator of $\mathbb{T}_i$ by σ_i and the delta operator by Δ_i.

Definition 6.139 Let a function $f : \mathbb{T}_1 \times \mathbb{T}_2 \times \cdots \times \mathbb{T}_n \to \mathbb{R}$ be given. The *directional delta derivative* of the function f at the point t^0 in the direction of the vector w is defined as the number

$$F^{\Delta}(x^0) = \frac{\partial f(t^0)}{\Delta w}.$$

Remark 6.140 If f is σ_i-completely delta differentiable at t^0, then, using Theorem 6.132, we have

$$\begin{aligned}
F^{\Delta}(x^0) &= f_{t_i}^{\Delta_i}(t^0)w_i \\
&\quad + f_{t_{i-1}}^{\Delta_{i-1}}(t_1^0, \dots, t_{i-1}^0, \sigma_i(t_i^0), t_{i+1}^0, \dots, t_n^0)w_{i-1} \\
&\quad + \cdots \\
&\quad + f_{t_1}^{\Delta_1}(t_1^0, \sigma_2(t_2^0), \dots, \sigma_i(t_i^0), t_{i+1}^0, \dots, t_n^0)w_1 \\
&\quad + f_{t_{i+1}}^{\Delta_{i+1}}(\sigma_1(t_1^0), \dots, \sigma_i(t_i^0), t_{i+1}^0, \dots, t_n^0)w_{i+1} \\
&\quad + \cdots \\
&\quad + f_{t_n}^{\Delta_n}(\sigma_1(t_1^0), \dots, \sigma_{n-1}(t_{n-1}^0), t_n^0)w_n.
\end{aligned}$$

Example 6.141 Let $\mathbb{T} = \mathbb{N}$, $x^0 = 1$, $w = (1, 0) \in \mathbb{R}^2$. We note that w is a unit vector in $\mathbb{R}^2$. Let $t^0 = (1, 1)$. We set

$$\mathbb{T}_1 = \{t_1 = 1 + (\xi - 1)1 : \xi \in \mathbb{T}\} = \mathbb{N},$$

$$\mathbb{T}_2 = \{t_2 = 1 + (\xi - 1)0 : \xi \in \mathbb{T}\} = \{1\}.$$

We consider $f(t) = t_1^2 + t_2^2$, $t \in \mathbb{T}_1 \times \mathbb{T}_2$. We have that f is σ_1-completely delta differentiable in $\mathbb{T}_1 \times \mathbb{T}_2$ and

$$\begin{aligned}
F^{\Delta}(1) &= \frac{\partial f}{\Delta w}(1, 1) \\
&= f_{t_1}^{\Delta_1}(1, 1)
\end{aligned}$$

$$= (\sigma_1(t_1) + t_1)\Big|_{t_1=1}$$

$$= (2t_1 + 1)\Big|_{t_1=1} = 3.$$

Example 6.142 Let $\mathbb{T} = \mathbb{Z}$, $x^0 = 0$, $w = \left(\frac{2}{\sqrt{13}}, \frac{3}{\sqrt{13}}\right) \in \mathbb{R}^2$. We note that w is a unit vector in $\mathbb{R}^2$. Let $t^0 = (t_1^0, t_2^0) = (0, 0)$ be a fixed point in $\mathbb{R}^2$. We set

$$\mathbb{T}_1 = \left\{t_1 = \frac{2}{\sqrt{13}}\xi : \xi \in \mathbb{Z}\right\} = \frac{2}{\sqrt{13}}\mathbb{Z},$$

$$\mathbb{T}_2 = \left\{t_2 = \frac{3}{\sqrt{13}}\xi : \xi \in \mathbb{Z}\right\} = \frac{3}{\sqrt{13}}\mathbb{Z}$$

so that

$$\sigma_1(t_1) = t_1 + \frac{2}{\sqrt{13}}, \quad t_1 \in \mathbb{T}_1, \quad \sigma_2(t_2) = t_2 + \frac{3}{\sqrt{13}}, \quad t_2 \in \mathbb{T}_2.$$

We consider

$$f(t) = t_1^3 + 3t_1^2 t_2 + t_2^2, \quad t = (t_1, t_2) \in \mathbb{T}_1 \times \mathbb{T}_2.$$

We note that f is σ_1-completely delta differentiable in $\mathbb{T}_1 \times \mathbb{T}_2$. Hence,

$$\frac{\partial f(0,0)}{\Delta w} = f_{t_1}^{\Delta_1}(0,0)\frac{2}{\sqrt{13}} + f_{t_2}^{\Delta_2}(\sigma_1(0), 0)\frac{3}{\sqrt{13}},$$

$$f_{t_1}^{\Delta_1}(t) = (\sigma_1(t_1))^2 + t_1\sigma_1(t_1) + t_1^2$$

$$+3(\sigma_1(t_1) + t_1)t_2$$

$$= \left(t_1 + \frac{2}{\sqrt{13}}\right)^2 + t_1\left(t_1 + \frac{2}{\sqrt{13}}\right) + t_1^2$$

$$+3\left(2t_1 + \frac{2}{\sqrt{13}}\right)t_2,$$

$$f_{t_1}^{\Delta_1}(0) = \left(\frac{2}{\sqrt{13}}\right)^2$$

$$= \frac{4}{13},$$

$$f_{t_2}^{\Delta_2}(t) = 3t_1^2 + \sigma_2(t_2) + t_2$$

$$= 3t_1^2 + 2t_2 + \frac{3}{\sqrt{13}},$$

$$f_{t_2}^{\Delta_2}(\sigma_1(0), 0) = f_{t_2}^{\Delta_2}\left(\frac{2}{\sqrt{13}}, 0\right)$$

$$= 3\left(\frac{2}{\sqrt{13}}\right)^2 + \frac{3}{\sqrt{13}}$$

$$= \frac{12}{13} + \frac{3}{\sqrt{13}}.$$

Hence,

$$\frac{\partial f(0,0)}{\Delta w} = \frac{4}{13} \cdot \frac{2}{\sqrt{13}} + \frac{3}{\sqrt{13}}\left(\frac{12}{13} + \frac{3}{\sqrt{13}}\right)$$

$$= \frac{8}{13\sqrt{13}} + \frac{36}{13\sqrt{13}} + \frac{9}{13}$$

$$= \frac{44}{13\sqrt{13}} + \frac{9}{13}.$$

Example 6.143 Let $\mathbb{T} = \mathbb{Z}$, $x^0 = 0$, $w = \left(-\frac{1}{\sqrt{3}}, \frac{2}{\sqrt{3}}\right) \in \mathbb{R}^2$. We note that w is a unit vector in $\mathbb{R}^2$. Let $t^0 = (0, 0)$. We set

$$\mathbb{T}_1 = \left\{t_1 = -\frac{\xi}{\sqrt{3}} : \xi \in \mathbb{T}\right\} = -\frac{1}{\sqrt{3}}\mathbb{Z},$$

$$\mathbb{T}_2 = \left\{t_2 = \frac{2\xi}{\sqrt{3}} : \xi \in \mathbb{T}\right\} = \frac{2}{\sqrt{3}}\mathbb{Z}.$$

We consider

$$f(t) = t_1^2 + 2t_1t_2, \quad t \in \mathbb{T}_1 \times \mathbb{T}_2.$$

We have that f is σ_1-completely delta differentiable in $\mathbb{T}_1 \times \mathbb{T}_2$ and

$$\sigma_1(t_1) = t_1 + \frac{1}{\sqrt{3}}, \quad t_1 \in \mathbb{T}_1, \quad \sigma_2(t_2) = t_2 + \frac{2}{\sqrt{3}}, \quad t_2 \in \mathbb{T}_2.$$

Hence,

$$\frac{\partial f(0,0)}{\Delta w} = f_{t_1}^{\Delta_1}(0,0)\left(-\frac{1}{\sqrt{3}}\right) + f_{t_2}^{\Delta_2}(\sigma_1(0), 0)\frac{2}{\sqrt{3}},$$

$$
\begin{aligned}
f_{t_1}^{\Delta_1}(t) &= \sigma_1(t_1) + t_1 + 2t_2\\
&= 2t_1 + \frac{1}{\sqrt{3}} + 2t_2,\\
f_{t_1}^{\Delta_1}(0,0) &= \frac{1}{\sqrt{3}},\\
f_{t_2}^{\Delta_2}(t) &= 2t_1,\\
f_{t_2}^{\Delta_2}(\sigma_1(0), 0) &= f_{t_2}^{\Delta_2}\left(\frac{1}{\sqrt{3}}, 0\right)\\
&= \frac{2}{\sqrt{3}},\\
\frac{\partial f(0,0)}{\Delta w} &= \frac{1}{\sqrt{3}}\left(-\frac{1}{\sqrt{3}}\right) + \frac{2}{\sqrt{3}} \cdot \frac{2}{\sqrt{3}}\\
&= -\frac{1}{3} + \frac{4}{3}\\
&= 1.
\end{aligned}
$$

Exercise 6.144 Let $\mathbb{T} = \mathbb{Z}$, $x^0 = 0$, $w = \left(\frac{1}{\sqrt{26}}, \frac{5}{\sqrt{26}}\right)$, $f(t) = t_1^2 + 2t_2$. Find

$$
\frac{\partial f(0,0)}{\Delta w}.
$$

Solution $\frac{1}{26} + \frac{10}{\sqrt{26}}$.

6.9 Implicit Functions

Let $\mathbb{T}_i$, $1 \le i \le n-1$, be time scales. We consider the equation

$$
f(t_1, t_2, \ldots, t_{n-1}, x) = 0 \tag{6.33}
$$

for $(t_1, t_2, \ldots, t_{n-1}, x) \in \mathbb{T}_1 \times \mathbb{T}_2 \times \mathbb{T}_{n-1} \times \mathbb{R}$.

Theorem 6.145 *Suppose an equation* (6.33) *satisfies the following conditions.*

(i) The function f is defined in a neighbourhood U of the point

$$
(t_1^0, t_2^0, \ldots, t_{n-1}^0, x) \in \mathbb{T}_1^{\kappa} \times \mathbb{T}_2^{\kappa} \times \cdots \times \mathbb{T}_{n-1}^{k} \times \mathbb{R}
$$

and is continuous in U together with its partial derivatives $f_{t_i}^{\Delta_i}(t_1, t_2, \ldots, t_{n-1}, x)$, $i \in \{1, \ldots, n-1\}$, and $\frac{\partial f}{\partial x}(t_1, \ldots, t_{n-1}, x)$.

(ii) $f(t_1^0, t_2^0, \ldots, t_{n-1}^0, x^0) = 0.$

(iii) $\frac{\partial f}{\partial x}(t_1^0, t_2^0, \ldots, t_{n-1}^0, x^0) \neq 0.$

Then the following statements are true.

(a) There is a "rectangle"

$$\mathscr{N} = \Big\{(t_1^0, t_2^0, \ldots, t_{n-1}^0, x) \in \mathbb{T}_1^\kappa \times \mathbb{T}_2^\kappa \times \cdots \times \mathbb{T}_{n-1}^k \times \mathbb{R} : \quad |t_i - t_i^0| < \delta_i, \quad i = 1, 2, \ldots, n-1, \quad |x - x^0| < \delta'\Big\} \tag{6.34}$$

belonging to U such that $\mathscr{M} \cap \mathscr{N}$ is described by a uniquely determined single-valued function

$$x = \psi(t_1, t_2, \ldots, t_{n-1}) \quad for \quad (t_1^0, t_2^0, \ldots, t_{n-1}^0, x) \in \mathscr{N}^0,$$

where

$$\mathscr{M} = \Big\{(t_1, t_2, \ldots, t_{n-1}, x) \in \mathbb{T}_1 \times \mathbb{T}_2 \times \ldots \times \mathbb{T}_{n-1} \times \mathbb{R} : \quad f(t_1, t_2, \ldots, t_{n-1}, x) = 0\Big\},$$

$$\mathscr{N}^0 = \Big\{(t_1, t_2, \ldots, t_{n-1}) \in \mathbb{T}_1 \times \mathbb{T}_2 \times \ldots \times \mathbb{T}_{n-1} : \quad |t_i - t_i^0| < \delta_i, \quad i = 1, 2, \ldots, n-1\Big\}.$$

(b) $x^0 = \psi(t_1^0, t_2^0, \ldots, t_{n-1}^0)$.

(c) The function $\psi(t_1, t_2, \ldots, t_{n-1})$ is continuous in $\mathscr{N}^0$.

(d) The function $\psi(t_1, t_2, \ldots, t_{n-1})$ has partial delta derivatives

$$\psi_{t_i}^{\Delta_i}(t_1, t_2, \ldots, t_{n-1}), \quad i = 1, 2, \ldots, n-1 \quad on \quad \mathscr{N}^0.$$

Proof (a) Without loss of generality, we can suppose that U is an open "rectangle" of the form

$$U = \Big\{(t_1, t_2, \ldots, t_{n-1}, x) \in \mathbb{T}_1 \times \mathbb{T}_2 \times \ldots \times \mathbb{T}_{n-1} \times \mathbb{R} : \quad |t_i - t_i^0| < \tilde{a}_i, \quad i = 1, 2, \ldots, n-1, \quad |x - x^0| < b\Big\},$$

and we can assume that

$$\frac{\partial}{\partial x} f(t_1^0, t_2^0, \ldots, t_{n-1}^0, x) > 0.$$

Because $\frac{\partial}{\partial x} f(t_1, t_2, \ldots, t_{n-1}, x)$ is continuous on U, we also have

$$\frac{\partial}{\partial x} f(t_1, t_2, \ldots, t_{n-1}, x) > 0 \tag{6.35}$$

in a small neighbourhood $U_1 \subset U$ of the point $(t_1^0, t_2^0, \ldots, t_{n-1}^0, x^0)$ and

$$U_1 = \Big\{(t_1, t_2, \ldots, t_{n-1}, x) \in \mathbb{T}_1 \times \mathbb{T}_2 \times \ldots \times \mathbb{T}_{n-1} \times \mathbb{R} :$$
$$|t_i - t_i^0| < a_i, \quad i = 1, 2, \ldots, n-1, \quad |x - x^0| < b_1\Big\}.$$

Since f is continuous in U, we have that f is continuous in U_1. Also, the function $f(t_1^0, t_2^0, \ldots, t_{n-1}^0, x)$ of the single variable x is continuous on the closed interval $[x^0 - b_1, x^0 + b_1]$. Hence, using (6.35), we have that $f(t_1^0, t_2^0, \ldots, t_{n-1}^0, x)$ is strongly increasing in $[x^0 - b_1, x^0 + b_1]$ and

$$f(t_1^0, t_2^0, \ldots, t_{n-1}^0, x^0) = 0.$$

Therefore,

$$f(t_1^0, t_2^0, \ldots, t_{n-1}^0, x_0 - b_1) < 0 \quad \text{and} \quad f(t_1^0, t_2^0, \ldots, t_{n-1}^0, x^0 + b_1) > 0. \tag{6.36}$$

By the continuity of f, there is a sufficiently small number $\delta > 0$ with $\delta < \min\{a_1, a_2, \ldots, a_{n-1}\}$ such that (6.36) holds for all

$$(t_1, t_2, \ldots, t_{n-1}) \in \mathscr{N}^0 = \Big\{(t_1, t_2, \ldots, t_{n-1}) \in \mathbb{T}_1 \times \mathbb{T}_2 \times \ldots \times \mathbb{T}_{n-1} :$$
$$|t_i - t_i^0| < \delta, \quad i = 1, 2, \ldots, n-1\Big\}.$$

Now, we choose an arbitrary $(t_1, t_2, \ldots, t_{n-1}) \in \mathscr{N}^0$, fix it temporarily, and consider the function $f(t_1, t_2, \ldots, t_{n-1}, x)$ on the real variable x in the interval $[x^0 - b_1, x^0 + b_1] \subset \mathbb{R}$. This function is continuous, strictly increasing, and assumes values of the opposite signs at the end points of this interval. Therefore, there is a single value $x \in (x^0 - b_1, x^0 + b_1)$, denoted by

$$x = \psi(t_1, t_2, \ldots, t_{n-1}),$$

for which

$$f(t_1, t_2, \ldots, t_{n-1}, \psi(t_1, t_2, \ldots, t_{n-1})) = 0.$$

Thus, letting $\delta' = b_1$, $\delta_i = \delta$, we see that in the neighbourhood $\mathscr{N}$ of the point

$$(t_1^0, t_2^0, \ldots, t_{n-1}^0, x^0),$$

defined by (6.34), (6.33) determines x as a unique function of $t_1, t_2, \ldots, t_{n-1}$:

$$x = \psi(t_1, t_2, \ldots, t_{n-1}).$$

(b) From the condition (ii), it follows that

$$x^0 = \psi(t_1^0, t_2^0, \ldots, t_{n-1}^0).$$

(c) We will show that the function ψ is a continuous function in $\mathscr{N}^0$. To do so, it is enough to prove that it is continuous at the point $(t_1^0, t_2^0, \ldots, t_{n-1}^0)$. Let $\varepsilon' \in (0, \delta')$ be arbitrarily chosen. There exists $\varepsilon \in (0, \delta)$ such that for the rectangle

$$\mathscr{N}_* = \Big\{(t_1, t_2, \ldots, t_{n-1}, x) \in \mathbb{T}_1 \times \mathbb{T}_2 \times \ldots \times \mathbb{T}_{n-1} \times \mathbb{R} :$$

$$|t_i - t_i^0| < \varepsilon, \quad i = 1, 2, \ldots, n-1, \quad |x - x^0| < \varepsilon'\Big\},$$

there exists a function $x = \psi_*(t_1, t_2, \ldots, t_{n-1})$ for

$$(t_1, t_2, \ldots, t_{n-1}) \in \mathscr{N}_*^* = \Big\{(t_1, t_2, \ldots, t_{n-1}) \in \mathbb{T}_1 \times \mathbb{T}_2 \times \ldots \times \mathbb{T}_{n-1} :$$

$$|t_i - t_i^0| < \varepsilon, \quad i = 1, 2, \ldots, n-1\Big\},$$

which describes the set $\mathscr{M} \cap \mathscr{N}_*$. Since $\mathscr{N}_* \subset \mathscr{N}$, we have that

$$\psi(t_1, t_2, \ldots, t_{n-1}) = \psi_*(t_1, t_2, \ldots, t_{n-1})$$

for $(t_1, t_2, \ldots, t_{n-1}) \in \mathscr{N}_*^*$. Hence, for any sufficiently small $\varepsilon' > 0$, there exists $\varepsilon > 0$ such that

$$|\psi(t_1, t_2, \ldots, t_{n-1}) - \psi(t_1^0, t_2^0, \ldots, t_{n-1}^0)| < \varepsilon'$$

provided that $|t_i - t_i^0| < \varepsilon, i \in \{1, 2, \ldots, n-1\}$, i.e., the function ψ is continuous at the point $(t_1^0, t_2^0, \ldots, t_{n-1}^0)$.

(d) We take $(t_1, t_2, \ldots, t_{n-1}) \in \mathscr{N}^0$. Let $i \in \{1, 2, \ldots, n-1\}$ be arbitrarily chosen.

a. First case. $\sigma_i(t_i) > t_i$. Since the function ψ is continuous at $(t_1, t_2, \ldots, t_{n-1})$, it has a partial delta derivative $\psi_{t_i}^{\Delta_i}(t_1, t_2, \ldots, t_{n-1})$ with

$$\psi_{t_i}^{\Delta_i}(t_1, t_2, \ldots, t_{n-1}) = \frac{\psi_i^{\sigma_i}(t) - \psi(t)}{\sigma_i(t_i) - t_i}.$$

b. Second case. $t_i = \sigma_i(t_i)$. Let

$$(t_1, t_2, \ldots, t_{i-1}, t_i', t_{i+1}, \ldots, t_n) \in \mathscr{N}^0, \quad t_i' \neq t_i.$$

We have that

$$f(t_1, t_2, \ldots, t_{i-1}, t_i, t_{i+1}, \ldots, t_{n-1}, \psi(t_1, t_2, \ldots, t_{i-1}, t_i, t_{i+1}, \ldots, t_n)) = 0,$$

$$f(t_1, t_2, \ldots, t_{i-1}, t_i', t_{i+1}, \ldots, t_{n-1}, \psi(t_1, t_2, \ldots, t_{i-1}, t_i', t_{i+1}, \ldots, t_n)) = 0.$$

Thus,

$$\begin{aligned}
& f(t_1, t_2, \ldots, t_{i-1}, t_i', t_{i+1}, \ldots, t_{n-1}, \psi(t_1, t_2, \ldots, t_{i-1}, t_i, t_{i+1}, \ldots, t_{n-1})) \\
& \quad - f(t_1, t_2, \ldots, t_{i-1}, t_i, t_{i+1}, \ldots, t_{n-1}, \psi(t_1, t_2, \ldots, t_{i-1}, t_i, t_{i+1}, \ldots, t_{n-1})) \\
& = f(t_1, t_2, \ldots, t_{i-1}, t_i', t_{i+1}, \ldots, t_{n-1}, \psi(t_1, t_2, \ldots, t_{i-1}, t_i, t_{i+1}, \ldots, t_{n-1})) \\
& \quad - f(t_1, t_2, \ldots, t_{i-1}, t_i', t_{i+1}, \ldots, t_{n-1}, \psi(t_1, t_2, \ldots, t_{i-1}, t_i', t_{i+1}, \ldots, t_{n-1})) \\
& = \frac{\partial f}{\partial x}(t_1, t_2, \ldots, t_{i-1}, t_i', t_{i+1}, \ldots, t_{n-1}, \theta) \\
& \quad \times \left(\psi(t_1, t_2, \ldots, t_{i-1}, t_i, t_{i+1}, \ldots, t_{n-1}) - \psi(t_1, t_2, \ldots, t_{i-1}, t_i', t_{i+1}, \ldots, t_{n-1})\right),
\end{aligned} \tag{6.37}$$

where θ is a real number between

$$\psi(t_1, t_2, \ldots, t_{i-1}, t_i', t_{i+1}, \ldots, t_{n-1}) \quad \text{and} \quad \psi(t_1, t_2, \ldots, t_{i-1}, t_i, t_{i+1}, \ldots, t_{n-1}).$$

Also, by the mean value theorem for delta derivatives, we have

$$f_{t_i}^{\Delta_i}(t_1, t_2, \ldots, t_{i-1}, \xi', t_{i+1}, \ldots, t_{n-1}, \psi(t_1, t_2, \ldots, t_{i-1}, t_i, t_{i+1}, \ldots, t_{n-1}))$$

$$\times(t_i' - t_i)$$

$$\leq f(t_1, t_2, \ldots, t_{i-1}, t_i', t_{i+1}, \ldots, t_{n-1}, \psi(t_1, t_2, \ldots, t_{i-1}, t_i, t_{i+1}, \ldots, t_{n-1}))$$

$$-f(t_1, t_2, \ldots, t_{i-1}, t_i, t_{i+1}, \ldots, t_{n-1}, \psi(t_1, t_2, \ldots, t_{i-1}, t_i, t_{i+1}, \ldots, t_{n-1}))$$

$$\leq f_{t_i}^{\Delta_i}(t_1, t_2, \ldots, t_{i-1}, \xi'', t_{i+1}, \ldots, t_{n-1}, \psi(t_1, t_2, \ldots, t_{i-1}, t_i, t_{i+1}, \ldots, t_{n-1}))$$

$$\times(t_i' - t_i).$$

Hence, using (6.37), we get

$$f_{t_i}^{\Delta_i}(t_1, t_2, \ldots, t_{i-1}, \xi', t_{i+1}, \ldots, t_{n-1}, \psi(t_1, t_2, \ldots, t_{i-1}, t_i, t_{i+1}, \ldots, t_{n-1}))$$

$$\times(t_i' - t_i)$$

$$\leq \frac{\partial}{\partial x} f(t_1, t_2, \ldots, t_{i-1}, t_i', t_{i+1}, \ldots, t_{n-1}, \theta)$$

$$\times\left(\psi(t_1, t_2, \ldots, t_{i-1}, t_i, t_{i+1}, \ldots, t_{n-1}) - \psi(t_1, t_2, \ldots, t_{i-1}, t_i', t_{i+1}, \ldots, t_{n-1})\right)$$

$$\leq f_{t_i}^{\Delta_i}(t_1, t_2, \ldots, t_{i-1}, \xi'', t_{i+1}, \ldots, t_{n-1}, \psi(t_1, t_2, \ldots, t_{i-1}, t_i, t_{i+1}, \ldots, t_{n-1}))$$

$$\times(t_i' - t_i).$$

Dividing the last inequality by

$$\frac{\partial}{\partial x} f(t_1, t_2, \ldots, t_{i-1}, t_i', t_{i+1}, \ldots, t_{n-1}, \theta)(t_i - t_i')$$

and using (6.35) and the continuity of $f_{t_i}^{\Delta_i}$ and $\frac{\partial f}{\partial x}$, we see that

$$\psi_{t_i}^{\Delta_i}(t_1, t_2, \ldots, t_{n-1})$$

$$= \lim_{t_i' \to t_i} \frac{\psi(t_1, t_2, \ldots, t_{i-1}, t_i', t_{i+1}, \ldots, t_{n-1}) - \psi(t_1, t_2, \ldots, t_{i-1}, t_i, t_{i+1}, \ldots, t_{n-1})}{t_i' - t_i}$$

$$= -\frac{f_{t_i}^{\Delta_i}(t_1, t_2, \ldots, t_{i-1}, t_i, t_{i+1}, \ldots, t_{n-1}, \psi(t_1, t_2, \ldots, t_{i-1}, t_i, t_{i+1}, \ldots, t_{n-1}))}{\frac{\partial f}{\partial x}(t_1, t_2, \ldots, t_{i-1}, t_i, t_{i+1}, \ldots, t_{n-1}, \psi(t_1, t_2, \ldots, t_{i-1}, t_i, t_{i+1}, \ldots, t_{n-1}))}.$$

The proof is complete. □

6.10 Advanced Practical Problems

Problem 6.146 Let $n = 2$ and $\Lambda^2 = \mathbb{N} \times 3^{\mathbb{N}}$. Find $\sigma(t)$.

Solution $(t_1 + 1, 3t_2)$, $(t_1, t_2) \in \mathbb{N} \times 3^{\mathbb{N}}$.

Problem 6.147 Let $n = 3$ and $\Lambda^3 = 3^{\mathbb{N}} \times 4^{\mathbb{N}} \times 5^{\mathbb{N}}$. Find $\rho(t)$.

Solution

$$\begin{cases} \left(\frac{t_1}{3}, \frac{t_2}{4}, \frac{t_3}{5}\right) & \text{if } (t_1, t_2, t_3) \in 3^{\mathbb{N}} \times 4^{\mathbb{N}} \times 5^{\mathbb{N}}, \quad t_1 \neq 3, \quad t_2 \neq 4, \quad t_3 \neq 5, \\ \left(3, \frac{t_2}{4}, \frac{t_3}{5}\right) & \text{if } (3, t_2, t_3) \in \Lambda^3, \quad t_2 \neq 4, \quad t_3 \neq 5, \\ \left(\frac{t_1}{3}, 4, \frac{t_3}{5}\right) & \text{if } (t_1, 4, t_3) \in \Lambda^3, \quad t_1 \neq 3, \quad t_3 \neq 5, \\ \left(\frac{t_1}{3}, \frac{t_2}{4}, 5\right) & \text{if } (t_1, t_2, 5) \in \Lambda^3, \quad t_1 \neq 3, \quad t_2 \neq 4, \\ \left(3, 4, \frac{t_3}{5}\right) & \text{if } t_1 = 3, \quad t_2 = 4, \quad t_3 \in \mathbb{T}_3 \setminus \{5\}, \\ \left(3, \frac{t_2}{4}, 5\right) & \text{if } t_1 = 3, \quad t_2 \in \mathbb{T}_2 \setminus \{4\}, t_3 = 5, \\ \left(\frac{t_1}{3}, 4, 5\right) & \text{if } t_1 \in \mathbb{T}_1 \setminus \{3\}, \quad t_2 = 4, \quad t_3 = 5, \\ (3, 4, 5) & \text{if } t_1 = 3, \quad t_2 = 4, \quad t_3 = 5. \end{cases}$$

Problem 6.148 Classify each point $t \in \Lambda^2 = \mathbb{N}_0^3 \times \mathbb{N}$ as strictly right-scattered, right-scattered, right-dense, strictly left-scattered, left-scattered, left-dense, strictly isolated, dense, and isolated, respectively.

Solution (t_1, t_2), $t_1 \neq 0$, $t_2 \neq 1$, are strictly isolated, $(t_1, 1)$, $t_1 \neq 0$, $(0, t_2)$, $t_2 \neq 1$, are isolated, $(0, 1)$ is dense.

Problem 6.149 Let $\Lambda^3 = 2^{\mathbb{N}} \times 3^{\mathbb{N}} \times 4^{\mathbb{N}}$. Find $\mu(t)$.

Solution $(t_1, 2t_2, 3t_3)$, $(t_1, t_2, t_3) \in \Lambda^3$.

Problem 6.150 Let $\Lambda^3 = \mathbb{N} \times 3\mathbb{Z} \times \mathbb{N}$ and define $f : \Lambda^3 \to \mathbb{R}$ by

$$f(t) = 2t_1 - t_2 + t_3, \quad t = (t_1, t_2, t_3) \in \Lambda^3.$$

Find

1. $f^{\sigma}(t)$,
2. $f_1^{\sigma_1}(t)$,
3. $f_2^{\sigma_2}(t)$,
4. $f_3^{\sigma_3}(t)$,
5. $f_{12}^{\sigma_1\sigma_2}(t)$,
6. $f_{13}^{\sigma_1\sigma_3}(t)$,
7. $f_{23}^{\sigma_2\sigma_3}(t)$,
8. $f^{\sigma}(t) - 3f_{13}^{\sigma_1\sigma_3}(t) + f_{23}^{\sigma_2\sigma_3}(t)$, $t \in \Lambda^3$.

Solution 1. $2t_1 - t_2 + t_3$,
2. $2t_1 - t_2 + t_3 + 2$,
3. $2t_1 - t_2 + t_3 - 3$,
4. $2t_1 - t_2 + t_3 + 1$,
5. $2t_1 - t_2 + t_3 - 1$,
6. $2t_1 - t_2 + t_3 + 3$,
7. $2t_1 - t_2 + t_3 - 2$,
8. $-2t_1 + t_2 - t_3 - 11$.

Problem 6.151 Let $\Lambda^2 = \mathbb{N} \times \mathbb{N}_0$ and define $f : \Lambda^2 \to \mathbb{R}$ by

$$f(t) = t_1^2 + t_2^2, \quad t = (t_1, t_2) \in \Lambda^2.$$

Find

1. $f^{\rho}(t)$,
2. $f_1^{\rho_1}(t)$,
3. $f_2^{\rho_2}(t)$,
4. $g(t) = f_1^{\rho_1}(t) + f_2^{\rho_2}(t)$, $t \in \Lambda^2$.

Solution Let $t = (t_1, t_2) \in \Lambda^2$, $t_1 \neq 1$, $t_2 \neq 0$.

1. $t_1^2 + t_2^2 - 2t_1 - 2t_2 + 2$,
2. $t_1^2 + t_2^2 - 2t_1 + 1$,
3. $t_1^2 + t_2^2 - 2t_2 + 1$,
4. $2(t_1^2 + t_2^2) - 2t_1 - 2t_2 + 2$.

Let $t = (t_1, t_2) \in \Lambda^2$, $t_1 = 1$, $t_2 \neq 0$.

1. $t_2^2 - 2t_2 + 2$,
2. $1 + t_2^2$,
3. $t_2^2 - 2t_2 + 2$,
4. $2t_2^2 - 2t_2 + 3$.

Let $t = (t_1, t_2) \in \Lambda^2$, $t_1 \neq 1$, $t_2 = 0$.

1. $t_1^2 - 2t_1 + 1$,
2. $t_1^2 - 2t_1 + 1$,

3. t_1^2,
4. $2t_1^2 - 2t_1 + 1$.

Let $t = (1, 0)$.

1. 1,
2. 1,
3. 1,
4. 2.

Problem 6.152 Let $f(t) = t_1t_2t_3 + t_1^2, t = (t_1, t_2, t_3) \in \Lambda^3$. Prove that

$$f_{t_1}^{\Delta_1}(t) = t_2t_3 + \sigma_1(t_1) + t_1.$$

Problem 6.153 Let $\Lambda^2 = \mathbb{Z} \times 2^{\mathbb{N}}$ and define $f : \Lambda^2 \to \mathbb{R}$ by

$$f(t) = t_1^4 + t_1t_2^4.$$

Find

$$f_{t_1}^{\Delta_1}(t), \quad t \in \Lambda_1^{\kappa_1 2} \quad \text{and} \quad f_{t_2}^{\Delta_2}(t), \quad t \in \Lambda_2^{\kappa_2 2}.$$

Solution $4t_1^3 + 6t_1^2 + 4t_1 + t_2^4 + 1$, $15t_1t_2^3$.

Problem 6.154 Let $\Lambda^2 = (2\mathbb{N}) \times \mathbb{Z}$ and define $f : \Lambda^2 \to \mathbb{R}$ by

$$f(t) = \sqrt{t_1} + t_1^2t_2, \quad t = (t_1, t_2) \in \Lambda^2.$$

Find $f_{t_1}^{\Delta_1}(2, t_2)$.

Solution $\frac{1}{2+\sqrt{2}} + 6t_2$.

Problem 6.155 Let $\Lambda^2 = 3^{\mathbb{N}} \times \mathbb{N}$ and define $h : \Lambda^2 \to \mathbb{R}$ by

$$h(t) = (t_1^2 + 2t_1t_2 + 3)(t_1^3 + t_1t_2 + t_2^2), \quad t \in \Lambda^2.$$

Find

1. $h_{t_1}^{\Delta_1}(t), t \in \Lambda_1^{\kappa_1 2}$,
2. $h_{t_2}^{\Delta_2}(t), t \in \Lambda_2^{\kappa_2 2}$.

Solution 1. $121t_1^4 + 80t_1^3t_2 + 13t_1^2(t_2 + 3) + 12t_1t_2^2 + 3t_2 + 2t_2^3, t \in \Lambda_1^{\kappa_1 2}$,

2. $2t_1^4 + t_1^3 + 3t_1^2(1 + 2t_2) + t_1(4t_2^2 + 6t_2 + 5) + 6t_2 + 3, t \in \Lambda_2^{\kappa_2 2}$.

Problem 6.156 Let $\Lambda^2 = \mathbb{N} \times \mathbb{N}_0$ and define $h : \Lambda^2 \to \mathbb{R}$ by

$$h(t) = \frac{t_1 - t_2}{t_1 + t_2 + 1}, \quad t \in \Lambda^2.$$

Find $h_{t_1}^{\Delta_1}(t)$ for $t \in \Lambda_1^{\kappa_1 2}$ and $h_{t_2}^{\Delta_2}(t)$ for $t \in \Lambda_2^{\kappa_2 2}$.

Solution

$$h_{t_1}^{\Delta_1}(t) = \frac{1+2t_2}{(t_1+t_2+1)(t_1+t_2+2)}, \quad h_{t_2}^{\Delta_2}(t) = \frac{-2t_1-1}{(t_1+t_2+1)(t_1+t_2+2)}.$$

Problem 6.157 Let $\Lambda^2 = \mathbb{N} \times \mathbb{R}$. Find $\sigma^2(t)$ and $\rho^2(t)$, $t \in \Lambda^2$.

Solution

$$\sigma^2(t) = (t_1+2, t_2), \quad \rho^2(t) = \begin{cases} (t_1-2, t_2) & \text{if } t_1 \geq 3, \\ (1, t_2) & \text{if } t_1 \in \{1, 2\}, \quad t \in \Lambda^2. \end{cases}$$

Problem 6.158 Let $\Lambda^3 = \mathbb{N}_0^2 \times \mathbb{N} \times \mathbb{R}$ and define $f : \Lambda^3 \to \mathbb{R}$ by

$$f(t) = \cos(t_1) + \sin(t_2) + t_1 t_2 t_3, \quad t \in \Lambda^3.$$

Find $f_{t_1 t_2 t_3}^{\Delta_1 \Delta_2 \Delta_3}(t)$, $t \in \Lambda^{\kappa 3}$.

Solution 1.

Problem 6.159 Let $\Lambda^2 = \mathbb{Z} \times \mathbb{R}$ and define $f : \Lambda^2 \to \mathbb{R}$ by

$$f(t) = t_1 t_2 + t_1^2 + t_2^2, \quad t \in \Lambda^2.$$

Prove that

$$f_{t_1}^{\nabla_1}(t) = 2t_1 + t_2 - 1, \quad t \in \Lambda_{1\kappa_1}^2, \quad f_{t_2}^{\nabla_2}(t) = t_1 + 2t_2, \quad t \in \Lambda_{2\kappa_2}^2.$$

Problem 6.160 Let $\Lambda^2 = \mathbb{Z} \times \mathbb{N}_0^2$ and define $f : \Lambda^2 \to \mathbb{R}$ by

$$f(t) = t_1 t_2^2, \quad t \in \Lambda^2.$$

Find $f_{t_2 t_1}^{\Delta_2 \nabla_1}(t)$, $t \in \Lambda_{21\kappa_1}^{\kappa_2 2}$.

Solution $2t_2 + \sqrt{t_2} + 1$.

Problem 6.161 Let $\Lambda^2 = \mathbb{N} \times 4^{\mathbb{N}}$ and define $f : \Lambda^2 \to \mathbb{R}$ by

$$f(t) = t_1 + 2t_2 - 3, \quad t \in \Lambda^2.$$

Prove that f is completely delta differentiable in Λ^2.

Problem 6.162 Let $\mathbb{T}_1 = [0, 1] \cup \{4\}$ and $\mathbb{T}_2 = [0, 1]$, where $[0, 1]$ is the real number interval. Define $f : \mathbb{T}_1 \times \mathbb{T}_2 \to \mathbb{R}$ by

$$f(t_1, t_2) = t_1 - 2t_2, \quad (t_1, t_2) \in [0, 1] \times [0, 1]$$

and

$$f(4, t_2) = a + 3t_2, \quad t_2 \in [0, 1],$$

where a is a real constant. Find a constant a so that f is completely delta differentiable in $(1, 1)$.

Solution $a = -1$.

Problem 6.163 Let $\mathbb{T}_1 = [0, 1] \cup \{5\}$ and $\mathbb{T}_2 = [0, 1]$, where $[0, 1]$ is the real number interval. Define $f : \mathbb{T}_1 \times \mathbb{T}_2 \to \mathbb{R}$ by

$$f(t) = t_1^2 + 5t_1 + 2t_2, \quad (t_1, t_2) \in [0, 1] \times [0, 1]$$

and

$$f(5, t_2) = a + 3t_2, \quad t_2 \in [0, 1],$$

where a is a real constant. Find a constant a so that the function f is completely delta differentiable, σ_1-completely delta differentiable, and σ_2-completely delta differentiable.

Solution $a = 33$.

Problem 6.164 Let $\mathbb{T}_1 = [0, 1] \times \{3\}$ and $\mathbb{T}_2 = [0, 1]$, where $[0, 1]$ is real number interval. Define $f : \mathbb{T}_1 \times \mathbb{T}_2 \to \mathbb{R}$ by

$$f(t) = t_2 + t_1^2 t_2 + 4t_1 - 2, \quad t \in [0, 1] \times [0, 1]$$

and

$$f(3, t_2) = t_2^4, \quad t_2 \in [0, 1].$$

Check if $f_{t_1 t_2}^{\Delta_1 \Delta_2}(1, t_2) = f_{t_2 t_1}^{\Delta_2 \Delta_1}(1, t_2)$, $t_2 \in [0, 1]$.

Problem 6.165 Let

$$\mathbb{T} = \mathbb{Z}, \quad x^0 = 0, \quad w = \left(\frac{1}{\sqrt{10}}, \frac{3}{\sqrt{10}}\right), \quad f(t) = t_1^2 + 3t_2.$$

Find $\frac{\partial f(0,0)}{\Delta w}$.

6.11 Notes and References

This chapter deals with differential calculus for multivariable functions on time scales and intends to prepare an instrument for introducing and investigating partial dynamic equations on time scales. Some papers related to this subject are [2, 8, 34]. There are a number of differences between the calculus of one and of several variables. Moreover, in this book, mainly partial delta derivatives are considered. Partial nabla derivatives and combinations of partial delta and nabla derivatives can be investigated in a similar manner. All results in this chapter are n-dimensional analogues of the two-dimensional results by Bohner and Guseinov [8].

Chapter 7
Multiple Integration on Time Scales

7.1 Multiple Riemann Integrals over Rectangles

Let $\mathbb{T}_i$, $i \in \{1, 2, \ldots, n\}$, be time scales. For $i \in \{1, 2, \ldots, n\}$, let σ_i, ρ_i, and Δ_i denote the forward jump operator, the backward jump operator, and the delta differentiation, respectively, on $\mathbb{T}_i$. Suppose $a_i < b_i$ are points in $\mathbb{T}_i$ and $[a_i, b_i)$ is the half-closed bounded interval in $\mathbb{T}_i$, $i \in \{1, \ldots, n\}$. Let us introduce a "rectangle" in $\Lambda^n = \mathbb{T}_1 \times \mathbb{T}_2 \times \ldots \times \mathbb{T}_n$ by

$$R = [a_1, b_1) \times [a_2, b_2) \times \ldots [a_n, b_n)$$

$$= \{(t_1, t_2, \ldots, t_n) : t_i \in [a_i, b_i),\ i = 1, 2, \ldots, n\}.$$

Let

$$a_i = t_i^0 < t_i^1 < \ldots < t_i^{k_i} = b_i.$$

Definition 7.1 We call the collection of intervals

$$P_i = \left\{[t_i^{j_i-1}, t_i^{j_i}) :\ j_i = 1, \ldots, k_i\right\}, \quad i = 1, 2, \ldots, n,$$

a *Δ_i-partition* of $[a_i, b_i)$ and denote the set of all Δ_i-partitions of $[a_i, b_i)$ by $P_i([a_i, b_i))$.

Definition 7.2 Let

$$\begin{aligned} R_{j_1 j_2 \ldots j_n} &= [t_1^{j_1-1}, t_1^{j_1}) \times [t_2^{j_2-1}, t_2^{j_2}) \times \ldots \times [t_n^{j_n-1}, t_n^{j_n}) \\ &1 \le j_i \le k_i, \quad i = 1, 2, \ldots, n. \end{aligned} \tag{7.1}$$

We call the collection

M. Bohner and S.G. Georgiev, *Multivariable Dynamic Calculus on Time Scales*,
DOI 10.1007/978-3-319-47620-9_7

$$P = \{R_{j_1 j_2 \ldots j_n} : 1 \le j_i \le k_i, \; i = 1, 2, \ldots, n\} \tag{7.2}$$

a *Δ-partition* of R, generated by the Δ_i-partitions P_i of $[a_i, b_i)$, and we write

$$P = P_1 \times P_2 \times \ldots \times P_n.$$

The set of all Δ-partitions of R is denoted by $\mathscr{P}(R)$. Moreover, for a bounded function $f : R \to \mathbb{R}$, we set

$$M = \sup\{f(t_1, t_2, \ldots, t_n) : (t_1, t_2, \ldots, t_n) \in R\},$$

$$m = \inf\{f(t_1, t_2, \ldots, t_n) : (t_1, t_2, \ldots, t_n) \in R\},$$

$$M_{j_1 j_2 \ldots j_n} = \sup\{f(t_1, t_2, \ldots, t_n) : (t_1, t_2, \ldots, t_n) \in R_{j_1 j_2 \ldots j_n}\},$$

$$m_{j_1 j_2 \ldots j_n} = \inf\{f(t_1, t_2, \ldots, t_n) : (t_1, t_2, \ldots, t_n) \in R_{j_1 j_2 \ldots j_n}\}.$$

Definition 7.3 The *upper Darboux Δ-sum* $U(f, P)$ and the *lower Darboux Δ-sum* $L(f, P)$ with respect to P are defined by

$$U(f, P) = \sum_{j_1=1}^{k_1} \sum_{j_2=1}^{k_2} \cdots \sum_{j_n=1}^{k_n} M_{j_1 j_2 \ldots j_n} (t_1^{j_1} - t_1^{j_1-1})(t_2^{j_2} - t_2^{j_2-1}) \ldots (t_n^{j_n} - t_n^{j_n-1})$$

and

$$L(f, P) = \sum_{j_1=1}^{k_1} \sum_{j_2=1}^{k_2} \cdots \sum_{j_n=1}^{k_n} m_{j_1 j_2 \ldots j_n} (t_1^{j_1} - t_1^{j_1-1})(t_2^{j_2} - t_2^{j_2-1}) \ldots (t_n^{j_n} - t_n^{j_n-1}).$$

Remark 7.4 We note that

$$\begin{aligned} U(f, P) &\le M \sum_{j_1=1}^{k_1} \sum_{j_2=1}^{k_2} \cdots \sum_{j_n=1}^{k_n} (t_1^{j_1} - t_1^{j_1-1})(t_2^{j_2} - t_2^{j_2-1}) \ldots (t_n^{j_n} - t_n^{j_n-1}) \\ &\le M(b_1 - a_1)(b_2 - a_2) \ldots (b_n - a_n) \end{aligned}$$

and

$$\begin{aligned} L(f, P) &\ge m \sum_{j_1=1}^{k_1} \sum_{j_2=1}^{k_2} \cdots \sum_{j_n=1}^{k_n} (t_1^{j_1} - t_1^{j_1-1})(t_2^{j_2} - t_2^{j_2-1}) \ldots (t_n^{j_n} - t_n^{j_n-1}) \\ &\ge M(b_1 - a_1)(b_2 - a_2) \ldots (b_n - a_n), \end{aligned}$$

i.e.,

$$\begin{aligned} m(b_1 - a_1)(b_2 - a_2)\dots(b_n - a_n) \leq & L(f, P) \\ \leq & U(f, P) \qquad (7.3) \\ \leq & M(b_1 - a_1)(b_2 - a_2)\dots(b_n - a_n). \end{aligned}$$

Definition 7.5 The *upper Darboux Δ-integral* $U(f)$ of f over R and the *lower Darboux Δ-integral* $L(f)$ of f over R are defined by

$$U(f) = \inf\{U(f, P) : P \in \mathscr{P}(R)\} \quad \text{and} \quad L(f) = \sup\{L(f, P) : P \in \mathscr{P}(R)\}.$$

From (7.3), it follows that $U(f)$ and $L(f)$ are finite real numbers.

Definition 7.6 We say that f is *Δ-integrable* over R provided $L(f) = U(f)$. In this case, we write

$$\int_R f(t_1, t_2, \dots, t_n)\Delta_1 t_1 \Delta_2 t_2 \dots \Delta_n t_n$$

for this common value. We call this integral the *Darboux Δ-integral.*

Remark 7.7 For a given rectangle

$$V = [c_1, d_1) \times [c_2, d_2) \times \dots \times [c_n, d_n) \subset \Lambda^n,$$

the "area" of V, i.e., $(d_1 - c_1)(d_2 - c_2)\dots(d_n - c_n)$, is denoted by $m(V)$.

Definition 7.8 Let $P, Q \in \mathscr{P}(R)$ and

$$P = P_1 \times P_2 \times \dots \times P_n, \quad Q = Q_1 \times Q_2 \times \dots \times Q_n,$$

where $P_i, Q_i \in \mathscr{P}([a_i, b_i))$. We say that Q is a *refinement* of P provided Q_i is a refinement of P_i for all $i \in \{1, 2, \dots, n\}$.

Theorem 7.9 *Let f be a bounded function on R. If P and Q are Δ-partitions of R and Q is a refinement of P, then*

$$L(f, P) \leq L(f, Q) \leq U(f, Q) \leq U(f, P),$$

i.e., refining of a partition increases the lower sum and decreases the upper sum.

Proof For every Δ-partition Q of R, we have

$$L(f, Q) \leq U(f, Q).$$

Now, we prove

$$L(f, P) \leq L(f, Q) \quad \text{and} \quad U(f, Q) \leq U(f, P).$$

To this end, let

$$P = \{R_1, R_2, \ldots, R_N\}.$$

Because Q is a refinement of P, there exists $k \in \{1, 2, \ldots, N\}$ such that

$$Q = \{R_1, R_2, \ldots, R_{k-1}, R'_k, R''_k, R_{k+1}, \ldots, R_N\},$$

where

$$R_k = R'_k \cup R''_k.$$

Define

$$\begin{aligned}
m_k &= \inf_{(t_1,t_2,\ldots,t_n)\in R_k} f(t_1, t_2, \ldots, t_n),\\
m_k^{(1)} &= \inf_{(t_1,t_2,\ldots,t_n)\in R'_k} f(t_1, t_2, \ldots, t_n),\\
m_k^{(2)} &= \inf_{(t_1,t_2,\ldots,t_n)\in R''_k} f(t_1, t_2, \ldots, t_n),\\
M_k &= \sup_{(t_1,t_2,\ldots,t_n)\in R_k} f(t_1, t_2, \ldots, t_n),\\
M_k^{(1)} &= \sup_{(t_1,t_2,\ldots,t_n)\in R'_k} f(t_1, t_2, \ldots, t_n),\\
M_k^{(2)} &= \sup_{(t_1,t_2,\ldots,t_n)\in R''_k} f(t_1, t_2, \ldots, t_n).
\end{aligned}$$

We note that

$$m_k \leq m_k^{(1)}, \quad m_k \leq m_k^{(2)}$$

and

$$M_k \geq M_k^{(1)}, \quad M_k \geq M_k^{(2)}.$$

Thus,

$$
\begin{aligned}
L(f, Q) - L(f, P) &= m_k^{(1)} m(R_k') + m_k^{(2)} m(R_k'') - m_k m(R_k) \\
&\geq m_k m(R_k') + m_k m(R_k'') - m_k m(R_k) \\
&= 0
\end{aligned}
$$

and

$$
\begin{aligned}
U(f, P) - U(f, Q) &= M_k m(R_k) - M_k^{(1)} m(R_k') - M_k^{(2)} m(R_k'') \\
&= M_k m(R_k') + M_k m(R_k'') - M_k^{(1)} m(R_k') - M_k^{(2)} m(R_k'') \\
&\geq M_k^{(1)} m(R_k') + M_k^{(2)} m(R_k'') - M_k^{(1)} m(R_k') - M_k^{(2)} m(R_k'') \\
&= 0,
\end{aligned}
$$

which completes the proof. □

Definition 7.10 Suppose

$$P = P_1 \times P_2 \times \ldots \times P_n \quad \text{and} \quad Q = Q_1 \times Q_2 \times \ldots \times Q_n,$$

where $P_i, Q_i \in \mathscr{P}([a_i, b_i))$, $i \in \{1, 2, \ldots, n\}$, are two Δ-partitions of

$$R = [a_1, b_1) \times [a_2, b_2) \times \ldots \times [a_n, b_n).$$

If P_i is generated by a set

$$\{t_i^0, t_i^1, \ldots, t_i^{k_i}\}, \quad \text{where} \quad a_i = t_i^0 < t_i^1 < \ldots < t_i^{k_i} = b_i,$$

and Q_i is generated by a set

$$\{\tau_i^0, \tau_i^2, \ldots, \tau_i^{p_i}\}, \quad \text{where} \quad a_i = \tau_i^0 < \tau_i^1 < \ldots < \tau_i^{p_i} = b_i,$$

then, by

$$P + Q = (P_1 + Q_1) \times (P_2 + Q_2) \times \ldots \times (P_n + Q_n),$$

we denote the Δ-partition of R generated by

$$P_i + Q_i = \{t_i^0, t_i^1, \ldots, t_i^{k_i}\} \cup \{\tau_i^0, \tau_i^1, \ldots, \tau_i^{p_i}\}, \quad i = 1, 2, \ldots, n.$$

Remark 7.11 Obviously, $P + Q$ is a refinement of both P and Q.

Theorem 7.12 *If f is a bounded function on R and if P and Q are any two Δ-partitions of R, then*

$$L(f, P) \leq U(f, Q),$$

i.e., every lower sum is less than or equal to every upper sum.

Proof Since $P + Q$ is a Δ-partition of R, which is a refinement of both P and Q, applying Theorem 7.9, we get

$$L(f, P) \leq L(f, P + Q) \leq U(f, P + Q) \leq U(f, Q),$$

i.e., $L(f, P) \leq U(f, Q)$. □

Theorem 7.13 *If f is a bounded function on R, then $L(f) \leq U(f)$.*

Proof Let $P \in \mathscr{P}(R)$. Then

$$L(f, P) \leq U(f, Q) \quad \text{for all} \quad Q \in \mathscr{P}(R).$$

Hence,

$$L(f, P) \leq \inf_{Q \in \mathscr{P}(R)} U(f, Q) = U(f).$$

Because $P \in \mathscr{P}(R)$ was arbitrarily chosen, we conclude that

$$\sup_{P \in \mathscr{P}(R)} L(f, P) \leq U(f),$$

i.e.,

$$L(f) \leq U(f),$$

completing the proof. □

Theorem 7.14 *If $L(f, P) = U(f, P)$ for some $P \in \mathscr{P}(R)$, then the function f is Δ-integrable over R and*

$$\int_R f(t_1, t_2, \ldots, t_n)\Delta_1 t_1 \Delta_2 t_2 \ldots \Delta_n t_n = L(f, P) = U(f, P).$$

Proof The result follows from the inequality

$$L(f, P) \leq L(f) \leq U(f) \leq U(f, P).$$

This completes the proof. □

Theorem 7.15 *A bounded function f on R is Δ-integrable if and only if for each $\varepsilon > 0$, there exists $P \in \mathscr{P}(R)$ such that*

$$U(f, P) - L(f, P) < \varepsilon. \tag{7.4}$$

Proof 1. Let f be Δ-integrable on R. Then

$$L(f) = U(f).$$

Using the definitions of $L(f)$ and $U(f)$, it follows that there exist $P, Q \in \mathscr{P}(R)$ such that

$$L(f, P) > L(f) - \frac{\varepsilon}{2} \quad \text{and} \quad U(f, Q) < U(f) + \frac{\varepsilon}{2}.$$

Let $S = P + Q$, which is a refinement of both P and Q. Thus, employing Theorem 7.9, we find

$$U(f, S) \leq U(f, Q) \quad \text{and} \quad L(f, S) \geq L(f, P)$$

and

$$\begin{aligned} U(f, S) - L(f, S) &\leq U(f, Q) - L(f, P) \\ &< U(f) + \frac{\varepsilon}{2} - L(f) + \frac{\varepsilon}{2} \\ &= \varepsilon. \end{aligned}$$

2. Conversely, suppose that for every $\varepsilon > 0$, the inequality (7.4) holds for some $P \in \mathscr{P}(R)$. Therefore,

$$\begin{aligned} U(f) &\leq U(f, P) \\ &= U(f, P) - L(f, P) + L(f, P) \\ &< \varepsilon + L(f, P) \\ &\leq \varepsilon + L(f). \end{aligned}$$

Since $\varepsilon > 0$ was arbitrarily chosen, we get

$$U(f) \leq L(f).$$

From the last inequality and from Theorem 7.13, we conclude that $U(f) = L(f)$, i.e., f is Δ-integrable on R.

The proof is complete. □

Remark 7.16 Let $\mathbb{T}$ be a time scale with forward jump operator σ. We note that for every $\delta > 0$, there exists at least one partition $P_1 \in \mathscr{P}([a, b])$ generated by a set

$$\{t_0, t_1, t_2, \ldots, t_n\} \subset [a, b], \quad \text{where} \quad a = t_0 < t_1 < \ldots < t_n = b,$$

such that for each $i \in \{1, 2, \ldots, n\}$ either

$$t_i - t_{i-1} < \delta$$

or

$$t_i - t_{i-1} > \delta \quad \text{and} \quad \sigma(t_{i-1}) = t_i.$$

Definition 7.17 We denote by $P_\delta([a, b))$ the set of all $P_1 \in \mathscr{P}([a, b))$ that possess the property indicated in Remark 7.16. Further, by $\mathscr{P}_\delta(R)$, we denote the set of all $P \in \mathscr{P}(R)$ such that

$$P = P_1 \times P_2 \times \ldots \times P_n, \quad \text{where} \quad P_i \in \mathscr{P}_\delta([a_i, b_i)), \quad i = 1, 2, \ldots, n.$$

Theorem 7.18 *Let $P^0 \in \mathscr{P}(R)$ be given by*

$$P^0 = P_1^0 \times P_2^0 \times \ldots \times P_n^0$$

in which $P_i^0 \in \mathscr{P}([a_i, b_i))$, $i \in \{1, 2, \ldots, n\}$, is generated by a set

$$A_i^0 = \{t_{i_0}^0, t_{i_1}^0, \ldots, t_{i_{n_i}}^0\} \subset [a_i, b_i], \quad \textit{where} \quad a_i = t_{i_0}^0 < t_{i_1}^0 < \ldots < t_{i_{n_i}}^0 = b_i.$$

Then, for each $P \in \mathscr{P}_\delta(R)$, we have

$$L(f, P^0 + P) - L(f, P) \le (M - m)D^{n-1}(n_1 + n_2 + \ldots + n_n - n)\delta$$

and

$$U(f, P) - U(f, P + P^0) \le (M - m)D^{n-1}(n_1 + n_2 + \ldots + n_n - n)\delta,$$

where $D = \max\limits_{i \in \{1,2,\ldots,n\}} \{b_i - a_i\}$, and M and m are defined as above.

Proof Suppose the partition P is given by

$$P = P_1 \times P_2 \times \ldots \times P_n$$

in which $P_i \in \mathscr{P}_\delta([a_i, b_i))$ is generated by a set

$$A_i = \{t_0^i, t_1^i, \ldots, t_{p_i}^i\} \subset [a_i, b_i],$$

where

$$a_i = t_0^i < t_1^i < \ldots < t_{p_i}^i = b_i, \quad i = 1, 2, \ldots, n.$$

Let $Q = P^0 + P = Q_1 \times Q_2 \times \ldots \times Q_n$, where $Q_i \in \mathscr{P}([a_i, b_i)), i = 1, 2, \ldots, n$, are generated by the sets

$$B_i = A_i^0 + A_i.$$

We suppose that there exists $i \in \{1, 2, \ldots, n\}$ such that B_i has one more point, say t', than A_i and $B_l = A_l$, $l \neq i$, $l \in \{1, 2, \ldots, n\}$. Then $t' \in (t^i_{k_i-1}, t^i_{k_i})$ for some $k_i \in \{1, 2, \ldots, p_i\}$, where $t^i_{k_i} - t^i_{k_i-1} \leq \delta$. If $t^i_{k_i} - t^i_{k_i-1} \geq \delta$, then, using

$$P_i \in \mathscr{P}_\delta([a_i, b_i)),$$

we have $\sigma(t^i_{k_i-1}) = t^i_{k_i}$ and $(t^i_{k_i-1}, t^i_{k_i}) = \emptyset$. Now, denoting by $m_{k_1k_2\ldots k_n}$, $m^{(1)}_{k_1k_2\ldots k_n}$, and $m^{(2)}_{k_1k_2\ldots k_n}$ the infimum of f on

$$R_{k_1k_2\ldots k_n} = [t_{k_1-1}, t_{k_1}) \times [t_{k_2-1}, t_{k_2}) \times \ldots \times [t_{k_n-1}, t_{k_n}),$$

$$R^{(1)}_{k_1k_2\ldots k_n} = [t_{k_1-1}, t_{k_1}) \times \ldots \times [t_{k_{i-1}-1}, t_{k_{i-1}}) \times [t_{k_i-1}, t') \times [t_{k_{i+1}-1}, t_{k_{i+1}}) \times \ldots \times [t_{k_n-1}, t_{k_n}),$$

$$R^{(2)}_{k_1k_2\ldots k_n} = [t_{k_1-1}, t_{k_1}) \times \ldots \times [t_{k_{i-1}-1}, t_{k_{i-1}}) \times [t', t_{k_i}) \times [t_{k_{i+1}-1}, t_{k_{i+1}}) \times \ldots \times [t_{k_n-1}, t_{k_n}),$$

respectively, we have

$$m^{(1)}_{k_1k_2\ldots k_n} \geq m_{k_1k_2\ldots k_n},$$

$$m^{(2)}_{k_1k_2\ldots k_n} \geq m_{k_1k_2\ldots k_n},$$

$$m^{(1)}_{k_1k_2\ldots k_n} - m_{k_1k_2\ldots k_n} \leq M - m,$$

$$m^{(2)}_{k_1k_2\ldots k_n} - m_{k_1k_2\ldots k_n} \leq M - m,$$

and

$$m(R_{k_1k_2\ldots k_n}) = m(R^{(1)}_{k_1k_2\ldots k_n}) + m(R^{(2)}_{k_1k_2\ldots k_n}),$$

so that

$$L(f,Q)-L(f,P)$$

$$=\sum_{j_{k_1}=1}^{p_1}\sum_{j_{k_2}=1}^{p_2}\cdots\sum_{j_{k_{i-1}}=1}^{p_{i-1}}\sum_{j_{k_{i+1}}=1}^{p_{i+1}}\cdots\sum_{j_{k_n}=1}^{p_n}\Big(m^{(1)}_{j_{k_1}j_{k_2}\ldots j_{k_n}}m(R^{(1)}_{j_{k_1}j_{k_2}\ldots j_{k_n}})$$

$$+m^{(2)}_{j_{k_1}j_{k_2}\ldots j_{k_n}}m(R^{(2)}_{j_{k_1}j_{k_2}\ldots j_{k_n}})-m_{j_{k_1}j_{k_2}\ldots j_{k_n}}m(R_{j_{k_1}j_{k_2}\ldots j_{k_n}})\Big)$$

$$=\sum_{j_{k_1}=1}^{p_1}\sum_{j_{k_2}=1}^{p_2}\cdots\sum_{j_{k_{i-1}}=1}^{p_{i-1}}\sum_{j_{k_{i+1}}=1}^{p_{i+1}}\cdots\sum_{j_{k_n}=1}^{p_n}\Big((m^{(1)}_{j_{k_1}j_{k_2}\ldots j_{k_n}}-m_{j_{k_1}j_{k_2}\ldots j_{k_n}})$$

$$\times m(R^{(1)}_{j_{k_1}j_{k_2}\ldots j_{k_n}})+(m^{(2)}_{j_{k_1}j_{k_2}\ldots j_{k_n}}-m_{j_{k_1}j_{k_2}\ldots j_{k_n}})m(R^{(2)}_{j_{k_1}j_{k_2}\ldots j_{k_n}})\Big)$$

$$\leq (M-m)\sum_{j_{k_1}=1}^{p_1}\sum_{j_{k_2}=1}^{p_2}\cdots\sum_{j_{k_{i-1}}=1}^{p_{i-1}}\sum_{j_{k_{i+1}}=1}^{p_{i+1}}\cdots\sum_{j_{k_n}=1}^{p_n}\Big(m(R^{(1)}_{j_{k_1}j_{k_2}\ldots j_{k_n}})$$

$$+m(R^{(2)}_{j_{k_1}j_{k_2}\ldots j_{k_n}})\Big)$$

$$=(M-m)(t_{k_i}-t_{k_i-1})$$

$$\times\sum_{j_{k_1}=1}^{p_1}\sum_{j_{k_2}=1}^{p_2}\cdots\sum_{j_{k_{i-1}}=1}^{p_{i-1}}\sum_{j_{k_{i+1}}}^{p_{i+1}}\cdots\sum_{j_{k_n}=1}^{p_n}(t_{j_{k_1}}-t_{j_{k_1-1}})(t_{j_{k_2}}-t_{j_{k_2-1}})\ldots(t_{j_{k_n}}-t_{j_{k_n-1}})$$

$$\leq (M-m)D^{n-1}\delta.$$

Since B_i has at most n_i-1 points that are not in A_i, an induction argument shows that

$$L(f,Q)-L(f,P)\leq (M-m)(n_1+n_2+\cdots+n_n-n)D^{n-1}\delta.$$

The proof of the other inequality is similar. □

Theorem 7.19 *A bounded function f on R is Δ-integrable if and only if for each $\varepsilon>0$, there exists $\delta>0$ such that*

$$P\in\mathscr{P}_\delta(R)\ \ \textit{implies}\ \ U(f,P)-L(f,P)<\varepsilon. \tag{7.5}$$

Proof 1. Suppose that for each $\varepsilon > 0$, there exists $\delta > 0$ such that (7.5) holds. Because $P \in \mathscr{P}(R)$, we have that (7.4) holds. Hence, using Theorem 7.15, we conclude that f is Δ-integrable on R.

2. Suppose that f is Δ-integrable over R. Let $\varepsilon > 0$ be arbitrarily chosen. Hence, by Theorem 7.15, it follows that there exists $P^0 \in \mathscr{P}(R)$ such that

$$U(f, P^0) - L(f, P^0) < \frac{\varepsilon}{2}.$$

Let D be as in Theorem 7.18 and P^0 be determined as in Theorem 7.18. We choose

$$\delta = \frac{\varepsilon}{4(M-m)D^{n-1}(n_1 + n_2 + \ldots + n_n - n)}.$$

Then, using Theorem 7.18, for each $P \in \mathscr{P}_\delta(R)$, we have

$$\begin{aligned} L(f, P^0 + P) - L(f, P) &\le (M-m)D^{n-1}(n_1 + n_2 + \cdots + n_n - n)\delta \\ &= \frac{\varepsilon}{4}, \\ U(f, P) - U(f, P^0 + P) &\le (M-m)D^{n-1}(n_1 + n_2 + \cdots + n_n - n)\delta \\ &= \frac{\varepsilon}{4}. \end{aligned}$$

Using this and

$$L(f, P^0) \le L(f, P^0 + P) \quad \text{and} \quad U(f, P^0 + P) \le U(f, P^0),$$

we obtain

$$L(f, P^0) - L(f, P) \le \frac{\varepsilon}{4} \quad \text{and} \quad U(f, P) - U(f, P^0) \le \frac{\varepsilon}{4},$$

i.e.,

$$-L(f, P) \le \frac{\varepsilon}{4} - L(f, P^0) \quad \text{and} \quad U(f, P) \le \frac{\varepsilon}{4} + U(f, P^0).$$

Therefore,

$$\begin{aligned} U(f, P) - L(f, P) &\le \frac{\varepsilon}{4} - L(f, P^0) + \frac{\varepsilon}{4} + U(f, P^0) \\ &= \frac{\varepsilon}{2} + U(f, P^0) - L(f, P^0) \end{aligned}$$

$$< \frac{\varepsilon}{2} + \frac{\varepsilon}{2}$$

$$= \varepsilon.$$

Thus, we have verified (7.5). □

Theorem 7.20 *For every bounded function f on R, the Darboux Δ-sums $L(f, P)$ and $U(f, P)$ evaluated for $P \in \mathscr{P}_\delta(R)$ have limits as $\delta \to 0$, uniformly with respect to P, and*

$$\lim_{\delta\to 0} L(f, P) = L(f) \quad \textit{and} \quad \lim_{\delta\to 0} U(f, P) = U(f).$$

Proof We fix $\varepsilon > 0$ and choose a partition $P^0 \in \mathscr{P}(R)$ so that

$$L(f) - L(f, P^0) < \varepsilon \quad \text{and} \quad U(f, P^0) - U(f) < \varepsilon.$$

Let P^0 be described as in Theorem 7.18. Then, for any $P \in \mathscr{P}_\delta(R)$, using Theorem 7.18, we have

$$L(f, P^0 + P) - L(f, P) \le (M - m)D^{n-1}(n_1 + n_2 + \cdots + n_n - n)\delta$$

and

$$U(f, P) - U(f, P^0 + P) \le (M - m)D^{n-1}(n_1 + n_2 + \cdots + n_n - n)\delta.$$

We take

$$\delta = \frac{\varepsilon}{(M - m)D^{n-1}(n_1 + n_2 + \cdots + n_n - n)}.$$

Because $P^0 + P$ is a refinement of P^0, we have

$$L(f, P^0) \le L(f, P^0 + P) \quad \text{and} \quad U(f, P^0 + P) \le U(f, P^0).$$

Thus,

$$L(f) - \varepsilon < L(f, P^0) \le L(f, P^0 + P)$$

$$\le L(f),$$

$$U(f) \le U(f, P^0 + P)$$

$$\le U(f, P^0)$$

$$< \varepsilon + U(f).$$

Hence,

$$L(f, P^0 + P) - L(f, P^0) < \varepsilon \quad \text{and} \quad U(f, P^0) - U(f, P^0 + P) < \varepsilon.$$

Therefore,

$$\begin{aligned}
|L(f) - L(f, P)| &= |L(f) - L(f, P^0) + L(f, P^0) - L(f, P^0 + P) \\
&\quad + L(f, P^0 + P) - L(f, P)| \\
&\leq |L(f) - L(f, P^0)| + |L(f, P^0) - L(f, P^0 + P)| \\
&\quad + |L(f, P + P^0) - L(f, P)| \\
&< \varepsilon + \varepsilon + (M - m)D^{n-1}(n_1 + n_2 + \cdots + n_n - n)\delta \\
&\leq \varepsilon + \varepsilon + \varepsilon \\
&= 3\varepsilon
\end{aligned}$$

and

$$\begin{aligned}
|U(f, P) - U(f)| &= |U(f, P) - U(f, P + P^0) + U(f, P + P^0) - U(f, P^0) \\
&\quad + U(f, P^0) - U(f)| \\
&\leq |U(f, P) - U(f, P + P^0)| + |U(f, P + P^0) - U(f, P^0)| \\
&\quad + |U(f, P^0) - U(f)| \\
&< (M - m)D^{n-1}(n_1 + n_2 + \cdots + n_n - n)\delta + \varepsilon + \varepsilon \\
&\leq \varepsilon + \varepsilon + \varepsilon \\
&= 3\varepsilon,
\end{aligned}$$

completing the proof. □

Definition 7.21 Let f be a bounded function on R and $P \in \mathscr{P}(R)$. In each "rectangle" $R_{j_1 j_2 \dots j_n}$, $1 \le j_i \le k_i$, $i = 1, 2, \dots, n$, choose a point $\xi_{j_1 j_2 \dots j_n}$ and form the sum

$$S = \sum_{i=1}^{n} \sum_{j_i=1}^{k_i} f(\xi_{j_1 j_2 \dots j_n})(t_1^{j_i} - t_1^{j_i - 1}) \dots (t_n^{j_n} - t_n^{j_n - 1}). \tag{7.6}$$

We call S a *Riemann Δ-sum* of f corresponding to $P \in \mathscr{P}(R)$. We say that f is *Riemann Δ-integrable* over R if there exists a number I such that, for each $\varepsilon > 0$, there exists $\delta > 0$ such that

$$|S - I| < \varepsilon$$

for every Riemann Δ-sum S of f corresponding to any $P \in \mathscr{P}_\delta(R)$, independent of the choice of the point $\xi_{j_1 j_2 \dots j_n} \in R_{j_1 j_2 \dots j_n}$ for $1 \le j_i \le k_i$, $i = 1, 2, \dots, n$. The number I is called the *Riemann Δ-integral* of f over R. We write

$$I = \lim_{\delta \to 0} S.$$

Theorem 7.22 *The Riemann Δ-integral is well defined.*

Proof Suppose that f is Riemann Δ-integrable over R and there are two numbers I_1 and I_2 such that for every $\varepsilon > 0$, there exists $\delta > 0$ so that

$$|S - I_1| < \frac{\varepsilon}{2} \quad \text{and} \quad |S - I_2| < \frac{\varepsilon}{2}$$

for every Riemann Δ-sum S of f corresponding to any $P \in \mathscr{P}_\delta(R)$, independent of the way in which $\xi_{j_1 j_2 \dots j_n} \in R_{j_1 j_2 \dots j_n}$ for $1 \le j_i \le k_i$, $i = 1, 2, \dots, n$, is chosen. Therefore,

$$\begin{aligned} |I_1 - I_2| &= |I_1 - S + S - I_2| \\ &\le |S - I_1| + |S - I_2| \\ &< \frac{\varepsilon}{2} + \frac{\varepsilon}{2} \\ &= \varepsilon. \end{aligned}$$

Consequently, $I_1 = I_2$. □

Remark 7.23 Note that in the Riemann definition of the integral, we need not assume the boundedness of f in advance. However, it follows that the Riemann integrability of a function f over R implies its boundedness on R.

Theorem 7.24 *A bounded function on R is Riemann Δ-integrable if and only if it is Darboux Δ-integrable, in which case the values of the integrals are equal.*

Proof 1. Suppose that f is Darboux Δ-integrable over R in the sense of Definition 7.6. Let $\varepsilon > 0$ and $\delta > 0$ be chosen so that (7.4) of Theorem 7.15 holds. Using the definition of S, we have

$$L(f, P) \leq S \leq U(f, P).$$

Also,

$$\begin{aligned} U(f, P) &< L(f, P) + \varepsilon \\ &\leq L(f) + \varepsilon \\ &= \int_R f(t_1, t_2, \ldots, t_n)\Delta_1 t_1 \Delta_2 t_2 \ldots \Delta_n t_n + \varepsilon, \\ L(f, P) &> U(f, P) - \varepsilon \\ &\geq U(f) - \varepsilon \\ &= \int_R f(t_1, t_2, \ldots, t_n)\Delta_1 t_1 \Delta_2 t_2 \ldots \Delta_n t_n - \varepsilon. \end{aligned}$$

Hence,

$$\begin{aligned} S - \int_R f(t_1, t_2, \ldots, t_n)\Delta_1 t_1 \Delta_2 t_2 \ldots \Delta_n t_n &\leq U(f, P) \\ &\quad - \int_R f(t_1, t_2, \ldots, t_n)\Delta_1 t_1 \Delta_2 t_2 \ldots \Delta_n t_n \\ &< \int_R f(t_1, t_2, \ldots, t_n)\Delta_1 t_1 \Delta_2 t_2 \ldots \Delta_n t_n + \varepsilon \\ &\quad - \int_R f(t_1, t_2, \ldots, t_n)\Delta_1 t_1 \Delta_2 t_2 \ldots \Delta_n t_n \\ &= \varepsilon \end{aligned}$$

and

$$\begin{aligned} S - \int_R f(t_1, t_2, \ldots, t_n)\Delta_1 t_1 \Delta_2 t_2 \ldots \Delta_n t_n &\geq L(f, P) \\ &\quad - \int_R f(t_1, t_2, \ldots, t_n)\Delta_1 t_1 \Delta_2 t_2 \ldots \Delta_n t_n \\ &> \int_R f(t_1, t_2, \ldots, t_n)\Delta_1 t_1 \Delta_2 t_2 \ldots \Delta_n t_n - \varepsilon \end{aligned}$$

$$- \int_R f(t_1, t_2, \ldots, t_n)\Delta_1 t_1 \Delta_2 t_2 \ldots \Delta_n t_n$$

$$= -\varepsilon.$$

Consequently,

$$\left| S - \int_R f(t_1, t_2, \ldots, t_n)\Delta_1 t_1 \Delta_2 t_2 \ldots \Delta_n t_n \right| < \varepsilon.$$

2. Now, we suppose that f is Riemann Δ-integrable in the sense of Definition 7.21. Select $P \in \mathscr{P}_\delta(R)$ of the type (7.1) and (7.2). For each $i \in \{1, 2, \ldots, n\}$ and $1 \le j_i \le k_i$, we choose $\xi_{j_1 j_2 \ldots j_n} \in R_{j_1 j_2 \ldots j_n}$ so that

$$M_{j_1 j_2 \ldots j_n} - \varepsilon < f(\xi_{j_1 j_2 \ldots j_n}) < m_{j_1 j_2 \ldots j_n} + \varepsilon.$$

The Riemann Δ-sum S for this choice of the points $\xi_{j_1 j_2 \ldots j_n}$ satisfies

$$U(f, P) - \varepsilon \prod_{i=1}^{n} (b_i - a_i) < S < L(f, P) + \varepsilon \prod_{i=1}^{n} (b_i - a_i)$$

as well as

$$-\varepsilon < S - I < \varepsilon.$$

Thus,

$$L(f) \ge L(f, P)$$

$$> S - \varepsilon \prod_{i=1}^{n} (b_i - a_i)$$

$$> I - \varepsilon - \varepsilon \prod_{i=1}^{n} (b_i - a_i)$$

and

$$U(f) \le U(f, P)$$

$$< S + \varepsilon \prod_{i=1}^{n} (b_i - a_i)$$

$$< I + \varepsilon + \varepsilon \prod_{i=1}^{n} (b_i - a_i).$$

Since $\varepsilon > 0$ was arbitrarily chosen, we conclude that

$$L(f) \geq I \quad \text{and} \quad U(f) \leq I,$$

i.e.,

$$I \leq L(f) \leq U(f) \leq I.$$

This completes the proof. □

Remark 7.25 In the definition of

$$\int_R f(t_1, t_2, \ldots, t_n) \Delta_1 t_1 \Delta_2 t_2 \ldots \Delta_n t_n$$

with $R = [a_1, b_1) \times \ldots \times [a_n, b_n)$, we assumed that $a_i < b_i$, $i \in \{1, \ldots, n\}$. We extend the definition to the case $a_i = b_i$ for some $i \in \{1, 2, \ldots, n\}$ by setting

$$\int_R f(t_1, t_2, \ldots, t_n) \Delta_1 t_1 \Delta_2 t_2 \ldots \Delta_n t_n = 0 \tag{7.7}$$

if $a_i = b_i$ for some $i \in \{1, \ldots, n\}$.

Theorem 7.26 *Let*

$$a = (a_1, \ldots, a_n) \in \Lambda^n \quad \textit{and} \quad b = (b_1, \ldots, b_n) \in \Lambda^n$$

with $a_i \leq b_i$ for all $i \in \{1, \ldots, n\}$. Every constant function

$$f(t_1, t_2, \ldots, t_n) = A \quad \textit{for} \quad (t_1, t_2, \ldots, t_n) \in R = [a_1, b_1) \times \ldots \times [a_n, b_n)$$

is Δ-integrable over R and

$$\int_R f(t_1, t_2, \ldots, t_n) \Delta_1 t_1 \Delta_2 t_2 \ldots \Delta_n t_n = A \prod_{i=1}^{n} (b_i - a_i). \tag{7.8}$$

Proof We assume that $a_i < b_i$ for all $i \in \{1, \ldots, n\}$. Consider a partition P of R of the type (7.1) and (7.2). Since

$$M_{j_1 j_2 \ldots j_n} = m_{j_1 j_2 \ldots j_n} = A \quad \text{for all} \quad 1 \leq j_i \leq k_i, \quad i \in \{1, \ldots, n\},$$

we have that

$$U(f, P) = L(f, P) = A \prod_{i=1}^{n} (b_i - a_i).$$

Hence, using Theorem 7.15, it follows that f is Δ-integrable and (7.8) holds. If $a_i = b_i$ for some $i \in \{1, \ldots, n\}$, then (7.8) follows by (7.7). Note that every Riemann Δ-sum of f associated with P is also equal to

$$A \prod_{i=1}^{n} (b_i - a_i).$$

This completes the proof. □

Theorem 7.27 *Let* $t^0 = (t_1^0, \ldots, t_n^0) \in \Lambda^n$. *Every function* $f : \Lambda^n \to \mathbb{R}$ *is* Δ-*integrable over*

$$R = R(t^0) = [t_1^0, \sigma_1(t_1^0)) \times \ldots \times [t_n^0, \sigma_n(t_n^0)),$$

and

$$\int_R f(t_1, t_2, \ldots, t_n) \Delta_1 t_1 \Delta_2 t_2 \ldots \Delta_n t_n = \prod_{i=1}^{n} \mu_i(t_i^0) f(t_1^0, \ldots, t_n^0). \tag{7.9}$$

Proof If $\mu_i(t_i^0) = 0$ for some $i \in \{1, \ldots, n\}$, then (7.9) is obvious as both sides of (7.9) are equal to zero in this case. If $\mu_i(t_i^0) > 0$ for all $i \in \{1, \ldots, n\}$, then a single partition P of $R(t^0)$ is

$$[t_1^0, \sigma_1(t_1^0)) \times \ldots \times [t_n^0, \sigma_n(t_n^0)) = \{(t_1^0, \ldots, t_n^0)\}.$$

Consequently, we have

$$U(f, P) = L(f, P) = \prod_{i=1}^{n} \mu_i(t_i^0) f(t_1^0, \ldots, t_n^0).$$

Therefore, Theorem 7.15 shows that f is Δ-integrable over $R(t^0)$ and (7.9) holds. Note that the Riemann Δ-sum associated with the above partition is also equal to the right-hand side of (7.9). □

Theorem 7.28 *Let*

$$a = (a_1, \ldots, a_n) \in \Lambda^n \quad \text{and} \quad b = (b_1, \ldots, b_n) \in \Lambda^n$$

with $a_i \leq b_i$ *for all* $i \in \{1, 2, \ldots, n\}$. *If* $\mathbb{T}_i = \mathbb{R}$ *for every* $i \in \{1, 2, \ldots, n\}$, *then every bounded function* f *on* $R = [a_1, b_1) \times [a_2, b_2) \times \ldots \times [a_n, b_n)$ *is* Δ-*integrable if and only if* f *is Riemann integrable on* R *in the classical sense, and in this case*

$$\int_R f(t_1, t_2, \ldots, t_n) \Delta_1 t_1 \Delta_2 t_2 \ldots \Delta_n t_n = \int_R f(t_1, t_2, \ldots, t_n) \mathrm{d}t_1 \mathrm{d}t_2 \ldots \mathrm{d}t_n,$$

where the integral on the right-hand side is the ordinary Riemann integral.

Proof Clearly, Definitions 7.6 and 7.21 of the Δ-integral coincide in the case $\mathbb{T}_i = \mathbb{R}$, $i \in \{1, 2, \ldots, n\}$, with the usual Darboux and Riemann definitions of the integral, respectively. Note that the classical definitions of Darboux's and Riemann's integral do not depend on whether the rectangles of the partition are taken closed, half-closed, or open. Moreover, if $\mathbb{T}_i = \mathbb{R}$, $i \in \{1, 2, \ldots, n\}$, then $\mathscr{P}_\delta(\mathbb{R})$ consists of all partitions of R with norm (mesh) less than or equal to $\delta\sqrt{n}$. □

Theorem 7.29 *Let*

$$a = (a_1, \ldots, a_n) \in \Lambda^n \quad \textit{and} \quad b = (b_1, \ldots, b_n) \in \Lambda^n$$

with $a_i \le b_i$ for all $i \in \{1, 2, \ldots, n\}$. If $\mathbb{T}_i = \mathbb{Z}$ for all $i \in \{1, 2, \ldots, n\}$, then every function defined on $R = [a_1, b_1) \times [a_2, b_2) \times \ldots \times [a_n, b_n)$ is Δ-integrable over R, and

$$\int_R f(t_1, t_2, \ldots, t_n)\Delta_1 t_1 \Delta_2 t_2 \ldots \Delta_n t_n$$

$$= \begin{cases} 0 & \textit{if} \quad a_i = b_i \quad \textit{for some} \quad i \in \{1, 2, \ldots, n\} \\ \sum\limits_{r_1=a_1}^{b_1-1} \sum\limits_{r_2=a_2}^{b_2-1} \cdots \sum\limits_{r_n=a_n}^{b_n-1} f(r_1, r_2, \ldots, r_n) & \textit{otherwise.} \end{cases} \tag{7.10}$$

Proof Let $b_i = a_i + p_i$, $p_i \in \mathbb{N}$, $i \in \{1, 2, \ldots, n\}$. Consider the partition P^* of R given by (7.1) and (7.2) with $k_i = p_i$, $i \in \{1, 2, \ldots, n\}$, and

$$t_i^0 = a_i, \quad t_i^1 = a_i + 1, \quad \ldots, \quad t_i^{k_i} = a_i + p_i.$$

Thus, $R_{j_1 j_2 \ldots j_n}$ contains the single point $(t_1^{j_1-1}, t_2^{j_2-1}, \ldots, t_n^{j_n-1})$. Therefore,

$$U(f, P^*) = L(f, P^*) = \sum_{r_1=a_1}^{b_1-1} \sum_{r_2=a_2}^{b_2-1} \cdots \sum_{r_n=a_n}^{b_n-1} f(r_1, r_2, \ldots, r_n).$$

Hence, Theorem 7.15 shows that f is Δ-integrable over R and (7.10) holds for $a_i < b_i$, $i \in \{1, 2, \ldots, n\}$. If $a_i = b_i$ for some $i \in \{1, 2, \ldots, n\}$, then the relation (7.7) shows the validity of (7.10). □

Example 7.30 Let $\mathbb{T}_1 = \mathbb{T}_2 = \mathbb{Z}$. We consider

$$I = \int_0^4 \int_1^8 t_2(2t_1 + 1)\Delta_1 t_1 \Delta_2 t_2.$$

Here,

$$f(t_1, t_2) = t_2(2t_1 + 1), \quad (t_1, t_2) \in \mathbb{T}_1 \times \mathbb{T}_2,$$

$$\sigma_1(t_1) = t_1 + 1, \quad t_1 \in \mathbb{T}_1, \quad \sigma_2(t_2) = t_2 + 1, \quad t_2 \in \mathbb{T}_2.$$

We note that, for $g(t_1, t_2) = t_2 t_1^2$, we have

$$\begin{aligned} g_{t_1}^{\Delta_1}(t_1, t_2) &= t_2(\sigma_1(t_1) + t_1) \\ &= t_2(t_1 + 1 + t_1) \\ &= t_2(2t_1 + 1). \end{aligned}$$

Therefore,

$$\int_1^8 t_2(2t_1 + 1)\Delta_1 t_1 = g(8, t_2) - g(1, t_2) = 64t_2 - t_2 = 63t_2.$$

Hence,

$$I = \int_0^4 63t_2 \Delta_2 t_2 = 63 \int_0^4 t_2 \Delta_2 t_2.$$

Since, for $h(t_2) = t_2^2$, we have

$$\begin{aligned} \frac{1}{2}(h^{\Delta_2}(t_2) - 1) &= \frac{1}{2}(\sigma_2(t_2) + t_2 - 1) \\ &= \frac{1}{2}(t_2 + 1 + t_2 - 1) \\ &= t_2, \end{aligned}$$

we get

$$\begin{aligned} I &= 63 \int_0^4 \frac{1}{2}(h^{\Delta_2}(t_2) - 1)\Delta_2 t_2 \\ &= \frac{63}{2} \int_0^4 h^{\Delta_2}(t_2)\Delta_2 t_2 - \frac{63}{2} \int_0^4 \Delta_2 t_2 \\ &= \frac{63}{2}(h(4) - h(0)) - 126 \end{aligned}$$

$$= 504 - 126$$

$$= 378.$$

Example 7.31 Let $\mathbb{T}_1 = \mathbb{Z}$ and $\mathbb{T}_2 = 2^{\mathbb{N}}$. We consider

$$I = \frac{1}{2} \sin \frac{1}{2} \int_0^3 \int_2^8 t_2 \cos \left(t_1 + \frac{1}{2} \right) \Delta_1 t_1 \Delta_2 t_2.$$

Here,

$$f(t_1, t_2) = \frac{1}{2} t_2 \cos \left(t_1 + \frac{1}{2} \right) \sin \frac{1}{2}, \quad (t_1, t_2) \in \mathbb{T}_1 \times \mathbb{T}_2,$$

$$\sigma_1(t_1) = t_1 + 1, \quad t_1 \in \mathbb{T}_1, \quad \sigma_2(t_2) = 2t_2, \quad t_2 \in \mathbb{T}_2.$$

Since, for $g(t_1) = \sin t_1$, we have

$$\begin{aligned} g^{\Delta_1}(t_1) &= \frac{\sin \sigma_1(t_1) - \sin t_1}{\sigma_1(t_1) - t_1} \\ &= \frac{\sin(t_1 + 1) - \sin t_1}{t_1 + 1 - t_1} \\ &= \frac{1}{2} \sin \frac{1}{2} \cos \left(t_1 + \frac{1}{2} \right), \end{aligned}$$

we get

$$\begin{aligned} \frac{1}{2} \sin \frac{1}{2} \int_2^8 t_2 \cos \left(t_1 + \frac{1}{2} \right) \Delta_1 t_1 &= t_2 \int_2^8 g^{\Delta_1}(t_1) \Delta_1 t_1 \\ &= t_2 (g(8) - g(2)) \\ &= t_2 (\sin 8 - \sin 2) \\ &= 2t_2 \sin 3 \cos 5. \end{aligned}$$

Therefore,

$$I = 2 \sin 3 \cos 5 \int_0^3 t_2 \Delta_2 t_2.$$

Because for $h(t_2) = t_2^2$, we have

$$h^{\Delta_2}(t_2) = \sigma_2(t_2) + t_2 = 2t_2 + t_2 = 3t_2,$$

we get

$$t_2 = \frac{1}{3} h^{\Delta_2}(t_2).$$

Consequently,

$$\begin{aligned} I &= 2 \sin 3 \cos 5 \int_0^3 \frac{1}{3} h^{\Delta_2}(t_2) \Delta_2 t_2 \\ &= \frac{2}{3} \sin 3 \cos 5 \int_0^3 h^{\Delta_2}(t_2) \Delta_2 t_2 \\ &= \frac{2(h(3) - h(0))}{3} \sin 3 \cos 5 \\ &= 6 \sin 3 \cos 5. \end{aligned}$$

Example 7.32 Let $\mathbb{T}_1 = 3\mathbb{Z}$ and $\mathbb{T}_2 = 3^{\mathbb{N}}$. We consider

$$I = \int_3^9 \int_{-3}^{12} (t_1 t_2 + 2t_1 + t_2 + 3) \Delta_1 t_1 \Delta_2 t_2.$$

Here,

$$f(t_1, t_2) = t_1 t_2 + 2t_1 + t_2 + 3, \quad (t_1, t_2) \in \mathbb{T}_1 \times \mathbb{T}_2,$$

$$\sigma_1(t_1) = t_1 + 3, \quad t_1 \in \mathbb{T}_1, \quad \sigma_2(t_2) = 3t_2, \quad t_2 \in \mathbb{T}_2.$$

We note that, for $g(t_1) = t_1^2$, we have

$$g^{\Delta_1}(t_1) = \sigma_1(t_1) + t_1 = 2t_1 + 3,$$

whereupon

$$t_1 = \frac{g^{\Delta_1}(t_1) - 3}{2},$$

$$t_1 t_2 + 2t_1 + t_2 + 3 = \frac{g^{\Delta_1}(t_1) - 3}{2}(t_2 + 2) + t_2 + 3,$$

and

$$\begin{aligned}
\int_{-3}^{12} f(t_1, t_2)\Delta_1 t_1 &= \int_{-3}^{12} \left(\frac{g^{\Delta_1}(t_1) - 3}{2}(t_2 + 2) + (t_2 + 3) \right) \Delta_1 t_1 \\
&= \frac{t_2 + 2}{2} \int_{-3}^{12} g^{\Delta_1}(t_1)\Delta_1 t_1 + \int_{-3}^{12} \left(-\frac{3}{2}(t_2 + 2) + t_2 + 3 \right) \Delta_1 t_1 \\
&= \frac{t_2 + 2}{2} t_1^2 (g(12) - g(-3)) - \frac{t_2}{2} \int_{-3}^{12} \Delta_1 t_1 \\
&= \frac{t_2 + 2}{2}(144 - 9) - \frac{15}{2} t_2 \\
&= \frac{135}{2}(t_2 + 2) - \frac{15}{2} t_2 \\
&= 60 t_2 + 135.
\end{aligned}$$

Therefore,

$$I = \int_3^9 (60 t_2 + 135)\Delta_2 t_2.$$

Since, for $h(t_2) = t_2^2$, we have

$$h^{\Delta_2}(t_2) = \sigma_2(t_2) + t_2 = 3t_2 + t_2 = 4t_2,$$

we get

$$t_2 = \frac{1}{4} h^{\Delta_2}(t_2).$$

Consequently,

$$\begin{aligned}
I &= \int_3^9 \left(60 \cdot \frac{1}{4} h^{\Delta_2}(t_2) + 135 \right) \Delta_2 t_2 \\
&= 15 \int_3^9 h^{\Delta_2}(t_2)\Delta_2 t_2 + 135 \int_3^9 \Delta_2 t_2 \\
&= 15(h(9) - h(3)) + 135 \cdot 6 \\
&= 15 \cdot (81 - 9) + 810 \\
&= 1890.
\end{aligned}$$

Exercise 7.33 Let $\mathbb{T}_1 = 2^{\mathbb{N}}$ and $\mathbb{T}_2 = 4\mathbb{Z}$. Compute the integral

$$-2\int_0^8 \int_2^8 \frac{t_2}{t_1} \sin\frac{t_1}{2} \sin\frac{3t_1}{2} \Delta_1 t_1 \Delta_2 t_2.$$

Solution $-32 \sin 3 \sin 5$.

7.2 Properties of Multiple Integrals over Rectangles

Note that Λ^n is a complete metric space with the metric d defined by

$$d(t,s) = \sqrt{\sum_{i=1}^{n} (t_i - s_i)^2} \quad \text{for} \quad t = (t_1, t_2, \ldots, t_n), \quad s = (s_1, s_2, \ldots, s_n) \in \Lambda^n,$$

and also with the equivalent metric

$$d(t,s) = \max_{i \in \{1,2,\ldots,n\}} \{|t_i - s_i|\}.$$

Definition 7.34 A function $f : \Lambda^n \to \mathbb{R}$ is said to be *continuous* at $t \in \Lambda^n$ if for every $\varepsilon > 0$, there exists $\delta > 0$ such that

$$|f(t) - f(s)| < \varepsilon$$

for all points $s \in \Lambda^n$ satisfying $d(t,s) < \delta$.

Remark 7.35 If t is an isolated point of Λ^n, then every function $f : \Lambda^n \to \mathbb{R}$ is continuous at t. In particular, if $\mathbb{T}_i = \mathbb{Z}$ for all $i \in \{1, 2, \ldots, n\}$, then every function $f : \Lambda^n \to \mathbb{R}$ is continuous at each point of Λ^n.

Theorem 7.36 *Every continuous function on* $K = [a_1, b_1] \times [a_2, b_2] \times \ldots \times [a_n, b_n]$ *is* Δ*-integrable over* $R = [a_1, b_1) \times [a_2, b_2) \times \ldots \times [a_n, b_n)$.

Proof Let $\varepsilon > 0$ be arbitrarily chosen. Since f is continuous, it is uniformly continuous on the compact subset K of Λ^n. Therefore, there exists $\delta > 0$ such that

$$\begin{cases} t = (t_1, t_2, \ldots, t_n), \quad t' = (t_1', t_2', \ldots, t_n') \in R \quad \text{and} \quad \max_{i \in \{1,2,\ldots,n\}} \{|t_i - t_i'|\} \le \delta \\ \quad \text{implies} \quad |f(t) - f(t')| < \frac{\varepsilon}{(2^n - 1)\prod_{i=1}^{n}(b_i - a_i + 1)}. \end{cases} \tag{7.11}$$

Consider $P \in \mathscr{P}(R)$ given by (7.1) and (7.2). Let

$$\tilde{R}_{j_1 j_2 \dots j_n} = [t_1^{j_1-1}, \sigma_1(t_1^{j_1-1})] \times [t_2^{j_2-1}, \sigma_2(t_2^{j_2-1})] \times \dots \times [t_n^{j_n-1}, \sigma_n(t_n^{j_n-1})],$$

$$\tilde{M}_{j_1 j_2 \dots j_n} = \sup\{f(t_1, t_2, \dots, t_n) : (t_1, t_2, \dots, t_n) \in \tilde{R}_{j_1 j_2 \dots j_n}\},$$

$$\tilde{m}_{j_1 j_2 \dots j_n} = \inf\{f(t_1, t_2, \dots, t_n) : (t_1, t_2, \dots, t_n) \in \tilde{R}_{j_1 j_2 \dots j_n}\}.$$

Then, since $R_{j_1 j_2 \dots j_n} \subset \tilde{R}_{j_1 j_2 \dots j_n}$, we have

$$\tilde{m}_{j_1 j_2 \dots j_n} \le m_{j_1 j_2 \dots j_n} \le M_{j_1 j_2 \dots j_n} \le \tilde{M}_{j_1 j_2 \dots j_n}$$

for $1 \le j_i \le k_i$, $i = 1, 2, \dots, n$. Therefore, taking into account that f assumes its maximum and minimum on each compact rectangle $\tilde{R}_{j_1 j_2 \dots j_n}$, (7.11) shows

$$U(f, P) - L(f, P) = \sum_{j_1=1}^{k_1} \sum_{j_2=1}^{k_2} \dots \sum_{j_n=1}^{k_n} (M_{j_1 j_2 \dots j_n} - m_{j_1 j_2 \dots j_n})$$

$$\times (t_1^{j_1} - t_1^{j_1-1})(t_2^{j_2} - t_2^{j_2-1}) \dots (t_n^{j_n} - t_n^{j_n-1})$$

$$\le \sum_{j_1=1}^{k_1} \sum_{j_2=1}^{k_2} \dots \sum_{j_n=1}^{k_n} (\tilde{M}_{j_1 j_2 \dots j_n} - \tilde{m}_{j_1 j_2 \dots j_n})$$

$$\times (t_1^{j_1} - t_1^{j_1-1})(t_2^{j_2} - t_2^{j_2-1}) \dots (t_n^{j_n} - t_n^{j_n-1})$$

$$= \sum_{t_1^{j_1} - t_1^{j_1-1} \le \delta} \sum_{t_2^{j_2} - t_2^{j_2-1} \le \delta} \dots \sum_{t_n^{j_n} - t_n^{j_n-1} \le \delta} (\tilde{M}_{j_1 j_2 \dots j_n} - \tilde{m}_{j_1 j_2 \dots j_n})$$

$$\times (t_1^{j_1} - t_1^{j_1-1})(t_2^{j_2} - t_2^{j_2-1}) \dots (t_n^{j_n} - t_n^{j_n-1})$$

$$+ \sum_{t_1^{j_1} - t_1^{j_1-1} \le \delta} \sum_{t_2^{j_2} - t_2^{j_2-1} \le \delta} \dots \sum_{t_n^{j_n} - t_n^{j_n-1} > \delta} (\tilde{M}_{j_1 j_2 \dots j_n} - \tilde{m}_{j_1 j_2 \dots j_n})$$

$$\times (t_1^{j_1} - t_1^{j_1-1})(t_2^{j_2} - t_2^{j_2-1}) \dots (t_n^{j_n} - t_n^{j_n-1})$$

$$+ \cdots$$

$$+ \sum_{t_1^{j_1}-t_1^{j_1-1}>\delta} \sum_{t_2^{j_2}-t_2^{j_2-1}>\delta} \cdots \sum_{t_n^{j_n}-t_n^{j_n-1}>\delta} (\tilde{M}_{j_1 j_2 \ldots j_n} - \tilde{m}_{j_1 j_2 \ldots j_n})$$

$$\times (t_1^{j_1} - t_1^{j_1-1})(t_2^{j_2} - t_2^{j_2-1}) \ldots (t_n^{j_n} - t_n^{j_n-1})$$

$$\leq \frac{\varepsilon(2^n - 1)}{(2^n - 1)\prod_{i=1}^n (b_i - a_i + 1)}$$

$$\times \sum_{j_1=1}^{k_1} \sum_{j_2=1}^{k_2} \cdots \sum_{j_n=1}^{k_n} (t_1^{j_1} - t_1^{j_1-1})(t_2^{j_2} - t_2^{j_2-1}) \ldots (t_n^{j_n} - t_n^{j_n-1})$$

$$= \frac{\varepsilon \prod_{i=1}^n (b_i - a_i)}{\prod_{i=1}^n (b_i - a_i + 1)}$$

$$< \varepsilon.$$

Thus, $U(f, P) - L(f, P) < \varepsilon$. Hence, Theorem 7.15 yields that f is Δ-integrable. □

Definition 7.37 We say that a function $\phi : [\alpha, \beta] \to \mathbb{R}$ satisfies the *Lipschitz condition* if there exists a constant $B > 0$, a so-called *Lipschitz constant*, such that

$$|\phi(u) - \phi(v)| \leq B|u - v| \quad \text{for all} \quad u, v \in [\alpha, \beta].$$

Example 7.38 Let $\phi : [0, 1] \to \mathbb{R}$ be defined by $\phi(x) = x^2 + 1$, $x \in [0, 1]$. Then, for $x, y \in [0, 1]$, we have

$$\begin{aligned} |\phi(x) - \phi(y)| &= |x^2 + 1 - y^2 - 1| \\ &= |x^2 - y^2| \\ &= |x - y||x + y| \\ &\leq |x - y|(|x| + |y|) \\ &\leq 2|x - y|, \end{aligned}$$

i.e., ϕ satisfies the Lipschitz condition with Lipschitz constant $B = 2$.

Example 7.39 Let $\phi : [0, \pi] \to \mathbb{R}$ be defined by $\phi(x) = \sin x$, $x \in [0, \pi]$. Then, for $x, y \in [0, \pi]$, we have

$$\begin{aligned}|\phi(x) - \phi(y)| &= |\sin x - \sin y| \\ &= 2\left|\sin\frac{x-y}{2}\cos\frac{x+y}{2}\right| \\ &= 2\left|\sin\frac{x-y}{2}\right|\left|\cos\frac{x+y}{2}\right| \\ &\le 2\frac{|x-y|}{2} \\ &= |x-y|,\end{aligned}$$

i.e., ϕ satisfies the Lipschitz condition with constant $B = 1$.

Example 7.40 Let $\phi : [0, 3] \to \mathbb{R}$ be defined by $\phi(x) = \frac{1}{x+4}$, $x \in [0, 3]$. Then, for $x, y \in [0, 3]$, we have

$$\begin{aligned}|\phi(x) - \phi(y)| &= \left|\frac{1}{x+4} - \frac{1}{y+4}\right| \\ &= \left|\frac{y+4-x-4}{(x+4)(y+4)}\right| \\ &= \frac{|x-y|}{(x+4)(y+4)} \\ &\le \frac{1}{16}|x-y|,\end{aligned}$$

i.e., ϕ satisfies the Lipschitz condition with Lipschitz constant $L = \frac{1}{16}$.

Exercise 7.41 Let $\phi : [0, 5] \to \mathbb{R}$ be defined by $\phi(x) = \frac{1}{x^2+10}$, $x \in [0, 5]$. Prove that ϕ satisfies the Lipschitz condition. Find a Lipschitz constant.

Solution $B = \frac{1}{10}$.

Theorem 7.42 *Let $\phi : [\alpha, \beta] \to \mathbb{R}$ be differentiable. Then ϕ satisfies the Lipschitz condition with Lipschitz constant B if and only if*

$$|\phi'(x)| \le B \quad \textit{for all} \quad x \in [\alpha, \beta].$$

Proof 1. Suppose ϕ satisfies the Lipschitz condition with Lipschitz constant B. Then, for every $x, y \in [\alpha, \beta]$, we have

$$|\phi(x) - \phi(y)| \leq B|x - y|,$$

whereupon

$$|\phi'(x)| \leq B \quad \text{for all} \quad x \in [\alpha, \beta].$$

2. Suppose $|\phi'(x)| \leq B$ for all $x \in [\alpha, \beta]$. Then, for $x, y \in [\alpha, \beta]$, using the mean value theorem, we have that there exists $\xi \in [\alpha, \beta]$ so that

$$|\phi(x) - \phi(y)| = |\phi'(\xi)||x - y| \leq B|x - y|,$$

i.e., ϕ satisfies the Lipschitz condition with Lipschitz constant B.

This completes the proof. □

Example 7.43 Let $\phi : [0, 2] \to \mathbb{R}$ be defined by $\phi(x) = \arctan x$, $x \in [0, 2]$. We note that ϕ is continuously differentiable on $[0, 2]$ and

$$\phi'(x) = \frac{1}{1 + x^2}, \quad |\phi'(x)| \leq 1 \quad \text{for all} \quad x \in [0, 2].$$

Consequently, ϕ satisfies the Lipschitz condition with Lipschitz constant $B = 1$.

Example 7.44 Let $\phi : [0, 1] \to \mathbb{R}$ be defined by $\phi(x) = \log(1 + x^2)$, $x \in [0, 1]$. We note that ϕ is continuously differentiable on $[0, 1]$ and

$$|\phi'(x)| = \left|\frac{2x}{1 + x^2}\right| \leq 2 \quad \text{for all} \quad x \in [0, 1].$$

Therefore, ϕ satisfies the Lipschitz condition with Lipschitz constant $B = 2$.

Example 7.45 Let

$$\phi(x) = \begin{cases} \frac{1}{x} & \text{if } x \in (0, 1], \\ 0 & \text{if } x = 0. \end{cases}$$

We assume that the function ϕ satisfies the Lipschitz condition with Lipschitz constant B. Then for all $x \in (0, 1]$ and $y = 0$, we have

$$\frac{1}{x} \leq B,$$

which is a contradiction.

Exercise 7.46 Let

$$\phi(x) = \begin{cases} \frac{1}{x^2 - 1} & \text{if } x \in [0, 1), \\ 10 & \text{if } x = 1. \end{cases}$$

Check if ϕ satisfies the Lipschitz condition.

Solution No.

Theorem 7.47 *Let f be bounded and Δ-integrable over*

$$R = [a_1, b_1) \times [a_2, b_2) \times \ldots \times [a_n, b_n)$$

and let M and m be its supremum and infimum over R, respectively. If $\phi : [m, M] \to \mathbb{R}$ is a function satisfying the Lipschitz condition, then the composite function $h = \phi \circ f$ is Δ-integrable over R.

Proof Let $\varepsilon > 0$ be arbitrarily chosen. Since f is Δ-integrable over R, there exists $P \in \mathscr{P}(R)$ given by (7.1) and (7.2) such that

$$U(f, P) - L(f, P) < \frac{\varepsilon}{B},$$

where B is a Lipschitz constant for ϕ. Let $M_{j_1 j_2 \ldots j_n}$ and $m_{j_1 j_2 \ldots j_n}$ be the supremum and infimum of f on $R_{j_1 j_2 \ldots j_n}$, respectively, and let $M^*_{j_1 j_2 \ldots j_n}$ and $m^*_{j_1 j_2 \ldots j_n}$ be the corresponding numbers for h. Then, for every

$$(t_1^{j_1}, t_2^{j_2}, \ldots, t_n^{j_n}), \quad (t_1^{\prime j_1}, t_2^{\prime j_2}, \ldots, t_n^{\prime j_n}) \in R_{j_1 j_2 \ldots j_n},$$

we have

$$\begin{aligned} & h(t_1^{j_1}, t_2^{j_2}, \ldots, t_n^{j_n}) - h(t_1^{\prime j_1}, t_2^{\prime j_2}, \ldots, t_n^{\prime j_n}) \\ & \leq |h(t_1^{j_1}, t_2^{j_2}, \ldots, t_n^{j_n}) - h(t_1^{\prime j_1}, t_2^{\prime j_2}, \ldots, t_n^{\prime j_n})| \\ & = |\phi(h(t_1^{j_1}, t_2^{j_2}, \ldots, t_n^{j_n})) - \phi(h(t_1^{\prime j_1}, t_2^{\prime j_2}, \ldots, t_n^{\prime j_n}))| \\ & \leq B|f(t_1^{j_1}, t_2^{j_2}, \ldots, t_n^{j_n}) - f(t_1^{\prime j_1}, t_2^{\prime j_2}, \ldots, t_n^{\prime j_n})| \\ & \leq B(M_{j_1 j_2 \ldots j_n} - m_{j_1 j_2 \ldots j_n}). \end{aligned}$$

Hence,

$$M^*_{j_1 j_2 \ldots j_n} - m^*_{j_1 j_2 \ldots j_n} \leq B(M_{j_1 j_2 \ldots j_n} - m_{j_1 j_2 \ldots j_n})$$

because there exist two sequences

$$(t_{1p}^{j_1}, t_{2p}^{j_2}, \ldots, t_{np}^{j_n}), (t_{1p}^{\prime j_1}, t_{2p}^{\prime j_2}, \ldots, t_{np}^{\prime j_n}) \in R_{j_1 j_2 \ldots j_n}$$

such that

$$h(t_{1p}^{j_1}, t_{2p}^{j_2}, \ldots, t_{np}^{j_n}) \to M^*_{j_1 j_2 \ldots j_n}, \quad h(t_{1p}^{\prime j_1}, t_{2p}^{\prime j_2}, \ldots, t_{np}^{\prime j_n}) \to m^*_{j_1 j_2 \ldots j_n}$$

as $p \to \infty$. Consequently,

$$U(h,P) - L(h,P) = \sum_{j_1=1}^{k_1}\sum_{j_2=1}^{k_2}\dots\sum_{j_n=1}^{k_n}(M^*_{j_1 j_2 \dots j_n} - m^*_{j_1 j_2 \dots j_n})$$

$$\times (t_1^{j_1} - t_1^{j_1-1})(t_2^{j_2} - t_2^{j_2-1})\dots(t_n^{j_n} - t_n^{j_n-1})$$

$$\le B\sum_{j_1=1}^{k_1}\sum_{j_2=1}^{k_2}\dots\sum_{j_n=1}^{k_n}(M_{j_1 j_2 \dots j_n} - m_{j_1 j_2 \dots j_n})$$

$$\times (t_1^{j_1} - t_1^{j_1-1})(t_2^{j_2} - t_2^{j_2-1})\dots(t_n^{j_n} - t_n^{j_n-1})$$

$$= B(U(f,P) - L(f,P))$$

$$< \varepsilon.$$

By Theorem 7.15, h is Δ-integrable. □

Theorem 7.48 *Let f be a bounded function that is Δ-integrable over*

$$R = [a_1, b_1) \times [a_2, b_2) \times \dots \times [a_n, b_n).$$

If $a_i', b_i' \in [a_i, b_i]$ with $a_i' < b_i'$ for all $i \in \{1, 2, \dots, n\}$, then f is Δ-integrable over $R' = [a_1', b_1') \times [a_2', b_2') \times \dots \times [a_n', b_n')$.

Proof Let $\varepsilon > 0$ be arbitrarily chosen. Since f is Δ-integrable over R, there exists $P \in \mathscr{P}(R)$ given by (7.1) and (7.2) so that

$$U(f,P) - L(f,P) < \varepsilon.$$

Let $P' \in \mathscr{P}(R)$ be such that

$$P' = P \cup \{\{a_1', b_1'\} \times \{a_2', b_2'\} \times \dots \times \{a_n', b_n'\}\}.$$

Then P' is a refinement of P. Therefore,

$$L(f,P) \le L(f,P') \le U(f,P') \le U(f,P).$$

Hence,

$$U(f,P') - L(f,P') \le U(f,P) - L(f,P) < \varepsilon.$$

Now, consider $P'' \in \mathscr{P}(R)$ consisting of all subrectangles of P' belonging to R'. If $\tilde{U}$ and $\tilde{L}$ are the upper and lower Δ-sums of f on R' associated with the partition P'', then

$$\tilde{U} - \tilde{L} \leq U(f, P') - L(f, P') < \varepsilon.$$

Hence, by Theorem 7.15, f is Δ-integrable over R'. □

Theorem 7.49 *Let f be a bounded function that is Δ-integrable on*

$$R = [a_1, b_1) \times [a_2, b_2) \times \ldots \times [a_n, b_n).$$

If $\alpha \in \mathbb{R}$, then αf is Δ-integrable on R and

$$\int_R \alpha f(t_1, t_2, \ldots, t_n)\Delta_1 t_1 \Delta_2 t_2 \ldots \Delta_n t_n = \alpha \int_R f(t_1, t_2, \ldots, t_n)\Delta_1 t_1 \Delta_2 t_2 \ldots \Delta_n t_n. \tag{7.12}$$

Proof If $\alpha = 0$, then (7.12) is obvious as both sides of (7.12) are equal to zero in this case. Let $\varepsilon > 0$ be arbitrarily chosen. We assume $\alpha \neq 0$.

1. Let $\alpha > 0$. Since f is Δ-integrable over R, there exists $P \in \mathscr{P}(R)$ given by (7.1) and (7.2) so that

$$U(f, P) - L(f, P) < \frac{\varepsilon}{\alpha}.$$

Thus,

$$U(\alpha f, P) - L(\alpha f, P) = \alpha U(f, P) - \alpha L(f, P) < \varepsilon.$$

Hence, by Theorem 7.15, αf is Δ-integrable over R. Also, we have

$$\alpha L(f, P) = L(\alpha f, P) \leq U(\alpha f, P) = \alpha U(f, P),$$

whereupon

$$\alpha L(f) = L(\alpha f) \leq U(\alpha f) = \alpha U(f).$$

From here, using that $L(f) = U(f)$, we conclude (7.12).

2. Let $\alpha < 0$. Since f is Δ-integrable over R, there exists $P \in \mathscr{P}(R)$ such that

$$\frac{\varepsilon}{\alpha} < U(f, P) - L(f, P) < -\frac{\varepsilon}{\alpha}.$$

Thus,

$$U(\alpha f, P) - L(\alpha f, P) \leq -\alpha(U(f, P) - L(f, P)) < \varepsilon,$$

and hence, by Theorem 7.15, αf is Δ-integrable over R. As in the previous case, we get (7.12).

The proof is complete. □

Theorem 7.50 *If f and g are bounded functions that are Δ-integrable over*

$$R = [a_1, b_1) \times [a_2, b_2) \times \ldots \times [a_n, b_n),$$

then $f + g$ is Δ-integrable over R and

$$\begin{aligned}&\int_R (f+g)(t_1, t_2, \ldots, t_n)\Delta_1 t_1 \Delta_2 t_2 \ldots \Delta_n t_n \\ &= \int_R f(t_1, t_2, \ldots, t_n)\Delta_1 t_1 \Delta_2 t_2 \ldots \Delta_n t_n + \int_R g(t_1, t_2, \ldots, t_n)\Delta_1 t_1 \Delta_2 t_2 \ldots \Delta_n t_n.\end{aligned} \tag{7.13}$$

Proof Let $\varepsilon > 0$ be arbitrarily chosen. Since f and g are Δ-integrable over R, there exists $P \in \mathscr{P}(R)$ given by (7.1) and (7.2) such that

$$U(f, P) - L(f, P) < \frac{\varepsilon}{2} \quad \text{and} \quad U(g, P) - L(g, P) < \frac{\varepsilon}{2}.$$

Because

$$U(f+g, P) \leq U(f, P) + U(g, P) \quad \text{and} \quad L(f+g, P) \geq L(f, P) + L(g, P), \tag{7.14}$$

we find

$$\begin{aligned}U(f+g, P) - L(f+g, P) &\leq U(f, P) + U(g, P) - L(f, P) - L(g, P) \\ &= U(f, P) - L(f, P) + U(g, P) - L(g, P) \\ &< \frac{\varepsilon}{2} + \frac{\varepsilon}{2} \\ &= \varepsilon.\end{aligned}$$

Hence, by Theorem 7.15, it follows that $f + g$ is Δ-integrable over R. From (7.14), we get

$$U(f+g) \leq U(f) + U(g) \quad \text{and} \quad L(f+g) \geq L(f) + L(g),$$

whereupon

$$L(f+g) \leq U(f+g) \leq U(f) + U(g) = L(f) + L(g) \leq L(f+g),$$

i.e., (7.13) holds. □

Corollary 7.51 *If f and g are bounded Δ-integrable over*

$$R = [a_1, b_1) \times [a_2, b_2) \times \ldots \times [a_n, b_n)$$

and $\alpha, \beta \in \mathbb{R}$, then $\alpha f + \beta g$ is Δ-integrable over R and

$$\int_R (\alpha f + \beta g)(t_1, t_2, \ldots, t_n)\Delta_1 t_1 \Delta_2 t_2 \ldots \Delta_n t_n$$
$$= \alpha \int_R f(t_1, t_2, \ldots, t_n)\Delta_1 t_1 \Delta_2 t_2 \ldots \Delta_n t_n + \beta \int_R g(t_1, t_2, \ldots, t_n)\Delta_1 t_1 \Delta_2 t_2 \ldots \Delta_n t_n.$$

Proof Since f and g are Δ-integrable over R, by Theorem 7.49, we get that αf and βg are Δ-integrable over R and

$$\int_R \alpha f(t_1, t_2, \ldots, t_n)\Delta_1 t_1 \Delta_2 t_2 \ldots \Delta_n t_n = \alpha \int_R f(t_1, t_2, \ldots, t_n)\Delta_1 t_1 \Delta_2 t_2 \ldots \Delta_n t_n,$$

$$\int_R \beta g(t_1, t_2, \ldots, t_n)\Delta_1 t_1 \Delta_2 t_2 \ldots \Delta_n t_n = \beta \int_R g(t_1, t_2, \ldots, t_n)\Delta_1 t_1 \Delta_2 t_2 \ldots \Delta_n t_n.$$

From here and from Theorem 7.50, we find that $\alpha f + \beta g$ is Δ-integrable over R and

$$\int_R (\alpha f + \beta g)(t_1, t_2, \ldots, t_n)\Delta_1 t_1 \Delta_2 t_2 \ldots \Delta_n t_n$$

$$= \int_R \alpha f(t_1, t_2, \ldots, t_n)\Delta_1 t_1 \Delta_2 t_2 \ldots \Delta_n t_n + \int_R \beta g(t_1, t_2, \ldots, t_n)\Delta_1 \Delta_2 t_2 \ldots \Delta_n t_n$$

$$= \alpha \int_R f(t_1, t_2, \ldots, t_n)\Delta_1 t_1 \Delta_2 t_2 \ldots \Delta_n t_n + \beta \int_R g(t_1, t_2, \ldots, t_n)\Delta_1 t_1 \Delta_2 t_2 \ldots \Delta_n t_n,$$

completing the proof. □

Theorem 7.52 *If f and g are bounded functions that are Δ-integrable over R with*

$$f(t_1, t_2, \ldots, t_n) \leq g(t_1, t_2, \ldots, t_n) \text{ for all } (t_1, t_2, \ldots, t_n) \in R,$$

then

$$\int_R f(t_1, t_2, \ldots, t_n)\Delta_1 t_1 \Delta_2 t_2 \ldots \Delta_n t_n \leq \int_R g(t_1, t_2, \ldots, t_n)\Delta_1 t_1 \Delta_2 t_2 \ldots \Delta_n t_n.$$

Proof By Corollary 7.51, we have that $g - f$ is Δ-integrable over R. Since $g - f$ is nonnegative on R, we have

$$L(g-f,P)\geq 0 \quad \text{for all} \quad P\in\mathscr{P}(R).$$

Hence, by Corollary 7.51, we get

$$\begin{aligned}0 &\leq L(g-f,P)\\ &\leq \int_R (g(t_1,t_2,\ldots,t_n)-f(t_1,t_2,\ldots,t_n))\Delta_1 t_1\Delta_2 t_2\ldots\Delta_n t_n\\ &= \int_R g(t_1,t_2,\ldots,t_n)\Delta_1 t_1\Delta_2 t_2\ldots\Delta_n t_n - \int_R f(t_1,t_2,\ldots,t_n)\Delta_1 t_1\Delta_2 t_2\ldots\Delta_n t_n,\end{aligned}$$

which completes the proof. □

Theorem 7.53 *If f is a bounded function that is Δ-integrable over R, then so is $|f|$, and*

$$\left|\int_R f(t_1,t_2,\ldots,t_n)\Delta_1 t_1\Delta_2 t_2\ldots\Delta_n t_n\right| \leq \int_R |f(t_1,t_2,\ldots,t_n)|\Delta_1 t_1\Delta_2 t_2\ldots\Delta_n t_n.$$

Proof Let $\varepsilon>0$ be arbitrarily chosen. Since f is Δ-integrable on R, there exists $P\in\mathscr{P}(R)$ given by (7.1) and (7.2) so that

$$U(f,P)-L(f,P)<\varepsilon.$$

Let

$$\overline{M}_{j_1 j_2\ldots j_n} = \sup\{|f(t_1,t_2,\ldots,t_n)| : (t_1,t_2,\ldots,t_n)\in R_{j_1 j_2\ldots j_n}\},$$

$$\overline{m}_{j_1 j_2\ldots j_n} = \inf\{|f(t_1,t_2,\ldots,t_n)| : (t_1,t_2,\ldots,t_n)\in R_{j_1 j_2\ldots j_n}\}.$$

Thus,

$$\begin{aligned}&\overline{M}_{j_1 j_2\ldots j_n} - \overline{m}_{j_1 j_2\ldots j_n}\\ &= \sup\{|f(t_1,t_2,\ldots,t_n)|-|f(t_1',t_2',\ldots,t_n')| : (t_1,t_2,\ldots,t_n),(t_1',t_2',\ldots,t_n')\in R_{j_1 j_2\ldots j_n}\}\\ &\leq \sup\{|f(t_1,t_2,\ldots,t_n)-f(t_1',t_2',\ldots,t_n')| : (t_1,t_2,\ldots,t_n),(t_1',t_2',\ldots,t_n')\in R_{j_1 j_2\ldots j_n}\}\\ &= M_{j_1 j_2\ldots j_n} - m_{j_1 j_2\ldots j_n}.\end{aligned}$$

Therefore,

$$U(|f|,P)-L(|f|,P)\leq U(f,P)-L(f,P)<\varepsilon.$$

Hence, using Theorem 7.15, we get that $|f|$ is Δ-integrable over R. Since

$$-f(t_1, t_2, \ldots, t_n) \le |f(t_1, t_2, \ldots, t_n)| \quad \text{and} \quad f(t_1, t_2, \ldots, t_n) \le |f(t_1, t_2, \ldots, t_n)|$$

for all $(t_1, t_2, \ldots, t_n) \in R$, using Theorem 7.52, we get

$$-\int_R f(t_1, t_2, \ldots, t_n)\Delta_1 t_1 \Delta_2 t_2 \ldots \Delta_n t_n \le \int_R |f(t_1, t_2, \ldots, t_n)|\Delta_1 t_1 \Delta_2 t_2 \ldots \Delta_n t_n$$

and

$$\int_R f(t_1, t_2, \ldots, t_n)\Delta_1 t_1 \Delta_2 t_2 \ldots \Delta_n t_n \le \int_R |f(t_1, t_2, \ldots, t_n)|\Delta_1 t_1 \Delta_2 t_2 \ldots \Delta_n t_n,$$

which completes the proof. □

Theorem 7.54 *If f is a bounded function that is Δ-integrable over R, then so is f^2.*

Proof Since $f^2 = |f||f|$, without loss of generality, we assume that f is nonnegative over R. Let

$$M_f = \sup\{f(t_1, t_2, \ldots, t_n) : (t_1, t_2, \ldots, t_n) \in R\}.$$

Let $\varepsilon > 0$ be arbitrarily chosen. Since f is Δ-integrable over R, there exists $P \in \mathscr{P}(R)$ given by (7.1) and (7.2) so that

$$U(f, P) - L(f, P) < \frac{\varepsilon}{2M_f + 1}.$$

Hence,

$$\begin{aligned}
&U(f^2, P) - L(f^2, P) \\
&= \sum_{j_1=1}^{k_1}\sum_{j_2=1}^{k_2}\cdots\sum_{j_n=1}^{k_n} M^2_{j_1 j_2 \ldots j_n}(t_1^{j_1} - t_1^{j_1-1})(t_2^{j_2} - t_2^{j_2-1})\ldots(t_n^{j_n} - t_n^{j_n-1}) \\
&\quad - \sum_{j_1=1}^{k_1}\sum_{j_2=1}^{k_2}\cdots\sum_{j_n=1}^{k_n} m^2_{j_1 j_2 \ldots j_n}(t_1^{j_1} - t_1^{j_1-1})(t_2^{j_2} - t_2^{j_2-1})\ldots(t_n^{j_n} - t_n^{j_n-1}) \\
&= \sum_{j_1=1}^{k_1}\sum_{j_2=1}^{k_2}\cdots\sum_{j_n=1}^{k_n} (M_{j_1 j_2 \ldots j_n} + m_{j_1 j_2 \ldots j_n})(M_{j_1 j_2 \ldots j_n} - m_{j_1 j_2 \ldots j_n}) \\
&\quad \times (t_1^{j_1} - t_1^{j_1-1})(t_2^{j_2} - t_2^{j_2-1})\ldots(t_n^{j_n} - t_n^{j_n-1})
\end{aligned}$$

$$\leq 2M_f \sum_{j_1=1}^{k_1}\sum_{j_2=1}^{k_2}\cdots\sum_{j_n=1}^{k_n}(M_{j_1 j_2 \ldots j_n} - m_{j_1 j_2 \ldots j_n})$$

$$\times (t_1^{j_1} - t_1^{j_1-1})(t_2^{j_2} - t_2^{j_2-1})\ldots(t_n^{j_n} - t_n^{j_n-1})$$

$$= 2M_f(U(f, P) - L(f, P))$$

$$< 2M_f \frac{\varepsilon}{2M_f + 1}$$

$$< \varepsilon,$$

which completes the proof. □

Theorem 7.55 *If f and g are Δ-integrable over R, then so is fg.*

Proof We have

$$fg = \frac{1}{4}\left((f+g)^2 - (f-g)^2\right).$$

Since f and g are Δ-integrable over R, by Theorem 7.50, we get that $f + g$ is Δ-integrable over R, and by Corollary 7.51, $f - g$ is Δ-integrable over R. Hence, using Theorem 7.54, we find that $(f + g)^2$ and $(f - g)^2$ are Δ-integrable over R. From here and from Corollary 7.51, we conclude that fg is Δ-integrable over R. □

Theorem 7.56 *Let the rectangle R be the union of two disjoint rectangles R_1 and R_2. If f is a bounded function that is Δ-integrable on each of R_1 and R_2, then f is Δ-integrable over R and*

$$\int_R f(t_1, t_2, \ldots, t_n)\Delta_1 t_1 \Delta_2 t_2 \ldots \Delta_n t_n = \int_{R_1} f(t_1, t_2, \ldots, t_n)\Delta_1 t_1 \Delta_2 t_2 \ldots \Delta_n t_n$$

$$+ \int_{R_2} f(t_1, t_2, \ldots, t_n)\Delta_1 t_1 \Delta_2 t_2 \ldots \Delta_n t_n.$$

Proof Let $\varepsilon > 0$ be arbitrarily chosen. Since f is Δ-integrable over R_1 and R_2, there exist $P_1 \in \mathscr{P}(R_1)$ and $P_2 \in \mathscr{P}(R_2)$ given by (7.1) and (7.2) so that

$$U(f, P_1) - L(f, P_1) < \frac{\varepsilon}{2} \quad \text{and} \quad U(f, P_2) - L(f, P_2) < \frac{\varepsilon}{2}.$$

We note that $P_1 \cup P_2 \in \mathscr{P}(R)$. Thus,

$$\begin{aligned} U(f, P_1 \cup P_2) - L(f, P_1 \cup P_2) &\leq U(f, P_1) + U(f, P_2) - L(f, P_1) - L(f, P_2) \\ &= U(f, P_1) - L(f, P_1) + U(f, P_2) - L(f, P_2) \\ &< \frac{\varepsilon}{2} + \frac{\varepsilon}{2} \\ &= \varepsilon. \end{aligned}$$

Hence, by Theorem 7.15, we conclude that f is Δ-integrable over R. From

$$\begin{aligned} L(f, P_3) + L(f, P_4) &\leq L(f, P_3 \cup P_4) \leq U(f, P_3 \cup P_4) \\ &\leq U(f, P_3) + U(f, P_4) \end{aligned}$$

for all $P_3 \in \mathscr{P}(R_1)$ and $P_4 \in \mathscr{P}(R_2)$, we get (7.13). □

Theorem 7.57 (Mean Value Theorem) *Let f and g be bounded functions that are Δ-integrable on R, and let g be nonnegative (or nonpositive) on R. If*

$$m = \inf\{f(t_1, t_2, \ldots, t_n) : (t_1, t_2, \ldots, t_n) \in R\}$$

and

$$M = \sup\{f(t_1, t_2, \ldots, t_n) : (t_1, t_2, \ldots, t_n) \in R\},$$

then there exists a real number $\Lambda \in [m, M]$ such that

$$\begin{aligned} &\int_R f(t_1, t_2, \ldots, t_n) g(t_1, t_2, \ldots, t_n) \Delta_1 t_1 \Delta_2 t_2 \ldots \Delta_n t_n \\ &\qquad = \Lambda \int_R g(t_1, t_2, \ldots, t_n) \Delta_1 t_1 \Delta_2 t_2 \ldots \Delta_n t_n. \end{aligned} \tag{7.15}$$

Proof We consider the case when g is nonnegative on R. The case when g is nonpositive on R is left to the reader. Since g is nonnegative on R, we have

$$m g(t_1, t_2, \ldots, t_n) \leq f(t_1, t_2, \ldots, t_n) g(t_1, t_2, \ldots, t_n) \leq M g(t_1, t_2, \ldots, t_n)$$

and

$$\int_R g(t_1, t_2, \ldots, t_n) \Delta_1 t_1 \Delta_2 t_2 \ldots \Delta_n t_n \geq 0.$$

We note that if $\int_R g(t_1, t_2, \ldots, t_n) \Delta_1 t_1 \Delta_2 t_2 \ldots \Delta_n t_n = 0$, then (7.15) holds for any $\Lambda \in [m, M]$. We assume that $\int_R g(t_1, t_2, \ldots, t_n) \Delta_1 t_1 \Delta_2 t_2 \ldots \Delta_n t_n > 0$. Hence, using Theorem 7.52, we get

$$m \int_R g(t_1, t_2, \ldots, t_n)\Delta_1 t_1 \Delta_2 t_2 \ldots \Delta_n t_n$$

$$\leq \int_R f(t_1, t_2, \ldots, t_n) g(t_1, t_2, \ldots, t_n)\Delta_1 t_1 \Delta_2 t_2 \ldots \Delta_n t_n$$

$$\leq M \int_R g(t_1, t_2, \ldots, t_n)\Delta_1 t_1 \Delta_2 t_2 \ldots \Delta_n t_n,$$

whereupon

$$m \leq \frac{\int_R f(t_1, t_2, \ldots, t_n) g(t_1, t_2, \ldots, t_n)\Delta_1 t_1 \Delta_2 t_2 \ldots \Delta_n t_n}{\int_R g(t_1, t_2, \ldots, t_n)\Delta_1 t_1 \Delta_2 t_2 \ldots \Delta_n t_n} \leq M.$$

Letting

$$\Lambda = \frac{\int_R f(t_1, t_2, \ldots, t_n) g(t_1, t_2, \ldots, t_n)\Delta_1 t_1 \Delta_2 t_2 \ldots \Delta_n t_n}{\int_R g(t_1, t_2, \ldots, t_n)\Delta_1 t_1 \Delta_2 t_2 \ldots \Delta_n t_n}$$

completes the proof. □

Theorem 7.58 *Let f be a bounded function that is Δ-integrable over $R = [a_1, b_1) \times [a_2, b_2)$ and assume that the single integral*

$$I(t_1) = \int_{a_2}^{b_2} f(t_1, t_2)\Delta_2 t_2 \tag{7.16}$$

exists for each $t_1 \in [a_1, b_1)$. Then the iterated integral

$$\int_{a_1}^{b_1} I(t_1)\Delta_1 t_1$$

exists and the equality

$$\int_R f(t_1, t_2)\Delta_1 t_1 \Delta_2 t_2 = \int_{a_1}^{b_1} \Delta_1 t_1 \int_{a_2}^{b_2} f(t_1, t_2)\Delta_2 t_2 \tag{7.17}$$

holds.

Proof Let $P \in \mathscr{P}(R)$ be given by (7.1) and (7.2) in the case $n = 2$. We have

$$m_{j_1 j_2} \leq f(t_1, t_2) \leq M_{j_1 j_2} \quad \text{on} \quad R_{j_1 j_2}. \tag{7.18}$$

Let $\xi_{j_1} \in [t_1^{j_1 - 1}, t_1^{j_1})$ for some $j_1 = 1, 2, \ldots, k_1$. We set $t_1 = \xi_{j_1}$ in (7.18). Thus,

$$m_{j_1 j_2} \le f(\xi_{j_1}, t_2) \le M_{j_1 j_2}.$$

We integrate this inequality from $t_2^{j_2-1}$ to $t_2^{j_2}$ for some $j_2 \in \{1, 2, \dots, k_2\}$ to obtain

$$m_{j_1 j_2}(t_2^{j_2} - t_2^{j_2-1}) \le \int_{t_2^{j_2-1}}^{t_2^{j_2}} f(\xi_{j_1}, t_2)\Delta_2 t_2 \le M_{j_1 j_2}(t_2^{j_2} - t_2^{j_2-1}). \tag{7.19}$$

We note that the integral in (7.19) exists because the existence of the integral in (7.16) is assumed over the entire interval $[a_2, b_2)$. Now, we multiply (7.19) by $t_1^{j_1} - t_1^{j_1-1}$ and get

$$\begin{aligned} m_{j_1 j_2}(t_1^{j_1} - t_1^{j_1-1})(t_2^{j_2} - t_2^{j_2-1}) &\le (t_1^{j_1} - t_1^{j_1-1}) \int_{t_2^{j_2-1}}^{t_2^{j_2}} f(\xi_{j_1}, t_2)\Delta_2 t_2 \\ &\le M_{j_1 j_2}(t_1^{j_1} - t_1^{j_1-1})(t_2^{j_2} - t_2^{j_2-1}), \end{aligned}$$

whereupon

$$\begin{aligned} &\sum_{j_1=1}^{k_1}\sum_{j_2=1}^{k_2} m_{j_1 j_2}(t_1^{j_1} - t_1^{j_1-1})(t_2^{j_2} - t_2^{j_2-1}) \\ &\le \sum_{j_1=1}^{k_1}\sum_{j_2=1}^{k_2} (t_1^{j_1} - t_1^{j_1-1}) \int_{t_2^{j_2-1}}^{t_2^{j_2}} f(\xi_{j_1}, t_2)\Delta_2 t_2 \\ &\le \sum_{j_1=1}^{k_1}\sum_{j_2=1}^{k_2} M_{j_1 j_2}(t_1^{j_1} - t_1^{j_1-1})(t_2^{j_2} - t_2^{j_2-1}), \end{aligned}$$

i.e.,

$$L(f, P) \le \sum_{j_1=1}^{k_1} (t_1^{j_1} - t_1^{j_1-1}) I(\xi_{j_1}) \le U(f, P). \tag{7.20}$$

Since f is Δ-integrable over R, we have

$$L(f, P) \le \int_R f(t_1, t_2)\Delta_1 t_1 \Delta_2 t_2 \le U(f, P). \tag{7.21}$$

By Theorem 7.19 and (7.21), there exists $\delta > 0$ such that $P \in \mathscr{P}_\delta(R)$ (in the case $n = 2$) implies

$$U(f, P) - L(f, P) < \frac{\varepsilon}{3},$$

$$\left|L(f, P) - \int_R f(t_1, t_2)\Delta_2 t_2\right| < \frac{\varepsilon}{3},$$

and

$$\left|U(f, P) - \int_R f(t_1, t_2)\Delta_1 t_1 \Delta_2 t_2\right| < \frac{\varepsilon}{3}.$$

For such a partition P, we have

$$U(f, P) - \sum_{j_1=1}^{k_1} (t_1^{j_1} - t_1^{j_1-1}) I(\xi_{j_1}) < \frac{\varepsilon}{3}$$

and

$$\sum_{j_1=1}^{k_1} (t_1^{j_1} - t_1^{j_1-1}) I(\xi_{j_1}) - L(f, P) < \frac{\varepsilon}{3}.$$

Thus,

$$\left|\sum_{j_1=1}^{k_1} (t_1^{j_1} - t_1^{j_1-1}) I(\xi_{j_1}) - \int_R f(t_1, t_2)\Delta_1 t_1 \Delta_2 t_2\right|$$

$$= \left|\sum_{j_1=1}^{k_1} (t_1^{j_1} - t_1^{j_1-1}) I(\xi_{j_1}) - L(f, P) + L(f, P) - U(f, P)\right.$$

$$\left. + U(f, P) - \int_R f(t_1, t_2)\Delta_1 t_1 \Delta_2 t_1\right|$$

$$\leq \left|\sum_{j_1=1}^{k_1} (t_1^{j_1} - t_1^{j_1-1}) I(\xi_{j_1}) - L(f, P)\right|$$

$$+ U(f, P) - L(f, P)$$

$$+\left|U(f,P)-\int_R f(t_1,t_2)\Delta_1 t_1\Delta_2 t_2\right|$$

$$<\frac{\varepsilon}{3}+\frac{\varepsilon}{3}+\frac{\varepsilon}{3}$$

$$=\varepsilon.$$

This means, by the Riemann definition of the single integral, that the function I is Δ-integrable from a_1 to b_1 and (7.17) holds. □

As above, one can prove the following result.

Theorem 7.59 *Let f be a bounded function that is Δ-integrable over $R=[a_1,b_1)\times[a_2,b_2)$ and assume that the single integral*

$$I(t_2)=\int_{a_1}^{b_1} f(t_1,t_2)\Delta_1 t_1 \tag{7.22}$$

exists for each $t_2\in[a_2,b_2)$. Then the iterated integral

$$\int_{a_2}^{b_2} I(t_2)\Delta_2 t_2$$

exists and the equality

$$\int_R f(t_1,t_2)\Delta_1 t_1\Delta_2 t_2=\int_{a_2}^{b_2}\Delta_2 t_2\int_{a_1}^{b_1} f(t_1,t_2)\Delta_1 t_1 \tag{7.23}$$

holds.

Remark 7.60 If, together with the double integral $\int_R f(t_1,t_2)\Delta_1 t_1\Delta_2 t_2$, there exist both integrals (7.16) and (7.22), then the formulas (7.17) and (7.23) hold simultaneously, i.e.,

$$\int_R f(t_1,t_2)\Delta_1 t_1\Delta_2 t_2=\int_{a_1}^{b_1}\Delta_1 t_1\int_{a_2}^{b_2} f(t_1,t_2)\Delta_2 t_2=\int_{a_2}^{b_2}\Delta_2 t_2\int_{a_1}^{b_1} f(t_1,t_2)\Delta_1 t_1.$$

Example 7.61 Let $\mathbb{T}_1=3^{\mathbb{N}_0}$ and $\mathbb{T}_2=2^{\mathbb{N}_0}$. Consider the integral

$$I=\int_2^4\int_1^3 (t_1^2+t_1t_2+t_2^2)\Delta_1 t_1\Delta_2 t_2.$$

We set

$$I_1(t_1) = \int_2^4 (t_1^2 + t_1 t_2 + t_2^2)\Delta_2 t_2, \quad I_2(t_2) = \int_1^3 (t_1^2 + t_1 t_2 + t_2^2)\Delta_1 t_1.$$

Here,

$$\sigma_1(t_1) = 3t_1, \quad t_1 \in \mathbb{T}_1, \quad \sigma_2(t_2) = 2t_2, \quad t_2 \in \mathbb{T}_2.$$

Thus, for $g_2(t_1) = t_1^2$, $g_3(t_1) = t_1^3$, $h_2(t_2) = t_2^2$, $h_3(t_2) = t_2^3$, we have

$$\begin{aligned} g_2^{\Delta_1}(t_1) &= \sigma_1(t_1) + t_1 \\ &= 3t_1 + t_1 \\ &= 4t_1, \\ g_3^{\Delta_1}(t_1) &= \sigma_1^2(t_1) + t_1\sigma_1(t_1) + t_1^2 \\ &= 9t_1^2 + 3t_1^2 + t_1^2 \\ &= 13t_1^2, \\ h_2^{\Delta_2}(t_2) &= \sigma_2(t_2) + t_2 \\ &= 2t_2 + t_2 \\ &= 3t_2, \\ h_3^{\Delta_2}(t_2) &= \sigma_2^2(t_2) + t_2\sigma_2(t_2) + t_2^2 \\ &= 4t_2^2 + 2t_2^2 + t_2^2 \\ &= 7t_2^2. \end{aligned}$$

Hence,

$$t_1 = \frac{1}{4} g_2^{\Delta_1}(t_1), \quad t_1^2 = \frac{1}{13} g_3^{\Delta_1}(t_1),$$

$$t_2 = \frac{1}{3} h_2^{\Delta_2}(t_2), \quad t_2^2 = \frac{1}{7} h_3^{\Delta_2}(t_2).$$

Therefore,

$$I_1(t_1) = \int_2^4 \left(t_1^2 + \frac{1}{3}t_1 h_2^{\Delta_2}(t_2) + \frac{1}{7}h_3^{\Delta_2}(t_2)\right)\Delta_2 t_2$$

$$= 2t_1^2 + \frac{1}{3}t_1(h_2(4) - h_2(1)) + \frac{1}{7}(h_3(4) - h_3(2))$$

$$= 2t_1^2 + 4t_1 + 8$$

$$= \frac{2}{13}g_3^{\Delta_1}(t_1) + g_2^{\Delta_1}(t_1) + 8,$$

$$I = \int_1^3 \left(\frac{2}{13}g_3^{\Delta_1}(t_1) + g_2^{\Delta_1}(t_1) + 8\right)\Delta_1 t_1$$

$$= \frac{2}{13}(g_3(3) - g_3(1)) + g_2(3) - g_2(1) + 16$$

$$= 4 + 8 + 16$$

$$= 28.$$

Also,

$$I_2(t_2) = \int_1^3 \left(\frac{1}{13}g_3^{\Delta_1}(t_1) + \frac{1}{4}g_2^{\Delta_1}(t_1)t_2 + t_2^2\right)\Delta_1 t_1$$

$$= \frac{1}{13}(g_3(3) - g_3(1)) + \frac{1}{4}t_2(g_2(3) - g_2(1)) + 2t_2^2$$

$$= 2 + 2t_2 + 2t_2^2$$

$$= 2 + \frac{2}{3}h_2^{\Delta_2}(t_2) + \frac{2}{7}h_3^{\Delta_2}(t_2),$$

$$I = \int_2^4 \left(2 + \frac{2}{3}h_2^{\Delta_2}(t_2) + \frac{2}{7}h_3^{\Delta_2}(t_2)\right)\Delta_2 t_2$$

$$= 4 + \frac{2}{3}(h_2(4) - h_2(2)) + \frac{2}{7}(h_3(4) - h_3(2))$$

$$= 4 + 8 + 16$$

$$= 28.$$

Consequently,

$$\int_2^4 \int_1^3 (t_1^2 + t_1t_2 + t_2^2)\Delta_1 t_1 \Delta_2 t_2 = \int_2^4 \Delta_2 t_2 \int_1^3 (t_1^2 + t_1t_2 + t_2^2)\Delta_1 t_1$$

$$= \int_1^3 \Delta_1 t_1 \int_2^4 (t_1^2 + t_1t_2 + t_2^2)\Delta_2 t_2.$$

Example 7.62 Let $\mathbb{T}_1 = \mathbb{N}$ and $\mathbb{T}_2 = 2^{\mathbb{N}_0}$. Let us consider the integral

$$I = \int_1^4 \int_1^2 (t_1 + t_2^2)\Delta_1 t_1 \Delta_2 t_2.$$

We set

$$I_1(t_1) = \int_1^4 (t_1 + t_2^2)\Delta_2 t_2, \quad I_2(t_2) = \int_1^2 (t_1 + t_2^2)\Delta_1 t_1.$$

Here,

$$\sigma_1(t_1) = t_1 + 1, \quad t_1 \in \mathbb{T}_1, \quad \sigma_2(t_2) = 2t_2, \quad t_2 \in \mathbb{T}_2.$$

Thus, using $g(t_1) = t_1^2$ and $h(t_2) = t_2^3$, we get

$$\begin{aligned} g^{\Delta_1}(t_1) &= \sigma_1(t_1) + t_1 \\ &= t_1 + 1 + t_1 \\ &= 2t_1 + 1, \\ h^{\Delta_2}(t_2) &= (\sigma_2(t_2))^2 + t_2\sigma_2(t_2) + t_2^2 \\ &= 4t_2^2 + 2t_2^2 + t_2^2 \\ &= 7t_2^2. \end{aligned}$$

Hence,

$$t_1 = \frac{1}{2}g^{\Delta_1}(t_1) - \frac{1}{2}, \quad t_2^2 = \frac{1}{7}h^{\Delta_2}(t_2).$$

Therefore,

$$
\begin{aligned}
I_1(t_1) &= \int_1^4 \left(t_1 + \frac{1}{7}h^{\Delta_2}(t_2)\right) \Delta_2 t_2 \\
&= 3t_1 + \frac{1}{7}(h(4) - h(1)) \\
&= 3t_1 + 9 \\
&= \frac{3}{2}g^{\Delta_1}(t_1) - \frac{3}{2} + 9 \\
&= \frac{3}{2}g^{\Delta_1}(t_1) + \frac{15}{2}, \\
I &= \int_1^2 \left(\frac{3}{2}g^{\Delta_1}(t_1) + \frac{15}{2}\right) \Delta_1 t_1 \\
&= \frac{3}{2}(g(2) - g(1)) + \frac{15}{2} \\
&= \frac{9}{2} + \frac{15}{2} \\
&= 12.
\end{aligned}
$$

Also,

$$
\begin{aligned}
I_2(t_2) &= \int_1^2 \left(\frac{1}{2}g^{\Delta_1}(t_1) - \frac{1}{2} + t_2^2\right) \Delta_1 t_1 \\
&= \frac{1}{2}(g(2) - g(1)) - \frac{1}{2} + t_2^2 \\
&= \frac{3}{2} - \frac{1}{2} + t_2^2 \\
&= 1 + \frac{1}{7}h^{\Delta_2}(t_2), \\
I &= \int_1^4 \left(1 + \frac{1}{7}h^{\Delta_2}(t_2)\right) \Delta_2 t_2 \\
&= 3 + \frac{1}{7}(h(4) - h(1))
\end{aligned}
$$

$$= 3 + 9$$

$$= 12.$$

Consequently,

$$\int_1^4 \int_1^2 (t_1 + t_2^2)\Delta_1 t_1 \Delta_2 t_2 = \int_1^4 \Delta_2 t_2 \int_1^2 (t_1 + t_2^2)\Delta_1 t_1$$
$$= \int_1^2 \Delta_1 t_1 \int_1^4 (t_1 + t_2^2)\Delta_2 t_2.$$

Exercise 7.63 Let $\mathbb{T}_1 = \mathbb{Z}$ and $\mathbb{T}_2 = 4^{\mathbb{N}_0}$. Prove that

$$\int_1^{16} \int_0^2 (t_1^3 + t_1 t_2 + t_2^3)\Delta_1 t_1 \Delta_2 t_2 = \int_1^{16} \Delta_2 t_2 \int_0^2 (t_1^3 + t_1 t_2 + t_2^3)\Delta_1 t_1$$
$$= \int_0^2 \Delta_1 t_1 \int_1^{16} (t_1^3 + t_1 t_2 + t_2^3)\Delta_2 t_2.$$

7.3 Multiple Integration over more General Sets

In this section, we extend the definition for the multiple Riemann Δ-integral over more general sets in Λ^n, called *Jordan Δ-measurable sets.*

Definition 7.64 Let $E \subset \Lambda^n$. A point $t = (t_1, t_2, \ldots, t_n) \in \Lambda^n$ is called a *boundary point* of E if every open ball $B(t, r) = \{x \in \Lambda^n : d(t, x) < r\}$ of radius r with center t contains at least one point of $E \setminus \{t\}$ and at least one point of $\Lambda^n \setminus E$. The set of all boundary points of E is called the *boundary* of E and is denoted by ∂E.

Definition 7.65 Let $E \subset \Lambda^n$. A point $t = (t_1, t_2, \ldots, t_n) \in \Lambda^n$ is called a *Δ-boundary point* of E if every rectangle of the form

$$V = [t_1, t_1') \times [t_2, t_2') \times \ldots \times [t_n, t_n') \subset \Lambda^n$$

with $t_i' \in \mathbb{T}_i$, $t_1' > t_i$, $i = 1, 2, \ldots, n$, contains at least one point of E and at least one point of $\Lambda^n \setminus E$. The set of all Δ-boundary points of E is called the *Δ-boundary* and is denoted by $\partial_\Delta E$.

For $i \in \{1, 2, \ldots, n\}$, we set $\mathbb{T}_i^0 = \mathbb{T}_i \setminus \{\max \mathbb{T}_i\}$. If $\mathbb{T}_i$ does not have a maximum, then $\mathbb{T}_i^0 = \mathbb{T}_i$, $i \in \{1, 2, \ldots, n\}$. Evidently, for every point $t_i \in \mathbb{T}_i^0$, there exists an interval $[\alpha_i, \beta_i) \subset \mathbb{T}_i$ that contains the point t_i, $i \in \{1, 2, \ldots, n\}$.

Definition 7.66 A point $t^0 = (t_1^0, t_2^0, \ldots, t_n^0) \in \Lambda^n$ is called *Δ-dense* if every rectangle of the form

$$V = [t_1^0, t_1) \times [t_2^0, t_2) \times \ldots \times [t_n^0, t_n) \subset \Lambda^n$$

with $t_i \in \mathbb{T}_i$, $t_i > t_i^0$, $i \in \{1, 2, \ldots, n\}$, contains at least one point distinct from t^0. Otherwise, the point t^0 is called *Δ-scattered.*

Remark 7.67 In fact, every Δ-dense point is a dense point in Λ^n and every Δ-scattered point is a scattered point in Λ^n.

Example 7.68 Let R be a rectangle in Λ^n. Then $\partial_\Delta R = \emptyset$.

Example 7.69 If $\mathbb{T}_i = \mathbb{Z}$, $i \in \{1, 2, \ldots, n\}$, then any set $E \subset \Lambda^n$ has neither any boundary nor any Δ-boundary points.

Example 7.70 Let $a, b, c, d \in \mathbb{R}$, $a < b$, $c < d$. Let

$$E_1 = [a, b) \times [c, d), \quad E_2 = (a, b] \times (c, d], \quad E_3 = [a, b] \times [c, d].$$

Here,

$$\partial E_1 = \partial E_2 = \partial E_3 = \{t_1 = a,\ c \le t_2 \le d\} \cup \{t_1 = b,\ c \le t_2 \le d_2\}$$

$$\cup \{a \le t_1 \le b,\ t_2 = c\} \cup \{a \le t_1 \le b,\ t_2 = d\},$$

$$\partial_\Delta E_1 = \emptyset,$$

$$\partial_\Delta E_2 = \{t_1 = a,\ c \le t_2 \le d\} \cup \{t_1 = b,\ c \le t_2 \le d_2\}$$

$$\cup \{a \le t_1 \le b,\ t_2 = c\} \cup \{a \le t_1 \le b,\ t_2 = d\},$$

$$\partial_\Delta E_3 = \{t_1 = b,\ c \le t_2 \le d\} \cup \{t_2 = d,\ a \le t_1 \le b\}.$$

Let $\Lambda_0^n = \mathbb{T}_1^0 \times \mathbb{T}_2^0 \times \ldots \times \mathbb{T}_n^0$.

Definition 7.71 Let $E \subset \Lambda_0^n$ be a bounded set and let $\partial_\Delta E$ be its boundary. Assume

$$R = [a_1, b_1) \times [a_2, b_2) \times \ldots \times [a_n, b_n)$$

is a rectangle in Λ^n such that $E \cup \partial_\Delta E \subset R$. Suppose $\mathscr{P}(R)$ denotes the set of all Δ-partitions of R of the type (7.1) and (7.2). For every $P \in \mathscr{P}(R)$, denote with $J_*(E, P)$ the sum of the areas of those subrectangles of P which are entirely contained in E, and let $J^*(E, P)$ be the sum of the areas of the subrectangles of P each of which contains at least one point of $E \cup \partial_\Delta E$. The numbers

$$J_*(E) = \sup\{J_*(E, P) : P \in \mathscr{P}(R)\} \quad \text{and} \quad J^*(E) = \inf\{J^*(E, P) : P \in \mathscr{P}(R)\}$$

are called the *inner* and *outer Jordan Δ-measure* of E, respectively. The set E is said to be *Jordan Δ-measurable* if $J_*(E) = J^*(E)$, in which case this common value is called the *Jordan Δ-measure* of E, denoted by $J(E)$.

We note that

$$0 \leq J_*(E) \leq J^*(E)$$

for every bounded set $E \subset \Lambda_0^n$. If E has Jordan Δ-measure zero, then $J_*(E) = J^*(E) = 0$. Therefore, we have the following statement.

Proposition 7.72 *A bounded set $E \subset \Lambda_0^n$ has Jordan Δ-measure zero if and only if for every $\varepsilon > 0$, the set E can be covered by a finite collection of rectangles of type*

$$V_j = [\alpha_1^j, \beta_1^j) \times [\alpha_2^j, \beta_2^j) \times \ldots \times [\alpha_n^j, \beta_n^j) \subset \Lambda^n, \quad j = 1, 2, \ldots, m,$$

the sum of whose areas is less than ε:

$$E \subset \bigcup_{j=1}^{m} V_j \quad \text{and} \quad \sum_{j=1}^{m} m(V_j) < \varepsilon.$$

If E is a set of Jordan Δ-measure zero, then so is any set $\tilde{E} \subset E$.

Proposition 7.73 *The union of a finite number of bounded subsets $E_1, E_2, \ldots, E_k \subset \Lambda_0^n$ each of which has Jordan Δ-measure zero is in turn a set of Jordan Δ-measure zero.*

Proof Let $\varepsilon > 0$ be arbitrarily chosen. We can construct, for each $l \in \{1, 2, \ldots, k\}$, a finite covering $\{V_j^{(l)}\}_{j=1}^{n_l}$ of E_l by rectangles of the needed type, the sum of whose areas is less than $\frac{\varepsilon}{2^l}$:

$$E_l \subset \bigcup_{j=1}^{k_l} V_j^{(l)} \quad \text{and} \quad \sum_{j=1}^{k_l} m(V_j^{(l)}) < \frac{\varepsilon}{2^l}$$

for all $l = 1, 2, \ldots, k$. The union of all these rectangles is a finite covering of $E = \bigcup_{l=1}^{k} E_l$ by rectangles, and the sum of the areas of all rectangles is less than $\sum_{l=1}^{\infty} \frac{\varepsilon}{2^l} = \varepsilon$. Since $\varepsilon > 0$ was arbitrarily chosen, we have that $J(E) = 0$, i.e., the set E is of Jordan Δ-measure zero. □

The empty set is a Jordan Δ-measurable set, and its Jordan Δ-measure is zero.

Theorem 7.74 *For each point $t^0 = (t_1^0, t_2^0, \ldots, t_n^0) \in \Lambda_0^n$, the single point set $\{t^0\}$ is Jordan Δ-measurable, and its Jordan Δ-measure is given by*

$$J(\{t^0\}) = (\sigma_1(t_1^0) - t_1^0)(\sigma_2(t_2^0) - t_2^0) \ldots (\sigma_n(t_n^0) - t_n^0). \tag{7.24}$$

Proof 1. If $\sigma_i(t_i^0) > t_i^0$ for all $i \in \{1, 2, \ldots, n\}$, then

$$\{t^0\} = [t_1^0, \sigma_1(t_1^0)) \times [t_2^0, \sigma_2(t_2^0)) \times \ldots \times [t_n^0, \sigma_n(t_n^0)).$$

Therefore, $\{t^0\}$ is Jordan Δ-measurable and (7.24) holds.

2. If $\sigma_l(t_l^0) = t_l^0$ for some $l \in \{1, 2, \ldots, n\}$, then there exist $t_k \in \mathbb{T}_k$ sufficiently close to t_k^0 and $t_k > t_k^0$, $k \in \{1, 2, \ldots, n\}$. If $\sigma_m(t_m^0) > t_m^0$ for some $m \in \{1, 2, \ldots, n\}$, then $t_m = \sigma_m(t_m^0)$. Therefore,

$$[t_1^0, t_1) \times [t_2^0, t_2) \times \ldots \times [t_n^0, t_n)$$

covers the single point set $\{t^0\}$ and has a sufficiently small area. Therefore, the single point set $\{t^0\}$ is Jordan Δ-measurable and has Jordan Δ-measure zero. In this case,

$$(\sigma_1(t_1^0) - t_1^0)(\sigma_2(t_2^0) - t_2^0) \ldots (\sigma_n(t_n^0) - t_n^0) = 0$$

because $\sigma_l(t_l^0) = t_l^0$.

This completes the proof. □

Corollary 7.75 *Every Δ-dense point of Λ_0^n has Jordan Δ-measure zero.*

Theorem 7.76 *If $E \subset \Lambda_0^n$ is a bounded set with Δ-boundary $\partial_\Delta E$, then*

$$J^*(\partial_\Delta E) = J^*(E) - J_*(E). \tag{7.25}$$

Moreover, E is Jordan Δ-measurable iff its Δ-boundary $\partial_\Delta E$ has Jordan Δ-measure zero.

Proof Let $R = [a_1, b_1) \times [a_2, b_2) \times \ldots \times [a_n, b_n)$ be a rectangle in Λ^n containing $E \cup \partial_\Delta E$. Then, for every $P \in \mathscr{P}(R)$, we have

$$J^*(\partial_\Delta E, P) = J^*(E, P) - J_*(E, P).$$

Therefore,

$$J^*(\partial_\Delta E, P) \geq J^*(E) - J_*(E),$$

whereupon

$$J^*(\partial_\Delta E) \geq J^*(E) - J_*(E). \tag{7.26}$$

Let $\varepsilon > 0$ be arbitrarily chosen. We choose $H, Q \in \mathscr{P}(R)$ such that

$$J_*(E, H) > J_*(E) - \frac{\varepsilon}{2} \quad \text{and} \quad J^*(E, Q) < J^*(E) + \frac{\varepsilon}{2}.$$

We set $P = H \cup Q$. Then P is a refinement of both H and Q. Therefore,

$$\begin{aligned} J^*(\partial_\Delta E) &\leq J^*(\partial_\Delta E, P) \\ &= J^*(E, P) - J_*(E, P) \\ &\leq J^*(E, Q) - J_*(E, H) \\ &< J^*(E) + \frac{\varepsilon}{2} - J_*(E) + \frac{\varepsilon}{2} \\ &= J^*(E) - J_*(E) + \varepsilon. \end{aligned}$$

Because $\varepsilon > 0$ was arbitrarily chosen, we conclude that

$$J^*(\partial_\Delta E) \leq J^*(E) - J_*(E).$$

By the last inequality and (7.26), we get (7.25). The set E is Jordan Δ-measurable iff $J^*(E) = J_*(E)$ iff $J^*(\partial_\Delta E) = 0$. □

Proposition 7.77 *Let $E_1, E_2 \in \Lambda^n$. Then we have the following relations.*

1. $\partial_\Delta(E_1 \cup E_2) \subset \partial_\Delta E_1 \cup \partial_\Delta E_2$,
2. $\partial_\Delta(E_1 \cap E_2) \subset \partial_\Delta E_1 \cup \partial_\Delta E_2$,
3. $\partial_\Delta(E_1 \setminus E_2) \subset \partial_\Delta E_1 \cup \partial_\Delta E_2$.

Proof 1. Let $t \in \partial_\Delta(E_1 \cup E_2)$ be arbitrarily chosen. Then every rectangle

$$V = [t_1, t_1') \times [t_2, t_2') \times \ldots \times [t_n, t_n')$$

contains a point $t'' \in E_1 \cup E_2$ and a point $t''' \in \Lambda^n \setminus (E_1 \cup E_2)$. Since $t'' \in E_1 \cup E_2$, we have $t'' \in E_1$ or $t'' \in E_2$.

a. Suppose $t'' \in E_1$. Therefore, every rectangle V contains a point $t'' \in E_1$ and a point $t''' \in \Lambda^n \setminus (E_1 \cup E_2) \subset \Lambda^2 \setminus E_1$. Consequently, $t \in \partial_\Delta E_1$ and hence

$$t \in \partial_\Delta E_1 \cup \partial_\Delta E_2. \tag{7.27}$$

b. Suppose $t'' \in E_2$. Therefore, every rectangle V contains a point $t'' \in E_2$ and a point $t''' \in \Lambda^n \setminus (E_1 \cup E_2) \subset \Lambda^2 \setminus E_2$. Consequently, $t \in \partial_\Delta E_2$ and hence (7.27) holds.

Because $t \in \partial_\Delta(E_1 \cup E_2)$ was arbitrarily chosen and due to (7.27), we get

$$\partial_\Delta(E_1 \cup E_2) \subset \partial_\Delta E_1 \cup \partial_\Delta E_2.$$

2. We have

$$\partial_\Delta(E_1 \cap E_2) \subset \partial_\Delta(E_1 \cup E_2) \subset \partial_\Delta E_1 \cup \partial_\Delta E_2.$$

3. We have

$$\partial_\Delta(E_1 \setminus E_2) \subset \partial_\Delta(E_1 \cup E_2) \subset \partial_\Delta E_1 \cup \partial_\Delta E_2.$$

The proof is complete. □

Theorem 7.78 *The union of a finite number of Jordan Δ-measurable sets is a Jordan Δ-measurable set.*

Proof If $E_1, E_2, \ldots, E_k$ are Jordan Δ-measurable sets, then

$$J(\partial_\Delta E_l) = 0 \quad \text{for all} \quad l \in \{1, 2, \ldots, k\}.$$

By Proposition 7.77, we get

$$\partial_\Delta(E_1 \cup E_2 \cup \ldots \cup E_k) \subset \partial_\Delta E_1 \cup \partial_\Delta E_2 \cup \ldots \cup \partial_\Delta E_k.$$

Therefore,

$$J(\partial_\Delta(E_1 \cup E_2 \cup \ldots \cup E_k)) \le J(\partial_\Delta E_1) + J(\partial_\Delta E_2) + \cdots + J(\partial_\Delta E_k) = 0.$$

Hence, by Theorem 7.76, we obtain that $E_1 \cup E_2 \cup \ldots \cup E_k$ is Jordan Δ-measurable. □

Theorem 7.79 *The intersection of a finite number of Jordan Δ-measurable sets is a Jordan Δ-measurable set.*

Proof If $E_1, E_2, \ldots, E_k$ are Jordan Δ-measurable sets, then

$$J(\partial_\Delta E_l) = 0 \quad \text{for all} \quad l \in \{1, 2, \ldots, k\}.$$

By Proposition 7.77, we get

$$\partial_\Delta(E_1 \cap E_2 \cap \ldots \cap E_k) \subset \partial_\Delta E_1 \cup \partial_\Delta E_2 \cup \ldots \cup \partial_\Delta E_k.$$

Therefore,

$$J(\partial_\Delta(E_1 \cap E_2 \cap \ldots \cap E_k)) \le J(\partial_\Delta E_1) + J(\partial_\Delta E_2) + \ldots + J(\partial_\Delta E_k) = 0.$$

Hence, using Theorem 7.76, we obtain that $E_1 \cup E_2 \cup \ldots \cup E_k$ is Jordan Δ-measurable. □

Theorem 7.80 *The difference of two Jordan Δ-measurable sets is a Jordan Δ-measurable set.*

Proof If E_1 and E_2 are two Jordan Δ-measurable sets, then

$$J(\partial_\Delta E_1) = J(\partial_\Delta E_2) = 0.$$

By Proposition 7.77, we get

$$\partial_\Delta(E_1 \setminus E_2) \subset \partial_\Delta E_1 \cup \partial_\Delta E_2.$$

Therefore,

$$J(\partial_\Delta(E_1 \setminus E_2)) \le J(\partial_\Delta E_1) + J(\partial_\Delta E_2) = 0.$$

Hence, by Theorem 7.76, we conclude that $E_1 \setminus E_2$ is Jordan Δ-measurable. □

Definition 7.81 Let f be defined and bounded on a bounded Jordan Δ-measurable set $E \subset \Lambda_0^n$. Let

$$R = [a_1, b_1) \times [a_2, b_2) \times \ldots \times [a_n, b_n) \subset \Lambda^n$$

be a rectangle containing E and put

$$K = [a_1, b_1] \times [a_2, b_2] \times \ldots \times [a_n, b_n].$$

Define f on K by

$$F(t) = \begin{cases} f(t) & \text{if } \ t \in E \\ 0 & \text{if } \ t \in K \setminus E. \end{cases} \tag{7.28}$$

(here $t = (t_1, t_2, \ldots, t_n)$). In this case, f is said to be *Riemann Δ-integrable* over E if F is Riemann Δ-integrable over R in the sense of Section 7.1. We write

$$\int_E f(t_1, t_2, \ldots, t_n)\Delta_1 t_1 \Delta_2 t_2 \ldots \Delta_n t_n = \int_R F(t_1, t_2, \ldots, t_n)\Delta_1 t_1 \Delta_2 t_2 \ldots \Delta_n t_n.$$

Definition 7.82 Assume that f is defined and bounded on a bounded Jordan Δ-measurable set $E \subset \Lambda_0^n$. Let $R = [a_1, b_1) \times [a_2, b_2) \times \ldots \times [a_n, b_n) \subset \Lambda^n$ be a rectangle such that $E \subset R$. Let $P \in \mathscr{P}(R)$ be given by (7.1) and (7.2). Some of the rectangles of P will be entirely within E, some will be outside of E, and some will be partly within and partly outside E. Let $P' = \{R_1, R_2, \ldots, R_k\}$ be the collection of subrectangles in P that lie completely within the set E. The collection P' is called the inner Δ-partition of the set E, determined by the partition P of the rectangle R. We choose an arbitrary point $\xi^l = (\xi_1^l, \xi_2^l, \ldots, \xi_n^l)$ in the subrectangle R_l of P' for $l \in \{1, 2, \ldots, k\}$. Let $m(R_l)$ denote the area of R_l. We set

$$S = \sum_{l=1}^{k} f(\xi^l) m(R_l).$$

We call S a *Riemann Δ-sum* of f corresponding to the partition $P \in \mathscr{P}(R)$. We say that f is *Riemann Δ-integrable* over $E \subset \Lambda_0^n$ if there exists a number I such that for each $\varepsilon > 0$, there exists $\delta > 0$ such that $|S - I| < \varepsilon$ for every Riemann Δ-sum S of f corresponding to any inner Δ-partition $P' = \{R_1, R_2, \ldots, R_k\}$ of E, determined by a partition $P \in \mathscr{P}_\delta(R)$, independent of the way in which $\xi^l \in R_l$ for $1 \leq l \leq k$ is chosen. The number I is called the *Riemann multiple Δ-integral* of f over E. We write $I = \lim_{\delta \to 0} S$.

Proposition 7.83 *Let $E \subset \Lambda_0^n$ be a bounded set with Δ-boundary $\partial_\Delta E$. If*

$$R = [a_1, b_1) \times [a_2, b_2) \times \ldots \times [a_n, b_n) \subset \Lambda^n$$

is a rectangle that contains $E \cup \partial_\Delta E$, then

$$\lim_{\delta \to 0} J_*(E, P) = J_*(E) \quad \textit{and} \quad \lim_{\delta \to 0} J^*(E, P) = J^*(E)$$

for any $P \in \mathscr{P}_\delta(R)$.

Proof Define the functions $g_1 : R \to \mathbb{R}$ and $g_2 : R \to \mathbb{R}$ by

$$g_1(t) = \begin{cases} 1 & \text{if } t \in E \\ 0 & \text{if } t \in R \setminus E \end{cases}$$

and

$$g_2(t) = \begin{cases} 1 & \text{if } t \in E \cup \partial_\Delta E \\ 0 & \text{if } t \in R \setminus (E \cup \partial_\Delta E). \end{cases}$$

Thus,

$$J_*(E, P) = L(g_1, P) \quad \text{and} \quad J^*(E, P) = U(g_2, P)$$

for all $P \in \mathscr{P}_\delta(R)$. Hence,

$$J_*(E) = L(g_1) \quad \text{and} \quad J^*(E) = U(g_2).$$

By Theorem 7.20, we have

$$\lim_{\delta \to 0} L(g_1, P) = L(g_1) \quad \text{and} \quad \lim_{\delta \to 0} U(g_2, P) = U(g_2),$$

which completes the proof. □

Proposition 7.84 *Let $\Gamma \subset \Lambda_0^n$ be a set of Jordan Δ-measure zero. Moreover, let*

$$R = [a_1, b_1) \times [a_2, b_2) \times \ldots \times [a_n, b_n)$$

be a rectangle in Λ^n that contains Γ. Then, for each $\varepsilon > 0$, there exists $\delta > 0$ such that for every partition $P \in \mathscr{P}_\delta(R)$, the sum of areas of subrectangles of P which have a common point with Γ is less than ε.

Proof By Proposition 7.83, we have that

$$\lim_{\delta \to 0} J_*(\Gamma, P) = \lim_{\delta \to 0} J^*(\Gamma, P) = 0.$$

This completes the proof. □

Theorem 7.85 *If $E \subset \Lambda_0^n$ is a bounded and Jordan Δ-measurable set and f is a bounded function on E, then Definitions 7.81 and 7.82 of the Riemann Δ-integrability of f over E are equivalent to each other.*

Proof Assume that $R = [a_1, b_1) \times [a_2, b_2) \times \ldots \times [a_n, b_n) \subset \Lambda^n$ contains E and let

$$K = [a_1, b_1] \times [a_2, b_2] \times \ldots \times [a_n, b_n].$$

Define f on K by (7.28). Let P be a Δ-partition of R into subrectangles $R_{j_1 j_2 \ldots j_n}$, $1 \le j_i \le k_i$, $i = 1, 2, \ldots, n$, defined by (7.1) and (7.2). Then, for any

$$\xi^{j_1 j_2 \ldots j_n} \in R_{j_1 j_2 \ldots j_n},$$

we have

$$\begin{aligned} &\sum_{j_1=1}^{k_1} \sum_{j_2=1}^{k_2} \cdots \sum_{j_n=1}^{k_n} F(\xi^{j_1 j_2 \ldots j_n}) m(R_{j_1 j_2 \ldots j_n}) \\ &= \sum_{(j_1, j_2, \ldots, j_n) \in A} f(\xi^{j_1 j_2 \ldots j_n}) m(R_{j_1 j_2 \ldots j_n}) + \sum_{(j_1, j_2, \ldots, j_n) \in B} F(\xi^{j_1 j_2 \ldots j_n}) m(R_{j_1 j_2 \ldots j_n}), \end{aligned} \tag{7.29}$$

where

$$\begin{aligned} A &= \{(j_1, j_2, \ldots, j_n) : \; R_{j_1 j_2 \ldots j_n} \subset E\}, \\ B &= \{(j_1, j_2, \ldots, j_n) : \; R_{j_1 j_2 \ldots j_n} \subset \Lambda^n \setminus E \quad \text{and} \quad R_{j_1 j_2 \ldots j_n} \cap \partial_\Delta E \neq \emptyset\}. \end{aligned} \tag{7.30}$$

By Proposition 7.84, the second sum of the right-hand side of (7.29) can be made sufficiently small for $P \in \mathscr{P}_\delta(R)$ as $\delta \to 0$ since $J(\partial_\Delta E) = 0$. □

Theorem 7.86 *If $E \subset \Lambda_0^n$ is a bounded and Jordan Δ-measurable set, then the integral*

$$\int_E 1\Delta_1 t_1 \Delta_2 t_2 \ldots \Delta_n t_n$$

exists. Moreover, we have

$$J(E) = \int_E 1\Delta_1 t_1 \Delta_2 t_2 \ldots \Delta_n t_n. \tag{7.31}$$

Proof Suppose $R = [a_1, b_1) \times [a_2, b_2) \times \ldots \times [a_n, b_n)$ contains E and

$$K = [a_1, b_1] \times [a_2, b_2] \times \ldots \times [a_n, b_n].$$

We set

$$F(t) = \begin{cases} 1 & \text{if } t \in E \\ 0 & \text{if } t \in K \setminus E, \end{cases}$$

where $t = (t_1, t_2, \ldots, t_n)$. Let P be a Δ-partition of R into subrectangles defined by (7.1) and (7.2), and let A and B be defined as in (7.30). If $(j_1, j_2, \ldots, j_n) \in A$, then

$$F(\xi^{j_1 j_2 \ldots j_n}) = 1$$

and (7.29) takes the form

$$\sum_{j_1=1}^{k_1} \sum_{j_2=1}^{k_2} \cdots \sum_{j_n=1}^{k_n} F(\xi^{j_1 j_2 \ldots j_n}) m(R_{j_1 j_2 \ldots j_n})$$
$$= J_*(E, P) + \sum_{(j_1, j_2, \ldots, j_n) \in B} F(\xi^{j_1 j_2 \ldots j_n}) m(R_{j_1 j_2 \ldots j_n}). \tag{7.32}$$

If $P \in \mathscr{P}_\delta(R)$ and $\delta \to 0$, then, by Proposition 7.83 and the Jordan Δ-measurability of E, we have

$$J_*(E, P) \to J(E). \tag{7.33}$$

Since E is Jordan Δ-measurable, we get $J(\partial_\Delta E) = 0$. Therefore,

$$\sum_{(j_1, j_2, \ldots, j_n) \in B} F(\xi^{j_1 j_2 \ldots j_n}) m(R_{j_1 j_2 \ldots j_n}) \to 0$$

as $\delta \to 0$. Hence, using (7.32) and (7.33), we get that 1 is integrable over E and (7.31) holds. □

Corollary 7.87 *Let $\mathbb{T}_i = \mathbb{Z}$, $i \in \{1, 2, \ldots, n\}$. Assume that $E \subset \Lambda^n$ is bounded. Then, for any function $f : E \to \mathbb{R}$, we have*

$$\int_E f(t_1, t_2, \ldots, t_n)\Delta_1 t_1 \Delta_2 t_2 \ldots \Delta_n t_n = \sum_{t\in E} f(t). \tag{7.34}$$

Proof We have that $\partial_\Delta E = \emptyset$. Therefore, E is Jordan Δ-measurable. Hence, utilizing Definition 7.81 and Theorem 7.29, we find that (7.34) holds. The Jordan Δ-measure of E coincides with the number of points of E. □

Example 7.88 Let $\mathbb{T}_1 = \mathbb{T}_2 = \mathbb{Z}$ and $E = [0, 2] \times [0, 5]$. Define $f : \Lambda^2 \to \mathbb{R}$ by

$$f(t) = t_1^2 + t_2^2, \quad t \in \Lambda^2.$$

We will compute

$$\int_E f(t)\Delta_1 t_1 \Delta_2 t_2.$$

We have $J(E) = 18$ and

$$\begin{aligned}
\int_E (t_1^2 + t_2^2)\Delta_1 t_1 \Delta_2 t_2 &= (t_1^2 + t_2^2)\Big|_{(t_1,t_2)=(0,0)} + (t_1^2 + t_2^2)\Big|_{(t_1,t_2)=(1,0)} \\
&\quad + (t_1^2 + t_2^2)\Big|_{(t_1,t_2)=(2,0)} + (t_1^2 + t_2^2)\Big|_{(t_1,t_2)=(0,1)} \\
&\quad + (t_1^2 + t_2^2)\Big|_{(t_1,t_2)=(1,1)} + (t_1^2 + t_2^2)\Big|_{(t_1,t_2)=(2,1)} \\
&\quad + (t_1^2 + t_2^2)\Big|_{(t_1,t_2)=(0,2)} + (t_1^2 + t_2^2)\Big|_{(t_1,t_2)=(1,2)} \\
&\quad + (t_1^2 + t_2^2)\Big|_{(t_1,t_2)=(2,2)} + (t_1^2 + t_2^2)\Big|_{(t_1,t_2)=(0,3)} \\
&\quad + (t_1^2 + t_2^2)\Big|_{(t_1,t_2)=(1,3)} + (t_1^2 + t_2^2)\Big|_{(t_1,t_2)=(2,3)} \\
&\quad + (t_1^2 + t_2^2)\Big|_{(t_1,t_2)=(0,4)} + (t_1^2 + t_2^2)\Big|_{(t_1,t_2)=(1,4)} \\
&\quad + (t_1^2 + t_2^2)\Big|_{(t_1,t_2)=(2,4)} + (t_1^2 + t_2^2)\Big|_{(t_1,t_2)=(0,5)} \\
&\quad + (t_1^2 + t_2^2)\Big|_{(t_1,t_2)=(1,5)} + (t_1^2 + t_2^2)\Big|_{(t_1,t_2)=(2,5)} \\
&= 0 + 1 + 4 + 1 + 2 + 5 + 4 + 5 + 8
\end{aligned}$$

$$+9+10+13+16+17+20+25+26+29$$

$$= 195.$$

Exercise 7.89 Let $\mathbb{T}_1 = \mathbb{T}_2 = \mathbb{Z}$, $E = [-1, 0] \times [1, 2]$. Find $J(E)$ and

$$\int_E (t_1 + t_2)\Delta_1 t_1 \Delta_2 t_2.$$

Solution 4, 4.

Theorem 7.90 *Let $E_1, E_2 \subset \Lambda_0^n$ be bounded Jordan Δ-measurable sets such that $J(E_1 \cap E_2) = 0$, and let $E = E_1 \cup E_2$. If $f : E \to \mathbb{R}$ is a bounded function which is Δ-integrable over each of E_1 and E_2, then f is Δ-integrable over E and*

$$\int_E f(t_1, t_2, \ldots, t_n)\Delta_1 t_1 \Delta_2 t_2 \ldots \Delta_n t_n = \int_{E_1} f(t_1, t_2, \ldots, t_n)\Delta_1 t_1 \Delta_2 t_2 \ldots \Delta_n t_n$$
$$+ \int_{E_2} f(t_1, t_2, \ldots, t_n)\Delta_1 t_1 \Delta_2 t_2 \ldots \Delta_n t_n.$$

Proof Suppose $R = [a_1, b_1) \times [a_2, b_2) \times \ldots \times [a_n, b_n)$ contains E and

$$K = [a_1, b_1] \times [a_2, b_2] \times \ldots \times [a_n, b_n].$$

Define F as in (7.28). Let $P = \{R_1, R_2, \ldots, R_k\}$ be a Δ-partition of R and form a Riemann Δ-sum

$$S(F, P) = \sum_{j=1}^{k} F(\xi^j)m(R_j).$$

If S_1 denotes the part of the sum arising from those rectangles containing only points of E_1, and if S_2 is similarly defined by E_2, then we can write

$$S(F, P) = S_1 + S_2 + S_3,$$

where S_3 contains those terms coming from subrectangles which contain points of $E_1 \cap E_2$. Since $J(E_1 \cap E_2) = 0$, we can make $|S_3|$ arbitrarily small when P is sufficiently fine. S_1 approximates the integral

$$\int_{E_1} f(t_1, t_2, \ldots, t_n)\Delta_1 t_1 \Delta_2 t_2 \ldots \Delta_n t_n,$$

and S_2 approximates the integral

$$\int_{E_2} f(t_1, t_2, \ldots, t_n)\Delta_1 t_1 \Delta_2 t_2 \ldots \Delta_n t_n,$$

which completes the proof. □

Exercise 7.91 Let f and g be Δ-integrable over E, and let $\alpha, \beta \in \mathbb{R}$. Prove that $\alpha f + \beta g$ is also Δ-integrable over E and

$$\begin{aligned}&\int_E (\alpha f + \beta g)(t_1, t_2, \ldots, t_n)\Delta_1 t_1 \Delta_2 t_2 \ldots \Delta_n t_n \\ &= \alpha \int_E f(t_1, t_2, \ldots, t_n)\Delta_1 t_1 \Delta_2 t_2 \ldots \Delta_n t_n + \beta \int_E g(t_1, t_2, \ldots, t_n)\Delta_1 t_1 \Delta_2 t_2 \ldots \Delta_n t_n.\end{aligned}$$

Exercise 7.92 Let f and g are Δ-integrable over E. Prove that so is their product fg.

Exercise 7.93 Let f be Δ-integrable over E. Prove that so is $|f|$ with

$$\left|\int_E f(t_1, t_2, \ldots, t_n)\Delta_1 t_1 \Delta_2 t_2 \ldots \Delta_n t_n\right| \leq \int_E |f(t_1, t_2, \ldots, t_n)|\Delta_1 t_1 \Delta_2 t_2 \ldots \Delta_n t_n.$$

Exercise 7.94 Let f and g be Δ-integrable over E and

$$f(t_1, t_2, \ldots, t_n) \leq g(t_1, t_2, \ldots, t_n) \quad \text{for all} \quad (t_1, t_2, \ldots, t_n) \in E.$$

Prove that

$$\int_E f(t_1, t_2, \ldots, t_n)\Delta_1 t_1 \Delta_2 t_2 \ldots \Delta_n t_n \leq \int_E g(t_1, t_2, \ldots, t_n)\Delta_1 t_1 \Delta_2 t_2 \ldots \Delta_n t_n.$$

Exercise 7.95 Let f and g be Δ-integrable over E and let g be nonnegative (or nonpositive) on E. Prove that there exists a real number

$$\Lambda \in [m, M]$$

such that

$$\begin{aligned}&\int_E f(t_1, t_2, \ldots, t_n) g(t_1, t_2, \ldots, t_n)\Delta_1 t_1 \Delta_2 t_2 \ldots \Delta_n t_n \\ &\qquad = \Lambda \int_E g(t_1, t_2, \ldots, t_n)\Delta_1 t_1 \Delta_2 t_2 \ldots \Delta_n t_n,\end{aligned}$$

where

$$m = \inf\{f(t_1, t_2, \ldots, t_n) : (t_1, t_2, \ldots, t_n) \in E\}$$

and

$$M = \sup\{f(t_1, t_2, \ldots, t_n) : (t_1, t_2, \ldots, t_n) \in E\}.$$

Lemma 7.96 *Let* $[a_i, b_i] \subset \mathbb{T}_i^0$, $i \in \{1, 2, \ldots, n-1\}$. *Suppose*

$$\phi : [a_1, b_1] \times [a_2, b_2] \times \ldots \times [a_{n-1}, b_{n-1}] \to \mathbb{T}_n^0$$

is a continuous function. Let Γ *be the set (graph of* ϕ*) in* Λ^n *given by*

$$\Gamma = \Big\{(t_1, t_2, \ldots, t_{n-1}, \phi(t_1, t_2, \ldots, t_{n-1})) :$$

$$(t_1, t_2, \ldots, t_{n-1}) \in [a_1, b_1) \times [a_2, b_2) \times \ldots \times [a_{n-1}, b_{n-1})\Big\}.$$

In this case, the subset Γ' *of* Γ *consisting of all* Δ*-dense points of* Γ *has Jordan* Δ*-measure zero in* Λ^n.

Proof Since ϕ is continuous on

$$[a_1, b_1] \times [a_2, b_2] \times \ldots \times [a_{n-1}, b_{n-1}],$$

it is uniformly continuous on it. Therefore, for each $\varepsilon > 0$, there exists $\delta > 0$ such that

$$t = (t_1, t_2, \ldots, t_{n-1}),\ t' = (t_1', t_2', \ldots, t_{n-1}') \in [a_1, b_1]\times[a_2, b_2]\times\ldots\times[a_{n-1}, b_{n-1}]$$

and

$$|t_i - t_i'| < \delta, \quad i = 1, 2, \ldots, n-1$$

implies

$$|\phi(t) - \phi(t')| < \frac{\varepsilon}{(b_1 - a_1)(b_2 - a_2)\ldots(b_{n-1} - a_{n-1})}.$$

Take a partition $P \in \mathscr{P}_\delta([a_1, b_1) \times [a_2, b_2) \times \ldots \times [a_{n-1}, b_{n-1}))$ determined by

$$a_i = t_i^0 < t_i^1 < \ldots < t_i^{k_i} = b_i, \quad i = 1, 2, \ldots, n-1.$$

For $l_i \in \{1, 2, \ldots, k_i\}$, $i = 1, 2, \ldots, n-1$, let us set

$$d_{l_1 l_2 \ldots l_{n-1}} = \min\Big\{\phi(t) :$$

$$t = (t_1, t_2, \ldots, t_{n-1}) \in [t_1^{l_1-1}, t_1^{l_1}] \times [t_2^{l_2-1}, t_2^{l_2}] \times \ldots \times [t_{n-1}^{l_{n-1}-1}, t_{n-1}^{l_{n-1}}]\Big\},$$

$$D_{l_1 l_2 \ldots l_{n-1}} = \max\Big\{\phi(t) :$$

$$t = (t_1, t_2, \ldots, t_{n-1}) \in [t_1^{l_1-1}, t_1^{l_1}] \times [t_2^{l_2-1}, t_2^{l_2}] \times \ldots \times [t_{n-1}^{l_{n-1}-1}, t_{n-1}^{l_{n-1}}]\Big\}.$$

Denote

$$I = \left\{(l_1, l_2, \ldots, l_{n-1}) : l_i \in \{1, 2, \ldots, k_i\},\ t_i^{l_i} - t_i^{l_i-1} \leq \delta,\ i = 1, 2, \ldots, n-1\right\},$$

$$I' = \left\{(l_1, l_2, \ldots, l_{n-1}) : l_i \in \{1, 2, \ldots, k_i\},\ t_i^{l_i} - t_i^{l_i-1} > \delta,\ i = 1, 2, \ldots, n-1\right\}.$$

Consider the rectangles $R_{l_1 l_2 \ldots l_{n-1}} \subset \Lambda^n$, $l_i \in \{1, 2, \ldots, k_i\}$, $i \in \{1, 2, \ldots, n-1\}$, defined by

$$R_{l_1 l_2 \ldots l_{n-1}} = [t_1^{l_1-1}, t_1^{l_1}) \times [t_2^{l_2-1}, t_2^{l_2}) \times \ldots \times [t_{n-1}^{l_{n-1}-1}, t_{n-1}^{l_{n-1}}) \times [d_{l_1 l_2 \ldots l_n}, D_{l_1 l_2 \ldots l_{n-1}}).$$

Obviously, all Δ-dense points of Γ may lie only in rectangles

$$R_{l_1 l_2 \ldots l_{n-1}} \quad \text{for} \quad (l_1, l_2, \ldots, l_{n-1}) \in I.$$

On the other hand,

$$\begin{aligned}
&\sum_{(l_1, l_2, \ldots, l_{n-1}) \in I} m(R_{l_1 l_2 \ldots l_{n-1}}) \\
&= \sum_{(l_1, l_2, \ldots, l_{n-1}) \in I} (t_1^{l_1} - t_1^{l_1-1})(t_2^{l_2} - t_2^{l_2-1}) \ldots (t_{n-1}^{l_{n-1}} - t_{n-1}^{l_{n-1}-1})(D_{l_1 l_2 \ldots l_{n-1}} - d_{l_1 l_2 \ldots l_{n-1}}) \\
&\leq \frac{\varepsilon}{(b_1 - a_1)(b_2 - a_2) \ldots (b_{n-1} - a_{n-1})} \\
&\quad \times \sum_{(l_1, l_2, \ldots, l_{n-1}) \in I} (t_1^{l_1} - t_1^{l_1-1})(t_2^{l_2} - t_2^{l_2-1}) \ldots (t_{n-1}^{l_{n-1}} - t_{n-1}^{l_{n-1}}) \\
&\leq \frac{\varepsilon}{(b_1 - a_1)(b_2 - a_2) \ldots (b_{n-1} - a_{n-1})} (b_1 - a_1)(b_2 - a_2) \ldots (b_{n-1} - a_{n-1}) \\
&= \varepsilon.
\end{aligned}$$

Since $\varepsilon > 0$ was arbitrarily chosen, this completes the proof. □

Theorem 7.97 *Let* $[a_i, b_i] \subset \mathbb{T}_i^0$, $i = 1, 2, \ldots, n-1$, *and let*

$$\phi : [a_1, b_1] \times [a_2, b_2] \times \ldots \times [a_{n-1}, b_{n-1}] \to \mathbb{T}_n^0$$

and

$$\psi : [a_1, b_1] \times [a_2, b_2] \times \ldots \times [a_{n-1}, b_{n-1}] \to \mathbb{T}_n^0$$

be continuous functions such that $\phi(t) < \psi(t)$ *for all*

$$t = (t_1, t_2, \dots, t_{n-1}) \in [a_1, b_1] \times [a_2, b_2] \times \dots \times [a_{n-1}, b_{n-1}].$$

Let E *be the bounded set in* Λ^n *given by*

$$E = \Big\{(t_1, t_2, \dots, t_n) \in \Lambda^n : \ a_i \le t_i < b_i, \ i \in \{1, 2, \dots, n-1\},$$

$$\phi(t_1, t_2, \dots, t_{n-1}) \le t_n < \psi(t_1, t_2, \dots, t_{n-1})\Big\}.$$

In this case, E *is Jordan* Δ*-measurable, and if* $f : E \to \mathbb{R}$ *is* Δ*-integrable over* E *and if the single integral*

$$\int_{\phi(t_1,t_2,\dots,t_{n-1})}^{\psi(t_1,t_2,\dots,t_{n-1})} f(t_1, t_2, \dots, t_{n-1}, t_n)\Delta_n t_n$$

exists for each $(t_1, t_2, \dots, t_{n-1}) \in [a_1, b_1) \times [a_2, b_2) \times \dots \times [a_{n-1}, b_{n-1})$*, then the iterated integral*

$$\int_{a_1}^{b_1}\int_{a_2}^{b_2}\dots\int_{a_{n-1}}^{b_{n-1}} \Delta_1 t_1 \Delta_2 t_2 \dots \Delta_{n-1} t_{n-1} \int_{\phi(t_1,t_2,\dots,t_{n-1})}^{\psi(t_1,t_2,\dots,t_{n-1})} f(t_1, t_2, \dots, t_n)\Delta_n t_n$$

exists. Moreover, we have

$$\int_E f(t_1, t_2, \dots, t_n)\Delta_1 t_1 \Delta_2 t_2 \dots \Delta_n t_n$$

$$= \int_{a_1}^{b_1}\int_{a_2}^{b_2}\dots\int_{a_{n-1}}^{b_{n-1}} \Delta_1 t_1 \Delta_2 t_2 \dots \Delta_{n-1} t_{n-1} \int_{\phi(t_1,t_2,\dots,t_{n-1})}^{\psi(t_1,t_2,\dots,t_{n-1})} f(t_1, t_2, \dots, t_n)\Delta_n t_n.$$

Proof By Lemma 7.96, it follows that $J(\partial_\Delta E) = 0$, and hence, E is Jordan Δ-measurable. Choose an interval

$$[a_n, b_n] \subset \mathbb{T}_n^0$$

such that the rectangle

$$R = [a_1, b_1) \times [a_2, b_2) \times \dots \times [a_{n-1}, b_{n-1}) \times [a_n, b_n)$$

contains E. Define the function f as in (7.28). Thus,

$$
\begin{aligned}
\int_{a_n}^{b_n} F(t_1, t_2, \dots, t_{n-1}, t_n)\Delta_n t_n &= \int_{a_n}^{\phi(t_1,t_2,\dots,t_{n-1})} F(t_1, t_2, \dots, t_{n-1}, t_n)\Delta_n t_n \\
&\quad + \int_{\phi(t_1,t_2,\dots,t_{n-1})}^{\psi(t_1,t_2,\dots,t_{n-1})} F(t_1, t_2, \dots, t_{n-1}, t_n)\Delta_n t_n \\
&\quad + \int_{\psi(t_1,t_2,\dots,t_{n-1})}^{b_n} F(t_1, t_2, \dots, t_{n-1}, t_n)\Delta_n t_n \\
&= \int_{\phi(t_1,t_2,\dots,t_{n-1})}^{\psi(t_1,t_2,\dots,t_{n-1})} F(t_1, t_2, \dots, t_{n-1}, t_n)\Delta_n t_n \\
&= \int_{\phi(t_1,t_2,\dots,t_{n-1})}^{\psi(t_1,t_2,\dots,t_{n-1})} f(t_1, t_2, \dots, t_{n-1}, t_n)\Delta_n t_n.
\end{aligned}
$$

Therefore,

$$
\begin{aligned}
&\int_R F(t_1, t_2, \dots, t_{n-1}, t_n)\Delta_1 t_1 \Delta_2 t_2 \dots \Delta_{n-1} t_{n-1} \Delta_n t_n \\
&= \int_{a_1}^{b_1} \int_{a_2}^{b_2} \dots \int_{a_{n-1}}^{b_{n-1}} \Delta_1 t_1 \Delta_2 t_2 \dots \Delta_{n-1} t_{n-1} \int_{a_n}^{b_n} F(t_1, t_2, \dots, t_{n-1}, t_n)\Delta_n t_n \\
&= \int_{a_1}^{b_1} \int_{a_2}^{b_2} \dots \int_{a_{n-1}}^{b_{n-1}} \Delta_1 t_1 \Delta_2 t_2 \dots \Delta_{n-1} t_{n-1} \int_{\phi(t_1,t_2,\dots,t_{n-1})}^{\psi(t_1,t_2,\dots,t_{n-1})} f(t_1, t_2, \dots, t_{n-1}, t_n)\Delta_n t_n.
\end{aligned}
$$

On the other hand, by Definition 7.81, we have

$$
\begin{aligned}
&\int_E f(t_1, t_2, \dots, t_{n-1}, t_n)\Delta_1 t_1 \Delta_2 t_2 \dots \Delta_{n-1} t_{n-1} \Delta_n t_n \\
&\qquad = \int_R f(t_1, t_2, \dots, t_{n-1}, t_n)\Delta_1 t_1 \Delta_2 t_2 \dots \Delta_{n-1} t_{n-1} \Delta_n t_n,
\end{aligned}
$$

so that the proof is complete. □

Example 7.98 Let $\mathbb{T}_1 = \mathbb{Z}$ and $\mathbb{T}_2 = \mathbb{Z}$. We will evaluate the integral

$$
I = \int_{-2}^{2} \int_{\sqrt{t_1}}^{\sqrt{3t_1}} (2t_2 + 1)\Delta_2 t_2 \Delta_1 t_1.
$$

We have

$$
\sigma_1(t_1) = t_1 + 1, \quad t_1 \in \mathbb{T}_1, \quad \sigma_2(t_2) = t_2 + 1, \quad t_2 \in \mathbb{T}_2.
$$

Also, with $g(t_1) = t_1^2$ and $h(t_2) = t_2^2$, we get

$$t_1 = \frac{1}{2}g^{\Delta_1}(t_1) - \frac{1}{2}, \quad t_1 \in \mathbb{T}_1, \quad t_2 = \frac{1}{2}h^{\Delta_2}(t_2) - \frac{1}{2}, \quad t_2 \in \mathbb{T}_2.$$

Hence,

$$\begin{aligned}
\int_{\sqrt{t_1}}^{\sqrt{3t_1}} (2t_2 + 1)\Delta_2 t_2 &= \int_{\sqrt{t_1}}^{\sqrt{3t_1}} h^{\Delta_2}(t_2)\Delta_2 t_2 \\
&= h\left(\sqrt{3t_1}\right) - h\left(\sqrt{t_1}\right) \\
&= 3t_1 - t_1 \\
&= 2t_1,
\end{aligned}$$

$$\begin{aligned}
I &= \int_{-2}^{2} \Delta_1 t_1 \int_{\sqrt{t_1}}^{\sqrt{3t_1}} (2t_2 + 1)\Delta_2 t_2 \\
&= 2\int_{-2}^{2} t_1 \Delta_1 t_1 \\
&= 2\int_{-2}^{2} \left(\frac{1}{2}g^{\Delta_1}(t_1) - \frac{1}{2}\right) \Delta_1 t_1 \\
&= \int_{-2}^{2} \left(g^{\Delta_1}(t_1) - 1\right) \Delta_1 t_1 \\
&= g(2) - g(-2) - 4 \\
&= 4 - 4 - 4 \\
&= -4.
\end{aligned}$$

Example 7.99 Let $\mathbb{T}_1 = \mathbb{T}_2 = \mathbb{N}_0$. We will evaluate the integral

$$I = \int_3^5 \int_0^{t_2} (2t_1 + 1)\sin_f(t_2, 1)\Delta_1 t_1 \Delta_2 t_2, \quad \text{where} \quad f(t_2) = t_2^2.$$

Here, with $g(t_1) = t_1^2$, we get

$$\sigma_1(t_1) = t_1 + 1, \quad 2t_1 + 1 = g^{\Delta_1}(t_1), \quad t_1 \in \mathbb{T}_1, \quad \sigma_2(t_2) = t_2 + 1, \quad t_2 \in \mathbb{T}_2$$

so that

$$\begin{aligned}
I &= \int_3^5 \sin_f(t_2, 1)\Delta_2 t_2 \int_0^{t_2} (2t_1 + 1)\Delta_1 t_1 \\
&= \int_3^5 \sin_f(t_2, 1)(g(t_2) - g(0))\Delta_2 t_2 \\
&= \int_3^5 f(t_2) \sin_f(t_2, 1)\Delta_2 t_2 \\
&= -\cos_f(t_2, 1)\Big|_{t_2=3}^{t_2=5} \\
&= \cos_f(3, 1) - \cos_f(5, 1).
\end{aligned}$$

Example 7.100 Let $\mathbb{T}_1 = \mathbb{T}_2 = 2^{\mathbb{N}_0} \cup \{0\}$. We will evaluate the integral

$$I = 7\int_1^2 \int_0^{t_2} t_1^2 \sinh_{t_2^3}(t_2, 1)\Delta_1 t_1 \Delta_2 t_2, \quad \text{where} \quad f(t_2) = t_2^3.$$

We have

$$\begin{aligned}
I &= 7\int_1^2 \sinh_f(t_2, 1)\Delta_2 t_2 \int_0^{t_2} t_1^2 \Delta_1 t_1 \\
&= \int_1^2 t_2^3 \sinh_f(t_2, 1)\Delta_2 t_2 \\
&= \int_1^2 \cosh_f^{\Delta_2}(t_2, 1)\Delta_2 t_2 \\
&= \cosh_f(t_2, 1)\Big|_{t_2=1}^{t_2=2} \\
&= \cosh_f(2, 1) - \cosh_f(1, 1).
\end{aligned}$$

Exercise 7.101 Let $\mathbb{T}_1 = \mathbb{T}_2 = 3^{\mathbb{N}_0} \cup \{0\}$. Compute the integral

$$40\int_1^3 \int_0^{t_1} t_2^3 \cosh_f(t_1, 1)\Delta_2 t_2 \Delta_1 t_1, \quad \text{where} \quad f(t_1) = t_1^4.$$

Solution $\sinh_f(3, 1) - 2$.

7.4 Advanced Practical Problems

Problem 7.102 Let $\mathbb{T}_1 = \mathbb{Z}$ and $\mathbb{T}_2 = 4\mathbb{Z}$. Compute

$$\int_{-4}^{4}\int_{4}^{8} t_2 \log\frac{t_1+1}{t_1}\Delta_1 t_1 \Delta_2 t_2.$$

Solution $-16\log 2$.

Problem 7.103 Let $\phi : [2, 8] \to \mathbb{R}$, $\phi(x) = x^3$. Prove that ϕ satisfies the Lipschitz condition. Find a Lipschitz constant.

Solution $B = 192$.

Problem 7.104 Let

$$\phi(x) = \begin{cases} \frac{1}{x^3-8} & \text{for } x \in [0, 2), \\ -1 & \text{for } x = 2. \end{cases}$$

Check if ϕ satisfies the Lipschitz condition.

Solution No.

Problem 7.105 Let $\mathbb{T}_1 = 2^{\mathbb{N}_0}$ and $\mathbb{T}_2 = \mathbb{Z}$. Prove that

$$\int_{-1}^{3}\int_{1}^{8}(t_1^3 + 3t_1^2 t_2 + t_1 t_2^2 + t_2^3)\Delta_1 t_1 \Delta_2 t_2 = \int_{-1}^{3}\Delta_2 t_2 \int_{1}^{8}(t_1^3 + 3t_1^2 t_2 + t_1 t_2^2 + t_2^3)\Delta_1 t_1$$
$$= \int_{1}^{8}\Delta_1 t_1 \int_{-1}^{3}(t_1^3 + 3t_1^2 t_2 + t_1 t_2^2 + t_2^3)\Delta_2 t_2.$$

Problem 7.106 Let

$$\mathbb{T}_1 = \mathbb{T}_2 = \mathbb{Z}, \quad E = [1, 2] \times [2, 4].$$

Find $J(E)$ and $\int_E (2t_1 - t_2)\Delta_1 t_1 \Delta_2 t_2$.

Solution 6, 0.

Problem 7.107 Let $\mathbb{T}_1 = \mathbb{T}_2 = 2^{\mathbb{N}_0}$. Compute the integral

$$\int_{1}^{3}\int_{t_2}^{2t_2}(t_1^2 + t_1 t_2 + 3t_2^2)\Delta_1 t_1 \Delta_2 t_2.$$

Solution $\frac{80}{3}$.

Problem 7.108 Let $\mathbb{T}_1 = \mathbb{T}_2 = 4^{\mathbb{N}_0} \cup \{0\}$. Compute the integral

$$\int_0^4 \int_0^{t_1} \left(\sinh_f(t_1, 1) - 5t_2 \cosh_{f^2}(t_1, 1)\right) \Delta_2 t_2 \Delta_1 t_1, \quad \text{where} \quad f(t_1) = t_1.$$

Solution $\cosh_f(4, 1) - \cosh_f(0, 1) - \sinh_{f^2}(4, 1) + \sinh_{f^2}(0, 1)$.

Problem 7.109 Let $\mathbb{T}_1 = 2^{\mathbb{N}_0} \cup \{0\}$, $\mathbb{T}_2 = 3^{\mathbb{N}_0} \cup \{0\}$. Compute the integral

$$\int_0^1 \int_0^{3t_1} (t_1 + 2t_2^2) \Delta_2 t_2 \Delta_1 t_1.$$

Solution $\frac{963}{1365}$.

Problem 7.110 Let $\mathbb{T}_1 = \mathbb{N}_0$ and $\mathbb{T}_2 = 2^{\mathbb{N}_0} \cup \{0\}$. Compute the integral

$$\int_0^2 \int_0^{t_1} \cosh_f(t_1, 1)) \Delta_2 t_2 \Delta_1 t_1, \quad \text{where} \quad f(t_1) = t_1.$$

Solution $\sinh_f(2, 1) - \sinh_f(0, 1)$.

Problem 7.111 Let $\mathbb{T}_1 = \mathbb{Z}$ and $\mathbb{T}_2 = 3^{\mathbb{N}_0} \cup \{0\}$. Compute the integral

$$\int_{-1}^1 \int_0^{t_1} e_f(t_1, 1) \Delta_2 t_2 \Delta_1 t_1, \quad \text{where} \quad f(t_1) = t_1.$$

Solution $-e_f(-1, 1)$.

Problem 7.112 Let $\mathbb{T}_1 = \mathbb{T}_2 = [0, 1] \cup \{3\}$, where $[0, 1]$ is the real number interval. Let $E = [0, 1) \times [0, 1)$. Find ∂E and $\partial_\Delta E$.

Solution $\partial E = \{t_1 \in \{0, 1\},\ 0 \le t_2 \le 1\} \cup \{t_2 \in \{0, 1\},\ 0 \le t_1 \le 1\}$, $\partial_\Delta E = \emptyset$.

Problem 7.113 Let $\mathbb{T}_1 = 2^{\mathbb{N}_0}$ and $\mathbb{T}_2 = 3^{\mathbb{N}_0}$. Find $J(\{(2, 3)\})$.

Solution 12.

7.5 Notes and References

In the original papers of Bernd Aulbach and Stefan Hilger [4, 32] on single-variable time scales calculus, the concept of integral was defined by means of an antiderivative (or pre-antiderivative) of a function and called the Cauchy integral. Next, Sonja Sailer [39] used the Darboux definition of the integral. Further, Riemann's definition of the integral on time scales was introduced in [6, 7, 21, 27–29], and a complete theory of integration for single variable time scales was developed. In [2], Calvin Ahlbrandt and Christina Morian introduced double integrals over rectangles on time scales as iterated integrals defined by using antiderivatives of single-variable functions, under the assumption that the order of integration in the iterated integral can be reversed. In [9], the Darboux and Riemann definitions of multiple integrals on time scales over arbitrary regions for functions of two variables are introduced. As before, this chapter considers only delta integrals. Nabla integrals and mixed integrals involving delta integration with respect to a part of the variables and nabla integration with respect to the other part of the variables can be investigated in a similar manner. The reader is also referred to [11] for a treatment of multiple Lebesgue integration on time scales. All results in this chapter are n-dimensional analogues of the two-dimensional results by Bohner and Guseinov [9].

Chapter 8
Line Integrals

8.1 Length of Time Scale Curves

Let $\mathbb{T}$ be a time scale with the forward jump operator σ and the delta operator Δ. Let $a, b \in \mathbb{T}$ with $a < b$. Assume that $\phi_i : [a, b] \to \mathbb{R}$ is continuous, $i \in \{1, \ldots, m\}$.

Definition 8.1 The system of functions

$$x_i = \phi_i(t), \quad t \in [a, b] \subset \mathbb{T}, \quad i \in \{1, \ldots, m\} \tag{8.1}$$

is said to define a time scale continuous *curve* Γ. The points $A(x_1, x_2, \ldots, x_m)$ with the coordinates $x_1, x_2, \ldots, x_m$ given by (8.1) are called the *points* of the curve. The set of all points of the curve will be referred to as the *curve*. The points

$$A_0(\phi_1(a), \ldots, \phi_m(a)) \quad \text{and} \quad A_1(\phi_1(b), \ldots, \phi_m(b))$$

are called the *initial point* and the *final point* of the curve, respectively. A_0 and A_1 are called the *end points* of the curve.

Example 8.2 The system

$$x_1 = t^2 + t, \quad x_2 = t - 2, \quad x_3 = t, \quad t \in [0, 2],$$

defines a continuous curve. Here,

$$\phi_1(t) = t^2 + t, \quad \phi_2(t) = t - 2, \quad \phi_3(t) = t.$$

Thus,

$$\phi_1(0) = 0, \quad \phi_2(0) = -2, \quad \phi_3(0) = 0,$$

$$\phi_1(2) = 6, \quad \phi_2(2) = 0, \quad \phi_3(2) = 2.$$

M. Bohner and S.G. Georgiev, *Multivariable Dynamic Calculus on Time Scales*,
DOI 10.1007/978-3-319-47620-9_8

Moreover, $A_0(0, -2, 0)$ is the initial point and $A_1(6, 0, 2)$ is the final point.

Definition 8.3 If the initial and final points coincide, then the curve is said to be *closed*.

Example 8.4 Consider the curve

$$x_1 = t^2 - 3t, \quad x_2 = t^3 - 4t^2 + 5t, \quad t \in [1, 2].$$

Here,

$$\phi_1(t) = t^2 - 3t, \quad \phi_2(t) = t^3 - 4t^2 + 5t.$$

Thus,

$$\phi_1(1) = -2, \quad \phi_2(1) = 2,$$

$$\phi_1(2) = -2, \quad \phi_2(2) = 2.$$

Hence, $A_1(-2, 2)$ is the initial point and $A_2(-2, 2)$ is the final point. The considered curve is closed.

Definition 8.5 The time scale parameter t is called the *parameter* of the curve.

Definition 8.6 The equations in (8.1) are called the *parametric equations* of the curve Γ.

Definition 8.7 We say that a curve Γ is an *oriented* curve in the sense that a point

$$(x_1', x_2', \ldots, x_m') = (\phi_1(t'), \phi_2(t'), \ldots \phi_m(t'))$$

is regarded as distinct from a point

$$(x_1'', x_2'', \ldots, x_m'') = (\phi_1(t''), \phi_2(t''), \ldots, \phi_m(t''))$$

if $t' \neq t''$ and as preceding $(x_1'', x_2'', \ldots, x_m'')$ if $t' < t''$. The oriented curve Γ is then said to be "traversed in the direction of increasing t."

Definition 8.8 Two curves Γ_1 and Γ_2 with equations

$$x_i = \phi_i(t), \quad i \in \{1, \ldots, m\}, \quad t \in \mathbb{T}_1,$$

and

$$x_i = \psi_i(t), \quad i \in \{1, \ldots, m\}, \quad t \in \mathbb{T}_2,$$

respectively, are regarded as *identical* if the equations of one curve can be transformed into the equations of the other curve by means of a continuous strictly increasing change of the parameter, i.e., if there is a continuous increasing function $\tau = \lambda(t)$, $t \in [a, b]$, with the range $[\alpha, \beta]$, such that

$$\psi_i(\lambda(t)) = \phi_i(t), \quad t \in [a, b].$$

We then say that the two curves have the same direction. We say that the two curves have opposite direction if the function λ is decreasing. In this case, the initial point of Γ_1 is the final point of Γ_2, and vice versa. The curve differing by Γ only by the direction in which it is traversed is denoted by $-\Gamma$.

Example 8.9 The curves

$$x_1 = t, \quad x_2 = t - 1, \quad x_3 = t^2 - 1, \quad t \in [0, 1],$$

and

$$x_1 = t^2, \quad x_2 = t^2 - 1, \quad x_3 = t^4 - 1, \quad t \in [0, 1],$$

are identical.

Example 8.10 The curves

$$x_1 = 2 - t, \quad x_2 = 3 - t, \quad x_3 = t, \quad t \in [0, 1],$$

and

$$x_1 = 2 - t^3, \quad x_2 = 3 - t^3, \quad x_3 = t^3, \quad t \in [0, 1],$$

are identical.

Example 8.11 The curves

$$x_1 = t, \quad x_2 = 2 - \cos t, \quad x_3 = 3 - \tan t, \quad t \in [0, 1],$$

and

$$x_1 = t^2, \quad x_2 = t^4, \quad x_3 = t^2, \quad t \in [0, 1],$$

are not identical.

Definition 8.12 If the same point $(x_1, \ldots, x_m)$ corresponds to more than one parameter value in the half-open interval $[a, b)$, then we say that $(x_1, \ldots, x_m)$ is a *multiple* point of the curve (8.1).

Definition 8.13 A curve with no multiple points is called a *simple curve* or *Jordan curve*.

Example 8.14 The curve

$$x_1 = t, \quad x_2 = 4 - t, \quad x_3 = 5 + t, \quad t \in [1, 4],$$

is a Jordan curve.

Example 8.15 Consider the curve

$$x_1 = t^2 - 5t, \quad x_2 = t^3 - 8t^2 + 21t, \quad t \in [0, 4].$$

Then

$$\phi_1(t) = t^2 - 5t, \quad \phi_2(t) = t^3 - 8t^2 + 21t.$$

Take the point $A(-6, 18)$. We can obtain it for $t = 2$ and $t = 3$. Therefore, the point A is a multiple point, and the considered curve is not a Jordan curve.

Exercise 8.16 Determine if the curve

$$x_1 = t^2 - 2t, \quad x_2 = t^3, \quad x_3 = t^2 + 1, \quad t \in [0, 2],$$

is a Jordan curve.

Exercise 8.17 Check if the curves

$$x_1 = t, \quad x_2 = 3 + 2t, \quad x_3 = t^2 - 1, \quad t \in [0, 2],$$

and

$$x_1 = t^2, \quad x_2 = t^4 - 3, \quad x_3 = t^3, \quad t \in [2, 4],$$

are identical.

Let Γ be a time scale continuous curve given by (8.1). Consider a partition of the interval $[a, b]$,

$$P = \{t_0, t_1, \ldots, t_n\} \subset [a, b], \quad \text{where} \quad a = t_0 < t_1 < \ldots < t_n = b. \tag{8.2}$$

For a given partition P, we denote by $A_0, A_1, \ldots, A_n$ the corresponding points of the curve Γ, i.e., $A_j(\phi_1(t_j), \phi_2(t_j), \ldots, \phi_m(t_j))$, $j \in \{0, 1, \ldots, n\}$. We set

$$l(\Gamma, P) = \sum_{j=1}^{n} d(A_{j-1}, A_j) = \sum_{j=1}^{n} \sqrt{\sum_{k=1}^{m} \left(\phi_k(t_j) - \phi_k(t_{j-1})\right)^2}, \tag{8.3}$$

where $d(A_{j-1}, A_j)$ denotes the distance from the point A_{j-1} to A_j.

Definition 8.18 The curve Γ is said to be *rectifiable* if

$$l(\Gamma) = \sup\{l(\Gamma, P) : \ P \quad \text{is a partition of} \quad [a, b]\} < \infty.$$

In this case, the nonnegative number $l = l(\Gamma)$ is called the *length* of the curve Γ. If the supremum does not exist, then the curve is said to be *nonrectifiable*.

Proposition 8.19 *If P and Q are partitions of $[a, b]$ and Q is a refinement of P, then*

$$l(\Gamma, P) \leq l(\Gamma, Q).$$

Proof We assume that P is given by (8.2). Without loss of generality, we suppose that Q has only one more point, say s, than P. Then

$$a = t_0 < t_1 < \ldots < t_{k-1} < s < t_k < \ldots < t_n = b$$

for some $k \in \{1, \ldots, n\}$. Let

$$B = B(\phi_1(s), \ldots, \phi_m(s)).$$

Thus,

$$l(\Gamma, Q) - l(\Gamma, P) = d(B, A_{k-1}) + d(A_k, B) - d(A_k, A_{k-1}) \geq 0,$$

where the triangle inequality for the distance between the points is used. □

Theorem 8.20 *If the curve Γ is rectifiable, then its length does not depend on the parameterization of this curve.*

Proof Suppose the curve Γ has two parameterizations and let t and τ be the parameters of these parameterizations defined on the intervals $[a, b]$ and $[\alpha, \beta]$, respectively. Note that, since t and τ are strictly increasing and continuous functions of each other, to each partition of $[a, b]$, there corresponds a definite partition P' of $[\alpha, \beta]$, and vice versa. We have

$$l(\Gamma, P) = l(\Gamma, P').$$

Hence, the lengths of Γ corresponding to the two parameterizations of Γ are equal. □

Theorem 8.21 *If a rectifiable curve Γ is split by means of a finite number of points $A_0, A_1, \ldots, A_n$ into a finite number of curves Γ_i and if the points A_i correspond to the values t_i of the parameter t and*

$$a = t_0 < t_1 < \ldots < t_n = b,$$

then each curve Γ_i is rectifiable and

$$l(\Gamma) = \sum_{i=1}^{n} l(\Gamma_i).$$

Proof Without loss of generality, we suppose that the curve Γ is split only into two curves Γ_1 and Γ_2 by means of a point A. Assume that the points of the curve

Γ_1 correspond to $[a, c]$, and the points of Γ_2 correspond to the interval $[c, b]$. Let P_1 and P_2 be arbitrary partitions of $[a, c]$ and $[c, b]$, respectively. Then their union $P = P_1 \cup P_2$ is a partition of $[a, b]$ and

$$l(\Gamma, P) = l(\Gamma_1, P_1) + l(\Gamma_2, P_2). \tag{8.4}$$

Since $l(\Gamma, P) < \infty$, we have that $l(\Gamma_1, P_1) < \infty$ and $l(\Gamma_2, P_2) < \infty$. Because P_1 and P_2 were arbitrarily chosen partitions of $[a, c]$ and $[c, b]$, respectively, we get

$$l(\Gamma_1) < \infty \quad \text{and} \quad l(\Gamma_2) < \infty,$$

i.e., the curves Γ_1 and Γ_2 are rectifiable. By (8.4), we obtain

$$l(\Gamma_1) + l(\Gamma_2) \le l(\Gamma). \tag{8.5}$$

Assume that

$$l(\Gamma_1) + l(\Gamma_2) < l(\Gamma).$$

Thus, there exists $\varepsilon > 0$ such that

$$\varepsilon = l(\Gamma) - l(\Gamma_1) - l(\Gamma_2). \tag{8.6}$$

There exists a partition P_0 of the interval $[a, b]$ such that

$$l(\Gamma) - l(\Gamma, P_0) < \varepsilon. \tag{8.7}$$

We add the point c to the partition P_0 and denote the obtained partition by P. Thus, P is a refinement of P_0, and by Proposition 8.19, we have that

$$l(\Gamma, P_0) \le l(\Gamma, P).$$

Therefore, using (8.7), we get

$$l(\Gamma) - l(\Gamma, P) \le l(\Gamma) - l(\Gamma, P_0) < \varepsilon. \tag{8.8}$$

Note that P can be represented as a union of some partitions P_1 and P_2 of the intervals $[a, c]$ and $[c, b]$. Thus,

$$l(\Gamma, P) = l(\Gamma, P_1) + l(\Gamma, P_2).$$

From here and from (8.8), we obtain

$$l(\Gamma) - l(\Gamma_1, P_1) - l(\Gamma_2, P_2) < \varepsilon.$$

Because

$$l(\Gamma_1, P_1) \le l(\Gamma_1) \quad \text{and} \quad l(\Gamma_2, P_2) \le l(\Gamma_2),$$

we obtain

$$l(\Gamma) - l(\Gamma_1) - l(\Gamma_2) < \varepsilon,$$

which contradicts (8.6). This concludes the proof. □

Lemma 8.22 *For any $l \in \mathbb{N}$, we have*

$$\left| \sqrt{\sum_{k=1}^{l} x_k^2} - \sqrt{\sum_{k=1}^{l} y_k^2} \right| \le \sum_{k=1}^{l} |x_k - y_k|. \tag{8.9}$$

Proof We will use induction.

1. For $l = 1$ and $l = 2$, the assertion is evidently true.
2. We assume that (8.9) holds for some $l \in \mathbb{N}$. We will prove the assertion for $l + 1$. We have

$$\begin{aligned}
\left| \sqrt{\sum_{k=1}^{l+1} x_k^2} - \sqrt{\sum_{k=1}^{l+1} y_k^2} \right| &= \left| \sqrt{\left(\sqrt{\sum_{k=1}^{l} x_k^2}\right)^2 + x_{l+1}^2} - \sqrt{\left(\sqrt{\sum_{k=1}^{l} y_k^2}\right)^2 + y_{l+1}^2} \right| \\
&\le \left| \sqrt{\sum_{k=1}^{l} x_k^2} - \sqrt{\sum_{k=1}^{l} y_k^2} \right| + |x_{l+1} - y_{l+1}| \\
&\le \sum_{k=1}^{l} |x_k - y_k| + |x_{l+1} - y_{l+1}| \\
&= \sum_{k=1}^{l+1} |x_k - y_k|.
\end{aligned}$$

This completes the proof. □

Theorem 8.23 *Assume the functions ϕ_i, $i \in \{1, \ldots, m\}$, are continuous on $[a, b]$ and Δ-differentiable on $[a, b)$. If their Δ-derivatives ϕ_i^Δ are bounded and Δ-integrable on $[a, b)$, then the curve Γ, defined by* (8.1), *is rectifiable, and its length $l(\Gamma)$ can be evaluated by the formula*

$$l(\Gamma) = \int_a^b \sqrt{\sum_{k=1}^{m} \left(\phi_k^\Delta(t)\right)^2} \Delta t.$$

Proof Let P be an arbitrary partition of $[a, b]$ of the form (8.2). Consider $l(\Gamma, P)$ defined by (8.3). Applying to each of the functions ϕ_k the mean value theorem (Theorem 2.36) on $[t_{j-1}, t_j]$ for $j \in \{1, \dots, n\}$, we get that there exist points ξ_j^k and η_j^k such that

$$\phi_k^\Delta(\xi_j^k)(t_j - t_{j-1}) \le \phi_k(t_j) - \phi_k(t_{j-1}) \le \phi_k^\Delta(\eta_j^k)(t_j - t_{j-1}). \tag{8.10}$$

We set

$$A_{kj} = \max\{|\phi_k^\Delta(\xi_j^k)|, |\phi_k^\Delta(\eta_j^k)|\}.$$

Thus,

$$|\phi_k(t_j) - \phi_k(t_{j-1})| \le A_{kj}(t_j - t_{j-1}).$$

Because $\phi_k^\Delta(t)$ are bounded on $[a, b]$, there exists a constant $C > 0$ such that

$$|\phi_k^\Delta(t)| \le C \quad \text{for all} \quad t \in [a, b], \quad k \in \{1, \dots, m\}.$$

Consequently,

$$|\phi_k(t_j) - \phi_k(t_{j-1})| \le C(t_j - t_{j-1})$$

for all $j \in \{1, \dots, n\}$ and all $k \in \{1, \dots, m\}$. Hence, using (8.3), we obtain

$$\begin{aligned} l(\Gamma, P) &= \sum_{j=1}^{n} \sqrt{\sum_{k=1}^{m} (\phi_k(t_j) - \phi_k(t_{j-1}))^2} \\ &\le \sum_{j=1}^{n} \sqrt{\sum_{k=1}^{m} C^2 (t_j - t_{j-1})^2} \\ &= C\sqrt{m} \sum_{j=1}^{n} (t_j - t_{j-1}) \\ &= C\sqrt{m}(b - a). \end{aligned}$$

Therefore, $l(\Gamma, P)$ is bounded by $C\sqrt{m}(b - a)$, which does not depend on the choice of the partition P of the interval $[a, b]$. From here, we conclude that $l(\Gamma) < \infty$, and then Γ is a rectifiable curve. Now, we consider the Riemann sum

$$S = \sum_{j=1}^{n} \sqrt{\sum_{k=1}^{m} \left(\phi_k^\Delta(\xi_j^k)\right)^2} (t_j - t_{j-1})$$

of the Δ-integrable function $\sum_{k=1}^{m} \left(\phi_k^{\Delta}(t)\right)^2$, corresponding to the partition P of $[a,b]$ given by (8.2) and the choice of intermediate points ξ_j^k defined by (8.10). For every $\delta > 0$, there exists at least one partition P of $[a,b]$ of the form (8.2) such that, for each $j \in \{1, \dots, n\}$, either $t_j - t_{j-1} \le \delta$ or $t_j - t_{j-1} > \delta$ and $\sigma(t_{j-1}) = t_j$. We denote by $\mathscr{P}_\delta([a,b])$ the set of all such partitions P of $[a,b]$. Let

$$I = \int_a^b \sqrt{\sum_{k=1}^{m} \left(\phi_k^{\Delta}(t)\right)^2}\,\Delta t.$$

From (8.10), we have that

$$\begin{aligned} 0 &\le \phi_k(t_j) - \phi_k(t_{j-1}) - \phi_k^{\Delta}(\xi_j^k)(t_j - t_{j-1}) \\ &\le \phi_k^{\Delta}(\eta_j^k)(t_j - t_{j-1}) - \phi_k^{\Delta}(\xi_j^k)(t_j - t_{j-1}) \\ &= \left(\phi_k^{\Delta}(\eta_j^k) - \phi_k^{\Delta}(\xi_j^k)\right)(t_j - t_{j-1}). \end{aligned}$$

Consequently,

$$\phi_k(t_j) - \phi_k(t_{j-1}) = \left(\phi_k^{\Delta}(\xi_j^k) + \alpha_{jk}\right)(t_j - t_{j-1}),$$

where

$$0 \le \alpha_{jk} \le \phi_k^{\Delta}(\eta_j^k) - \phi_k^{\Delta}(\xi_j^k) \le M_j^k - m_j^k$$

and

$$M_j^k = \sup_{t \in [t_{j-1}, t_j]} \phi_k^{\Delta}(t), \quad m_j^k = \inf_{t \in [t_{j-1}, t_j]} \phi_k^{\Delta}(t).$$

Using Lemma 8.22, we obtain

$$\begin{aligned} \left| \sqrt{\sum_{k=1}^{m} \left(\phi_k^{\Delta}(\xi_j^k) + \alpha_{jk}\right)^2} - \sqrt{\sum_{k=1}^{m} \left(\phi_k^{\Delta}(\xi_j^k)\right)^2} \right| &\le \sum_{k=1}^{m} |\phi_k^{\Delta}(\xi_j^k) + \alpha_{jk} - \phi_k^{\Delta}(\xi_j^k)| \\ &= \sum_{k=1}^{m} |\alpha_{jk}| \\ &\le \sum_{k=1}^{m} (M_j^k - m_j^k). \end{aligned}$$

Therefore,

$$\begin{aligned}|l(\Gamma, P) - S| &= \Bigg|\sum_{j=1}^{n}\sqrt{\sum_{k=1}^{m}(\phi_k(t_j)-\phi_k(t_{j-1}))^2} \\ &\quad - \sum_{j=1}^{n}\sqrt{\sum_{k=1}^{m}\left(\phi_k^{\Delta}(\xi_j^k)\right)^2}(t_j - t_{j-1})\Bigg| \\ &= \Bigg|\sum_{j=1}^{n}\sqrt{\sum_{k=1}^{m}\left(\phi_k^{\Delta}(\xi_j^k)+\alpha_{jk}\right)^2}(t_j - t_{j-1}) \\ &\quad - \sum_{j=1}^{n}\sqrt{\sum_{k=1}^{m}\left(\phi_k^{\Delta}(\xi_j^k)\right)^2}(t_j - t_{j-1})\Bigg| \\ &\leq \sum_{j=1}^{n}\left(\sum_{k=1}^{m}(M_j^k - m_j^k)\right)(t_j - t_{j-1}) \\ &= \sum_{k=1}^{m}\sum_{j=1}^{n}(M_j^k - m_j^k)(t_j - t_{j-1}) \\ &= \sum_{k=1}^{m}\left(U(\phi_k^{\Delta}, P) - L(\phi_k^{\Delta}, P)\right),\end{aligned}$$

where U and L denote the upper and lower Darboux Δ-sums, respectively, of ϕ_k^{Δ}. Since the functions $\sqrt{\sum_{k=1}^{m}\left(\phi_k^{\Delta}\right)^2}$ and ϕ_k^{Δ}, $k \in \{1, \ldots, m\}$, are Δ-integrable over $[a, b)$, for every $\varepsilon > 0$, there exists $\delta > 0$ such that

$$|S - I| < \frac{\varepsilon}{4}, \quad U(\phi_k^{\Delta}, P) - L(\phi_k^{\Delta}, P) < \frac{\varepsilon}{4m}$$

for all $P \in \mathscr{P}_\delta([a, b])$. Therefore,

$$\begin{aligned}|l(\Gamma, P) - I| &= |l(\Gamma, P) - S + S - I| \\ &\leq |l(\Gamma, P) - S| + |S - I| \\ &< \frac{\varepsilon}{4} + \sum_{k=1}^{m}\frac{\varepsilon}{4m} \\ &= \frac{\varepsilon}{2}.\end{aligned}$$

On the other hand, among the partitions $P \in \mathscr{P}_\delta([a,b])$, there exists a partition P such that

$$0 < l(\Gamma) - l(\Gamma, P) < \frac{\varepsilon}{2}.$$

Consequently,

$$\begin{aligned}|l(\Gamma) - I| &= |l(\Gamma) - l(\Gamma, P) + l(\Gamma, P) - I|\\ &\le |l(\Gamma) - l(\Gamma, P)| + |l(\Gamma, P) - I|\\ &< \frac{\varepsilon}{2} + \frac{\varepsilon}{2}\\ &= \varepsilon.\end{aligned}$$

Since $\varepsilon > 0$ was arbitrarily chosen, we have that $l(\Gamma) = I$, and the proof is completed. □

Example 8.24 Let $\mathbb{T} = \mathbb{Z}$. Consider the curve

$$\Gamma = \begin{cases} x_1 = t^2 + 2t \\ x_2 = t^3 - 3 \\ x_3 = t, \quad t \in [0, 2]. \end{cases}$$

Here,

$$\sigma(t) = t + 1, \quad \mu(t) = 1$$

and

$$\phi_1(t) = t^2 + 2t, \quad \phi_2(t) = t^3 - 3, \quad \phi_3(t) = t.$$

Thus,

$$\begin{aligned}\phi_1^\Delta(t) &= \sigma(t) + t + 2\\ &= t + 1 + t + 2\\ &= 2t + 3,\\ \phi_2^\Delta(t) &= (\sigma(t))^2 + t\sigma(t) + t^2\end{aligned}$$

$$= (t+1)^2 + t(t+1) + t^2$$

$$= t^2 + 2t + 1 + t^2 + t + t^2$$

$$= 3t^2 + 3t + 1,$$

$$\phi_3^{\Delta}(t) = 1.$$

Hence,

$$(\phi_1^{\Delta}(t))^2 + (\phi_2^{\Delta}(t))^2 + (\phi_3^{\Delta}(t))^2 = (2t+3)^2 + (3t^2+3t+1)^2 + 1^2$$

$$= 4t^2 + 12t + 9 + 9t^4 + 9t^2 + 1$$

$$+8t^3 + 6t^2 + 6t + 1$$

$$= 9t^4 + 8t^3 + 19t^2 + 18t + 11.$$

Consequently,

$$l(\Gamma) = \int_0^2 \sqrt{(\phi_1^{\Delta}(t))^2 + (\phi_2^{\Delta}(t))^2 + (\phi_3^{\Delta}(t))^2}\Delta t$$

$$= \int_0^2 \sqrt{9t^4 + 8t^3 + 19t^2 + 18t + 11}\Delta t$$

$$= \sqrt{9t^4 + 8t^3 + 19t^2 + 18t + 11}\Big|_{t=0} + \sqrt{9t^4 + 8t^3 + 19t^2 + 18t + 11}\Big|_{t=1}$$

$$= \sqrt{11} + 5\sqrt{3}.$$

Example 8.25 Let $\mathbb{T} = 2^{\mathbb{N}_0} \cup \{0\}$. Consider the curve

$$\Gamma = \begin{cases} x_1(t) = e_f(t,1), \quad \text{where} \quad f(t) = t^2 \\ x_2(t) = t^3 \\ x_3(t) = t^2 + 2t, \quad t \in [1,4]. \end{cases}$$

Here,

$$\sigma(t) = 2t, \quad \mu(t) = t$$

and

$$\phi_1(t) = e_f(t,1), \quad \phi_2(t) = t^3, \quad \phi_3(t) = t^2 + 2t.$$

Thus,

$$\begin{aligned}
\phi_1^{\Delta}(t) &= f(t)e_f(t,1),\\
\phi_2^{\Delta}(t) &= (\sigma(t))^2 + t\sigma(t) + t^2\\
&= 4t^2 + 2t^2 + t^2\\
&= 7t^2,\\
\phi_3^{\Delta}(t) &= \sigma(t) + t + 2\\
&= 2t + t + 2\\
&= 3t + 2.
\end{aligned}$$

Therefore,

$$\begin{aligned}
\left(\phi_1^{\Delta}(t)\right)^2 + \left(\phi_2^{\Delta}(t)\right)^2 + \left(\phi_3^{\Delta}(t)\right)^2 &= \left(t^2 e_f(t,1)\right)^2 + (7t^2)^2 + (3t+2)^2\\
&= t^4 e_f^2(t,1) + 49t^4 + 9t^2 + 12t + 4.
\end{aligned}$$

Hence,

$$\begin{aligned}
l(\Gamma) &= \int_1^4 \sqrt{\left(\phi_1^{\Delta}(t)\right)^2 + \left(\phi_2^{\Delta}(t)\right)^2 + \left(\phi_3^{\Delta}(t)\right)^2}\,\Delta t\\
&= \int_1^4 \sqrt{t^4 e_f^2(t,1) + 49t^4 + 9t^2 + 12t + 4}\,\Delta t\\
&= t\sqrt{t^4 e_f^2(t,1) + 49t^4 + 9t^2 + 12t + 4}\Big|_{t=1}\\
&\quad + t\sqrt{t^4 e_f^2(t,1) + 49t^4 + 9t^2 + 12t + 4}\Big|_{t=2}\\
&= \sqrt{e_f^2(1,1) + 49 + 9 + 12 + 4} + 2\sqrt{16 e_f(2,1) + 784 + 36 + 24 + 4}
\end{aligned}$$

$$= \sqrt{75} + 2\sqrt{16e_f(2,1) + 848}$$

$$= 5\sqrt{3} + 2\sqrt{16e_f(2,1) + 848}.$$

Example 8.26 Let $\mathbb{T} = \mathbb{N}_0^2$. Consider the curve

$$\Gamma = \begin{cases} x_1(t) = t + 2 \\ x_2(t) = t^2 + 3, \quad t \in [1, 4]. \end{cases}$$

Here,

$$\sigma(t) = (1 + \sqrt{t})^2, \quad \mu(t) = 1 + 2\sqrt{t}$$

and

$$\phi_1(t) = t + 2, \quad \phi_2(t) = t^2 + 3.$$

Thus,

$$\phi_1^\Delta(t) = 1,$$

$$\begin{aligned} \phi_2^\Delta(t) &= t + \sigma(t) \\ &= t + (1 + \sqrt{t})^2 \\ &= t + t + 2\sqrt{t} + 1 \\ &= 2t + 2\sqrt{t} + 1. \end{aligned}$$

Therefore,

$$\begin{aligned} \left(\phi_1^\Delta(t)\right)^2 + \left(\phi_2^\Delta(t)\right)^2 &= 1 + (1 + 2t + 2\sqrt{t})^2 \\ &= 1 + 4t^2 + 4t + 1 + 8t\sqrt{t} + 4t + 4\sqrt{t} \\ &= 4t^2 + 8t\sqrt{t} + 8t + 4\sqrt{t} + 2. \end{aligned}$$

Hence,

$$
\begin{aligned}
l(\Gamma) &= \int_1^4 \sqrt{\left(\phi_1^{\Delta}(t)\right)^2 + \left(\phi_2^{\Delta}(t)\right)^2}\Delta t \\
&= \int_1^4 \sqrt{4t^2 + 8t\sqrt{t} + 8t + 4\sqrt{t} + 2}\Delta t \\
&= (1 + 2\sqrt{t})\sqrt{4t^2 + 8t\sqrt{t} + 8t + 4\sqrt{t} + 2}\Big|_{t=1} \\
&= 3\sqrt{4 + 8 + 8 + 4 + 2} \\
&= 3\sqrt{26}.
\end{aligned}
$$

Example 8.27 Let $\mathbb{T} = [-3, 1] \cup \{2\}$, where $[-3, 1]$ is the real line interval. Consider the curve

$$
\Gamma = \begin{cases} x_1 = t^3 \\ x_2 = t^3 + 2, \quad t \in \mathbb{T}. \end{cases}
$$

Here,

$$
\sigma(t) = \begin{cases} t & \text{for } t \in [-3, 1) \\ 2 & \text{for } t = 1, \end{cases} \qquad \mu(t) = \begin{cases} 0 & \text{for } t \in [-3, 1) \\ 1 & \text{for } t = 1 \end{cases}
$$

and

$$
\phi_1(t) = t^3, \quad \phi_2(t) = t^3 + 2.
$$

We have

$$
\phi_1^{\Delta}(t) = \phi_2^{\Delta}(t) = \begin{cases} 3t^2 & \text{for } t \in [-3, 1) \\ 7 & \text{for } t = 1. \end{cases}
$$

Let

$$
h(t) = \sqrt{\left(\phi_1^{\Delta}(t)\right)^2 + \left(\phi_2^{\Delta}(t)\right)^2}.
$$

Thus,

$$
h(t) = \begin{cases} \sqrt{9t^4 + 9t^4} & \text{for } t \in [-3, 1) \\ 7\sqrt{2} & \text{for } t = 1. \end{cases}
$$

Consequently,

$$\begin{aligned}
l(\Gamma) &= \int_{-3}^{2} h(t)\Delta t \\
&= \int_{-3}^{1} h(t)\Delta t + \int_{1}^{2} h(t)\Delta t \\
&= 3\sqrt{2}\int_{-3}^{1} t^2 \mathrm{d}t + h(1) \\
&= \sqrt{2}t^3\Big|_{t=-3}^{t=1} + 7\sqrt{2} \\
&= 28\sqrt{2} + 7\sqrt{2} \\
&= 35\sqrt{2}.
\end{aligned}$$

Exercise 8.28 Let $\mathbb{T} = (-\infty, 0] \cup \mathbb{N}$, where $(-\infty, 0]$ is the real line interval. Find $l(\Gamma)$, where

$$\Gamma = \begin{cases} x_1 = t^3 \\ x_2 = t^2, \quad t \in [-1, 0] \cup \{1, 2, 3\}. \end{cases}$$

Solution $\frac{1}{27}(8 - 13\sqrt{13}) + \sqrt{2} + \sqrt{58} + \sqrt{386}.$

8.2 Line Integrals of the First Kind

Let Γ be a curve defined by (8.1). Put

$$A = (\phi_1(a), \ldots, \phi_m(a)) \quad \text{and} \quad B = (\phi_1(b), \ldots, \phi_m(b)).$$

Suppose that the function $f(x_1, \ldots, x_m)$ is defined and continuous on the curve Γ. Let

$$P = \{t_0, t_1, \ldots, t_n\} \subset [a, b],$$

where

$$a = t_0 < t_1 < \ldots < t_n = b,$$

be a partition of the interval $[a, b]$, and put

$$A_i = (\phi_1(t_i), \ldots, \phi_m(t_i)), \quad i \in \{0, 1, \ldots, n\}.$$

Take any $\tau_k \in [t_{k-1}, t_k)$, $k \in \{1, \ldots, n\}$. Denote by l_k the length of the piece of the curve Γ between its points A_{k-1} and A_k. Then, using Theorem 8.23, we have

$$l_k = \int_{t_{k-1}}^{t_k} \sqrt{\sum_{l=1}^{m} \left(\phi_l^{\Delta}(t)\right)^2}\, \Delta t.$$

We introduce

$$S_1 = \sum_{k=1}^{n} f(\phi_1(\tau_k), \ldots, \phi_m(\tau_k)) l_k.$$

Definition 8.29 We say that a complex number I_1 is the line delta integral of the *first kind* of the function f along the curve Γ if for each $\varepsilon > 0$, there exists $\delta > 0$ such that

$$|S_1 - I_1| < \varepsilon$$

for every integral sum S_1 of f corresponding to a partition $P \in \mathscr{P}_\delta([a, b])$ independent of the way in which $\tau_k \in [t_{k-1}, t_k)$ for $k \in \{1, 2, \ldots, n\}$ is chosen. We denote the number I_1 symbolically by

$$\int_\Gamma f(x_1, \ldots, x_m) \Delta t \quad \text{or} \quad \int_{AB} f(x_1, \ldots, x_m) \Delta l, \tag{8.11}$$

where A and B are the initial and final points of the curve Γ, respectively.

Theorem 8.30 *Suppose that the curve Γ is given by* (8.1)*, where ϕ_i, $i \in \{1, \ldots, m\}$, are continuous on $[a, b]$ and Δ-differentiable on $[a, b)$. If ϕ_i^{Δ}, $i \in \{1, \ldots, m\}$, are bounded and Δ-integrable over $[a, b)$ and if the function f is continuous on Γ, then the line integral* (8.11) *exists and can be computed by the formula*

$$\int_\Gamma f(x_1, \ldots, x_m) \Delta l = \int_a^b f(\phi_1(t), \ldots, \phi_m(t)) \sqrt{\sum_{k=1}^{m} \left(\phi_k^{\Delta}(t)\right)^2}\, \Delta t.$$

Proof Let $\varepsilon > 0$ be arbitrarily chosen. We have

$$\begin{aligned} S_1 &= \sum_{k=1}^{n} f(\phi_1(\tau_k), \ldots, \phi_m(\tau_k)) \int_{t_{k-1}}^{t_k} \sqrt{\sum_{l=1}^{m} \left(\phi_l^{\Delta}(t)\right)^2}\, \Delta t \\ &= \sum_{k=1}^{n} \int_{t_{k-1}}^{t_k} f(\phi_1(\tau_k), \ldots, \phi_m(\tau_k)) \sqrt{\sum_{l=1}^{m} \left(\phi_l^{\Delta}(t)\right)^2}\, \Delta t. \end{aligned}$$

Let

$$I=\sum_{k=1}^{n}\int_{t_{k-1}}^{t_k} f(\phi_1(t),\dots,\phi_m(t))\sqrt{\sum_{l=1}^{m}\left(\phi_l^{\Delta}(t)\right)^2}\,\Delta t.$$

Hence,

$$S_1-I=\sum_{k=1}^{n}\int_{t_{k-1}}^{t_k}\left(f(\phi_1(\tau_k),\dots,\phi_m(\tau_k))-f(\phi_1(t),\dots,\phi_m(t))\right)\sqrt{\sum_{l=1}^{m}\left(\phi_l^{\Delta}(t)\right)^2}\,\Delta t.$$

Since f is continuous on Γ, there exists $\delta>0$ such that $t,\tau\in[a,b]$ and $|t-\tau|\le\delta$ imply

$$|f(\phi_1(t),\dots,\phi_m(t))-f(\phi_1(\tau),\dots,\phi_m(\tau))|<\frac{\varepsilon}{l(\Gamma)}.$$

Thus,

$$\begin{aligned}
|S_1-I| &= \left|\sum_{k=1}^{n}\int_{t_{k-1}}^{t_k}\left(f(\phi_1(\tau_k),\dots,\phi_m(\tau_k))-f(\phi_1(t),\dots,\phi_m(t))\right)\sqrt{\sum_{l=1}^{m}\left(\phi_l^{\Delta}(t)\right)^2}\,\Delta t\right|\\
&\le \sum_{k=1}^{n}\int_{t_{k-1}}^{t_k}|f(\phi_1(\tau_k),\dots,\phi_m(\tau_k))-f(\phi_1(t),\dots,\phi_m(t))|\sqrt{\sum_{l=1}^{m}\left(\phi_l^{\Delta}(t)\right)^2}\,\Delta t\\
&= \sum_{|t_k-t_{k-1}|\le\delta}\int_{t_{k-1}}^{t_k}|f(\phi_1(\tau_k),\dots,\phi_m(\tau_k))-f(\phi_1(t),\dots,\phi_m(t))|\sqrt{\sum_{l=1}^{m}\left(\phi_l^{\Delta}(t)\right)^2}\,\Delta t\\
&\quad+\sum_{|t_k-t_{k-1}|>\delta}\int_{t_{k-1}}^{t_k}|f(\phi_1(\tau_k),\dots,\phi_m(\tau_k))-f(\phi_1(t),\dots,\phi_m(t))|\sqrt{\sum_{l=1}^{m}\left(\phi_l^{\Delta}(t)\right)^2}\,\Delta t\\
&\le \frac{\varepsilon}{l(\Gamma)}\sum_{|t_k-t_{k-1}|<\delta}\int_{t_{k-1}}^{t_k}\sqrt{\sum_{l=1}^{m}\left(\phi_l^{\Delta}(t)\right)^2}\,\Delta t\\
&\le \frac{\varepsilon}{l(\Gamma)}\sum_{k=1}^{n}\int_{t_{k-1}}^{t_k}\sqrt{\sum_{l=1}^{m}\left(\phi_l^{\Delta}(t)\right)^2}\,\Delta t\\
&= \frac{\varepsilon}{l(\Gamma)}l(\Gamma)\\
&= \varepsilon,
\end{aligned}$$

which completes the proof. □

Example 8.31 Let $\mathbb{T} = \mathbb{Z}$ and

$$\Gamma = \begin{cases} x_1 = t^2 + 2t + 1 \\ x_2 = t^3 - 2t^2 \\ x_3 = t, \quad t \in [0, 3]. \end{cases}$$

We will compute

$$I = \int_\Gamma (x_1 + x_2)\Delta l.$$

Here,

$$\sigma(t) = t + 1, \quad \mu(t) = 1$$

and

$$\phi_1(t) = t^2 + 2t + 1, \quad \phi_2(t) = t^3 - 2t^2, \quad \phi_3(t) = t.$$

Thus,

$$\begin{aligned} \phi_1^\Delta(t) &= \sigma(t) + t + 2 \\ &= t + 1 + t + 2 \\ &= 2t + 3, \\ \phi_2^\Delta(t) &= (\sigma(t))^2 + t\sigma(t) + t^2 - 2(\sigma(t) + t) \\ &= (t+1)^2 + t(t+1) + t^2 - 2(t + 1 + t) \\ &= t^2 + 2t + 1 + t^2 + t + t^2 - 4t - 2 \\ &= 3t^2 - t - 1, \\ \phi_3^\Delta(t) &= 1, \end{aligned}$$

$$\left(\phi_1^\Delta(t)\right)^2 + \left(\phi_2^\Delta(t)\right)^2 + \left(\phi_3^\Delta(t)\right)^2 = (2t+3)^2 + (3t^2 - t - 1)^2 + 1$$

$$= 4t^2 + 12t + 9 + 9t^4 + t^2 + 1 - 6t^3 - 6t^2 + 2t + 1$$

$$= 9t^4 - 6t^3 - t^2 + 14t + 11.$$

Therefore,

$$\begin{aligned}
I &= \int_0^3 (t^2 + 2t + 1 + t^3 - 2t^2)\sqrt{9t^4 - 6t^3 - t^2 + 14t + 11}\Delta t \\
&= \int_0^3 (t^3 - t^2 + 2t + 1)\sqrt{9t^4 - 6t^3 - t^2 + 14t + 11}\Delta t \\
&= (t^3 - t^2 + 2t + 1)\sqrt{9t^4 - 6t^3 - t^2 + 14t + 11}\Big|_{t=0} \\
&\quad +(t^3 - t^2 + 2t + 1)\sqrt{9t^4 - 6t^3 - t^2 + 14t + 11}\Big|_{t=1} \\
&\quad +(t^3 - t^2 + 2t + 1)\sqrt{9t^4 - 6t^3 - t^2 + 14t + 11}\Big|_{t=2} \\
&= \sqrt{11} + 9\sqrt{3} + 9\sqrt{131}.
\end{aligned}$$

Example 8.32 Let $\mathbb{T} = 3^{\mathbb{N}_0}$ and

$$\Gamma = \begin{cases} x_1 = t + 2 \\ x_2 = t^2 + t \\ x_3 = t^3, \quad t \in [1, 3]. \end{cases}$$

We will compute

$$I = \int_\Gamma x_1 x_2 \Delta l.$$

Here,

$$\sigma(t) = 3t, \quad \mu(t) = 2t$$

and

$$\phi_1(t) = t + 2, \quad \phi_2(t) = t^2 + t, \quad \phi_3(t) = t^3.$$

Thus,

$$\phi_1^\Delta(t) = 1,$$

$$
\begin{aligned}
\phi_2^{\Delta}(t) &= t + \sigma(t) + 1 \\
&= t + 3t + 1 \\
&= 4t + 1, \\
\phi_3^{\Delta}(t) &= (\sigma(t))^2 + t\sigma(t) + t^2 \\
&= 9t^2 + 3t^2 + t^2 \\
&= 13t^2, \\
\left(\phi_1^{\Delta}(t)\right)^2 + \left(\phi_2^{\Delta}(t)\right)^2 + \left(\phi_3^{\Delta}(t)\right)^2 &= 1 + (4t+1)^2 + 169t^4 \\
&= 1 + 16t^2 + 8t + 1 + 169t^4 \\
&= 169t^4 + 16t^2 + 8t + 2.
\end{aligned}
$$

Therefore,

$$
\begin{aligned}
I &= \int_1^3 (t+2)(t^2+t)\sqrt{169t^4 + 16t^2 + 8t + 2}\,\Delta t \\
&= (t+2)(t^2+t)\sqrt{169t^4 + 16t^2 + 8t + 2}\,\mu(t)\Big|_{t=1} \\
&= 12\sqrt{195}.
\end{aligned}
$$

Example 8.33 Let $\mathbb{T} = [-4, 0] \cup \{2, 4\}$, where $[-4, 0]$ is the real line interval. Define Γ by $x_1 + x_2 = 4$. We will compute

$$
I = \int_{\Gamma} x_1 x_2 \Delta l.
$$

Here,

$$
\sigma(t) = \begin{cases} t & \text{for } t \in [-4, 0] \\ 4 & \text{for } t = 2, \end{cases} \qquad \mu(t) = \begin{cases} 0 & \text{for } t \in [-4, 0] \\ 2 & \text{for } t = 2. \end{cases}
$$

Let $x_1 = t$. Then $x_2 = 4 - t$ and

$$\Gamma = \begin{cases} x_1 = t \\ x_2 = 4 - t, \quad t \in [-4, 0] \cup \{2, 4\}. \end{cases}$$

Therefore,

$$\phi_1(t) = t, \quad \phi_2(t) = 4 - t.$$

Hence,

$$\phi_1^{\Delta}(t) = 1, \quad \phi_2^{\Delta}(t) = -1, \quad \left(\phi_1^{\Delta}(t)\right)^2 + \left(\phi_2^{\Delta}(t)\right)^2 = 2.$$

Consequently,

$$\begin{aligned} I &= \int_{-4}^{0} t(4-t)\sqrt{2}\Delta t + \int_{2}^{4} t(4-t)\sqrt{2}\Delta t \\ &= \sqrt{2}\int_{-4}^{0} t(4-t)\mathrm{d}t + t(4-t)\sqrt{2}\mu(t)\Big|_{t=2} \\ &= 4\sqrt{2}\int_{-4}^{0} t\mathrm{d}t - \sqrt{2}\int_{-4}^{0} t^2\mathrm{d}t + 8\sqrt{2} \\ &= 2\sqrt{2}t^2\Big|_{t=-4}^{t=0} - \frac{\sqrt{2}}{3}t^3\Big|_{t=-4}^{t=0} + 8\sqrt{2} \\ &= -32\sqrt{2} - \frac{64\sqrt{2}}{3} + 8\sqrt{2} \\ &= -\frac{136}{3}\sqrt{2}. \end{aligned}$$

Exercise 8.34 Let $\mathbb{T} = 2^{\mathbb{N}_0} \cup \{0\}$ and

$$\Gamma = \begin{cases} x_1 = t \\ x_2 = t^2 + t \\ x_3 = t - 2 \\ x_4 = t^3, \quad t \in [0, 2]. \end{cases}$$

Find

$$I = \int_{\Gamma} \left(x_1^2 + x_3^2 - \sqrt{x_2 x_4}\right) \Delta l.$$

Solution $I = (2 - \sqrt{2})\sqrt{67}$.

Theorem 8.35 (Linearity) *Let $\alpha, \beta \in \mathbb{R}$. If the functions f and g are Δ-integrable along the curve Γ, then $\alpha f + \beta g$ is also Δ-integrable along the curve Γ and*

$$\int_\Gamma (\alpha f + \beta g)(x_1, \ldots, x_m)\Delta l = \alpha \int_\Gamma f(x_1, \ldots, x_m)\Delta l + \beta \int_\Gamma g(x_1, \ldots, x_m)\Delta l.$$

Proof We have

$$\begin{aligned}\alpha \int_\Gamma f(x_1, \ldots, x_m)\Delta l &= \alpha \int_a^b f(\phi_1(t), \ldots, \phi_m(t)) \sqrt{\sum_{k=1}^m \left(\phi_k^\Delta(t)\right)^2} \Delta t \\ &= \int_a^b \alpha f(\phi_1(t), \ldots, \phi_m(t)) \sqrt{\sum_{k=1}^m \left(\phi_k^\Delta(t)\right)^2} \Delta t,\end{aligned}$$

$$\begin{aligned}\beta \int_\Gamma g(x_1, \ldots, x_m)\Delta l &= \beta \int_a^b g(\phi_1(t), \ldots, \phi_m(t)) \sqrt{\sum_{k=1}^m \left(\phi_k^\Delta(t)\right)^2} \Delta t \\ &= \int_a^b \beta g(\phi_1(t), \ldots, \phi_m(t)) \sqrt{\sum_{k=1}^m \left(\phi_k^\Delta(t)\right)^2} \Delta t.\end{aligned}$$

Hence,

$$\begin{aligned}&\alpha \int_\Gamma f(x_1, \ldots, x_m)\Delta l + \beta \int_\Gamma g(x_1, \ldots, x_m)\Delta l \\ &= \int_a^b \alpha f(\phi_1(t), \ldots, \phi_m(t)) \sqrt{\sum_{k=1}^m \left(\phi_k^\Delta(t)\right)^2} \Delta t \\ &\quad + \int_a^b \beta g(\phi_1(t), \ldots, \phi_m(t)) \sqrt{\sum_{k=1}^m \left(\phi_k^\Delta(t)\right)^2} \Delta t \\ &= \int_a^b (\alpha f(\phi_1(t), \ldots, \phi_m(t)) + \beta g(\phi_1(t), \ldots, \phi_m(t))) \sqrt{\sum_{k=1}^m \left(\phi_k^\Delta(t)\right)^2} \Delta t,\end{aligned}$$

which completes the proof. □

Theorem 8.36 (Additivity) *If the curve AB consists of two parts AC and CB and if the function f is Δ-integrable along the curve AB, then it is Δ-integrable along each of the curves AC and CB and*

$$\int_{AB} f(x_1, \dots, x_m)\Delta l = \int_{AC} f(x_1, \dots, x_m)\Delta l + \int_{CB} f(x_1, \dots, x_m)\Delta l.$$

Proof Let

$$A(\phi_1(a), \dots, \phi_m(a)), \quad B(\phi_1(b), \dots, \phi_m(b)), \quad C(\phi_1(c), \dots, \phi_m(c)),$$

where $a \leq c \leq b$. Hence,

$$\begin{aligned}\int_{AB} f(x_1, \dots, x_m)\Delta l &= \int_a^b f(\phi_1(t), \dots, \phi_m(t)) \sqrt{\sum_{k=1}^m \left(\phi_k^\Delta(t)\right)^2}\Delta t \\ &= \int_a^c f(\phi_1(t), \dots, \phi_m(t)) \sqrt{\sum_{k=1}^m \left(\phi_k^\Delta(t)\right)^2}\Delta t \\ &\quad + \int_c^b f(\phi_1(t), \dots, \phi_m(t)) \sqrt{\sum_{k=1}^m \left(\phi_k^\Delta(t)\right)^2}\Delta t \\ &= \int_{AC} f(x_1, \dots, x_m)\Delta l + \int_{CB} f(x_1, \dots, x_m)\Delta l,\end{aligned}$$

which completes the proof. □

Theorem 8.37 (Existence of Modulus of Integral) *If f is Δ-integrable along the curve Γ, then so is $|f|$ and*

$$\left|\int_\Gamma f(x_1, \dots, x_m)\Delta l\right| \leq \int_\Gamma |f(x_1, \dots, x_m)|\Delta l. \tag{8.12}$$

Proof Because f is Δ-integrable along the curve Γ, we have

$$\int_\Gamma f(x_1, \dots, x_m)\Delta l = \int_a^b f(\phi_1(t), \dots, \phi_m(t)) \sqrt{\sum_{k=1}^m \left(\phi_k^\Delta(t)\right)^2}\Delta t. \tag{8.13}$$

Therefore, the function

$$f(\phi_1(t), \dots, \phi_m(t)) \sqrt{\sum_{k=1}^m \left(\phi_k^\Delta(t)\right)^2}$$

is Δ-integrable on $[a, b)$. Hence,

$$\left| f(\phi_1(t), \dots, \phi_m(t)) \sqrt{\sum_{k=1}^{m} \left(\phi_k^{\Delta}(t)\right)^2} \right|$$

is Δ-integrable on $[a, b)$ and

$$\begin{aligned}
&\left| \int_a^b f(\phi_1(t), \dots, \phi_m(t)) \sqrt{\sum_{k=1}^{m} \left(\phi_k^{\Delta}(t)\right)^2 \Delta t} \right| \\
&\le \int_a^b \left| f(\phi_1(t), \dots, \phi_m(t)) \sqrt{\sum_{k=1}^{m} \left(\phi_k^{\Delta}(t)\right)^2} \right| \Delta t \\
&= \int_a^b |f(\phi_1(t), \dots, \phi_m(t))| \sqrt{\sum_{k=1}^{m} \left(\phi_k^{\Delta}(t)\right)^2 \Delta t} \\
&= \int_\Gamma |f(x_1, \dots, x_m)| \Delta l.
\end{aligned}$$

From the last inequality and from (8.13), we get (8.12). □

Theorem 8.38 (Mean Value Theorem) *Suppose f is bounded and Δ-integrable along the curve Γ. Let us set*

$$m = \inf\{f(x_1, \dots, x_m) : (x_1, \dots, x_m) \in \Gamma\}$$

and

$$M = \sup\{f(x_1, \dots, x_m) : (x_1, \dots, x_m) \in \Gamma\}.$$

Then there exists a real number $\Lambda \in [m, M]$ such that

$$\int_\Gamma f(x_1, \dots, x_m) \Delta l = \Lambda l(\Gamma). \tag{8.14}$$

Proof Since f is Δ-integrable along the curve Γ, we have

$$\int_\Gamma f(x_1, \dots, x_m) \Delta l = \int_a^b f(\phi_1(t), \dots, \phi_m(t)) \sqrt{\sum_{k=1}^{m} \left(\phi_k^{\Delta}(t)\right)^2 \Delta t}. \tag{8.15}$$

Because $f(\phi_1(t), \dots, \phi_m(t))$ and

$$\sqrt{\sum_{k=1}^{m} \left(\phi_k^{\Delta}(t)\right)^2}$$

are bounded Δ-integrable on $[a, b)$ and

$$\sqrt{\sum_{k=1}^{m} \left(\phi_k^{\Delta}(t)\right)^2}$$

is nonnegative on $[a, b]$, using Theorem 7.57, there exists $\Lambda \in [m, M]$ such that

$$\int_a^b f(\phi_1(t), \ldots, \phi_m(t)) \sqrt{\sum_{k=1}^{m} \left(\phi_k^{\Delta}(t)\right)^2} \Delta t = \Lambda \int_a^b \sqrt{\sum_{k=1}^{m} \left(\phi_k^{\Delta}(t)\right)^2} \Delta t.$$

Hence, utilizing Theorem 8.23, we obtain

$$\int_a^b f(\phi_1(t), \ldots, \phi_m(t)) \sqrt{\sum_{k=1}^{m} \left(\phi_k^{\Delta}(t)\right)^2} \Delta t = \Lambda l(\Gamma).$$

From here and from (8.15), we obtain (8.14). □

8.3 Line Integrals of the Second Kind

Here, we will use the same notation as in the previous section. Assume that g_i, $i \in \{1, \ldots, m\}$, are defined and continuous on the curve Γ. We introduce the integral sums

$$S_{2i} = \sum_{k=1}^{m} g_i(\phi_1(\tau_k), \ldots, \phi_m(\tau_k))(\phi_i(t_k) - \phi_i(t_{k-1})), \quad i = 1, \ldots, m.$$

Definition 8.39 We say that a complex number I_i, $i \in \{1, \ldots, m\}$, is the line delta integral of the *second kind* of the function g_i along the curve Γ if for each $\varepsilon > 0$, there exists $\delta > 0$ such that

$$|S_{2i} - I_i| < \varepsilon$$

for every integral sum S_{2i} of g_i corresponding to a partition

$$P \in \mathscr{P}_\delta([a, b])$$

independent of the way in which $\tau_k \in [t_{k-1}, t_k)$ for $k \in \{1, \ldots, n\}$ is chosen. We denote the number I_i symbolically by

$$\int_\Gamma g_i(x_1, \ldots, x_m)\Delta_i x_i \quad \text{or} \quad \int_{AB} g_i(x_1, \ldots, x_m)\Delta_i x_i. \tag{8.16}$$

The sum

$$\int_{\Gamma} g_1(x_1, \ldots, x_m)\Delta_1 x_1 + \cdots + \int_{\Gamma} g_m(x_1, \ldots, x_m)\Delta_m x_m$$

is called a general line delta integral of the second kind, and it is denoted by

$$\int_{\Gamma} g_1(x_1, \ldots, x_m)\Delta_1 x_1 + \cdots + g_m(x_1, \ldots, x_m)\Delta_m x_m.$$

Theorem 8.40 *Suppose that the curve Γ is given by* (8.1)*, where ϕ_i are continuous on $[a, b]$ and Δ-integrable on $[a, b)$. If ϕ_i^{Δ} are bounded and Δ-integrable on $[a, b)$ and if the functions g_i are continuous on Γ, then the line integrals* (8.16) *exist and can be computed by the formula*

$$\int_{\Gamma} g_i(x_1, \ldots, x_m)\Delta_i x_i = \int_a^b g_i(\phi_1(t), \ldots, \phi_m(t))\phi_i^{\Delta}(t)\Delta t.$$

Proof We have

$$S_{2i} - I_{2i} = \sum_{k=1}^{n} \int_{t_{k-1}}^{t_k} (g_i(\phi_1(\tau_k), \ldots, \phi_m(\tau_k)) - g_i(\phi_1(t), \ldots, \phi_m(t)))\, \phi_i^{\Delta}(t)\Delta t.$$

Because ϕ_i^{Δ} are bounded and Δ-integrable on $[a, b)$, there exists a positive constant M such that

$$|\phi_i^{\Delta}(t)| \leq M \quad \text{for} \quad t \in [t_{k-1}, t_k) \quad \text{and for all} \quad k \in \{1, \ldots, n\}.$$

For arbitrary $\varepsilon > 0$, there exists $\delta > 0$ such that $t, \tau \in [a, b]$ and $|t - \tau| \leq \delta$ imply

$$|g_i(\phi_1(t), \ldots, \phi_m(t)) - g_i(\phi_1(\tau), \ldots, \phi_m(\tau))| < \frac{\varepsilon}{M(b-a)}.$$

Now, assuming that $P \in \mathscr{P}_{\delta}([a, b])$, we have

$$\begin{aligned}
|S_{2i} - I_{2i}| &= \left| \sum_{k=1}^{n} \int_{t_{k-1}}^{t_k} (g_i(\phi_1(\tau_k), \ldots, \phi_m(\tau_k)) - g_i(\phi_1(t), \ldots, \phi_m(t)))\phi_i^{\Delta}(t)\Delta t \right| \\
&\leq \sum_{k=1}^{n} \int_{t_{k-1}}^{t_k} |g_i(\phi_1(\tau_k), \ldots, \phi_m(\tau_k)) - g_i(\phi_1(t), \ldots, \phi_m(t))||\phi_i^{\Delta}(t)|\Delta t \\
&= \sum_{t_k - t_{k-1} \leq \delta} \int_{t_{k-1}}^{t_k} |g_i(\phi_1(\tau_k), \ldots, \phi_m(\tau_k)) - g_i(\phi_1(t), \ldots, \phi_m(t))||\phi_i^{\Delta}(t)|\Delta t
\end{aligned}$$

$$+ \sum_{t_k - t_{k-1} > \delta} \int_{t_{k-1}}^{t_k} |g_i(\phi_1(\tau_k), \ldots, \phi_m(\tau_k)) - g_i(\phi_1(t), \ldots, \phi_m(t))||\phi_i^{\Delta}(t)|\Delta t$$

$$= \sum_{t_k - t_{k-1} \le \delta} \int_{t_{k-1}}^{t_k} |g_i(\phi_1(\tau_k), \ldots, \phi_m(\tau_k)) - g_i(\phi_1(t), \ldots, \phi_m(t))||\phi_i^{\Delta}(t)|\Delta t$$

$$\le \frac{\varepsilon}{M(b-a)} M \sum_{t_k - t_{k-1} \le \delta} \int_{t_{k-1}}^{t_k} \Delta t$$

$$\le \frac{\varepsilon}{b-a} \sum_{k=1}^{n} \int_{t_{k-1}}^{t_k} \Delta t$$

$$= \varepsilon,$$

which completes the proof. □

Example 8.41 Let $\mathbb{T} = \mathbb{Z}$ and

$$\Gamma = \begin{cases} x_1 = t^2 \\ x_2 = t + 2 \\ x_3 = t, \quad t \in [0, 2]. \end{cases}$$

We will compute

$$I = \int_{\Gamma} x_1 \Delta_1 x_1 + x_2 \Delta_2 x_2 + x_3 \Delta_3 x_3.$$

Let

$$I_1 = \int_{\Gamma} x_1 \Delta_1 x_1, \quad I_2 = \int_{\Gamma} x_2 \Delta_2 x_2, \quad I_3 = \int_{\Gamma} x_3 \Delta_3 x_3.$$

Here,

$$\sigma(t) = t + 1, \quad \mu(t) = 1$$

and

$$\phi_1(t) = t^2, \quad \phi_2(t) = t + 2, \quad \phi_3(t) = t.$$

Thus,

$$\phi_1^{\Delta}(t) = \sigma(t) + t = 2t + 1,$$

$$\phi_2^{\Delta}(t) = \phi_3^{\Delta}(t) = 1.$$

Hence,

$$
\begin{aligned}
I_1 &= \int_0^2 t^2\phi_1^{\Delta}(t)\Delta t \\
&= \int_0^2 t^2(2t+1)\Delta t \\
&= \int_0^2 (2t^3+t^2)\Delta t \\
&= (2t^3+t^2)\Big|_{t=0} + (2t^3+t^2)\Big|_{t=1} \\
&= 3,
\end{aligned}
$$

$$
\begin{aligned}
I_2 &= \int_0^2 (t+2)\phi_2^{\Delta}(t)\Delta t \\
&= \int_0^2 (t+2)\Delta t \\
&= (t+2)\Big|_{t=0} + (t+2)\Big|_{t=1} \\
&= 5,
\end{aligned}
$$

$$
\begin{aligned}
I_3 &= \int_0^2 t\phi_3^{\Delta}(t)\Delta t \\
&= \int_0^2 t\Delta t \\
&= t\Big|_{t=0} + t\Big|_{t=1} \\
&= 1.
\end{aligned}
$$

Consequently,

$$
I = I_1 + I_2 + I_3 = 3 + 5 + 1 = 9.
$$

Example 8.42 Let $\mathbb{T} = 2^{\mathbb{N}_0} \cup \{0\}$ and

$$\Gamma = \begin{cases} x_1 = t^2 + 2 \\ x_2 = t^3 + 1 \\ x_3 = t^2, \quad t \in [0, 4]. \end{cases}$$

We will compute

$$I = \int_\Gamma x_1 x_2 \Delta_1 x_1 + x_2 \Delta_2 x_2 + x_3 \Delta_3 x_3.$$

Let

$$I_1 = \int_\Gamma x_1 x_2 \Delta_1 x_1, \quad I_2 = \int_\Gamma x_2 \Delta_2 x_2, \quad I_3 = \int_\Gamma x_3 \Delta_3 x_3.$$

Here, $\sigma(0) = 1$, $\mu(0) = 1$,

$$\sigma(t) = 2t, \quad \mu(t) = t \quad \text{for} \quad t \neq 0$$

and

$$\phi_1(t) = t^2 + 2, \quad \phi_2(t) = t^3 + 1, \quad \phi_3(t) = t^2.$$

Thus,

$$\begin{aligned} \phi_1^\Delta(t) &= \phi_3^\Delta(t) \\ &= t + \sigma(t) \\ &= t + 2t \\ &= 3t, \\ \phi_2^\Delta(t) &= (\sigma(t))^2 + t\sigma(t) + t^2 \\ &= 4t^2 + 2t^2 + t^2 \\ &= 7t^2. \end{aligned}$$

From here,

$$
\begin{aligned}
I_1 &= \int_0^4 (t^2+2)(t^3+1)\phi_1^{\Delta}(t)\Delta t \\
&= \int_0^4 (t^2+2)(t^3+1)3t\Delta t \\
&= 3t(t^2+2)(t^3+1)\mu(t)\Big|_{t=0} + 3t^2(t^2+2)(t^3+1)\Big|_{t=1} \\
&\quad + 3t^2(t^2+2)(t^3+1)\Big|_{t=2} \\
&= 18+648 \\
&= 666,
\end{aligned}
$$

$$
\begin{aligned}
I_2 &= \int_0^4 (t^3+1)\phi_2^{\Delta}(t)\Delta t \\
&= \int_0^4 (t^3+1)7t^2\Delta t \\
&= 7t^2(t^3+1)\mu(t)\Big|_{t=0} + 7t^3(t^3+1)\Big|_{t=1} + 7t^3(t^3+1)\Big|_{t=2} \\
&= 14+504 \\
&= 518,
\end{aligned}
$$

$$
\begin{aligned}
I_3 &= \int_0^4 t^2\phi_3^{\Delta}(t)\Delta t \\
&= \int_0^2 3t^3\Delta t \\
&= 3t^3\mu(t)\Big|_{t=0} + 3t^4\Big|_{t=1} + 3t^4\Big|_{t=2} \\
&= 3+48 \\
&= 51.
\end{aligned}
$$

Consequently,

$$
I = I_1 + I_2 + I_3 = 666 + 518 + 51 = 1235.
$$

Example 8.43 Let $\mathbb{T} = 3^{\mathbb{N}_0}$ and

$$\Gamma = \begin{cases} x_1 = t^2 \\ x_2 = t^3, \quad t \in [1, 3]. \end{cases}$$

We will compute

$$I = \int_\Gamma x_1^2 \Delta_1 x_1 + x_2^3 \Delta_2 x_2.$$

Let

$$I_1 = \int_\Gamma x_1^2 \Delta_1 x_1, \quad I_2 = \int_\Gamma x_2^3 \Delta_2 x_2.$$

Here,

$$\sigma(t) = 3t, \quad \mu(t) = 2t$$

and

$$\phi_1(t) = t^2, \quad \phi_2(t) = t^3.$$

Thus,

$$\begin{aligned} \phi_1^\Delta(t) &= t + \sigma(t) \\ &= t + 3t \\ &= 4t, \\ \phi_2^\Delta(t) &= (\sigma(t))^2 + t\sigma(t) + t^2 \\ &= 9t^2 + 3t^2 + t^2 \\ &= 13t^2. \end{aligned}$$

Hence,

$$\begin{aligned} I_1 &= \int_1^3 t^4 \phi_1^\Delta(t) \Delta t \\ &= \int_1^3 4t^5 \Delta t \end{aligned}$$

$$= 8t^6\Big|_{t=1}$$

$$= 8,$$

$$I_2 = \int_1^3 t^9 \phi_2^\Delta(t)\Delta t$$

$$= 13\int_1^3 t^{11}\Delta t$$

$$= 26t^{12}\Big|_{t=1}$$

$$= 26.$$

Consequently,

$$I = I_1 + I_2 = 8 + 26 = 34.$$

Exercise 8.44 Let $\mathbb{T} = \mathbb{Z}$ and

$$\Gamma = \begin{cases} x_1 = t + 2 \\ x_2 = 2t^2 - 3t + 2 \\ x_3 = t^2 + 1, \quad t \in [-1, 1]. \end{cases}$$

Compute

$$\int_\Gamma x_1^2 \Delta_1 x_1 + x_2^2 \Delta_2 x_2 + x_3^2 \Delta_3 x_3.$$

Solution -35.

Theorem 8.45 (Linearity) *Let $\alpha, \beta \in \mathbb{R}$. If f_i and g_i, $i \in \{1, \dots, m\}$, are line Δ-integrable of the second kind along the curve Γ, then $\alpha f_i + \beta g_i$ is line Δ-integrable along the curve Γ and*

$$\int_\Gamma (\alpha f_i + \beta g_i)(x_1, \dots, x_m)\Delta_i x_i$$
$$= \alpha \int_\Gamma f_i(x_1, \dots, x_m)\Delta_i x_i + \beta \int_\Gamma g_i(x_1, \dots, x_m)\Delta_i x_i.$$

Proof We have

$$
\begin{aligned}
&\alpha \int_{\Gamma} f_i(x_1, \ldots, x_m)\Delta_i x_i + \beta \int_{\Gamma} g_i(x_1, \ldots, x_m)\Delta_i x_i \\
&= \alpha \int_a^b f_i(\phi_1(t), \ldots, \phi_m(t))\phi_i^{\Delta}(t)\Delta t + \beta \int_a^b g_i(\phi_1(t), \ldots, \phi_m(t))\phi_i^{\Delta}(t)\Delta t \\
&= \int_a^b (\alpha f_i(\phi_1(t), \ldots, \phi_m(t)) + \beta g_i(\phi_1(t), \ldots, \phi_m(t)))\phi_i^{\Delta}(t)\Delta t \\
&= \int_{\Gamma} (\alpha f_i(x_1, \ldots, x_m) + \beta g_i(x_1, \ldots, x_m))\Delta_i x_i,
\end{aligned}
$$

which completes the proof. □

Theorem 8.46 (Additivity) *If the curve AB consists of two curves AC and CB and if the functions g_i, $i \in \{1, \ldots, m\}$, are line Δ-integrable of the second kind along the curve AB, then they are line Δ-integrable of the second kind along each of the curves AC and CB and*

$$
\int_{AB} g_i(x_1, \ldots, x_m)\Delta_i x_i = \int_{AC} g_i(x_1, \ldots, x_m)\Delta_i x_i + \int_{CB} g_i(x_1, \ldots, x_m)\Delta_i x_i.
$$

Proof Let

$$
A(\phi_1(a), \ldots, \phi_m(a)), \quad B(\phi_1(b), \ldots, \phi_m(b)), \quad C(\phi_1(c), \ldots, \phi_m(c)).
$$

Hence,

$$
\begin{aligned}
\int_{AB} g_i(x_1, \ldots, x_m)\Delta_i x_i &= \int_a^b g_i(\phi_1(t), \ldots, \phi_m(t))\phi_i^{\Delta}(t)\Delta t \\
&= \int_a^c g_i(\phi_1(t), \ldots, \phi_m(t))\phi_i^{\Delta}(t)\Delta t \\
&\quad + \int_c^b g_i(\phi_1(t), \ldots, \phi_m(t))\phi_i^{\Delta}(t)\Delta t \\
&= \int_{AC} g_i(x_1, \ldots, x_m)\Delta_i x_i + \int_{CB} g_i(x_1, \ldots, x_m)\Delta_i x_i,
\end{aligned}
$$

which completes the proof. □

Theorem 8.47 (Mean Value Theorem) *Suppose g_i, $i \in \{1, \ldots, m\}$, are bounded and line Δ-integrable of the second kind along the curve Γ. Suppose $\phi_i^{\Delta}(t)$ is nonnegative on $[a, b]$. Let us set*

$$
m_i = \inf\{g_i(x_1, \ldots, x_m) : (x_1, \ldots, x_m) \in \Gamma\}
$$

and

$$M_i = \sup\{g_i(x_1, \ldots, x_m) : (x_1, \ldots, x_m) \in \Gamma\}.$$

Then there exists a real number $\Lambda_i \in [m_i, M_i]$ *such that*

$$\int_\Gamma g_i(x_1, \ldots, x_m)\Delta_i x_i = \Lambda_i(\phi_i(b) - \phi_i(a)).$$

Proof By Theorem 8.40, we have

$$\int_\Gamma g_i(x_1, \ldots, x_m)\Delta_i x_i = \int_a^b g_i(\phi_1(t), \ldots, \phi_m(t))\phi_i^\Delta(t)\Delta t.$$

Since $g_i(\phi_1(t), \ldots, \phi_m(t))$ are bounded Δ-integrable on $[a, b]$ and $\phi_i^\Delta(t)$ is nonnegative on $[a, b]$, there exists $\Lambda_i \in [m_i, M_i]$ such that

$$\begin{aligned}\int_a^b g_i(\phi_1(t), \ldots, \phi_m(t))\phi_i^\Delta(t)\Delta t &= \Lambda_i \int_a^b \phi_i^\Delta(t)\Delta t \\ &= \Lambda_i \phi_i(t)\Big|_{t=a}^{t=b} \\ &= \Lambda_i(\phi_i(b) - \phi_i(a)),\end{aligned}$$

which completes the proof. □

8.4 Green's Formula

Let $\mathbb{T}_1$ and $\mathbb{T}_2$ be two time scales. Since $\mathbb{T}_1$ and $\mathbb{T}_2$ are closed subsets of $\mathbb{R}$, the set $\mathbb{T}_1 \times \mathbb{T}_2$ is a complete metric space with metric D defined by

$$d((x_1, x_2), (x_1', x_2')) = \sqrt{(x_1 - x_1')^2 + (x_2 - x_2')^2}.$$

Definition 8.48 If $\mathscr{M}$ is a metric space, then any continuous mapping $h : [a, b] \to \mathscr{M}$ is called a *continuous curve* in $\mathscr{M}$.

Definition 8.49 Let $[a, b] \subset \mathbb{T}_1$ with $a, b \in \mathbb{T}_1$ and $y_0 \in \mathbb{T}_2$. The set

$$\{(x, y_0) : x \in [a, b]\}$$

is called a *horizontal line segment* in $\mathbb{T}_1 \times \mathbb{T}_2$ and denoted by AB, where $A = (a, y_0)$ and $B = (b, y_0)$. We take $x_0 \in \mathbb{T}_1$ and $[c, d] \subset \mathbb{T}_2$ and define a *vertical line segment* in $\mathbb{T}_1 \times \mathbb{T}_2$ by

$$\{(x_0, y) : y \in [c, d]\}$$

and denote it by CD, where $C = (x_0, c)$ and $D = (x_0, d)$.

Definition 8.50 A finite sequence $A_1B_1, A_2B_2, \ldots, A_nB_n$, each of whose term A_kB_k, $k \in \{1, \ldots, n\}$, is a horizontal or vertical line segment in $\mathbb{T}_1 \times \mathbb{T}_2$, is said to form a *polygonal path* (or *broken line*) in $\mathbb{T}_1 \times \mathbb{T}_2$ with terminal points A_1 and B_n if

$$B_1 = A_2, \quad B_2 = A_3, \quad \ldots, \quad B_{n-1} = A_n.$$

Definition 8.51 A set of points of $\mathbb{T}_1 \times \mathbb{T}_2$ is said to be *connected* if any two of its points are terminal points of a polygonal path of points contained in the set.

Definition 8.52 A *component* of a set $\Omega \subset \mathbb{T}_1 \times \mathbb{T}_2$ is a nonempty maximal connected subset of Ω.

Definition 8.53 A nonempty open connected set of points of $\mathbb{T}_1 \times \mathbb{T}_2$ is called a *domain*.

Definition 8.54 A *closed domain* is a subset of $\mathbb{T}_1 \times \mathbb{T}_2$ which is the closure of a domain in $\mathbb{T}_1 \times \mathbb{T}_2$.

For the rectangle R in $\mathbb{T}_1 \times \mathbb{T}_2$ defined by

$$R = [a_1, b_1) \times [a_2, b_2), \tag{8.17}$$

we set

$$L_1 = \{(x, a_2) : x \in [a_1, b_1]\}, \quad L_2 = \{(b_1, y) : y \in [a_2, b_2]\},$$

$$L_3 = \{(x, b_2) : x \in [a_1, b_1]\}, \quad L_4 = \{(a_1, y) : y \in [a_2, b_2]\}.$$

Each of L_j, $j \in \{1, 2, 3, 4\}$, is an oriented line segment.

Definition 8.55 The closed curve

$$\Gamma = L_1 \cup L_2 \cup (-L_3) \cup (-L_4) \tag{8.18}$$

is called the *positively oriented fence* of R, i.e., the rectangle Γ remains on the "left" side along the fence curve Γ.

Definition 8.56 We say that the set $E \subset \mathbb{T}_1 \times \mathbb{T}_2$ is a set of *type ω* if it is a connected set in $\mathbb{T}_1 \times \mathbb{T}_2$ which is the union of a finite number of rectangles of the form (8.17) that are pairwise disjoint and adjoining to each other.

Let $E \subset \mathbb{T}_1 \times \mathbb{T}_2$ be a set of type ω so that

$$E = \bigcup_{k=1}^{m} R_k,$$

where

$$R_k = [a_1^k, b_1^k) \times [a_2^k, b_2^k) \subset \mathbb{T}_1 \times \mathbb{T}_2, \quad k \in \{1, \dots, m\},$$

and $R_1, R_2, \dots, R_m$ are pairwise disjoint and adjoining to each other. Let Γ_k be the positively oriented fence of the rectangle R_k. We set

$$X = \bigcup_{k=1}^{m} \Gamma_k.$$

Let X_0 consist of a finite number of line segments each of which serves as a common part of fences of two adjoining rectangles belonging to $\{R_1, R_2, \dots, R_m\}$.

Definition 8.57 The set

$$\Gamma = X \setminus X_0$$

forms a positively oriented closed "polygonal curve", which is called the *positively oriented fence* of the set E.

Theorem 8.58 (Green's Formula) *Let $E \subset \mathbb{T}_1 \times \mathbb{T}_2$ be a set of type ω and let Γ be its positively oriented fence. If the functions M and N are continuous and have continuous partial delta derivatives $M_{y_2}^{\Delta_2}$ and $N_{x_1}^{\Delta_1}$ on $E \cup \Gamma$, then*

$$\int\int_E \left(N_{x_1}^{\Delta_1}(x_1, x_2) - M_{x_2}^{\Delta_2}(x_1, x_2)\right) \Delta_1 x_1 \Delta_2 x_2 = \int_\Gamma M(x_1, x_2)\Delta_1 x_1 + N(x_1, x_2)\Delta_2 x_2.$$

Proof Let E be a rectangle R of the form (8.17). Let L_1, L_2, L_3, L_4 be the oriented line segments defined above, and let Γ be the positively oriented fence of R defined by (8.18). Hence,

$$\begin{aligned}
&\int\int_E \left(N_{x_1}^{\Delta_1}(x_1, x_2) - M_{x_2}^{\Delta_2}(x_1, x_2)\right) \Delta_1 x_1 \Delta_2 x_2 \\
&= \int_{a_1}^{b_1} \int_{a_2}^{b_2} \left(N_{x_1}^{\Delta_1}(x_1, x_2) - M_{x_2}^{\Delta_2}(x_1, x_2)\right) \Delta_2 x_2 \Delta_1 x_1 \\
&= \int_{a_1}^{b_1} \int_{a_2}^{b_2} N_{x_1}^{\Delta_1}(x_1, x_2)\Delta_2 x_2 \Delta_1 x_1 - \int_{a_1}^{b_1} \int_{a_2}^{b_2} M_{x_2}^{\Delta_2}(x_1, x_2)\Delta_2 x_2 \Delta_1 x_1 \\
&= \int_{a_2}^{b_2} \int_{a_1}^{b_1} N_{x_1}^{\Delta_1}(x_1, x_2)\Delta_1 x_1 \Delta_2 x_2 - \int_{a_1}^{b_1} M(x_1, x_2)\Big|_{x_2=a_2}^{x_2=b_2} \Delta_1 x_1 \\
&= \int_{a_2}^{b_2} N(x_1, x_2)\Big|_{x_1=a_1}^{x_1=b_1} \Delta_2 x_2 - \int_{a_1}^{b_1} (M(x_1, b_2) - M(x_1, a_2))\Delta_1 x_1
\end{aligned}$$

$$= \int_{a_2}^{b_2} (N(b_1, x_2) - N(a_1, x_2))\Delta_2 x_2 - \int_{a_1}^{b_1} (M(x_1, b_2) - M(x_1, a_2))\Delta_1 x_1$$

$$= \int_{a_2}^{b_2} N(b_1, x_2)\Delta_2 x_2 - \int_{a_2}^{b_2} N(a_1, x_2)\Delta_2 x_2 - \int_{a_1}^{b_1} M(x_1, b_2)\Delta_1 x_1$$

$$+ \int_{a_1}^{b_1} M(x_1, a_2)\Delta_1 x_1$$

$$= \int_{L_1} M(x_1, x_2)\Delta_1 x_1 + \int_{(-L_3)} M(x_1, x_2)\Delta_1 x_1$$

$$+ \int_{L_2} N(x_1, x_2)\Delta_2 x_2 + \int_{(-L_4)} N(x_1, x_2)\Delta_2 x_2.$$

We note that

$$\int_{L_2} M(x_1, x_2)\Delta_1 x_1 = \int_{(-L_4)} M(x_1, x_2)\Delta_1 x_1 = 0,$$

$$\int_{L_1} N(x_1, x_2)\Delta_2 x_2 = \int_{(-L_3)} N(x_1, x_2)\Delta_2 x_2 = 0.$$

Therefore,

$$\int\int_E \left(N_{x_1}^{\Delta_1}(x_1, x_2) - M_{x_2}^{\Delta_2}(x_1, x_2)\right) \Delta_1 x_1 \Delta_2 x_2$$

$$= \int_{L_1} M(x_1, x_2)\Delta_1 x_1 + \int_{L_2} M(x_1, x_2)\Delta_1 x_1 + \int_{(-L_3)} M(x_1, x_2)\Delta_1 x_1$$

$$+ \int_{(-L_4)} M(x_1, x_2)\Delta_1 x_1 + \int_{L_1} N(x_1, x_2)\Delta_2 x_2 + \int_{L_2} N(x_1, x_2)\Delta_2 x_2$$

$$+ \int_{(-L_3)} N(x_1, x_2)\Delta_2 x_2 + \int_{(-L_4)} N(x_1, x_2)\Delta_2 x_2$$

$$= \int_{\Gamma} M(x_1, x_2)\Delta_1 x_1 + N(x_1, x_2)\Delta_2 x_2.$$

Now, we assume that E is a set of type ω, i.e.,

$$E = \bigcup_{k=1}^{m} R_k,$$

where $R_k = [a_1^k, b_1^k) \times [a_2^k, b_2^k)$, $k \in \{1, \ldots, m\}$, and $R_1, \ldots, R_m$ are pairwise disjoint. Let Γ_k, $k \in \{1, \ldots, m\}$, be the positively oriented fence of the rectangle R_k. Thus,

$$\begin{aligned}
&\int\int_E \left(N_{x_1}^{\Delta_1}(x_1, x_2) - M_{x_2}^{\Delta_2}(x_1, x_2)\right) \Delta_1 x_1 \Delta_2 x_2 \\
&= \sum_{k=1}^{m} \int\int_{R_k} \left(N_{x_1}^{\Delta_1}(x_1, x_2) - M_{x_2}^{\Delta_2}(x_1, x_2)\right) \Delta_1 x_1 \Delta_2 x_2 \\
&= \sum_{k=1}^{m} \int_{\Gamma_k} M(x_1, x_2)\Delta_1 x_1 + N(x_1, x_2)\Delta_2 x_2 \\
&= \int_{\Gamma} M(x_1, x_2)\Delta_1 x_1 + N(x_1, x_2)\Delta_2 x_2.
\end{aligned}$$

This completes the proof. □

Example 8.59 Let $\mathbb{T}_1 = \mathbb{T}_2 = \mathbb{Z}$ and

$$\begin{aligned}
\Gamma &= \{(x_1, 1) : \ x_1 \in [1, 2]\} \cup \{(x_1, 2) : \ x_1 \in [1, 2]\} \\
&\quad \cup \{(1, x_2) : \ x_2 \in [1, 2]\} \cup \{(2, x_2) : \ x_2 \in [1, 2]\}.
\end{aligned}$$

Consider the line integral

$$I = \int_{\Gamma} (x_1^2 + 3x_1 x_2)\Delta_1 x_1 + (x_1^3 - 3x_1 + x_2)\Delta_2 x_2.$$

Here,

$$\sigma_1(x_1) = x_1 + 1, \quad x_1 \in \mathbb{T}_1, \quad \sigma_2(x_2) = x_2 + 1, \quad x_2 \in \mathbb{T}_2$$

and

$$M(x_1, x_2) = x_1^2 + 3x_1 x_2, \quad N(x_1, x_2) = x_1^3 - 3x_1 + x_2.$$

Thus,

$$\begin{aligned}
M_{x_2}^{\Delta_2}(x_1, x_2) &= 3x_1, \\
N_{x_1}^{\Delta_1}(x_1, x_2) &= (\sigma_1(x_1))^2 + x_1 \sigma_1(x_1) + x_1^2 - 3 \\
&= (x_1 + 1)^2 + x_1(x_1 + 1) + x_1^2 - 3
\end{aligned}$$

$$= x_1^2 + 2x_1 + 1 + x_1^2 + x_1 + x_1^2 - 3$$

$$= 3x_1^2 + 3x_1 - 2.$$

Then, using Green's formula, Theorem 8.58, we transform the line integral I as

$$\begin{aligned} I &= \int_1^2 \int_1^2 \left(N_{x_1}^{\Delta_1}(x_1, x_2) - M_{x_2}^{\Delta_2}(x_1, x_2)\right) \Delta_1 x_1 \Delta_2 x_2 \\ &= \int_1^2 \int_1^2 (3x_1^2 + 3x_1 - 2 - 3x_1)\Delta_1 x_1 \Delta_2 x_2 \\ &= \int_1^2 \int_1^2 (3x_1^2 - 2)\Delta_1 x_1 \Delta_2 x_2 \\ &= \int_1^2 (3x_1^2 - 2)\Big|_{x_1=1} \Delta_2 x_2 \\ &= \int_1^2 \Delta_2 x_2 \\ &= 1. \end{aligned}$$

Example 8.60 Let $\mathbb{T}_1 = \mathbb{T}_2 = 2^{\mathbb{N}_0}$. Let Γ be the positively oriented fence of the set

$$E = \left\{(x_1, x_2) \in \mathbb{T}_1 \times \mathbb{T}_2 : \ -1 \le x_1^3 - 2x_1^2 x_2 \le 1\right\}.$$

Consider the line integral

$$I = \int_\Gamma (x_1^4 - 3x_1 x_2^2)\Delta_1 x_1 + (x_2^3 - x_1^2)\Delta_2 x_2.$$

Here,

$$\sigma_1(x_1) = 2x_1, \quad x_1 \in \mathbb{T}_1, \quad \sigma_2(x_2) = 2x_2, \quad x_2 \in \mathbb{T}_2$$

and

$$M(x_1, x_2) = x_1^4 - 3x_1 x_2^2, \quad N(x_1, x_2) = x_2^3 - x_1^2.$$

Thus,

$$\begin{aligned} M_{x_2}^{\Delta_2}(x_1, x_2) &= -3x_1(\sigma_2(x_2) + x_2) \\ &= -3x_1(3x_2) \end{aligned}$$

$$= -9x_1x_2,$$

$$N_{x_1}^{\Delta_1}(x_1, x_2) = -(\sigma_1(x_1) + x_1)$$

$$= -(2x_1 + x_1)$$

$$= -3x_1.$$

Using Green's formula, Theorem 8.58, we obtain

$$I = \int\int_E \left(N_{x_1}^{\Delta_1}(x_1, x_2) - M_{x_2}^{\Delta_2}(x_1, x_2)\right) \Delta_1 x_1 \Delta_2 x_2$$

$$= \int\int_E (-3x_1 + 9x_1x_2)\Delta_1 x_1 \Delta_2 x_2.$$

Example 8.61 Let $\mathbb{T}_1 = 3^{\mathbb{N}_0} \cup \{0\}$ and $\mathbb{T}_2 = 2^{\mathbb{N}_0} \cup \{0\}$. Let Γ be the positively oriented fence of the set

$$E = \left\{(x_1, x_2) \in \mathbb{T}_1 \times \mathbb{T}_2 : \ 0 \leq 2x_1^4 + x_2^5 \leq 40\right\}.$$

Consider the line integral

$$I = \int_\Gamma (x_1^3 + 2x_1^2 + x_2^3)\Delta_1 x_1 + (x_1 + x_2^3 + x_1^3 x_2^2)\Delta_2 x_2.$$

Here,

$$\sigma_1(x_1) = 3x_1, \quad x_1 \in \mathbb{T}_1 \setminus \{0\}, \quad \sigma_1(0) = 1,$$

$$\sigma_2(x_2) = 2x_2, \quad x_2 \in \mathbb{T}_2 \setminus \{0\}, \quad \sigma_2(0) = 1,$$

and

$$M(x_1, x_2) = x_1^3 + 2x_1^2 + x_2^3, \quad N(x_1, x_2) = x_1 + x_2^3 + x_1^3 x_2^2.$$

Thus,

$$
\begin{aligned}
M_{x_2}^{\Delta_2}(x_1, x_2) &= (\sigma_2(x_2))^2 + x_2\sigma_2(x_2) + x_2^2 \\
&= (2x_2)^2 + x_2(2x_2) + x_2^2 \\
&= 4x_2^2 + 2x_2^2 + x_2^2 \\
&= 7x_2^2, \\
N_{x_1}^{\Delta_1}(x_1, x_2) &= 1 + ((\sigma_1(x_1))^2 + x_1\sigma_1(x_1) + x_1^2)x_2^2 \\
&= 1 + ((3x_1)^2 + x_1(3x_1) + x_1^2)x_2^2 \\
&= 1 + (9x_1^2 + 3x_1^2 + x_1^2)x_2^2 \\
&= 1 + 13x_1^2x_2^2.
\end{aligned}
$$

Using Green's formula, Theorem 8.58, we get

$$
\begin{aligned}
I &= \int\int_E \left(N_{x_1}^{\Delta_1}(x_1, x_2) - M_{x_2}^{\Delta_2}(x_1, x_2)\right) \Delta_1 x_1 \Delta_2 x_2 \\
&= \int\int_E (1 + 13x_1^2x_2^2 - 7x_2^2)\Delta_1 x_1 \Delta_2 x_2.
\end{aligned}
$$

Exercise 8.62 Let $\mathbb{T}_1 = \mathbb{Z}$ and $\mathbb{T}_2 = 2^{\mathbb{N}_0} \cup \{0\}$. Let Γ be the positively oriented fence of the set

$$
E = \left\{(x_1, x_2) \in \mathbb{T}_1 \times \mathbb{T}_2 : \; x_1^2 + x_2^2 \leq 4\right\}.
$$

Using Green's formula, Theorem 8.58, transform the line integral

$$
\int_\Gamma (x_1^4 + 3x_1^2 - x_1x_2)\Delta_1 x_1 + (x_1^2 - x_1x_2)\Delta_2 x_2.
$$

Solution $\int\int_E (3x_1 - x_2 + 1)\Delta_1 x_1 \Delta_2 x_2$.

Example 8.63 Let $\mathbb{T}_1 = \mathbb{T}_2 = \mathbb{Z}$. Let Γ be the positively oriented fence of the rectangle R defined by (8.17). Assume that M and N are continuous and have continuous partial delta derivatives $M_{x_2}^{\Delta_2}(x_1, x_2)$ and $N_{x_1}^{\Delta_1}(x_1, x_2)$ on $R \cup \Gamma$. Then, using Green's formula, we put

$$
I = \int_\Gamma M(x_1, x_2)\Delta_1 x_1 + N(x_1, x_2)\Delta_2 x_2,
$$

and thus obtain

$$\begin{aligned}
I &= \int_{a_1}^{b_1} \int_{a_2}^{b_2} \left(N_{x_1}^{\Delta_1}(x_1, x_2) - M_{x_2}^{\Delta_2}(x_1, x_2) \right) \Delta_2 x_2 \Delta_1 x_1 \\
&= \int_{a_2}^{b_2} \int_{a_1}^{b_1} N_{x_1}^{\Delta_1}(x_1, x_2) \Delta_1 x_1 \Delta_2 x_2 - \int_{a_1}^{b_1} \int_{a_2}^{b_2} M_{x_2}^{\Delta_2}(x_1, x_2) \Delta_2 x_2 \Delta_1 x_1 \\
&= \int_{a_2}^{b_2} (N(b_1, x_2) - N(a_1, x_2)) \Delta_2 x_2 - \int_{a_1}^{b_1} (M(x_1, b_2) - M(x_1, a_2)) \Delta_1 x_1 \\
&= \sum_{x_2 \in [a_2, b_2)} (N(b_1, x_2) - N(a_1, x_2)) - \sum_{x_1 \in [a_1, b_1)} (M(x_1, b_2) - M(x_1, a_2)).
\end{aligned}$$

Exercise 8.64 Let $\mathbb{T}_1 = \mathbb{T}_2 = 2^{\mathbb{N}_0}$. Let Γ be the positively oriented fence of the rectangle R defined by (8.17). Assume that M and N are continuous and have continuous partial delta derivatives $M_{x_2}^{\Delta_2}(x_1, x_2)$ and $N_{x_1}^{\Delta_1}(x_1, x_2)$ on $R \cup \Gamma$. Using Green's formula, compute

$$I = \int_\Gamma M(x_1, x_2) \Delta_1 x_1 + N(x_1, x_2) \Delta_2 x_2.$$

Solution

$$I = \sum_{x_2 \in [a_2, b_2)} (N(b_1, x_2) - N(a_1, x_2)) x_2 - \sum_{x_1 \in [a_1, b_1)} (M(x_1, b_2) - M(x_1, a_2)) x_1.$$

8.5 Advanced Practical Problems

Problem 8.65 Let $\mathbb{T} = \mathbb{Z}$ and

$$\Gamma = \begin{cases} x_1 = t^2 + 2t \\ x_2 = t \\ x_3 = t^2 \\ x_4 = t^3, \quad t \in [0, 3]. \end{cases}$$

Find $l(\Gamma)$.

Solution $2\sqrt{3} + 2\sqrt{21} + 2\sqrt{109}$.

Problem 8.66 Let $\mathbb{T} = 3^{\mathbb{N}_0}$ and

$$\Gamma = \begin{cases} x_1 = t^2 \\ x_2 = t^3 \\ x_3 = t, \quad t \in [1, 3]. \end{cases}$$

Find

$$I = \int_\Gamma (x_1 + x_2 + x_3)\Delta l.$$

Solution $I = 6\sqrt{186}$.

Problem 8.67 Let $\mathbb{T} = 2^{\mathbb{N}_0}$ and

$$\Gamma = \begin{cases} x_1 = t^2 + 2t \\ x_2 = t^2 - 2 \\ x_3 = t^2 + 1, \quad t \in [1, 2]. \end{cases}$$

Compute

$$\int_\Gamma x_1^2 \Delta_1 x_1 + x_2^2 \Delta_2 x_2 + x_3^2 \Delta_3 x_3.$$

Solution 60.

Problem 8.68 Let $\mathbb{T}_1 = \mathbb{T}_2 = 3^{\mathbb{N}_0}$. Let Γ be the positively oriented fence of the set

$$E = \left\{(x_1, x_2) \in \mathbb{T}_1 \times \mathbb{T}_2 : \ 1 \le x_1^3 + x_2^3 \le 27\right\}.$$

Using Green's formula, transform the line integral

$$\int_\Gamma (x_1^2 + e_f(x_1, 1) + x_2^3)\Delta_1 x_1 + (e_f(x_1, 1) + x_1^3 x_2^3)\Delta_2 x_2, \quad \text{where} \quad f(x_1) = x_1.$$

Solution $\int\int_E (13x_1^2 x_2^3 + x_1 e_f(x_1, 1) - 13x_2^2)\Delta_1 x_1 \Delta_2 x_2$.

Problem 8.69 Let $\mathbb{T}_1 = 3^{\mathbb{N}_0} \cup \{0\}$ and $\mathbb{T}_2 = 5^{\mathbb{N}_0}$. Let Γ be the positively oriented fence of the rectangle R defined by (8.17). Assume that M and N are continuous and have continuous partial delta derivatives $M_{x_2}^{\Delta_2}(x_1, x_2)$ and $N_{x_1}^{\Delta_1}(x_1, x_2)$ on $R \cup \Gamma$. Using Green's formula, compute

$$I = \int_\Gamma M(x_1, x_2)\Delta_1 x_1 + N(x_1, x_2)\Delta_2 x_2.$$

Solution

$$I = 4 \sum_{x_2 \in [a_2, b_2)} (N(b_1, x_2) - N(a_1, x_2))x_2 - 2 \sum_{x_1 \in [a_1, b_1)} (M(x_1, b_2) - M(x_1, a_2))x_1.$$

Problem 8.70 Let $\mathbb{T}_1 = \mathbb{Z}$ and $\mathbb{T}_2 = 3^{\mathbb{N}_0}$. Let Γ be the positively oriented fence of the rectangle R defined by (8.17). Assume that M and N are continuous and have continuous partial delta derivatives $M_{x_2}^{\Delta_2}(x_1, x_2)$ and $N_{x_1}^{\Delta_1}(x_1, x_2)$ on $R \cup \Gamma$. Using Green's formula, compute

$$I = \int_{\Gamma} M(x_1, x_2)\Delta_1 x_1 + N(x_1, x_2)\Delta_2 x_2.$$

Solution

$$I = 2 \sum_{x_2 \in [a_2, b_2)} (N(b_1, x_2) - N(a_1, x_2))x_2 - \sum_{x_1 \in [a_1, b_1)} (M(x_1, b_2) - M(x_1, a_2)).$$

Problem 8.71 Let Γ be the positively oriented fence of the set

$$E = \left\{(x_1, x_2) \in \mathbb{T}_1 \times \mathbb{T}_2 : \ 1 \le x_1 + x_1^2 x_2 + x_2 \le 10\right\}.$$

Using Green's formula, compute the following line integrals.

1. $\int_{\Gamma} (x_1^2 - 3x_2)\Delta_1 x_1 + (x_2^3 - 3x_1 x_2)\Delta_2 x_2$, $\mathbb{T}_1 = \mathbb{T}_2 = \mathbb{Z}$,
2. $\int_{\Gamma} x_1^3 x_2 \Delta_1 x_1 + x_1 x_2^2 \Delta_2 x_2$, $\mathbb{T}_1 = \mathbb{T}_2 = \mathbb{N}$,
3. $\int_{\Gamma} (x_1 - 2x_2)\Delta_1 x_1 + x_1^3 x_2^4 \Delta_2 x_2$, $\mathbb{T}_1 = \mathbb{N}$, $\mathbb{T}_2 = 2^{\mathbb{N}_0} \cup \{0\}$,
4. $\int_{\Gamma} (x_1^2 + x_1 x_2 + x_2^2)\Delta_1 x_1 + (x_1^2 - x_1 x_2 + x_2^2)\Delta_2 x_2$, $\mathbb{T}_1 = \mathbb{T}_2 = 3^{\mathbb{N}_0} \cup \{0\}$,
5. $\int_{\Gamma} x_1 x_2 \Delta_1 x_1 - (x_1^3 - x_2^4)\Delta_2 x_2$, $\mathbb{T}_1 = \mathbb{Z}$, $\mathbb{T}_2 = \mathbb{N}$,
6. $\int_{\Gamma} (x_1 + 2x_2 - x_2^4)\Delta_1 x_1 + x_2^5 \Delta_2 x_2$, $\mathbb{T}_1 = \mathbb{T}_2 = \mathbb{Z}$.

8.6 Notes and References

This chapter continues the developments from Chapters 6 and 7 and develops line integration along time scale curves. The length of time scale curves is defined. Rectifiable curves on time scales are introduced, and some of their properties are given. Line delta integrals of the first and second kinds are defined. Sufficient conditions for the existence of the line integrals are presented. A mean value result for line integrals of the first kind is offered, and a time scale version of Green's formula is given. Green's formula establishes a connection between double and line integrals. The length of time scales curves and the definitions and properties of line integrals of the first and second kind are n-dimensional analogues of the two-dimensional definitions and results by Bohner and Guseinov [14], while Green's formula (Theorem 8.58) is directly taken from [14].

Chapter 9
Surface Integrals

9.1 Surface Areas

Let $\mathbb{T}_i, i \in \{1, \ldots, n\}$, be time scales. Suppose $\Omega \subset \mathbb{T}_1 \times \ldots \times \mathbb{T}_n$. Let $\phi_i : \Omega \to \mathbb{R}$ be continuous functions on Ω.

Definition 9.1 The system of functions

$$x_i = \phi_i(y), \quad y \in \Omega, \quad i \in \{1, \ldots, m\}, \quad m = n + 1, \tag{9.1}$$

is said to define a time scale continuous *surface* S. The points $(x_1, \ldots, x_m)$ with x_i, $i \in \{1, \ldots, m\}$, defined by (9.1), are called the points of the surface S.

Example 9.2 Let $\mathbb{T}_1 = \mathbb{T}_2 = \mathbb{Z}$ and

$$\Omega = \{(y_1, y_2) \in \mathbb{T}_1 \times \mathbb{T}_2 : \; -3 \le y_1 \le 4, \; 0 \le y_2 \le 5\}.$$

Thus,

$$\begin{cases} x_1 = y_1^2 + y_2^2 \\ x_2 = y_1 - y_2 \\ x_3 = y_1 \quad (y_1, y_2) \in \Omega, \end{cases}$$

is a time scale continuous surface.

Let S be a surface given by (9.1). We can rewrite the system (9.1) in the form

$$\begin{aligned} \mathbf{r} &= \mathbf{r}(y_1, \ldots, y_n) \\ &= \phi_1(y_1, \ldots, y_n)\mathbf{e}_1 + \cdots + \phi_m(y_1, \ldots, y_n)\mathbf{e}_m, \end{aligned}$$

M. Bohner and S.G. Georgiev, *Multivariable Dynamic Calculus on Time Scales*,
DOI 10.1007/978-3-319-47620-9_9

where

$$\mathbf{e}_1 = (1, 0, \ldots, 0), \quad \mathbf{e}_2 = (0, 1, \ldots, 0), \quad \ldots, \quad \mathbf{e}_m = (0, 0, \ldots, 1).$$

Let

$$R_p = [y_1^p, y_1'^p) \times \ldots \times [y_n^p, y_n'^p),$$

$$y_j^p < y_j'^p, \quad j \in \{1, \ldots, n\},$$

$$(y_1^p, \ldots, y_n^p), (y_1'^p, \ldots, y_n'^p) \in \Omega, \quad p \in \{1, \ldots, k\},$$

and $P' = \{R_1, \ldots, R_k\}$, i.e., P' is an arbitrary inner Δ-partition of Ω. Let

$$\begin{aligned}
\mathbf{u}_j^p &= \mathbf{r}(y_1^p, \ldots, y_{j-1}^p, y_j'^p, y_{j+1}^p, \ldots, y_n^p) \\
&\quad -\mathbf{r}(y_1^p, \ldots, y_{j-1}^p, y_j^p, y_{j+1}^p, \ldots, y_n^p) \\
&= \phi_1(y_1^p, \ldots, y_{j-1}^p, y_j'^p, y_{j+1}^p, \ldots, y_n^p)\mathbf{e}_1 \\
&\quad +\phi_2(y_1^p, \ldots, y_{j-1}^p, y_j'^p, y_{j+1}^p, \ldots, y_n^p)\mathbf{e}_2 \\
&\quad +\cdots \\
&\quad +\phi_m(y_1^p, \ldots, y_{j-1}^p, y_j'^p, y_{j+1}^p, \ldots, y_n^p)\mathbf{e}_m \\
&\quad -\phi_1(y_1^p, \ldots, y_{j-1}^p, y_j^p, y_{j+1}^p, \ldots, y_n^p)\mathbf{e}_1 \\
&\quad -\phi_2(y_1^p, \ldots, y_{j-1}^p, y_j^p, y_{j+1}^p, \ldots, y_n^p)\mathbf{e}_2 \\
&\quad -\cdots \\
&\quad -\phi_m(y_1^p, \ldots, y_{j-1}^p, y_j^p, y_{j+1}^p, \ldots, y_n^p)\mathbf{e}_m \\
&= \left(\phi_1(y_1^p, \ldots, y_{j-1}^p, y_j'^p, y_{j+1}^p, \ldots, y_n^p) - \phi_1(y_1^p, \ldots, y_{j-1}^p, y_j^p, y_{j+1}^p, \ldots, y_n^p)\right) \mathbf{e}_1 \\
&\quad + \left(\phi_2(y_1^p, \ldots, y_{j-1}^p, y_j'^p, y_{j+1}^p, \ldots, y_n^p) - \phi_2(y_1^p, \ldots, y_{j-1}^p, y_j^p, y_{j+1}^p, \ldots, y_n^p)\right) \mathbf{e}_2 \\
&\quad +\cdots \\
&\quad + \left(\phi_m(y_1^p, \ldots, y_{j-1}^p, y_j'^p, y_{j+1}^p, \ldots, y_n^p) - \phi_m(y_1^p, \ldots, y_{j-1}^p, y_j^p, y_{j+1}^p, \ldots, y_n^p)\right) \mathbf{e}_m
\end{aligned}$$

and

$$u_j^{lp} = \phi_l(y_1^p, \dots, y_{j-1}^p, y_j'^p, y_{j+1}^p, \dots, y_n^p)$$

$$-\phi_l(y_1^p, \dots, y_{j-1}^p, y_j^p, y_{j+1}^p, \dots, y_n^p), \quad j \in \{1, \dots, n\}.$$

Thus,

$$\mathbf{u}_j^p = \sum_{l=1}^{m} u_j^{lp} \mathbf{e}_l.$$

The area of the parallelogram spanned by the vectors

$$\mathbf{u}_1^p, \quad \mathbf{u}_2^p, \quad \dots, \quad \mathbf{u}_m^p$$

is equal to

$$|\mathbf{u}_1^p \times \mathbf{u}_2^p \times \dots \times \mathbf{u}_m^p| = \left| \begin{vmatrix} \mathbf{e}_1 & \mathbf{e}_2 & \dots & \mathbf{e}_m \\ u_1^{1p} & u_1^{2p} & \dots & u_1^{mp} \\ u_2^{1p} & u_2^{2p} & \dots & u_2^{mp} \\ \dots & \dots & \dots & \dots \\ u_m^{1p} & u_m^{2p} & \dots & u_m^{mp} \end{vmatrix} \right|.$$

Let us set

$$A(S, P') = \sum_{p=1}^{k} |\mathbf{u}_1^p \times \mathbf{u}_2^p \times \dots \times \mathbf{u}_m^p|.$$

Definition 9.3 The surface S is said to be *rectifiable* or *squarable* if

$$\sup\left\{A(S, P') : P' \quad \text{is an inner } \Delta\text{-partition of} \quad \Omega\right\} = A(S) < \infty.$$

The nonnegative real number $A = A(S)$ is called the *area* of the surface S. If the supremum (finite) does not exist, then the surface is said to be *nonsquarable* or *nonrectifiable*.

Theorem 9.4 *Let P' and Q' be inner Δ-partitions of Ω and suppose that Q' is a refinement of P'. Then*

$$A(S, P') \le A(S, Q').$$

Proof Without loss of generality, we may assume that Q' has only one more element than P'. If P' is given by

$$P' = \{R_1, R_2, \dots, R_k\},$$

then there exists $j \in \{1, \dots, k\}$ such that Q' is given by

$$Q' = \{R_1, \dots, R_{j-1}, R_j^1, R_j^2, R_{j+1}, \dots, R_k\},$$

where $R_j = R_j^1 \cup R_j^2$. Assume that

$$R_j = [y_1^j, y_1^{\prime j}) \times \ldots \times [y_n^j, y_n^{\prime j}),$$

$$R_j^1 = [y_1^j, y_1^{\prime 1j}) \times \ldots \times [y_n^j, y_n^{\prime 1j}),$$

$$R_j^2 = [y_1^{\prime 1j}, y_1^{\prime j}) \times \ldots \times [y_n^{\prime 1j}, y_n^{\prime j}),$$

where

$$y_k^{\prime 1j} \in \mathbb{T}_k, \quad y_k^j < y_k^{\prime 1j} < y_k^{\prime j}.$$

Thus,

$$A(S, Q') = \sum_{p=1}^{j-1} |\mathbf{u}_1^p \times \mathbf{u}_2^p \times \ldots \times \mathbf{u}_m^p| + \sum_{p=j+1}^{k} |\mathbf{u}_1^p \times \mathbf{u}_2^p \times \ldots \times \mathbf{u}_m^p|$$

$$+ |\mathbf{u}_1^{1j} \times \ldots \times \mathbf{u}_m^{1j}| + |\mathbf{u}_1^{2j} \times \ldots \times \mathbf{u}_m^{2j}|,$$

$$A(S, P') = \sum_{p=1}^{j-1} |\mathbf{u}_1^p \times \ldots \times \mathbf{u}_m^p| + \sum_{p=j+1}^{k} |\mathbf{u}_1^p \times \ldots \times \mathbf{u}_m^p|$$

$$+ |\mathbf{u}_1^j \times \ldots \times \mathbf{u}_m^j|,$$

where

$$\mathbf{u}_l^{1j} = \mathbf{r}(y_1^j, \ldots, y_{l-1}^j, y_l^{\prime 1j}, y_{l+1}^j, \ldots, y_n^j) - \mathbf{r}(y_1^j, \ldots, y_{l-1}^j, y_l^j, y_{l+1}^j, \ldots, y_n^j),$$

$$\mathbf{u}_l^{2j} = \mathbf{r}(y_1^{\prime 1j}, \ldots, y_{l-1}^{\prime 1j}, y_l^{\prime j}, y_{l+1}^{\prime 1j}, \ldots, y_n^{\prime 1j}) - \mathbf{r}(y_1^{\prime 1j}, \ldots, y_{l-1}^{\prime 1j}, y_l^{\prime 1j}, y_{l+1}^{\prime 1j}, \ldots, y_n^{\prime 1j}),$$

$$\mathbf{u}_l^j = \mathbf{r}(y_1^j, \ldots, y_{l-1}^j, y_l^{\prime j}, y_{l+1}^j, \ldots, y_n^j) - \mathbf{r}(y_1^j, \ldots, y_{l-1}^j, y_l^j, y_{l+1}^j, \ldots, y_n^j),$$

$l \in \{1, \ldots, m\}$. Hence,

$$A(S, Q') - A(S, P') = |\mathbf{u}_1^{1j} \times \ldots \times \mathbf{u}_m^{1j}| + |\mathbf{u}_1^{2j} \times \ldots \times \mathbf{u}_m^{2j}| - |\mathbf{u}_1^j \times \ldots \times \mathbf{u}_m^j| \geq 0,$$

which completes the proof. □

Theorem 9.5 *Let the functions ϕ_l, $l \in \{1, \ldots, m\}$, be continuous and have continuous first-order partial derivatives in the closure of the region $\Omega \subset \mathbb{T}_1 \times \ldots \times \mathbb{T}_n$. If the region Ω is bounded and Jordan Δ-measurable, then the surface S defined by the parametric equations in* (9.1) *is squarable, and its area $A(S)$ can be evaluated* *by the formula*

$$A(S)=\int_{\Omega}|\mathbf{r}^{\Delta_1}(y_1,\ldots,y_n)\times\ldots\times\mathbf{r}^{\Delta_n}(y_1,\ldots,y_n)|\Delta_1 y_1\ldots\Delta_n y_n.$$

Proof First, we will show that the surface S is squarable. Let $P'=\{R_1,\ldots,R_k\}$ be an arbitrary Δ-partition of Ω and put

$$R_p=[y_1^p,y_1'^p)\times\ldots\times[y_n^p,y_n'^p),$$

$$u_j^{lp}=\phi_l(y_1^p,\ldots,y_{j-1}^p,y_j'^p,y_{j+1}^p,\ldots,y_n^p)$$
$$-\phi_l(y_1^p,\ldots,y_{j-1}^p,y_j^p,y_{j+1}^p,\ldots,y_n^p),$$

$$\mathbf{u}_j^p=\sum_{l=1}^m u_j^{lp}\mathbf{e}_l,\quad j\in\{1,\ldots,n\}.$$

Thus,

$$\begin{aligned}|\mathbf{u}_1^p\times\mathbf{u}_2^p\times\ldots\times\mathbf{u}_n^p|^2&\le|\mathbf{u}_1^p|^2|\mathbf{u}_2^p|^2\ldots|\mathbf{u}_n^p|^2\\&=\prod_{q=1}^n|\mathbf{u}_q^p|^2\\&=\prod_{q=1}^n\left|\sum_{l=1}^m u_q^{lp}\mathbf{e}_l\right|^2\\&=\prod_{q=1}^n\sum_{l=1}^m|u_q^{lp}|^2.\end{aligned}\tag{9.2}$$

Applying the mean value theorem to the functions u_q^{lp}, we get that there exist ξ_q^{lp} and η_q^{lp} such that

$$\begin{aligned}\phi_l^{\Delta_q}(y_1^p,\ldots,y_{q-1}^p,\eta_q^{lp},y_{q+1}^p,\ldots,y_n^p)(y_q'^p-y_q^p)&\le u_q^{lp}\\&\le\phi_l^{\Delta_q}(y_1^p,\ldots,y_{q-1}^p,\xi_q^{lp},y_{q+1}^p,\ldots,y_n^p)(y_q'^p-y_q^p).\end{aligned}\tag{9.3}$$

Since $\phi_l^{\Delta_q}$ have continuous first-order partial delta derivatives on the closure of the region Ω, there exists a positive constant C such that

$$|\phi_l^{\Delta_q}| \le C \quad \text{on} \quad \Omega.$$

Consequently,

$$|u_q^{lp}| \le C|y_q'^p - y_q^p|.$$

Hence, using (9.2), we obtain

$$|\mathbf{u}_1^p \times \ldots \times \mathbf{u}_n^p|^2 \le \prod_{q=1}^n \sum_{l=1}^m |u_q^{lp}|^2$$

$$\le C^2 \prod_{q=1}^n \sum_{l=1}^m |y_q'^p - y_q^p|^2 \tag{9.4}$$

$$= C^2 m \prod_{q=1}^n |y_q'^p - y_q^p|^2.$$

Assuming that

$$\Omega \subset R = [a_1, b_1) \times \ldots \times [a_n, b_n)$$

and using (9.4), we get

$$|\mathbf{u}_1^p \times \ldots \times \mathbf{u}_n^p|^2 \le C^2 m \prod_{q=1}^n (b_q - a_q).$$

This shows that the set

$$\{A(S, P') : P' \quad \text{is an inner } \Delta \text{-partition of} \quad \Omega\}$$

is bounded. Therefore, S is squarable. Let

$$I = \int_\Omega |\mathbf{r}^{\Delta_1}(y_1, \ldots, y_n) \times \ldots \times \mathbf{r}^{\Delta_n}(y_1, \ldots, y_n)| \Delta_1 y_1 \ldots \Delta_n y_n.$$

Consider the Riemann Δ-sum

$$\Lambda = \sum_{p=1}^k |\mathbf{r}^{\Delta_1}(\xi_1, \ldots, \xi_n) \times \ldots \times \mathbf{r}^{\Delta_n}(\xi_1, \ldots, \xi_n)| m(R_p)$$

of the Δ-integrable function $|\mathbf{r}^{\Delta_1} \times \ldots \times \mathbf{r}^{\Delta_n}|$, corresponding to the inner Δ-partition $P' = \{R_1, \ldots, R_k\}$ of Ω and any choice of the points $(\xi_1, \ldots, \xi_n) \in R_p$, where $m(R_p)$ denotes the area of R_p. Let $\varepsilon > 0$ be arbitrarily chosen. Thus,

$$A(S, P) - \Lambda = \sum_{p=1}^{n} \Big\{ |\mathbf{u}_1^p \times \ldots \times \mathbf{u}_n^p|$$

$$- |\mathbf{r}^{\Delta_1}(\xi_1, \ldots, \xi_n) \times \ldots \times \mathbf{r}^{\Delta_n}(\xi_1, \ldots, \xi_n)| m(R_p) \Big\}.$$

From (9.3), we have

$$0 \le \phi_l(y_1^p, \ldots, y_{q-1}^p, y_q'^p, y_{q+1}^p, \ldots, y_n^p)$$

$$-\phi_l(y_1^p, \ldots, y_{q-1}^p, y_q^p, y_{q+1}^p, \ldots, y_n^p)$$

$$-\phi_l^{\Delta_q}(y_1^p, \ldots, y_{q-1}^p, \eta_q^{lp}, y_{q+1}^p, \ldots, y_n^p)(y_q'^p - y_q^p)$$

$$\le \phi_l^{\Delta_q}(y_1^p, \ldots, y_{q-1}^p, \xi_q^{lp}, y_{q+1}^p, \ldots, y_n^p)(y_q'^p - y_q^p)$$

$$-\phi_l^{\Delta_q}(y_1^p, \ldots, y_{q-1}^p, \eta_q^{lp}, y_{q+1}^p, \ldots, y_n^p)(y_q'^p - y_q^p)$$

and

$$u_q^{lp} = \left(\phi_l^{\Delta_q}(y_1^p, \ldots, y_{q-1}^p, \eta_q^{lp}, y_{q+1}^p, \ldots, y_n^p) + \alpha_{lpq}\right)(y_q'^p - y_q^p).$$

Let

$$\overline{u}_q^{lp} = \phi_l^{\Delta_q}(y_1^p, \ldots, y_{q-1}^p, \eta_q^{lp}, y_{q+1}^p, \ldots, y_n^p) + \alpha_{lpq}.$$

Thus,

$$u_q^{lp} = \overline{u}_q^{lp}(y_q'^p - y_q^p)$$

and

$$\alpha_{lpq} \le \phi_l^{\Delta_q}(y_1^p, \ldots, y_{q-1}^p, \xi_q^{lp}, y_{q+1}^p, \ldots, y_n^p)$$

$$-\phi_l^{\Delta_q}(y_1^p, \ldots, y_{q-1}^p, \eta_q^{lp}, y_{q+1}^p, \ldots, y_n^p)$$

$$\le M_{lpq} - m_{lpq},$$

where M_{lpq} and m_{lpq} are the supremum and infimum of $\phi_l^{\Delta_q}$ on R_p, respectively. Hence,

$$\mathbf{u}_q^p = \sum_{l=1}^{m} u_q^{lp} \mathbf{e}_l$$

$$= \sum_{l=1}^{m} \overline{u}_q^{lp} (y_q^{\prime p} - y_q^p) \mathbf{e}_l.$$

Moreover,

$$A(S, P') - \Lambda = \sum_{p=1}^{k} \Big\{ |\overline{\mathbf{u}}_\mathbf{1}^\mathbf{p} \times \ldots \times \overline{\mathbf{u}}_\mathbf{n}^\mathbf{p}|$$

$$- |\mathbf{r}^{\Delta_1}(\xi_1, \ldots, \xi_n) \times \ldots \times \mathbf{r}^{\Delta_n}(\xi_1, \ldots, \xi_n)| \Big\} m(R_p),$$

where

$$\overline{\mathbf{u}}_\mathbf{q}^\mathbf{p} = \sum_{l=1}^{m} \overline{u}_q^{lp} \mathbf{e}_l.$$

Therefore,

$$|A(S, P') - \Lambda| < \frac{\varepsilon}{4} \tag{9.5}$$

for every inner Δ-partition $P' = \{R_1, \ldots, R_k\}$ of Ω, determined by a partition $P \in \mathscr{P}_\delta(R)$. Using the definition of the Δ-integral, diminishing δ if necessary, we may assume that for the same partitions P' for which (9.5) is satisfied, we have

$$|\Lambda - I| < \frac{\varepsilon}{4}. \tag{9.6}$$

Also, there is an inner Δ-partition P_0' of Ω such that

$$0 \le A(S) - A(S, P_0') < \frac{\varepsilon}{2}. \tag{9.7}$$

We refine the partition P_0' so that an inner Δ-partition P' is obtained, which is determined by a partition $P \in \mathscr{P}_\delta(R)$. Hence, by Theorem 9.4, we get

$$A(S, P_0') \le A(S, P').$$

From the last inequality and from (9.7), we obtain

$$\begin{aligned}0 &\le A(S) - A(S, P')\\ &\le A(S) - A(S, P'_0)\\ &< \frac{\varepsilon}{2}.\end{aligned}$$

Hence, using (9.5) and (9.6), we get

$$\begin{aligned}|A(S) - I| &= |A(S) - A(S, P') + A(S, P') - \Lambda + \Lambda - I|\\ &\le A(S) - A(S, P') + |A(S, P') - \Lambda| + |\Lambda - I|\\ &< \frac{\varepsilon}{2} + \frac{\varepsilon}{4} + \frac{\varepsilon}{4}\\ &= \varepsilon.\end{aligned}$$

Since $\varepsilon > 0$ was arbitrarily chosen, we obtain $A(S) = I$, which completes the proof. □

Remark 9.6 Let $m = 3$ and $n = 2$, i.e.,

$$S : \begin{cases} x_1 = \phi_1(y_1, y_2)\\ x_2 = \phi_2(y_1, y_2)\\ x_3 = \phi_3(y_1, y_2).\end{cases}$$

Thus,

$$\mathbf{r}(y_1, y_2) = \phi_1(y_1, y_2)\mathbf{e}_1 + \phi_2(y_1, y_2)\mathbf{e}_2 + \phi_3(y_1, y_2)\mathbf{e}_3.$$

Hence,

$$\mathbf{r}_{y_1}^{\Delta_1} = \phi_{1y_1}^{\Delta_1}(y_1, y_2)\mathbf{e}_1 + \phi_{2y_1}^{\Delta_1}(y_1, y_2)\mathbf{e}_2 + \phi_{3y_1}^{\Delta_1}(y_1, y_2)\mathbf{e}_3,$$

$$\mathbf{r}_{y_2}^{\Delta_2} = \phi_{1y_2}^{\Delta_2}(y_1, y_2)\mathbf{e}_1 + \phi_{2y_2}^{\Delta_2}(y_1, y_2)\mathbf{e}_2 + \phi_{3y_2}^{\Delta_2}(y_1, y_2)\mathbf{e}_3,$$

and

$$\mathbf{r}_{y_1}^{\Delta_1}(y_1, y_2) \times \mathbf{r}_{y_2}^{\Delta_2}(y_1, y_2) = \begin{vmatrix} \mathbf{e}_1 & \mathbf{e}_2 & \mathbf{e}_3\\ \phi_{1y_1}^{\Delta_1} & \phi_{2y_1}^{\Delta_1} & \phi_{3y_1}^{\Delta_1}\\ \phi_{1y_2}^{\Delta_2} & \phi_{2y_2}^{\Delta_2} & \phi_{3y_2}^{\Delta_2}\end{vmatrix}$$

$$= \begin{vmatrix} \phi_{2y_1}^{\Delta_1} & \phi_{3y_1}^{\Delta_1} \\ \phi_{2y_2}^{\Delta_2} & \phi_{3y_2}^{\Delta_2} \end{vmatrix} \mathbf{e}_1 - \begin{vmatrix} \phi_{1y_1}^{\Delta_1} & \phi_{3y_1}^{\Delta_1} \\ \phi_{1y_2}^{\Delta_2} & \phi_{3y_2}^{\Delta_2} \end{vmatrix} \mathbf{e}_2 + \begin{vmatrix} \phi_{1y_1}^{\Delta_1} & \phi_{2y_1}^{\Delta_1} \\ \phi_{1y_2}^{\Delta_2} & \phi_{2y_2}^{\Delta_2} \end{vmatrix} \mathbf{e}_3$$

$$= \left(\phi_{2y_1}^{\Delta_1}\phi_{3y_2}^{\Delta_2} - \phi_{2y_2}^{\Delta_2}\phi_{3y_1}^{\Delta_1}\right)\mathbf{e}_1 - \left(\phi_{1y_1}^{\Delta_1}\phi_{3y_2}^{\Delta_2} - \phi_{1y_2}^{\Delta_2}\phi_{3y_1}^{\Delta_1}\right)\mathbf{e}_2$$

$$+ \left(\phi_{1y_1}^{\Delta_1}\phi_{2y_2}^{\Delta_2} - \phi_{1y_2}^{\Delta_2}\phi_{2y_1}^{\Delta_1}\right)\mathbf{e}_3,$$

$$|\mathbf{r}_{y_1}^{\Delta_1}(y_1, y_2) \times \mathbf{r}_{y_2}(y_1, y_2)|^2 = \left(\phi_{2y_1}^{\Delta_1}\phi_{3y_2}^{\Delta_2} - \phi_{2y_2}^{\Delta_2}\phi_{3y_1}^{\Delta_1}\right)^2$$

$$+ \left(\phi_{1y_1}^{\Delta_1}\phi_{3y_2}^{\Delta_2} - \phi_{1y_2}^{\Delta_2}\phi_{3y_1}^{\Delta_1}\right)^2 + \left(\phi_{1y_1}^{\Delta_1}\phi_{2y_2}^{\Delta_2} - \phi_{1y_2}^{\Delta_2}\phi_{2y_1}^{\Delta_1}\right)^2$$

$$= \left(\phi_{2y_1}^{\Delta_1}\right)^2\left(\phi_{3y_2}^{\Delta_2}\right)^2 - 2\phi_{2y_1}^{\Delta_1}\phi_{3y_2}^{\Delta_2}\phi_{2y_2}^{\Delta_2}\phi_{3y_1}^{\Delta_1} + \left(\phi_{2y_2}^{\Delta_2}\right)^2\left(\phi_{3y_1}^{\Delta_1}\right)^2$$

$$+ \left(\phi_{1y_1}^{\Delta_1}\right)^2\left(\phi_{3y_2}^{\Delta_2}\right)^2 - 2\phi_{1y_1}^{\Delta_1}\phi_{1y_2}^{\Delta_2}\phi_{3y_1}^{\Delta_1}\phi_{3y_2}^{\Delta_2} + \left(\phi_{3y_1}^{\Delta_1}\right)^2\left(\phi_{1y_2}^{\Delta_2}\right)^2$$

$$+ \left(\phi_{1y_1}^{\Delta_1}\right)^2\left(\phi_{2y_2}^{\Delta_2}\right)^2 - 2\phi_{1y_1}^{\Delta_1}\phi_{1y_2}^{\Delta_2}\phi_{2y_1}^{\Delta_1}\phi_{2y_2}^{\Delta_2} + \left(\phi_{1y_2}^{\Delta_2}\right)^2\left(\phi_{2y_1}^{\Delta_1}\right)^2$$

$$= \left(\phi_{2y_1}^{\Delta_1}\right)^2\left(\left(\phi_{1y_2}^{\Delta_2}\right)^2 + \left(\phi_{3y_2}^{\Delta_2}\right)^2\right) + \left(\phi_{1y_1}^{\Delta_1}\right)^2\left(\left(\phi_{2y_2}^{\Delta_2}\right)^2 + \left(\phi_{3y_2}^{\Delta_2}\right)^2\right)$$

$$+ \left(\phi_{3y_1}^{\Delta_1}\right)^2\left(\left(\phi_{1y_2}^{\Delta_2}\right)^2 + \left(\phi_{2y_2}^{\Delta_2}\right)^2\right) - 2\phi_{2y_1}^{\Delta_1}\phi_{3y_2}^{\Delta_2}\phi_{2y_2}^{\Delta_2}\phi_{3y_1}^{\Delta_1}$$

$$- 2\phi_{1y_1}^{\Delta_1}\phi_{1y_2}^{\Delta_2}\phi_{3y_1}^{\Delta_1}\phi_{3y_2}^{\Delta_2} - 2\phi_{1y_1}^{\Delta_1}\phi_{1y_2}^{\Delta_2}\phi_{2y_1}^{\Delta_1}\phi_{2y_2}^{\Delta_2}$$

$$= \left(\phi_{2y_1}^{\Delta_1}\right)^2\left(\left(\phi_{1y_2}^{\Delta_2}\right)^2 + \left(\phi_{2y_2}^{\Delta_2}\right)^2 + \left(\phi_{3y_2}^{\Delta_2}\right)^2\right) - \left(\phi_{2y_1}^{\Delta_1}\right)^2\left(\phi_{2y_2}^{\Delta_2}\right)^2$$

$$+ \left(\phi_{1y_1}^{\Delta_1}\right)^2\left(\left(\phi_{1y_2}^{\Delta_2}\right)^2 + \left(\phi_{2y_2}^{\Delta_2}\right)^2 + \left(\phi_{3y_2}^{\Delta_2}\right)^2\right) - \left(\phi_{1y_1}^{\Delta_1}\right)^2\left(\phi_{1y_2}^{\Delta_2}\right)^2$$

$$+\left(\phi_{3y_1}^{\Delta_1}\right)^2\left(\left(\phi_{1y_2}^{\Delta_2}\right)^2+\left(\phi_{2y_2}^{\Delta_2}\right)^2+\left(\phi_{3y_2}^{\Delta_2}\right)^2\right)-\left(\phi_{3y_1}^{\Delta_1}\right)^2\left(\phi_{3y_2}^{\Delta_2}\right)^2$$

$$-2\phi_{2y_1}^{\Delta_1}\phi_{3y_2}^{\Delta_2}\phi_{2y_2}^{\Delta_2}\phi_{3y_1}^{\Delta_1}-2\phi_{1y_1}^{\Delta_1}\phi_{1y_2}^{\Delta_2}\phi_{3y_1}^{\Delta_1}\phi_{3y_2}^{\Delta_2}$$

$$-2\phi_{1y_1}^{\Delta_1}\phi_{1y_2}^{\Delta_2}\phi_{2y_1}^{\Delta_1}\phi_{2y_2}^{\Delta_2}$$

$$=\left(\left(\phi_{1y_1}^{\Delta_1}\right)^2+\left(\phi_{2y_1}^{\Delta_1}\right)^2+\left(\phi_{3y_1}^{\Delta_1}\right)^2\right)\left(\left(\phi_{1y_2}^{\Delta_2}\right)^2+\left(\phi_{2y_2}^{\Delta_2}\right)^2+\left(\phi_{3y_2}^{\Delta_2}\right)^2\right)$$

$$-\left(\phi_{1y_1}^{\Delta_1}\phi_{1y_2}^{\Delta_2}+\phi_{2y_1}^{\Delta_1}\phi_{2y_2}^{\Delta_2}+\phi_{3y_1}^{\Delta_1}\phi_{3y_2}^{\Delta_2}\right)^2.$$

Let

$$E=\left(\phi_{1y_1}^{\Delta_1}\right)^2+\left(\phi_{2y_1}^{\Delta_1}\right)^2+\left(\phi_{3y_1}^{\Delta_1}\right)^2,$$

$$F=\phi_{1y_1}^{\Delta_1}\phi_{1y_2}^{\Delta_2}+\phi_{2y_1}^{\Delta_1}\phi_{2y_2}^{\Delta_2}+\phi_{3y_1}^{\Delta_1}\phi_{3y_2}^{\Delta_2},$$

$$G=\left(\phi_{1y_2}^{\Delta_2}\right)^2+\left(\phi_{2y_2}^{\Delta_2}\right)^2+\left(\phi_{3y_2}^{\Delta_2}\right)^2.$$

Then

$$|\mathbf{r}_{y_1}^{\Delta_1}(y_1,y_2)\times\mathbf{r}_{y_2}^{\Delta_2}(y_1,y_2)|=\sqrt{EG-F^2}$$

and

$$A(S)=\int\int_{\Omega}\sqrt{EG-F^2}\Delta_1 y_1\Delta_2 y_2. \tag{9.8}$$

Remark 9.7 Let

$$S:\begin{cases}x_1=y_1\\ x_2=y_2\\ x_3=\phi(y_1,y_2)\\ (y_1,y_2)\in\Omega\subset\mathbb{T}_1\times\mathbb{T}_2.\end{cases}$$

Here,

$$\phi_1(y_1, y_2) = y_1, \quad \phi_2(y_1, y_2) = y_2, \quad \phi_3(y_1, y_2) = \phi(y_1, y_2),$$

$$\phi_{1y_1}^{\Delta_1} = 1, \quad \phi_{1y_2}^{\Delta_2} = 0,$$

$$\phi_{2y_1}^{\Delta_1} = 0, \quad \phi_{2y_2}^{\Delta_2} = 1,$$

$$\phi_{3y_1}^{\Delta_1} = \phi_{y_1}^{\Delta_1}, \quad \phi_{3y_2}^{\Delta_2} = \phi_{y_2}^{\Delta_2},$$

and

$$\begin{aligned} E &= \left(\phi_{1y_1}^{\Delta_1}\right)^2 + \left(\phi_{2y_1}^{\Delta_1}\right)^2 + \left(\phi_{3y_1}^{\Delta_1}\right)^2 \\ &= 1 + \left(\phi_{y_1}^{\Delta_1}\right)^2, \\ F &= \phi_{1y_1}^{\Delta_1}\phi_{1y_2}^{\Delta_2} + \phi_{2y_1}^{\Delta_1}\phi_{2y_2}^{\Delta_2} + \phi_{3y_1}^{\Delta_1}\phi_{3y_2}^{\Delta_2} \\ &= \phi_{y_1}^{\Delta_1}\phi_{y_2}^{\Delta_2}, \\ G &= \left(\phi_{1y_2}^{\Delta_1}\right)^2 + \left(\phi_{2y_2}^{\Delta_1}\right)^2 + \left(\phi_{3y_2}^{\Delta_1}\right)^2 \\ &= 1 + \left(\phi_{y_2}^{\Delta_2}\right)^2. \end{aligned}$$

Therefore,

$$\begin{aligned} EG - F^2 &= \left(1 + \left(\phi_{y_1}^{\Delta_1}\right)^2\right)\left(1 + \left(\phi_{y_2}^{\Delta_2}\right)^2\right) - \left(\phi_{y_1}^{\Delta_1}\right)^2\left(\phi_{y_2}^{\Delta_2}\right)^2 \\ &= 1 + \left(\phi_{y_1}^{\Delta_1}\right)^2 + \left(\phi_{y_2}^{\Delta_2}\right)^2 + \left(\phi_{y_1}^{\Delta_1}\right)^2\left(\phi_{y_2}^{\Delta_2}\right)^2 - \left(\phi_{y_1}^{\Delta_1}\right)^2\left(\phi_{y_2}^{\Delta_2}\right)^2 \\ &= 1 + \left(\phi_{y_1}^{\Delta_1}\right)^2 + \left(\phi_{y_2}^{\Delta_2}\right)^2. \end{aligned}$$

From here and (9.8), we obtain

$$A(S) = \int\int_{\Omega} \sqrt{1 + \left(\phi_{y_1}^{\Delta_1}\right)^2 + \left(\phi_{y_2}^{\Delta_2}\right)^2}\, \Delta_1 y_1 \Delta_2 y_2.$$

Example 9.8 Let $\mathbb{T}_1 = \mathbb{T}_2 = \mathbb{Z}$ and

$$\Omega = \{(y_1, y_2) \in \mathbb{T}_1 \times \mathbb{T}_2 : \ 0 \le y_1, y_2 \le 3\}.$$

Consider

$$S: \begin{cases} x_1 = y_1 + 2 \\ x_2 = y_2 - 2 \\ x_3 = y_1 + y_2. \end{cases}$$

Here,

$$\phi_1(y_1, y_2) = y_1 + 2, \quad \phi_2(y_1, y_2) = y_2 - 2, \quad \phi_3(y_1, y_2) = y_1 + y_2.$$

Then

$$\phi_{1y_1}^{\Delta_1}(y_1, y_2) = 1, \quad \phi_{1y_2}^{\Delta_2}(y_1, y_2) = 0,$$

$$\phi_{2y_1}^{\Delta_1}(y_1, y_2) = 0, \quad \phi_{2y_2}^{\Delta_2}(y_1, y_2) = 1,$$

$$\phi_{3y_1}^{\Delta_1}(y_1, y_2) = 1, \quad \phi_{3y_2}^{\Delta_2}(y_1, y_2) = 1.$$

Hence,

$$\begin{aligned} E &= \left(\phi_{1y_1}^{\Delta_1}(y_1, y_2)\right)^2 + \left(\phi_{2y_1}^{\Delta_1}(y_1, y_2)\right)^2 + \left(\phi_{3y_1}^{\Delta_1}(y_1, y_2)\right)^2 \\ &= 2, \\ F &= \phi_{1y_1}^{\Delta_1}(y_1, y_2)\phi_{1y_2}^{\Delta_2}(y_1, y_2) + \phi_{2y_1}^{\Delta_1}(y_1, y_2)\phi_{2y_2}^{\Delta_2}(y_1, y_2) \\ &\quad + \phi_{3y_1}^{\Delta_1}(y_1, y_2)\phi_{3y_2}^{\Delta_2}(y_1, y_2) \\ &= 1, \\ G &= \left(\phi_{1y_2}^{\Delta_2}(y_1, y_2)\right)^2 + \left(\phi_{2y_2}^{\Delta_2}(y_1, y_2)\right)^2 + \left(\phi_{3y_2}^{\Delta_2}(y_1, y_2)\right)^2 \\ &= 2 \end{aligned}$$

so that

$$EG - F^2 = 3$$

and

$$A(S) = \int\int_{\Omega} \sqrt{3}\Delta_1 y_1 \Delta_2 y_2 = 9\sqrt{3}.$$

Example 9.9 Let $\mathbb{T}_1 = \mathbb{Z}$, $\mathbb{T}_2 = 2^{\mathbb{N}_0}$,

$$\Omega = \{(y_1, y_2) : 0 \le y_1 \le 1, \quad 1 \le y_2 \le 2\},$$

and

$$S : \begin{cases} x_1 = y_1^2 + y_1 y_2^2 \\ x_2 = y_1^2 + y_2^2 \\ x_3 = y_1 + y_2. \end{cases}$$

Here,

$$\sigma_1(y_1) = y_1 + 1, \quad \mu_1(y_1) = 1, \quad y_1 \in \mathbb{T}_1,$$

$$\sigma_2(y_2) = 2y_2, \quad \mu_2(y_2) = y_2, \quad y_2 \in \mathbb{T}_2,$$

$$\phi_1(y_1, y_2) = y_1^2 + y_1 y_2^2, \quad \phi_2(y_1, y_2) = y_1^2 + y_2^2, \quad \phi_3(y_1, y_2) = y_1 + y_2.$$

Thus,

$$\begin{aligned} \phi_{1y_1}^{\Delta_1}(y_1, y_2) &= \sigma_1(y_1) + y_1 + y_2^2 \\ &= y_1 + 1 + y_1 + y_2^2 \\ &= 2y_1 + 1 + y_2^2, \\ \phi_{1y_2}^{\Delta_2}(y_1, y_2) &= y_1(\sigma_2(y_2) + y_2) \\ &= y_1(2y_2 + y_2) \\ &= 3y_1 y_2, \\ \phi_{2y_1}^{\Delta_1}(y_1, y_2) &= \sigma_1(y_1) + y_1 \\ &= y_1 + 1 + y_1 \end{aligned}$$

$$= 2y_1 + 1,$$

$$\phi_{2y_2}^{\Delta_2}(y_1, y_2) = \sigma_2(y_2) + y_2$$

$$= 2y_2 + y_2$$

$$= 3y_2,$$

$$\phi_{3y_1}^{\Delta_1}(y_1, y_2) = 1,$$

$$\phi_{3y_2}^{\Delta_2}(y_1, y_2) = 1$$

so that

$$E = \left(\phi_{1y_1}^{\Delta_1}(y_1, y_2)\right)^2 + \left(\phi_{2y_1}^{\Delta_1}(y_1, y_2)\right)^2 + \left(\phi_{3y_1}^{\Delta_1}(y_1, y_2)\right)^2$$

$$= (2y_1 + 1 + y_2^2)^2 + (2y_1 + 1)^2 + 1$$

$$= 4y_1^2 + 1 + y_2^4 + 4y_1 + 4y_1y_2^2 + 2y_2^2 + 4y_1^2 + 4y_1 + 1 + 1$$

$$= 8y_1^2 + y_2^4 + 2y_2^2 + 4y_1y_2^2 + 8y_1 + 3,$$

$$F = \phi_{1y_1}^{\Delta_1}(y_1, y_2)\phi_{1y_2}^{\Delta_2}(y_1, y_2) + \phi_{2y_1}^{\Delta_1}(y_1, y_2)\phi_{2y_2}^{\Delta_2}(y_1, y_2)$$

$$+\phi_{3y_1}^{\Delta_1}(y_1, y_2)\phi_{3y_2}^{\Delta_2}(y_1, y_2)$$

$$= 3y_1y_2(1 + 2y_1 + y_2^2) + (2y_1 + 1)3y_2 + 1$$

$$= 3y_1y_2 + 6y_1^2y_2 + 3y_1y_2^3 + 6y_1y_2 + 3y_2 + 1$$

$$= 6y_1^2y_2 + 3y_1y_2^3 + 9y_1y_2 + 3y_2 + 1,$$

$$G = \left(\phi_{1y_2}^{\Delta_2}(y_1, y_2)\right)^2 + \left(\phi_{2y_2}^{\Delta_2}(y_1, y_2)\right)^2 + \left(\phi_{3y_2}^{\Delta_2}(y_1, y_2)\right)^2$$

$$= 9y_1^2y_2^2 + 9y_2^2 + 1.$$

Let

$$h(y_1, y_2) = EG - F^2$$

$$= (8y_1^2 + y_2^4 + 2y_2^2 + 4y_1y_2^2 + 8y_1 + 2)(9y_1^2y_2^2 + 9y_2^2 + 1)$$

$$-(6y_1^2y_2 + 3y_1y_2^3 + 9y_1y_2 + 3y_2 + 1)^2,$$

$$l(y_2) = \sqrt{h(y_1, y_2)\Big|_{y_1=0}}$$

$$= \sqrt{(y_2^4 + 2y_2^2 + 3)(9y_2^2 + 1) - (3y_2 + 1)^2}.$$

Hence,

$$A(S) = \int_1^2 \int_0^1 \sqrt{h(y_1, y_2)}\Delta_1 y_1 \Delta_2 y_2$$

$$= \int_1^2 \sqrt{h(y_1, y_2)}\mu_1(y_1)\Big|_{y_1=0}\Delta_2 y_2$$

$$= \int_1^2 l(y_2)\Delta_2 y_2$$

$$= l(y_2)\mu_2(y_2)\Big|_{y_2=1}$$

$$= \sqrt{44}.$$

Example 9.10 Let $\mathbb{T}_1 = \mathbb{T}_2 = \mathbb{T}_3 = 2^{\mathbb{N}_0}$,

$$\Omega = \{(y_1, y_2, y_3) \in \mathbb{T}_1 \times \mathbb{T}_2 \times \mathbb{T}_3 : \ 1 \le y_1, y_2, y_3 \le 2\},$$

and

$$S : \begin{cases} x_1 = 2y_1^2 \\ x_2 = y_1 + y_2^2 + y_3^2 \\ x_3 = y_1 - y_3 \\ x_4 = y_3^2. \end{cases}$$

Here,

$$\sigma_1(y_1) = 2y_1, \quad \mu_1(y_1) = y_1, \quad y_1 \in \mathbb{T}_1,$$

$$\sigma_2(y_2) = 2y_2, \quad \mu_2(y_2) = y_2, \quad y_2 \in \mathbb{T}_2,$$

$$\sigma_3(y_3) = 2y_3, \quad \mu_3(y_3) = y_3, \quad y_3 \in \mathbb{T}_3,$$

$$\phi_1(y_1, y_2, y_3) = 2y_1^2, \quad \phi_2(y_1, y_2, y_3) = y_1 + y_2^2 + y_3^2,$$

$$\phi_3(y_1, y_2, y_3) = y_1 - y_3, \quad \phi_4(y_1, y_2, y_3) = y_3^2.$$

Then

$$\begin{aligned}
\phi_{1y_1}^{\Delta_1}(y_1, y_2, y_3) &= 2(\sigma_1(y_1) + y_1) \\
&= 2(2y_1 + y_1) \\
&= 6y_1, \\
\phi_{1y_2}^{\Delta_2}(y_1, y_2, y_3) &= 0, \\
\phi_{1y_3}^{\Delta_3}(y_1, y_2, y_3) &= 0, \\
\phi_{2y_1}^{\Delta_1}(y_1, y_2, y_3) &= 1, \\
\phi_{2y_2}^{\Delta_2}(y_1, y_2, y_3) &= \sigma_2(y_2) + y_2 \\
&= 2y_2 + y_2 \\
&= 3y_2, \\
\phi_{2y_3}^{\Delta_3}(y_1, y_2, y_3) &= \sigma_3(y_3) + y_3 \\
&= 2y_3 + y_3 \\
&= 3y_3,
\end{aligned}$$

$$\phi_{3y_1}^{\Delta_1}(y_1, y_2, y_3) = 1,$$

$$\phi_{3y_2}^{\Delta_2}(y_1, y_2, y_3) = 0,$$

$$\phi_{3y_3}^{\Delta_3}(y_1, y_2, y_3) = -1,$$

$$\phi_{4y_1}^{\Delta_1}(y_1, y_2, y_3) = 0,$$

$$\phi_{4y_2}^{\Delta_2}(y_1, y_2, y_3) = 0,$$

$$\begin{aligned}\phi_{4y_3}^{\Delta_3}(y_1, y_2) &= \sigma_3(y_3) + y_3\\ &= 2y_3 + y_3\\ &= 3y_3.\end{aligned}$$

Therefore,

$$\mathbf{r}_{y_1}^{\Delta_1}(y_1, y_2, y_3) \times \mathbf{r}_{y_2}^{\Delta_2}(y_1, y_2, y_3) \times \mathbf{r}_{y_3}^{\Delta_3}(y_1, y_2, y_3)$$

$$= \begin{vmatrix} \mathbf{e}_1 & \mathbf{e}_2 & \mathbf{e}_3 & \mathbf{e}_4 \\ \phi_{1y_1}^{\Delta_1}(y_1, y_2, y_3) & \phi_{2y_1}^{\Delta_1}(y_1, y_2, y_3) & \phi_{3y_1}^{\Delta_1}(y_1, y_2, y_3) & \phi_{4y_1}^{\Delta_1}(y_1, y_2, y_3) \\ \phi_{1y_2}^{\Delta_2}(y_1, y_2, y_3) & \phi_{2y_2}^{\Delta_2}(y_1, y_2, y_3) & \phi_{3y_2}^{\Delta_2}(y_1, y_2, y_3) & \phi_{4y_2}^{\Delta_2}(y_1, y_2, y_3) \\ \phi_{1y_3}^{\Delta_3}(y_1, y_2, y_3) & \phi_{2y_3}^{\Delta_3}(y_1, y_2, y_3) & \phi_{3y_3}^{\Delta_3}(y_1, y_2, y_3) & \phi_{4y_3}^{\Delta_3}(y_1, y_2, y_3) \end{vmatrix}$$

$$= \begin{vmatrix} \mathbf{e}_1 & \mathbf{e}_2 & \mathbf{e}_3 & \mathbf{e}_4 \\ 6y_1 & 1 & 1 & 0 \\ 0 & 3y_2 & 0 & 0 \\ 0 & 3y_3 & -1 & 3y_3 \end{vmatrix}$$

$$= \begin{vmatrix} 1 & 1 & 0 \\ 3y_2 & 0 & 0 \\ 3y_3 & -1 & 3y_3 \end{vmatrix}\mathbf{e}_1 - \begin{vmatrix} 6y_1 & 1 & 0 \\ 0 & 0 & 0 \\ 0 & -1 & 3y_3 \end{vmatrix}\mathbf{e}_2$$

$$+ \begin{vmatrix} 6y_1 & 1 & 0 \\ 0 & 3y_2 & 0 \\ 0 & 3y_3 & 3y_3 \end{vmatrix}\mathbf{e}_3 - \begin{vmatrix} 6y_1 & 1 & 1 \\ 0 & 3y_2 & 0 \\ 0 & 3y_3 & -1 \end{vmatrix}\mathbf{e}_4$$

$$= -9y_2y_3\mathbf{e}_1 + 54y_1y_2y_3\mathbf{e}_3 + 18y_1y_2\mathbf{e}_4,$$

$$|\mathbf{r}^{\Delta_1}_{y_1}(y_1, y_2, y_3) \times \mathbf{r}^{\Delta_2}_{y_2}(y_1, y_2, y_3) \times \mathbf{r}^{\Delta_3}_{y_3}(y_1, y_2, y_3)| = \sqrt{9^2 y_2^2 y_3^2 + 6^2 9^2 y_1^2 y_2^2 y_3^2 + 9^2 2^2 y_1^2 y_2^2}$$

$$= 9\sqrt{y_2^2 y_3^2 + 36 y_1^2 y_2^2 y_3^2 + 4 y_1^2 y_2^2}.$$

Hence,

$$\begin{aligned}
A(S) &= 9 \int_1^2 \int_1^2 \int_1^2 \sqrt{y_2^2 y_3^2 + 36 y_1^2 y_2^2 y_3^2 + 4 y_1^2 y_2^2}\, \Delta_3 y_3 \Delta_2 y_2 \Delta_1 y_1 \\
&= 9 \int_1^2 \int_1^2 \sqrt{y_2^2 y_3^2 + 36 y_1^2 y_2^2 y_3^2 + 4 y_1^2 y_2^2}\, \mu_3(y_3)\Big|_{y_3=1} \Delta_2 y_2 \Delta_1 y_1 \\
&= 9 \int_1^2 \int_1^2 \sqrt{y_2^2 + 36 y_1^2 y_2^2 + 4 y_1^2 y_2^2}\, \Delta_2 y_2 \Delta_1 y_1 \\
&= 9 \int_1^2 \int_1^2 \sqrt{y_2^2 + 40 y_1^2 y_2^2}\, \Delta_2 y_2 \Delta_1 y_1 \\
&= 9 \int_1^2 \sqrt{y_2^2 + 40 y_1^2 y_2^2}\, \mu_2(y_2)\Big|_{y_2=1} \Delta_1 y_1 \\
&= 9 \int_1^2 \sqrt{1 + 40 y_1^2}\, \Delta_1 y_1 \\
&= 9\sqrt{1 + 40 y_1^2}\, \mu_1(y_1)\Big|_{y_1=1} \\
&= 9\sqrt{41}.
\end{aligned}$$

Exercise 9.11 Find $A(S)$, where

1. $\mathbb{T}_1 = \mathbb{T}_2 = \mathbb{Z}$,

$$\Omega = \{(y_1, y_2) \in \mathbb{T}_1 \times \mathbb{T}_2 : -1 \le y_1 \le 1,\ 0 \le y_2 \le 2\},$$

and

$$S : \begin{cases} x_1 = 2y_1 - 3y_2 \\ x_2 = y_2 \\ x_3 = y_1^2 + y_2^2, \end{cases}$$

2. $\mathbb{T}_1 = \mathbb{Z}$, $\mathbb{T}_2 = 2^{\mathbb{N}_0} \cup \{0\}$,

$$\Omega = \{(y_1, y_2) \in \mathbb{T}_1 \times \mathbb{T}_2 : \ -2 \le y_1 \le 2, \ 0 \le y_2 \le 4\},$$

and

$$S : \begin{cases} x_1 = y_1 \\ x_2 = y_2 \\ x_3 = y_1^2 + y_1 y_2 + y_2^2, \end{cases}$$

3. $\mathbb{T}_1 = \mathbb{Z}$, $\mathbb{T}_2 = 3^{\mathbb{N}_0} \cup \{0\}$, $\mathbb{T}_3 = 2^{\mathbb{N}_0} \cup \{0\}$,

$$S : \begin{cases} x_1 = y_1 + y_2 + y_3 \\ x_2 = y_1 - y_3 \\ x_3 = y_1^2 + y_2^2 + y_3^2 \\ x_4 = y_1 - y_3. \end{cases}$$

9.2 Surface Δ-Integrals

Let $\mathbb{T}_1, \ldots, \mathbb{T}_n$ be time scales. For $i \in \{1, \ldots, n\}$, let σ_i and Δ_i denote the forward jump operator and the delta operator, respectively, on $\mathbb{T}_i$. Let $\Omega \subset \mathbb{T}_1 \times \ldots \times \mathbb{T}_n$ be a bounded and Jordan Δ-measurable set. Suppose S is a time continuous surface defined by the parametric equations

$$x_i = \phi_i(y_1, \ldots, y_n), \quad i \in \{1, \ldots, m\}, \quad m = n + 1, \quad (y_1, \ldots, y_n) \in \Omega, \quad (9.9)$$

where ϕ_i are continuous and have continuous first-order partial delta derivatives in the closure of the region Ω. Let $P' = \{R_1, \ldots, R_k\}$ be an arbitrary Δ-partition of Ω. Denote by A_i the area of the piece of the surface S corresponding to the piece R_i of Ω. From Theorem 9.5, we get

$$A_i = \int_{R_i} |\mathbf{r}^{\Delta_1}(y_1, \ldots, y_n) \times \ldots \times \mathbf{r}^{\Delta_n}(y_1, \ldots, y_n)| \Delta_1 y_1 \ldots \Delta_n y_n.$$

Let $h(x_1, \ldots, x_m)$ be a function that is defined and continuous on the closure $\overline{S}$ of the surface S. Take any $(\xi_1^l, \ldots, \xi_n^l) \in R_l$, $l \in \{1, \ldots, n\}$, and consider the integral sum

$$\Sigma = \sum_{l=1}^{k} h(\phi_1(\xi_1^l, \ldots, \xi_n^l), \ldots, \phi_m(\xi_1^l, \ldots, \xi_n^l))A_l.$$

Definition 9.12 We say that a number I is the surface Δ-integral of the function h over the surface S if for every $\varepsilon > 0$, there exists $\delta > 0$ such that

$$|\Sigma - I| < \varepsilon$$

for every integral sum Σ of h corresponding to any inner Δ-partition

$$P' = \{R_1, \ldots, R_k\}$$

of Ω, determined by a partition $P \in \mathscr{P}_\delta(\mathbb{R})$, independent of the choice of the points $(\xi_1^l, \ldots, \xi_n^l) \in R_l$ for $l \in \{1, \ldots, k\}$, where

$$R = [a_1, b_1) \times [a_2, b_2) \times \ldots \times [a_n, b_n) \subset \mathbb{T}_1 \times \mathbb{T}_2 \times \ldots \times \mathbb{T}_n$$

is a rectangle that contains Ω. We denote the number I, symbolically, by

$$\int_S h(x_1, \ldots, x_m)\Delta A. \tag{9.10}$$

Theorem 9.13 *Suppose that the surface S is given by the parametric equations (9.9), where the region $\Omega \subset \mathbb{T}_1 \times \ldots \times \mathbb{T}_n$ is bounded and Jordan Δ-measurable, the functions ϕ_l, $l \in \{1, \ldots, m\}$, are continuous and have continuous first-order partial delta derivatives in the closure $\overline{\Omega}$ of Ω, and the function h is continuous on the closure $\overline{S}$ of the surface S. Then the surface integral (9.10) exists and can be computed by*

$$\int_S h(x_1, \ldots, x_m)\Delta A = \int_\Omega h(\phi_1(y_1, \ldots, y_n), \ldots, \phi_m(y_1, \ldots, y_n)) \times |\mathbf{r}_{y_1}^{\Delta_1}(y_1, \ldots, y_n) \times \ldots \times \mathbf{r}_{y_n}^{\Delta_n}(y_1, \ldots, y_n)|\Delta_1 y_1 \ldots \Delta_n y_n. \tag{9.11}$$

Proof Since h is continuous on $\overline{S}$ and ϕ_l, $l \in \{1, \ldots, m\}$, are continuous on $\overline{\Omega}$, the multiple Δ-integral on the right-hand side of (9.11) exists. Let

$$R = [a_1, b_1) \times \ldots \times [a_n, b_n) \subset \mathbb{T}_1 \times \ldots \times \mathbb{T}_n$$

be a rectangle that contains $\Omega \cup \partial_\Delta \Omega$. Let $\varepsilon > 0$ be arbitrarily chosen. Since Ω is Jordan Δ-measurable, there exists $\delta > 0$ such that for every partition $P \in \mathscr{P}_\delta(R)$, the sum of the areas of subrectangles of R which have a common point with $\partial_\Delta \Omega$ is less than ε. Let $R_1, \ldots, R_k$ be all rectangles of the partition P that are entirely within Ω and $R_{k+1}, \ldots, R_N$ be all subrectangles of P that are not entirely within Ω and each of which has a common point with $\partial_\Delta \Omega$. Note that the collection $P' = \{R_1, \dots, R_k\}$ forms an inner Δ-partition of Ω and

$$\sum_{l=k+1}^{N} m(R_l) < \varepsilon.$$

Let

$$\Sigma = \sum_{l=1}^{k} H(\xi_1^l, \dots, \xi_n^l) \int_{R_l} \Phi(y_1, \dots, y_n) \Delta_1 y_1 \dots \Delta_n y_n,$$

where

$$H(\xi_1, \dots, \xi_n) = h(\phi_1(\xi_1, \dots, \xi_n), \dots, \phi_m(\xi_1, \dots, \xi_n)),$$

$$\Phi(y_1, \dots, y_n) = |\mathbf{r}_{y_1}^{\Delta_1}(y_1, \dots, y_n) \times \dots \times \mathbf{r}_{y_n}^{\Delta_n}(y_1, \dots, y_n)|$$

for $(\xi_1, \dots, \xi_n) \in \Omega$ and

$$H(\xi_1, \dots, \xi_n) = \Phi(\xi_1, \dots, \xi_n) = 0$$

for $(\xi_1, \dots, \xi_n) \in R \setminus \Omega$. Therefore,

$$\begin{aligned} I &= \int_{\Omega} H(\xi_1, \dots, \xi_n) \Phi(\xi_1, \dots, \xi_n) \Delta_1 \xi_1 \dots \Delta_n \xi_n \\ &= \sum_{l=1}^{k} \int_{R_l} H(\xi_1, \dots, \xi_n) \Phi(\xi_1, \dots, \xi_n) \Delta_1 \xi_1 \dots \Delta_n \xi_n \\ &\quad + \sum_{l=k+1}^{n} \int_{R_l} H(\xi_1, \dots, \xi_n) \Phi(\xi_1, \dots, \xi_n) \Delta_1 \xi_1 \dots \Delta_n \xi_n. \end{aligned}$$

Consider the difference

$$\begin{aligned} \Sigma - I = &\sum_{l=1}^{k} \int_{R_l} H(\xi_1^l, \dots, \xi_n^l) \Phi(\xi_1, \dots, \xi_n) \Delta_1 \xi_1 \dots \Delta_n \xi_n \\ &- \sum_{l=1}^{k} \int_{R_l} H(\xi_1, \dots, \xi_n) \Phi(\xi_1, \dots, \xi_n) \Delta_1 \xi_1 \dots \Delta_n \xi_n \\ &- \sum_{l=k+1}^{n} \int_{R_l} H(\xi_1, \dots, \xi_n) \Phi(\xi_1, \dots, \xi_n) \Delta_1 \xi_1 \dots \Delta_n \xi_n \end{aligned}$$

$$= \sum_{l=1}^{k} \int_{R_l} \left(H(\xi_1^l, \ldots, \xi_n^l) - H(\xi_1, \ldots, \xi_n)\right) \Phi(\xi_1, \ldots, \xi_n) \Delta_1 \xi_1 \ldots \Delta_n \xi_n$$

$$- \sum_{l=k+1}^{n} \int_{R_l} H(\xi_1, \ldots, \xi_n) \Phi(\xi_1, \ldots, \xi_n) \Delta_1 \xi_1 \ldots \Delta_n \xi_n.$$

We have

$$\begin{aligned} |\Sigma - I| =& \Big| \sum_{l=1}^{k} \int_{R_l} \left(H(\xi_1^l, \ldots, \xi_n^l) - H(\xi_1, \ldots, \xi_n)\right) \Phi(\xi_1, \ldots, \xi_n) \Delta_1 \xi_1 \ldots \Delta_n \xi_n \\ & - \sum_{l=k+1}^{n} \int_{R_l} H(\xi_1, \ldots, \xi_n) \Phi(\xi_1, \ldots, \xi_n) \Delta_1 \xi_1 \ldots \Delta_n \xi_n \Big| \\ \leq & \sum_{l=1}^{k} \int_{R_l} |H(\xi_1^l, \ldots, \xi_n^l) - H(\xi_1, \ldots, \xi_n)| \Phi(\xi_1, \ldots, \xi_n) \Delta_1 \xi_1 \ldots \Delta_n \xi_n \\ & + \sum_{l=k+1}^{N} \int_{R_l} |H(\xi_1, \ldots, \xi_n)| \Phi(\xi_1, \ldots, \xi_n) \Delta_1 \xi_1 \ldots \Delta_n \xi_n. \end{aligned} \tag{9.12}$$

Since h, ϕ_l, $l \in \{1, \ldots, m\}$, are continuous on $\overline{\Omega}$, we have that H and Φ are continuous on $\overline{\Omega}$. Consequently, the functions H and Φ are bounded and uniformly continuous on $\overline{\Omega}$. Hence,

$$\sup\{|H(\xi_1, \ldots, \xi_n)| \Phi(\xi_1, \ldots, \xi_n) : (\xi_1, \ldots, \xi_n) \in \Omega\} = M,$$

and for given $\varepsilon > 0$ and $\delta > 0$, we can diminish δ if necessary,

$$(\xi_1, \ldots, \xi_n), \ (\xi_1', \ldots, \xi_n') \in \overline{\Omega} \quad \text{and} \quad |\xi_l - \xi_l'| < \delta, \quad l \in \{1, \ldots, n\}$$

implies

$$|H(\xi_1, \ldots, \xi_n) - H(\xi_1', \ldots, \xi_n')| < \varepsilon. \tag{9.13}$$

Consequently,

$$\sum_{l=k+1}^{N}\int_{R_l}|H(\xi_1,\ldots,\xi_n)|\Phi(\xi_1,\ldots,\xi_n)\Delta_1\xi_1\ldots\Delta_n\xi_n \le M\sum_{l=k+1}^{N}\int_{R_l}\Delta_1\xi_1\ldots\Delta_n\xi_n$$

$$= M\sum_{l=k+1}^{N} m(R_l)$$

$$\le M\varepsilon.$$

Also,

$$\sum_{l=1}^{k}\int_{R_l}|H(\xi_1^l,\ldots,\xi_n^l)-H(\xi_1,\ldots,\xi_n)|\Phi(\xi_1,\ldots,\xi_n)\Delta_1\xi_1\ldots\Delta_n\xi_n$$

$$= \sum_{\substack{|x_1'-\xi_1|\le\delta\\ |\xi_2'-\xi_2|\le\delta\\ \ldots\\ |\xi_n'-\xi_n|\le\delta}}\int_{R_l}|H(\xi_1^l,\ldots,\xi_n^l)-H(\xi_1,\ldots,\xi_n)|\Phi(\xi_1,\ldots,\xi_n)\Delta_1\xi_1\ldots\Delta_n\xi_n$$

$$+ \sum_{\substack{|x_1'-\xi_1|>\delta\\ |\xi_2'-\xi_2|\le\delta\\ \ldots\\ |\xi_n'-\xi_n|\le\delta}}\int_{R_l}|H(\xi_1^l,\ldots,\xi_n^l)-H(\xi_1,\ldots,\xi_n)|\Phi(\xi_1,\ldots,\xi_n)\Delta_1\xi_1\ldots\Delta_n\xi_n$$

$$+\cdots$$

$$+ \sum_{\substack{|x_1'-\xi_1|>\delta\\ |\xi_2'-\xi_2|>\delta\\ \ldots\\ |\xi_n'-\xi_n|>\delta}}\int_{R_l}|H(\xi_1^l,\ldots,\xi_n^l)-H(\xi_1,\ldots,\xi_n)|\Phi(\xi_1,\ldots,\xi_n)\Delta_1\xi_1\ldots\Delta_n\xi_n.$$

Using (9.13), we have

$$\sum_{\substack{|x_1'-\xi_1|\le\delta\\ |\xi_2'-\xi_2|\le\delta\\ \ldots\\ |\xi_n'-\xi_n|\le\delta}}\int_{R_l}|H(\xi_1^l,\ldots,\xi_n^l)-H(\xi_1,\ldots,\xi_n)|\Phi(\xi_1,\ldots,\xi_n)\Delta_1\xi_1\ldots\Delta_n\xi_n$$

$$< \varepsilon \sum_{\substack{|x_1' - \xi_1| \le \delta \\ |\xi_2' - \xi_2| \le \delta \\ \dots \\ |\xi_n' - \xi_n| \le \delta}} \int_{R_l} \Phi(\xi_1, \dots, \xi_n)\Delta_1\xi_1 \dots \Delta_n\xi_n$$

$$\le \varepsilon \sum_{l=1}^{k} \int_{R_l} \Phi(\xi_1, \dots, \xi_n)\Delta_1\xi_1 \dots \Delta_n\xi_n$$

$$= \varepsilon A(S).$$

If $\xi_1' - \xi_1 > \delta, 0 < \xi_p' - \xi_p \le \delta$, $p \in \{2, \dots, n\}$, then $\xi_1' = \sigma_1(\xi_1)$ and

$$\int_{R_l} |H(\xi_1^l, \dots, \xi_n^l) - H(\xi_1, \dots, \xi_n)|\Phi(\xi_1, \dots, \xi_n)\Delta_1\xi_1 \dots \Delta_n\xi_n$$

$$= \int_{\xi_1}^{\sigma_1(\xi_1)} \int_{\xi_2}^{\xi_2'} \dots \int_{\xi_n}^{\xi_n'} |H(\xi_1^l, \dots, \xi_n^l) - H(\xi_1, \dots, \xi_n)|\Phi(\xi_1, \dots, \xi_n)\Delta_1\xi_1 \dots \Delta_n\xi_n$$

$$= \int_{\xi_2}^{\xi_2'} \dots \int_{\xi_n}^{\xi_n'} |H(\xi_1, \xi_2^l, \dots, \xi_n^l) - H(\xi_1, \xi_2, \dots, \xi_n)|\Phi(\xi_1, \xi_2, \dots, \xi_n)\times$$

$$\times (\sigma_1(\xi_1) - \xi_1)\Delta_2\xi_2 \dots \Delta_n\xi_n$$

$$< \varepsilon \int_{\xi_2}^{\xi_2'} \dots \int_{\xi_n}^{\xi_n'} \Phi(\xi_1, \xi_2, \dots, \xi_n)(\sigma_1(\xi_1) - \xi_1)\Delta_2\xi_2 \dots \Delta_n\xi_n$$

$$= \varepsilon \int_{R_l} \Phi(\xi_1, \dots, \xi_n)\Delta_1\xi_1 \dots \Delta_n\xi_n.$$

Hence,

$$\sum_{\substack{|x_1' - \xi_1| > \delta \\ |\xi_2' - \xi_2| \le \delta \\ \dots \\ |\xi_n' - \xi_n| \le \delta}} \int_{R_l} |H(\xi_1^l, \dots, \xi_n^l) - H(\xi_1, \dots, \xi_n)|\Phi(\xi_1, \dots, \xi_n)\Delta_1\xi_1 \dots \Delta_n\xi_n$$

$$\leq \varepsilon \sum_{l=1}^{k} \int_{R_l} \Phi(\xi_1, \ldots, \xi_n) \Delta_1 \xi_1 \ldots \Delta_n \xi_n$$

$$\leq \varepsilon A(S),$$

and so on, and finally

$$\sum_{\substack{|x_1' - \xi_1| > \delta \\ |\xi_2' - \xi_2| > \delta \\ \ldots \\ |\xi_n' - \xi_n| > \delta}} \int_{R_l} |H(\xi_1^l, \ldots, \xi_n^l) - H(\xi_1, \ldots, \xi_n)| \Phi(\xi_1, \ldots, \xi_n) \Delta_1 \xi_1 \ldots \Delta_n \xi_n$$

$$\leq \varepsilon A(S).$$

From here and from (9.12),

$$|\Sigma - I| \leq \sum_{l=1}^{k} \varepsilon A(S) + M\varepsilon \leq k\varepsilon A(S) + M\varepsilon.$$

Since $\varepsilon > 0$ was arbitrarily chosen, the proof is complete. □

Example 9.14 Let $\mathbb{T}_1 = \mathbb{T}_2 = \mathbb{Z}$,

$$S : \begin{cases} \frac{1}{3}x_1 + \frac{1}{2}x_2 + x_3 = 1 \\ \\ x_1 \geq 0, \quad x_2 \geq 0, \quad x_3 \geq 0 \end{cases}.$$

We will compute

$$I = \int_S (x_1 - x_2 - x_3) \Delta A.$$

Let

$$x_1 = y_1, \quad x_2 = y_2.$$

Then

$$x_3 = 1 - \frac{1}{3}x_1 - \frac{1}{2}x_2 = 1 - \frac{1}{3}y_1 - \frac{1}{2}y_2.$$

Hence,

$$0 \leq 1 - \frac{1}{3}y_1 - \frac{1}{2}y_2,$$

i.e.,

$$0 \le y_2 \le 2 - \frac{2}{3}y_1,$$

whereupon

$$0 \le y_1 \le 3.$$

Therefore,

$$S : \begin{cases} x_1 = y_1 \\ x_2 = y_2 \\ x_3 = 1 - \frac{1}{3}y_1 - \frac{1}{2}y_2 \\ 0 \le y_1 \le 3, \quad 0 \le y_2 \le 2 - \frac{2}{3}y_1 \end{cases}$$

and

$$\begin{aligned} x_1 - x_2 - x_3 &= y_1 - y_2 - \left(1 - \frac{1}{3}y_1 - \frac{1}{2}y_2\right) \\ &= y_1 - y_2 - 1 + \frac{1}{3}y_1 + \frac{1}{2}y_2 \\ &= -1 + \frac{4}{3}y_1 - \frac{1}{2}y_2. \end{aligned}$$

Here,

$$\sigma_1(y_1) = y_1 + 1, \quad \mu_1(y_1) = 1, \quad y_1 \in \mathbb{T}_1,$$

$$\sigma_2(y_2) = y_2 + 1, \quad \mu_2(y_2) = 1, \quad y_2 \in \mathbb{T}_2,$$

$$\phi_1(y_1, y_2) = y_1, \quad \phi_2(y_1, y_2) = y_2, \quad \phi_3(y_1, y_2) = 1 - \frac{1}{3}y_1 - \frac{1}{2}y_2.$$

Thus,

$$\phi_{1y_1}^{\Delta_1}(y_1, y_2) = 1, \quad \phi_{1y_2}^{\Delta_2}(y_1, y_2) = 0,$$

$$\phi_{2y_1}^{\Delta_1}(y_1, y_2) = 0, \quad \phi_{2y_2}^{\Delta_2}(y_1, y_2) = 1,$$

$$\phi_{3y_1}^{\Delta_1}(y_1, y_2) = -\frac{1}{3}, \quad \phi_{3y_2}^{\Delta_2}(y_1, y_2) = -\frac{1}{2}$$

and

$$\begin{aligned}
E &= \left(\phi_{1y_1}^{\Delta_1}(y_1, y_2)\right)^2 + \left(\phi_{2y_1}^{\Delta_1}(y_1, y_2)\right)^2 + \left(\phi_{3y_1}^{\Delta_1}(y_1, y_2)\right)^2 \\
&= 1 + \frac{1}{9} \\
&= \frac{10}{9}, \\
F &= \phi_{1y_1}^{\Delta_1}(y_1, y_2)\phi_{1y_2}^{\Delta_2}(y_1, y_2) + \phi_{2y_1}^{\Delta_1}(y_1, y_2)\phi_{2y_2}^{\Delta_2}(y_1, y_2) \\
&\quad + \phi_{3y_1}^{\Delta_1}(y_1, y_2)\phi_{3y_2}^{\Delta_2}(y_1, y_2) \\
&= 0 + 0 + \frac{1}{6} \\
&= \frac{1}{6}, \\
G &= \left(\phi_{1y_2}^{\Delta_2}(y_1, y_2)\right)^2 + \left(\phi_{2y_2}^{\Delta_2}(y_1, y_2)\right)^2 + \left(\phi_{3y_2}^{\Delta_2}(y_1, y_2)\right)^2 \\
&= 0 + 1 + \frac{1}{4} \\
&= \frac{5}{4}
\end{aligned}$$

so that

$$\sqrt{EG - F^2} = \sqrt{\frac{25}{18} - \frac{1}{36}} = \sqrt{\frac{49}{36}} = \frac{7}{6}.$$

Hence, (here we use the notation $\lfloor x \rfloor = \max\{y \in \mathbb{Z} : y \leq x\}$ for $x \in \mathbb{R}$)

$$\begin{aligned}
I &= \int_0^3 \int_0^{\lfloor 2 - \frac{2}{3}y_1 \rfloor} \left(-1 + \frac{4}{3}y_1 - \frac{1}{2}y_2\right) \frac{7}{6} \Delta_2 y_2 \Delta_1 y_1 \\
&= -\frac{7}{6} \int_0^2 \left(1 + \frac{1}{2}y_2\right) \Delta_2 y_2 + \frac{7}{6} \int_0^1 \left(\frac{1}{3} - \frac{1}{2}y_2\right) \Delta_2 y_2 \\
&= -\frac{7}{6}\left(1 \cdot 2 + \frac{1}{2} \cdot \frac{1 \cdot 2}{2}\right) + \frac{7}{6}\left(\frac{1}{3} \cdot 1 - \frac{1}{2} \cdot \frac{0 \cdot 1}{2}\right)
\end{aligned}$$

$$= -\frac{91}{36}.$$

Example 9.15 Let $\mathbb{T}_1 = \mathbb{Z}$, $\mathbb{T}_2 = 2^{\mathbb{N}_0}$,

$$S : \begin{cases} x_1 = y_1^2 \\ x_2 = y_2 \\ x_3 = y_1^2 + y_2^2, \quad 0 \le y_1 \le 2, \quad 1 \le y_2 \le 2. \end{cases}$$

We will compute

$$I = \int_S (x_1 + x_2 + x_3) \Delta A.$$

Here,

$$\sigma_1(y_1) = y_1 + 1, \quad \mu_1(y_1) = 1, \quad y_1 \in \mathbb{T}_1,$$

$$\sigma_2(y_2) = 2y_2, \quad \mu_2(y_2) = y_2, \quad y_2 \in \mathbb{T}_2,$$

$$\phi_1(y_1, y_2) = y_1^2, \quad \phi_2(y_1, y_2) = y_2, \quad \phi_3(y_1, y_2) = y_1^2 + y_2^2.$$

Then

$$\begin{aligned} \phi_{1y_1}^{\Delta_1}(y_1, y_2) &= \sigma_1(y_1) + y_1 \\ &= y_1 + 1 + y_1 \\ &= 2y_1 + 1, \end{aligned}$$

$$\phi_{1y_2}^{\Delta_2}(y_1, y_2) = 0,$$

$$\phi_{2y_1}^{\Delta_1}(y_1, y_2) = 0,$$

$$\phi_{2y_2}^{\Delta_2}(y_1, y_2) = 1,$$

$$\phi_{3y_1}^{\Delta_1}(y_1, y_2) = \sigma_1(y_1) + y_1$$

$$= y_1 + 1 + y_1$$

$$= 1 + 2y_1,$$

$$\phi_{3y_2}^{\Delta_2}(y_1, y_2) = \sigma_2(y_2) + y_2$$

$$= 2y_2 + y_2$$

$$= 3y_2.$$

Next,

$$x_1 + x_2 + x_3 = y_1^2 + y_2 + y_1^2 + y_2^2 = 2y_1^2 + y_2 + y_2^2$$

and

$$E = \left(\phi_{1y_1}^{\Delta_1}(y_1, y_2)\right)^2 + \left(\phi_{2y_1}^{\Delta_1}(y_1, y_2)\right)^2 + \left(\phi_{3y_1}^{\Delta_1}(y_1, y_2)\right)^2$$

$$= (2y_1 + 1)^2 + 0 + (1 + 2y_1)^2$$

$$= 2 + 8y_1 + 8y_1^2,$$

$$F = \phi_{1y_1}^{\Delta_1}(y_1, y_2)\phi_{1y_2}^{\Delta_2}(y_1, y_2) + \phi_{2y_1}^{\Delta_1}(y_1, y_2)\phi_{2y_2}^{\Delta_2}(y_1, y_2)$$

$$+\phi_{3y_1}^{\Delta_1}(y_1, y_2)\phi_{3y_2}^{\Delta_2}(y_1, y_2)$$

$$= 0 + 0 + (1 + 2y_1)3y_2$$

$$= 3y_2 + 6y_1y_2,$$

$$G = \left(\phi_{1y_2}^{\Delta_2}(y_1, y_2)\right)^2 + \left(\phi_{2y_2}^{\Delta_2}(y_1, y_2)\right)^2 + \left(\phi_{3y_2}^{\Delta_2}(y_1, y_2)\right)^2$$

$$= 0 + 1 + (3y_2)^2$$

$$= 1 + 9y_2^2$$

so that

$$EG - F^2 = (2 + 8y_1 + 8y_1^2)(1 + 9y_2^2) - (3y_2 + 6y_1y_2)^2$$

$$= 2 + 18y_2^2 + 8y_1 + 72y_1y_2^2 + 8y_1^2 + 72y_1^2y_2^2$$

$$-9y_2^2 - 36y_1y_2^2 - 36y_1^2y_2^2$$

$$= 2 + 8y_1 + 8y_1^2 + 9y_2^2 + 36y_1y_2^2 + 36y_1^2y_2^2.$$

Therefore,

$$I = \int_0^2 \int_1^2 (2y_1^2 + y_2 + y_2^2)\sqrt{2 + 8y_1 + 8y_1^2 + 9y_2^2 + 36y_1y_2^2 + 36y_1^2y_2^2}\Delta_2 y_2 \Delta_1 y_1$$

$$= \int_0^2 (2y_1^2 + y_2 + y_2^2)\sqrt{2 + 8y_1 + 8y_1^2 + 9y_2^2 + 36y_1y_2^2 + 36y_1^2y_2^2}\mu_2(y_2)\Big|_{y_2=1} \Delta_1 y_1$$

$$= \int_0^2 2(y_1^2 + 1)\sqrt{2 + 8y_1 + 8y_1^2 + 9 + 36y_1 + 36y_1^2}\Delta_1 y_1$$

$$= 2(y_1^2 + 1)\sqrt{11 + 44y_1 + 44y_1^2}\mu_1(y_1)\Big|_{y_1=0}$$

$$+2(y_1^2 + 1)\sqrt{11 + 44y_1 + 44y_1^2}\mu_1(y_1)\Big|_{y_1=1}$$

$$= 2\sqrt{11} + 4\sqrt{11 + 44 + 44}$$

$$= 14\sqrt{11}.$$

Example 9.16 Let $\mathbb{T}_1 = \mathbb{T}_2 = \mathbb{T}_3 = 3^{\mathbb{N}_0} \cup \{0\}$,

$$S : \begin{cases} x_1 = y_1 + y_2 + y_3 \\ x_2 = y_1 - y_3 \\ x_3 = y_3 \\ x_4 = y_1 + y_3, \quad 0 \le y_1 \le 3, \quad 0 \le y_2 \le 3, \quad 0 \le y_3 \le 9. \end{cases}$$

We will compute

$$\int_S (x_1 - x_2 + 2x_3)\Delta A.$$

Here,

$$\sigma_1(y_1) = 3y_1, \quad \mu_1(y_1) = 2y_1, \quad y_1 \in \mathbb{T}_1 \setminus \{0\}, \quad \sigma_1(0) = 1, \quad \mu_1(0) = 1,$$

$$\sigma_2(y_2) = 3y_2, \quad \mu_2(y_2) = 2y_2, \quad y_2 \in \mathbb{T}_2 \setminus \{0\}, \quad \sigma_2(0) = 1, \quad \mu_2(0) = 1,$$

$$\sigma_3(y_3) = 3y_3, \quad \mu_3(y_3) = 2y_3, \quad y_3 \in \mathbb{T}_3 \setminus \{0\}, \quad \sigma_3(0) = 1, \quad \mu_3(0) = 1,$$

$$\phi_1(y_1, y_2, y_3) = y_1 + y_2 + y_3, \quad \phi_2(y_1, y_2, y_3) = y_1 - y_3,$$

$$\phi_3(y_1, y_2, y_3) = y_3, \quad \phi_4(y_1, y_2, y_3) = y_1 + y_3.$$

Thus,

$$\phi_{1y_1}^{\Delta_1}(y_1, y_2, y_3) = \phi_{1y_2}^{\Delta_2}(y_1, y_2, y_3) = \phi_{1y_3}^{\Delta_3}(y_1, y_2, y_3) = 1,$$

$$\phi_{2y_1}^{\Delta_1}(y_1, y_2, y_3) = 1, \quad \phi_{2y_2}^{\Delta_2}(y_1, y_2, y_3) = 0, \quad \phi_{2y_3}^{\Delta_3}(y_1, y_2, y_3) = -1,$$

$$\phi_{3y_1}^{\Delta_1}(y_1, y_2, y_3) = 0, \quad \phi_{3y_2}^{\Delta_2}(y_1, y_2, y_3) = 0, \quad \phi_{3y_3}^{\Delta_3}(y_1, y_2, y_3) = 1,$$

$$\phi_{4y_1}^{\Delta_1}(y_1, y_2, y_3) = 1, \quad \phi_{4y_2}^{\Delta_2}(y_1, y_2, y_3) = 0, \quad \phi_{4y_3}^{\Delta_3}(y_1, y_2, y_3) = 1.$$

Next,

$$x_1 - x_2 + 2x_3 = y_1 + y_2 + y_3 - y_1 + y_3 + 2y_3 = y_2 + 4y_3$$

and

$$\mathbf{r}_{y_1}^{\Delta_1}(y_1, y_2, y_3) \times \mathbf{r}_{y_2}^{\Delta_2}(y_1, y_2, y_3) \times \mathbf{r}_{y_3}^{\Delta_3}(y_1, y_2, y_3)$$

$$= \begin{vmatrix} \mathbf{e}_1 & \mathbf{e}_2 & \mathbf{e}_3 & \mathbf{e}_4 \\ \phi_{1y_1}^{\Delta_1}(y_1, y_2, y_3) & \phi_{2y_1}^{\Delta_1}(y_1, y_2, y_3) & \phi_{3y_1}^{\Delta_1}(y_1, y_2, y_3) & \phi_{4y_1}^{\Delta_1}(y_1, y_2, y_3) \\ \phi_{1y_2}^{\Delta_2}(y_1, y_2, y_3) & \phi_{2y_2}^{\Delta_2}(y_1, y_2, y_3) & \phi_{3y_2}^{\Delta_2}(y_1, y_2, y_3) & \phi_{4y_2}^{\Delta_2}(y_1, y_2, y_3) \\ \phi_{1y_3}^{\Delta_3}(y_1, y_2, y_3) & \phi_{2y_3}^{\Delta_3}(y_1, y_2, y_3) & \phi_{3y_3}^{\Delta_3}(y_1, y_2, y_3) & \phi_{4y_3}^{\Delta_3}(y_1, y_2, y_3) \end{vmatrix}$$

$$= \begin{vmatrix} \mathbf{e}_1 & \mathbf{e}_2 & \mathbf{e}_3 & \mathbf{e}_4 \\ 1 & 1 & 0 & 1 \\ 1 & 0 & 0 & 0 \\ 1 & -1 & 1 & 1 \end{vmatrix}$$

$$= \begin{vmatrix} 1 & 0 & 1 \\ 0 & 0 & 0 \\ -1 & 1 & 1 \end{vmatrix} \mathbf{e}_1 - \begin{vmatrix} 1 & 0 & 1 \\ 1 & 0 & 0 \\ 1 & 1 & 1 \end{vmatrix} \mathbf{e}_2 + \begin{vmatrix} 1 & 1 & 1 \\ 1 & 0 & 0 \\ 1 & -1 & 1 \end{vmatrix} \mathbf{e}_3 - \begin{vmatrix} 1 & 1 & 0 \\ 1 & 0 & 0 \\ 1 & -1 & 1 \end{vmatrix} \mathbf{e}_4$$

$$= \mathbf{e}_2 - 2\mathbf{e}_3 + \mathbf{e}_4.$$

Hence,

$$|\mathbf{r}_{y_1}^{\Delta_1}(y_1, y_2, y_3) \times \mathbf{r}_{y_2}^{\Delta_2}(y_1, y_2, y_3) \times \mathbf{r}_{y_3}^{\Delta_3}(y_1, y_2, y_3)| = \sqrt{1+4+1} = \sqrt{6}$$

and

$$\begin{aligned}
I &= \sqrt{6} \int_0^3 \int_0^3 \int_0^9 (y_2 + 4y_3) \Delta_3 y_3 \Delta_2 y_2 \Delta_1 y_1 \\
&= \sqrt{6} \int_0^3 \int_0^3 \Big((y_2 + 4y_3)\mu_3(y_3)\Big|_{y_3=0} + (y_2 + 4y_3)\mu_3(y_3)\Big|_{y_3=1} \\
&\quad + (y_2 + 4y_3)\mu_3(y_3)\Big|_{y_3=3} \Big) \Delta_2 y_2 \Delta_1 y_1 \\
&= \sqrt{6} \int_0^3 \int_0^3 (y_2 + 2(y_2+4) + 6(y_2+12))\, \Delta_2 y_2 \Delta_1 y_1 \\
&= \sqrt{6} \int_0^3 \int_0^3 (y_2 + 2y_2 + 8 + 6y_2 + 72) \Delta_2 y_2 \Delta_1 y_1 \\
&= \sqrt{6} \int_0^3 \int_0^3 (9y_2 + 80) \Delta_2 y_2 \Delta_1 y_1 \\
&= \sqrt{6} \int_0^3 \left((9y_2 + 80)\mu_2(y_2)\Big|_{y_2=0} + (9y_2 + 80)\mu_2(y_2)\Big|_{y_2=1} \right) \Delta_1 y_1 \\
&= \sqrt{6} \int_0^3 (80 + 178) \Delta_1 y_1 \\
&= 774\sqrt{6}.
\end{aligned}$$

Exercise 9.17 Let $\mathbb{T}_1 = \mathbb{T}_2 = \mathbb{T}_3 = \mathbb{Z}$,

$$S : \begin{cases} x_1 = y_1 - 2y_2 + y_3^2 \\ x_2 = y_1 + y_2^2 \\ x_3 = y_3^2 \\ x_4 = y_1^2 + y_2 + y_3, \quad 0 \le y_1 \le 4, \quad 0 \le y_2 \le 2, \quad 0 \le y_3 \le 1. \end{cases}$$

Compute

$$\int_S (x_1^2 + x_2) \Delta A.$$

9.3 Advanced Practical Problems

Problem 9.18 Find $A(S)$, where

1. $\mathbb{T}_1 = \mathbb{T}_2 = \mathbb{Z}$,

$$\Omega = \{(y_1, y_2) \in \mathbb{T}_1 \times \mathbb{T}_2 : -3 \le y_1 \le 3, \ 0 \le y_2 \le 1\},$$

and

$$S : \begin{cases} x_1 = y_1 \\ x_2 = y_2 \\ x_3 = y_1^2 + y_2^2, \end{cases}$$

2. $\mathbb{T}_1 = \mathbb{T}_2 = 2^{\mathbb{N}_0}$,

$$\Omega = \{(y_1, y_2) \in \mathbb{T}_1 \times \mathbb{T}_2 : 2 \le y_1 \le 4, \ 2 \le y_2 \le 8\},$$

and

$$S : \begin{cases} x_1 = y_1^2 - y_2^2 \\ x_2 = y_1 + y_2 \\ x_3 = y_1^3 + y_2^2, \end{cases}$$

3. $\mathbb{T}_1 = 2^{\mathbb{N}_0}$, $\mathbb{T}_2 = \mathbb{Z}$, $\mathbb{T}_3 = \mathbb{Z}$,

$$\Omega = \{(y_1, y_2, y_3) \in \mathbb{T}_1 \times \mathbb{T}_2 \times \mathbb{T}_3 : \ 1 \le y_1 \le 4, \ -1 \le y_2 \le 1, \ 0 \le y_3 \le 2\},$$

and

$$S: \begin{cases} x_1 = y_1 \\ x_2 = y_2 \\ x_3 = y_3 \\ x_4 = y_1^2 + y_2^2 + y_3^2. \end{cases}$$

Problem 9.19 Let $\mathbb{T}_1 = \mathbb{Z}$, $\mathbb{T}_2 = \mathbb{N}_0$, $\mathbb{T}_3 = 2^{\mathbb{N}_0}$,

$$S: \begin{cases} x_1 = y_1 + y_2 + y_3 \\ x_2 = y_1^2 + y_2^2 \\ x_3 = y_2^2 - y_3^2 \\ x_4 = y_1 - 3y_2 + y_3^2, \quad 0 \le y_1 \le 2, \quad 0 \le y_2 \le 4, \quad 1 \le y_3 \le 8. \end{cases}$$

Compute

$$\int_S (x_1 - x_2^2 + 2x_3 + x_4) \Delta A.$$

9.4 Notes and References

In this chapter, the theory of surface integrals is extended to general time scales. The concept of a surface parameterized by time scale parameters is introduced, and an integral formula for the computation of its area is established. Surface delta integrals on time scales are defined, and some sufficient conditions for the existence of these integrals are offered. Moreover, a formula for their evaluation is presented. All results in this chapter are n-dimensional analogues of the three-dimensional results by Bohner and Guseinov [17].

References

1. R.P. Agarwal, M. Bohner, Basic calculus on time scales and some of its applications. Results Math. **35**(1–2), 3–22 (1999)
2. C.D. Ahlbrandt, C. Morian, Partial differential equations on time scales. J. Comput. Appl. Math. **141**(1–2):35–55 (2002). (Dynamic equations on time scales)
3. F.M. Atici, G.Sh. Guseinov, On Green's functions and positive solutions for boundary value problems on time scales. J. Comput. Appl. Math. **141**(1–2):75–99 (2002). (Dynamic equations on time scales)
4. B. Aulbach, S. Hilger, *Linear dynamic processes with inhomogeneous time scale, in Nonlinear Dynamics and Quantum Dynamical Systems (Gaussig, 1990), Mathematical Research*, vol. 59 (Akademie-Verlag, Berlin, 1990), pp. 9–20
5. B. Aulbach, S. Hilger, *A unified approach to continuous and discrete dynamics, in Qualitative Theory of Differential Equations (Szeged, 1988), Colloquia Mathematical Society János Bolyai*, vol. 53 (North-Holland, Amsterdam, 1990), pp. 37–56
6. M. Bohner, G.Sh. Guseinov, Improper integrals on time scales. Dyn. Syst. Appl. **12**(1–2), 45–65 (2003). (Special issue: dynamic equations on time scales)
7. M. Bohner, G.Sh. Guseinov, *Riemann and Lebesgue integration, in Advances in Dynamic Equations on Time Scales* (Birkhäuser Boston, Boston, MA, 2003), pp. 117–163
8. M. Bohner, G.Sh. Guseinov, Partial differentiation on time scales. Dyn. Syst. Appl. **13**(3–4), 351–379 (2004)
9. M. Bohner, G.Sh. Guseinov, Multiple integration on time scales. Dyn. Syst. Appl. **14**(3–4), 579–606 (2005)
10. M. Bohner, G.Sh. Guseinov, An introduction to complex functions on products of two time scales. J. Differ. Equ. Appl. **12**(3–4), 369–384 (2006)
11. M. Bohner, G.Sh. Guseinov, Multiple Lebesgue integration on time scales. Adv. Differ. Equ. Art. ID **26391**, 12 (2006)
12. M. Bohner, G.Sh. Guseinov, The convolution on time scales. Abstr. Appl. Anal. Art. ID **58373**, 24 (2007)
13. M. Bohner, G.Sh. Guseinov, Double integral calculus of variations on time scales. Comput. Math. Appl. **54**(1), 45–57 (2007)
14. M. Bohner, G.Sh. Guseinov, Line integrals and Green's formula on time scales. J. Math. Anal. Appl. **326**(2), 1124–1141 (2007)
15. M. Bohner, G.Sh. Guseinov, The h-Laplace and q-Laplace transforms. J. Math. Anal. Appl. **365**(1), 75–92 (2010)
16. M. Bohner, G.Sh. Guseinov, The Laplace transform on isolated time scales. Comput. Math. Appl. **60**(6), 1536–1547 (2010)
17. M. Bohner, G.Sh. Guseinov, Surface areas and surface integrals on time scales. Dyn. Syst. Appl. **19**(3–4), 435–453 (2010)

M. Bohner and S.G. Georgiev, *Multivariable Dynamic Calculus on Time Scales*,
DOI 10.1007/978-3-319-47620-9

18. M. Bohner, G.Sh. Guseinov, B. Karpuz, Properties of the Laplace transform on time scales with arbitrary graininess. Integr. Trans. Spec. Funct. **22**(11), 785–800 (2011)
19. M. Bohner, G.Sh. Guseinov, B. Karpuz, Further properties of the Laplace transform on time scales with arbitrary graininess. Integr. Trans. Spec. Funct. **24**(4), 289–301 (2013)
20. M. Bohner, G.Sh. Guseinov, A. Peterson, *Introduction to the time scales calculus, in Advances in Dynamic Equations on Time Scales* (Birkhäuser Boston, Boston, MA, 2003), pp. 1–15
21. M. Bohner, A. Peterson, *Dynamic Equations on Time Scales* (Birkhäuser Boston, Inc., Boston, MA, 2001). (An introduction with applications)
22. M. Bohner, A. Peterson, First and second order linear dynamic equations on time scales. J. Differ. Equ. Appl. **7**(6), 767–792 (2001). (On the occasion of the 60th birthday of Calvin Ahlbrandt)
23. M. Bohner, A. Peterson, A survey of exponential function on times scales. Cubo Mat. Educ. **3**(2), 285–301 (2001)
24. M. Bohner, A. Peterson, Laplace transform and Z-transform: unification and extension. Methods Appl. Anal. **9**(1), 151–157 (2002)
25. M. Bohner, A. Peterson (eds.), *Advances in Dynamic Equations on Time Scales* (Birkhäuser Boston Inc, Boston, MA, 2003)
26. J. Dieudonné, *Foundations of Modern Analysis* (Academic Press, New York, 1969). (Enlarged and corrected printing, Pure and Applied Mathematics, Vol. 10-I)
27. G.Sh. Guseinov, Integration on time scales. J. Math. Anal. Appl. **285**(1), 107–127 (2003)
28. G.Sh. Guseinov, B. Kaymakçalan, Basics of Riemann delta and nabla integration on time scales. J. Differ. Equ. Appl. **8**(11), 1001–1017 (2002). (Special issue in honour of Professor Allan Peterson on the occasion of his 60th birthday, Part I)
29. G.Sh. Guseinov, B. Kaymakçalan, On the Riemann integration on time scales, in *Proceedings of the Sixth International Conference on Difference Equations* (CRC, Boca Raton, FL, 2004), pp. 289–298
30. G.Sh. Guseinov, E. Özyılmaz, Tangent lines of generalized regular curves parametrized by time scales. Turkish J. Math. **25**(4), 553–562 (2001)
31. S. Hilger, *Ein Maßkettenkalkül mit Anwendung auf Zentrumsmannigfaltigkeiten*. PhD thesis, Universität Würzburg (1988)
32. S. Hilger, Analysis on measure chains-a unified approach to continuous and discrete calculus. Results Math. **18**(1–2), 18–56 (1990)
33. S. Hilger, Special functions, Laplace and Fourier transform on measure chains. Dyn. Syst. Appl. **8**(3–4), 471–488 (1999)
34. J. Hoffacker, Basic partial dynamic equations on time scales. J. Differ. Equ. Appl. **8**(4), 307–319 (2002). (In honor of Professor Lynn Erbe)
35. S. Keller, *Asymptotisches Verhalten invarianter Faserbündel bei Diskretisierung und Mittelwertbildung im Rahmen der Analysis auf Zeitskalen*. PhD thesis, Universität Augsburg (1999)
36. L. Pang, K. Wang, Function series theory of time scales. Comput. Math. Appl. **62**(9), 3427–3437 (2011)
37. Z. Pospíšil, Hyperbolic sine and cosine functions on measure chains, in *Proceedings of the Third World Congress of Nonlinear Analysts, Part 2 (Catania, 2000)*, vol. 47 (2001), pp. 861–872
38. C. Pötzsche, Chain rule and invariance principle on measure chains. J. Comput. Appl. Math. **141**(1–2), 249–254 (2002). (Dynamic equations on time scales)
39. S. Sailer, Riemann–Stieltjes Integrale auf Zeitmengen. Technical report, Universität Augsburg, October 5, 1992. (Schriftliche Hausarbeit)
40. C. Zhang, K. Wang, Parameter-dependent integral on time scales. Indian J. Pure Appl. Math. **45**(2), 139–163 (2014)

Index

M. Bohner and S.G. Georgiev, *Multivariable Dynamic Calculus on Time Scales*,
DOI 10.1007/978-3-319-47620-9

MIX
Papier aus verantwortungsvollen Quellen
Paper from responsible sources
FSC® C105338

If you have any concerns about our products,
you can contact us on
ProductSafety@springernature.com

In case Publisher is established outside the EU,
the EU authorized representative is
Springer Nature Customer Service Center GmbH
Europaplatz 3, 69115 Heidelberg, Germany

Printed by Libri Plureos GmbH
in Hamburg, Germany

MIX
Papier aus verantwortungsvollen Quellen
Paper from responsible sources
FSC® C105338

If you have any concerns about our products,
you can contact us on
ProductSafety@springernature.com

In case Publisher is established outside the EU,
the EU authorized representative is:
Springer Nature Customer Service Center GmbH
Europaplatz 3, 69115 Heidelberg, Germany

Printed by Libri Plureos GmbH
in Hamburg, Germany